"Our knowledge of pollen, the gold dust that carries the male germ of flowering plants and is vital for sexual reproduction and seed formation, has 'come of age' with this book. Here, for the first time in a single volume, are all the ideas and techniques developed in the last two decades concerning the manipulation of pollen and pollen tubes in plant breeding and biotechnology. Pollen has never been an easy topic to come to grips with, with its variable and often inexplicable terminolology that has made it a more difficult field in which to work. This book will remedy that, with its overview of pollen biology and pollen–pistil interactions that explains terms and concepts of the male function of pollen in a way that is readily understandable."

– From the foreword by R. Bruce Knox

POLLEN BIOTECHNOLOGY
FOR CROP PRODUCTION
AND IMPROVEMENT

POLLEN BIOTECHNOLOGY FOR CROP PRODUCTION AND IMPROVEMENT

Edited by

K. R. SHIVANNA
University of Delhi

V. K. SAWHNEY
University of Saskatchewan

PUBLISHED BY THE PRESS SYNDICATE OF THE UNIVERSITY OF CAMBRIDGE
The Pitt Building, Trumpington Street, Cambridge CB2 1RP, United Kingdom

CAMBRIDGE UNIVERSITY PRESS
The Edinburgh Building, Cambridge CB2 2RU, United Kingdom
40 West 20th Street, New York, NY 10011-4211, USA
10 Stamford Road, Oakleigh, Melbourne 3166, Australia

First published 1997

Printed in the United States of America

Typeset in Times Roman

Library of Congress Cataloging-in-Publication Data

Pollen Biotechnology for Crop Production and Improvement / edited by
K. R. Shivanna, V. K. Sawhney.
p.   cm.
ISBN 0-521-47180-X (hbk.)
1. Pollen – Biotechnology. 2. Plant breeding. 3. Crops – Genetic
engineering. 4. Crop improvement. I. Shivanna, K. R.
II. Sawhney, V. K.
SB106.B56P65   1996
631.5'3–dc20   96-14026
                                                          CIP

A catalog record for this book is available from the British Library

ISBN 0 521 47180 X hardback

Dedicated to
Pramila and Ramma

Contents

Contributors   ix
Foreword   xi
Preface   xiii

1. Pollen biology and pollen biotechnology: an introduction
   K. R. SHIVANNA and V. K. SAWHNEY   1

Part I. Pollen biology: an overview   **13**
2. Pollen development and pollen–pistil interaction
   K. R. SHIVANNA, M. CRESTI, and F. CIAMPOLINI   15
3. Gene expression during pollen development
   D. A. HAMILTON and J. P. MASCARENHAS   40
4. Pollination biology and plant breeding systems
   P. G. KEVAN   59

Part II. Pollen biotechnology and optimization of crop yield   **85**
5. Pollination efficiency of insects
   A. R. DAVIS   87
6. Pollination constraints and management of pollinating insects
   for crop production
   R. W. CURRIE   121

Part III. Pollen biotechnology and hybrid seed production   **153**
7. Cytoplasmic male sterility
   P. B. E. MCVETTY   155
8. Genic male sterility
   V. K. SAWHNEY   183
9. Self-incompatibility
   A. MCCUBBIN and H. G. DICKINSON   199
10. Chemical induction of male sterility
    J. W. CROSS and P. J. SCHULZ   218
11. Male sterility through recombinant DNA technology
    M. E. WILLIAMS, J. LEEMANS, and F. MICHIELS   237

vii

Part IV. Pollen biotechnology and plant breeding   **259**

12. Barriers to hybridization
    K. R. SHIVANNA   261
13. Methods for overcoming interspecific crossing barriers
    J. M. VAN TUYL and M. J. DE JEU   273
14. Storage of pollen
    B. BARNABÁS and G. KOVÁCS   293
15. Mentor effects in pistil-mediated pollen–pollen interactions
    M. VILLAR and M. GAGET-FAUROBERT   315
16. Pollen tube growth and pollen selection
    M. SARI-GORLA and C. FROVA   333
17. Isolation and manipulation of sperm cells
    D. D. CASS   352
18. Isolation and micromanipulation of embryo sac and
    egg cell in maize
    E. MATTHYS-ROCHON, R. MÒL, J. E. FAURE, C. DIGONNET,
    and C. DUMAS   363
19. In vitro fertilization with single isolated gametes
    E. KRANZ   377
20. Pollen embryos
    C. E. PALMER and W. A. KELLER   392
21. Use of pollen in gene transfer
    H. MORIKAWA and M. NISHIHARA   423

Index   438

Contributors

B. BARNABÁS *Cell Biology Department, Agricultural Research Institute of the Hungarian Academy of Sciences, H-2462 Martonvásár, Hungary*

D. D. CASS *Department of Biological Sciences, University of Alberta, Edmonton, Alberta T6G 2E9, Canada*

F. CIAMPOLINI *Department of Environmental Biology, University of Siena, Via P.A. Mattioli 4, 53100 Siena, Italy*

M. CRESTI *Department of Environmental Biology, University of Siena, Via P.A. Mattioli 4, 53100 Siena, Italy*

J. W. CROSS *7759 Inversham Drive, Suite 245, Falls Church, VA 22042, U.S.A.*

R. W. CURRIE *Department of Entomology, University of Manitoba, Winnipeg, Manitoba R3T 2N2, Canada*

A. R. DAVIS *Department of Biology, University of Saskatchewan, 112 Science Place, Saskatoon, Saskatchewan S7N 5E2, Canada*

H. G. DICKINSON *Department of Plant Sciences, University of Oxford, South Parks Road, Oxford OX1 3RB, United Kingdom*

C. DIGONNET *Ecole Normale Supérieure de Lyon, UMR 9938 CNRS-INRA-ENS, 69634 Lyon Cedex 07, France*

C. DUMAS *Ecole Normale Supérieure de Lyon, UMR INRA 9938 CNRS-INRA-ENS, 69634 Lyon Cedex 07, France*

J. E. FAURE *Ecole Normale Supérieure de Lyon, UMR 9938 CNRS-INRA-ENS, 69634 Lyon Cedex 07, France*

C. FROVA *Department of Genetics and Microbiology, University of Milan, Via Celoria 2b, Milan 20133, Italy*

M. GAGET-FAUROBERT *INRA, Station de recherches fruitières mediterranéenes, F-84140 Montfavet, France*

D. A. HAMILTON *Department of Biological Sciences and Center for Molecular Genetics, State University of New York, Albany NY 12222, U.S.A.*

M. J. DE JEU *Department of Plant Breeding, Wageningen Agricultural University (WAU), P.O. Box 386, NL-6700 AJ Wageningen, The Netherlands*

W. A. KELLER *Plant Biotechnology Institute (NRC), 110 Gymnasium Place, Saskatoon, Saskatchewan S7N 0W9, Canada*

P. G. KEVAN *Department of Environmental Biology, University of Guelph, Guelph, Ontario N1G 2W1, Canada*

G. KOVÁCS *Cell Biology Department, Agricultural Research Institute of the Hungarian Academy of Sciences, 2462 Martonvásár, Hungary.*

E. KRANZ *Institute für Allgemeine Botanik, Angewandte Molekularbiologie der Pflanzen II, Universität Hamburg, Ohnhorststrasse 18, D-22609 Hamburg, Germany*

J. LEEMANS *Plant Genetic Systems NV, J. Plateaustraat 22, B-9000 Gent, Belgium*

J. P. MASCARENHAS *Department of Biological Sciences and Center for Molecular Genetics, State University of New York, Albany, NY 12222, U.S.A.*

E. MATTHYS-ROCHON *Ecole Normale Supérieure de Lyon, UMR 9938 CNRS-INRA-ENS, 69364 Lyon Cedex 07, France*

A. MCUBBIN *Department of Biochemistry and Molecular Biology, Eberly College of Science, The Pennsylvania State University, University Park, PA 16802, U.S.A.*

P. B. E. McVETTY *Department of Plant Science, University of Manitoba, Winnipeg, Manitoba R3T 2N2, Canada*

F. MICHIELS *Plant Genetic Systems NV, J. Plateaustraat 22, B-9000 Gent, Belgium*

R. MÒL *Laboratory of Botany, Adam Mickiewicz University, 61 813 Poznan, Poland*

H. MORIKAWA *Department of Gene Science, Faculty of Science, Hiroshima University, Higashi-Hiroshima 724, Japan*

M. NISHIHARA *Department of Gene Science, Faculty of Science, Hiroshima University, Higashi-Hiroshima 724, Japan*

C. E. PALMER *Department of Plant Science, University of Manitoba, Winnipeg, Manitoba R3T 2N2, Canada*

M. SARI-GORLA *Department of Genetics and Microbiology, University of Milan, Via Celoria 26, Milan 20133, Italy*

V. K. SAWHNEY *Department of Biology, University of Saskactchewan, 112 Science Place, Saskatoon, Saskatchewan S7N 5E2, Canada*

P . J. SCHULZ *Biology Department, University of San Francisco, San Francisco, CA 94117, U.S.A.*

K. R. SHIVANNA *Department of Botany, University of Delhi, Delhi 110 007, India*

J. M. VAN TUYL *Centre for Plant Breeding and Reproduction Research (CPRO-DLO), P.O. Box 16, NL-6700 AA, Wageningen, The Netherlands*

M. VILLAR *INRA, Station d'Amélioration des Arbres Forestiers, F-45160 Ardon, France*

M. E. WILLIAMS *Du Pont Agricultural Products, E. I. duPont de Nemours and Co. Stine-Haskell Research Center, P. O. Box 30, Newark, DE 19714-0030, U.S.A.*

Foreword

Our knowledge of pollen, the gold dust that carries the male germ line of flowering plants and is vital for sexual reproduction and seed formation, has "come of age" with the publication of this book. Here, for the first time in a single volume, are all the ideas and techniques developed in the last two decades concerning the manipulation of pollen and pollen tubes in plant breeding and biotechnology. Pollen has never been an easy topic to come to grips with, with its variable and often inexplicable terminology that has made it a difficult field in which to work. This book will remedy that, with its overview of pollen biology and pollen–pistil interactions that explains terms and concepts of the male function of pollen in a way that is readily understandable.

This new biotechnology of pollen had its origins 40 years ago in developments in plant tissue culture, physiology, and electron microscopy, and, more recently, with the advent of molecular genetics. In reviewing pollen developmental processes and genetic defects in the early 1970s, Professor Jack Heslop-Harrison FRS showed how the opportunities for manipulation might be achieved:

On the one hand, are developmental faults involving deviation from the presumptive behaviour of a spore in the anther, with the production of a female gametophyte or even a sporophyte instead of a pollen grain. On the other hand, are failures of differentiation and various forms of abortion that result in death or gross malfunction. Events of the first category . . . illustrate the totipotency of the spore nucleus in an immediate and dramatic way, and show that the determination of the fate of the spore depends upon influence acting upon it soon after meiosis. Aberrations of the second kind are important because they form the basis of pollen sterility. (J. Heslop-Harrison 1972, Sexuality of angiosperms; in *Plant Physiology 6C*, ed. F. C. Steward, p. 249; New York: Academic Press.)

Arising from these approaches have come such developments as anther and microspore culture for production of haploid plants; in vitro culture of anthers for genetic transformation; isolation of sperm and egg cells; in vitro fertilization and embryo production; male sterility; and hybrid seed production. Of especial interest here are the refinements being made in the control of cytoplasmic male sterility through the control of the mitochondrial genome (Chapter 7), sterility caused by nuclear gene mutants (Chapter 8), and chemically induced male sterility (Chapter 10).

The question of the natural function of genes, which in the mutant form block anther or microspore development and thereby cause pollen sterility, is a challenge for the future, when some of these genes will have been sequenced. An exciting approach to this question has been to study a gene that is expressed exclusively in the anther and so is likely to be important for pollen development (Chapter 11). The gene *Bcp1* from *Brassica* pollen encodes a 12-kd protein that is expressed in both tapetum and pollen. Down regulation of this gene using antisense RNA in transgenic *Arabidopsis* blocked microspore development, showing that this gene is essential for maintenance of male fertility. Manipulation of this gene provides a potential tool for induction of male sterility in crop plants (H. Xu et al. 1995, *Proc. Nat. Acad. Sci. USA* 92:2106–2110).

The dramatic progress that has been made toward in vitro fertilization is reviewed in Chapters 17, 18, and 19. First, we learn how sperm and egg cells are isolated, and how long their viability can be retained in a test tube. Fertilization, a process that normally occurs deep within the ovule, can now be carried out in a test tube with the induced fusion of the male and female gametes. This has been achieved by means of an electrical stimulus to produce a zygote (and, subsequently, embryos and plants of maize) or by the more natural process of manipulating the Ca^{2+} concentrations of fusion media, whereby zygotes have been produced, but as yet, apparently no embryos or plants! This is a new frontier for the future.

The authors (and editors) of this book are well-known and internationally respected pollen biologists who have been at the forefront of developing new techniques of pollen biotechnology themselves. This book is a considerable achievement, and it gives me great personal pleasure to congratulate them and wish them well for a successful volume that will go on to many new editions in the future.

R. Bruce Knox
Melbourne, Australia

Preface

Pollen biotechnology offers a powerful tool for (a) the optimization of crop production, (b) exploitation of hybrid vigor, and (c) crop improvement through breeding. A large number of papers on various aspects of pollen biology and biotechnology are being published in a variety of journals. There is no book that brings together various aspects related to pollen biotechnology. Because most researchers in pollen biology and biotechnology work in specialized areas, their interests tend to be confined to journals in their fields. This has resulted in a wide gap between biologists interested in fundamental knowledge on pollen biology and those interested in the application of this knowledge to crop production and improvement.

This volume aims to provide a comprehensive account, written by international authorities, of different aspects of pollen biotechnology. We hope that this volume will help bridge the gap between pollen biologists, and agri-horticulturists, foresters, and, particularly, plant breeders. The information contained should also help pollen biologists focus their research in areas directly relevant to crop production and improvement.

The introductory chapter gives a general account of the scope of pollen biology and biotechnology. The remaining chapters are grouped under four sections. The first section gives an overview of pollen biology and includes pollen development, pollination, and pollen–pistil interaction. This is intended to provide, particularly for nonspecialists, the basic information on pollen biology necessary for a better understanding of different aspects of pollen biotechnology covered in other sections. The remaining three sections include chapters pertaining to the role of pollen biotechnology in (a) optimization of crop production, (b) hybrid seed production, and (c) plant breeding. In each chapter, an attempt is made to provide a brief review of the current status of the field and its application to crop production and improvement. It is hoped that this volume will promote an appreciation on the part of pollen biologists and plant breeders of the potentials of pollen biotechnology, and its application to crop production and improvement. In a volume like this, in which each chapter is meant to be complete, some overlap of information is inevitable. Nonetheless, we, with the help of other authors, have tried to minimize the overlap as much as possible. To keep the volume to a reasonable size, it has not been possible to include chapters on other interesting areas, such as the role of pollen in the honey and pharmaceutical industries, pollen allergy, and the use of pollen in monitoring cytotoxicity. However, some literature on these aspects is included in the introductory chapter.

The volume essentially covers interdisciplinary areas and caters primarily to pollen biologists involved in basic and applied aspects of research for crop production and improvement. The book may also serve as a supplementary text for senior undergraduate and graduate courses on plant biotechnology, pollen biology, pollination biology, reproductive biology, and plant breeding.

We are grateful to Professor R. B. Knox for writing the foreword for this volume. We sincerely thank all the contributors who readily accepted our invitation and sent their chapters within the time frame. We also thank the authors for their cooperation with revisions. The help and advice of Dr. Robin Smith, Life Sciences Editor of Cambridge University Press, throughout the course of this project is gratefully acknowledged. Thanks are due also to Ms. Kathy Weiman and Ms. Joan Ryan for their help with correspondence and with word processing of final copies, and to Ms. Diane Davis for advice and assistance with the index.

K. R. Shivanna

V. K. Sawhney

1

Pollen biology and pollen biotechnology: an introduction

K. R. SHIVANNA and V. K. SAWHNEY

Pollen biology   1
Pollen biotechnology   3
    Optimization of crop yield   3
    Hybrid seed production   3
    Plant breeding   5
    Production of other economic products   9
References   9

Pollen biology

Pollen grains embody the male partners in sexual reproduction. They are generally shed in a desiccated condition and their moisture level is less than 20%. At the time of shedding, pollen grains are either two-celled – a large vegetative cell enclosing a generative cell; or three-celled – a vegetative cell and two sperm cells formed by the division of the generative cell. There is considerable variation in the size and shape of pollen grains (Erdtman 1966; Moore and Webb 1978; Iwanami et al. 1988; Faegri and Iversen 1989; Cresti et al. 1992). Although a majority of pollen grains are spherical, in some marine angiosperms, such as *Amphibolis* and *Zostera*, they are filiform (up to 5 mm) (Ducker et al. 1978). The wall of the pollen grain is made up of two layers: an outer, acetolysis-resistant exine composed of sporopollenin and an inner pectocellulosic intine. One of the conspicuous structural features of pollen grains is the ornamentation of the wall formed by the outer part of the exine (Cresti et al. 1992).

Pollen biology involves a comprehensive understanding of the structural and functional aspects of pollen grains. The main function of the pollen is to discharge male gametes in the embryo sac for fertilization and for subsequent seed and fruit development. This function depends on the successful completion of a number of sequential events. The following are considered the major events in pollen biology:

- Pollen development
- Free dispersed phase
- Pollination
- Pollen–pistil interaction
- Fertilization

1

Pollen grains develop inside the anther and are dispersed by dehiscence of the anther. After dispersal, pollen grains remain as independent functional units and are exposed to the prevailing environmental conditions for varying periods. Depending on the period and severity of the environment, the quality of pollen grains, particularly their viability and vigor, may be affected during this prepollination phase. Eventually, pollen grains are deposited on the stigma (pollination) through biotic or abiotic agents.

Plants have evolved an amazing range of adaptations to achieve pollination (Free 1970; McGregor 1976; Real 1983). Pollination is a critical event and often a major constraint in crop productivity. Although studies on pollination biology have been, and still are, largely from the point of view of academic interest, there is a growing realization of the commercial rewards of such studies in increasing the pollination efficiency of our crops. In recent years, the possibility of the escape of engineered genes from the crop plants to their wild species has also increased the interest in pollination biology (Ellstrand 1988; Dale 1992; Raynolds and Gray 1993).

Following successful pollination, pollen grains germinate on the stigma and the resulting pollen tubes grow through the tissues of the stigma and style and ultimately enter the embryo sac. The processes from pollination until the entry of pollen tubes into the embryo sac are referred to as pollen–pistil inter-action or the progamic phase. This phase plays an important role in determining the breeding system of the species/population (Richards 1986). During pollen–pistil interaction, pollen is selected for quality and compatibility (Shivanna and Johri 1985).

Pollen grains represent the gametophytic generation. A large number of genes, some unique to the pollen and others common to the sporophytic generation, are expressed during pollen development as well as during the post-germination phase of the pollen. Studies on the expression of these genes are important for the application of pollen selection, for recombinant DNA technology, and for the induction of pollen embryos.

In addition to these areas of pollen biology, which have direct relevance to pollen function, other areas that are equally important include pollen in relation to taxonomy and phylogeny (Erdtman 1966), fossil palynology (Faegri and Eversen 1989), aeropalynology and pollen allergy (Stanley and Linskens 1974; Knox 1979; Singh et al. 1991; Mohapatra and Knox 1996), and the use of pollen to analyze the effects of ecotoxic chemicals (Kappler and Kristen 1987; Wolters and Martens 1987; Strube et al. 1991; Pfahler 1992).

The effective exploitation of the potentials of pollen biotechnology depends on the extent of basic information available on various aspects of pollen biology, particularly those relevant to its function. The main objective of the chapters included in "Pollen Biology: An Overview," Part I, is to present a summary of pollen biology so that the reader can appreciate the potentials and limitations of different aspects of pollen biotechnology presented in other parts of the book.

Pollen biotechnology

Pollen biotechnology is the manipulation of pollen development and function, including alteration of its genome, for the production and improvement of crops and related products. Pollen biotechnology has direct application in the following four main areas:

- Optimization of crop yield
- Hybrid seed production
- Plant breeding
- Production of other economic products

Optimization of crop yield

In the majority of crop plants, seeds and/or fruits are the economic products. As mentioned earlier, for effective fruit and seed set, pollination is a prerequisite. Most of the crop species are cross-pollinated. Pollination of crop plants is often a major constraint due to one or more of the following factors:

- (i) Habitat degradation of pollinators, and extensive use of pesticides with drastic effects on native pollinator population
- (ii) High density of plants in monoculture cropping system, compared to that in the wild, limiting the availability of native pollinators
- (iii) Crops grown in regions where natural pollinators are absent

Insufficient pollination is one of the important causative factors for low yield in many of the field and orchard species. Effective management of pollinating agents can overcome pollination constraints and thus improve the yield. Until recently, management of pollinating agents was confined to honeybees, and largely to field and orchard crops (Jay 1986). Intensive studies carried out during recent years have clearly shown the potential of many other insects for increasing pollination efficiency not only for field and orchard crops, but also for glass-house–grown crops (Torchio 1990). In Europe and Canada, the use of bumblebee colonies for greenhouse pollination of tomatoes is of increasing importance (Kearns and Inouye 1993). A range of options are now available for the crop grower for pollinator management. The selection of an effective option, however, requires a thorough knowledge of both the crop plants and the pollinators. Such studies are presently gathering momentum. The two chapters in Part II highlight some of the pollination constraints and the present status of the management of pollinating insects for improving pollination efficiency.

Hybrid seed production

Development of hybrid seed technology for crop production has been one of the most important advances in the history of agriculture. The foundation of hybrid seed technology goes back to the second part of the 19th century (Allard 1960; Gardner 1968), during which time many investigators showed that hybrid corn had more vigor than inbred lines and open-pollinated varieties. In the 1930s, intensive investigations for practical utilization of this observation (heterosis, or hybrid vigor) led to the commercial production of hybrid seeds (Shull 1909; Anderson 1938). By 1940 hybrid maize was grown in more than half the acreage in the United States, and by 1968 the figure had increased to 99% (Gardner 1968). Presently, hybrid corn is grown extensively in almost all corn-growing countries. Hybrid seed technology has also been extended to many other grain and vegetable crops, such as sorghum, pearl millet, rice, tomato, onion, and vegetable brassicas. The application of hybrid seed technology is considered one of the important approaches for the improvement of yield in crops. Commercial production of hybrid seeds requires the following:

(i)   Development and maintenance of inbred lines
(ii) Prevention of self- and inter-line pollination in hybrid seed production plots
(iii) Effective cross-pollination between two inbred lines

Of these, prevention of self- and inter-line pollination is the most critical, and the cost of hybrid seed production is determined by the technology used to achieve this objective. The architecture of the maize plant, in which the staminate (tassel) and the pistillate (ear) inflorescences are separated from each other in space, has made this species uniquely suited for controlled pollination. As maize is a cross-pollinated crop, prevention of self-pollination and promotion of cross-pollination are easily achieved by removing the tassels of the female parent before they shed pollen. In species with bisexual flowers, controlled pollination requires labor-intensive emasculation and/or manual pollination; this increases the cost of hybrid seed production enormously and thus makes this technology cost prohibitive. However, different types of male sterility and self-incompatibility systems have provided useful tools for extending hybrid seed technology to a range of crop species. Presently, the following pollination control systems are available for use in hybrid seed production:

(i)   Nuclear (genic) male sterility
(ii) Cytoplasmic male sterility
(iii) Self-incompatibility
(iv) Gametocides or chemical hybridizing agents
(v)   Recombinant DNA (R-DNA) technology

Cytoplasmic male sterility is the most convenient and is being used extensively for many crops. Nuclear male sterility and self-incompatibility, although in use for some crops, particularly vegetable crops, have major limitations. The use of chemical hybridizing agents for commercial production of hybrid seeds is in an advanced stage of testing and is likely to soon be an available choice for wheat.

Recently, the use of R-DNA technology (the introduction of chimeric ribonuclease gene and antisense RNA) for the effective control of pollinating systems has raised considerable excitement. In rapeseed, male-sterile and restorer transgenic lines are in an advanced stage of development in Europe and Canada. The chapters included in "Pollen Biotechnology and Hybrid Seed Production," Part III, summarize the available information on different pollination control systems and highlight their advantages and limitations for commercial hybrid seed production.

Plant breeding

A majority of present-day cultivars of crop species are the result of extensive breeding. The major aim of the breeding program has been to maximize yield, quality, uniformity, and resistance to biotic and abiotic stresses. However, until recently, the breeding work has largely concentrated on yield and genetic uniformity, and it has achieved remarkable success in these areas. Nevertheless, this has greatly eroded genetic variability in crops. The dependence of agriculture on a narrow genetic base has dramatically increased the susceptibility of cultivars to biotic and abiotic stresses.

Although present-day cultivars have the potential for high yield under optimal agronomic conditions, their vulnerability to biotic and abiotic stresses results in a significant reduction in yield, often up to 50–80% of their genetic potential (Boyer 1982). Also, the environmental aspects of agriculture (excessive use of agrochemicals), so far ignored by farmers and breeders, have become increasingly important in recent years. Attempts are now being made to breed varieties that have low requirements for chemical fertilizers, pesticides, insecticides, and other environmentally unfriendly agents. Thus, new adaptive traits, particularly for resistance to biotic and abiotic stresses, have to be constantly incorporated into cultivars to maintain and improve yield and quality. Furthermore, a large area of agricultural land is becoming marginal because of salinity, alkalinity, and contamination with industrial chemicals. This has necessitated the breeding of crops that can be grown in such marginal lands.

Realization of the objectives just enumerated is a difficult task. The future plant-breeding programs, therefore, are going to be more complex and challenging. Because of the constant erosion of genetic diversity, genes imparting resistance or tolerance to stresses are no longer available within the cultivated species. A large number of the wild and weedy species of crop plants form a good repository for such desirable genes. Because many of these traits are polygenic, R-DNA technology is not readily amenable to the transfer of these traits. Plant breeders have to depend largely on sexual reproductive systems for gene transfer and will need to extend the breeding programs to distantly related species and, in some cases, genera. A large number of examples of suc-

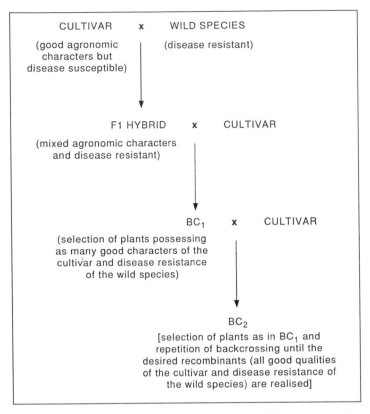

Figure 1.1 Protocol for transfer of desirable genes from wild species to cultivars through wide hybridization. Transfer of disease resistance is used as an example.

cessful gene transfer through wide hybridization are already available (Hawkes 1977; Stalker 1980; Goodman et al. 1987; Kalloo 1992).

The following are the essential steps for transferring genes through a conventional breeding program (Chopra and Sharma 1989):

(i)   Identification of accessions that possess desirable genes
(ii) Production of hybrids between the parents having desirable gene(s) and the cultivar (Figure 1.1)
(iii) Screening of a large number of plants in each of the backcross generations and identification of the required recombinants
(iv) Stabilization of the recombinants

Each of these steps is laborious and time-consuming. Apart from the time factor, a conventional breeding program is generally ineffective in crosses involving distantly related species. Application of appropriate pollen biotechnology to one or more of the steps just listed greatly reduces the time and cost of breeding (Table 1.1).

Table 1.1. *Integration of pollen biotechnology into conventional breeding program*

Major constraints	Biotechnological approaches
Identification of accessions with desirable genes	Use of pollen or other cells/ tissues for screening
Spatial/temporal isolation of parental species	Pollen storage
Presence of pre- and post-fertilization barriers	Application of various methods to overcome the barriers
Multiplication of hybrids	Use of tissue culture techniques
Hybrid sterility	Induction of amphiploidy
Production of backcross (BC) generation	Use of F_2 and F_3 generations Application of embryo rescue
Handling of BC generations	Use of pollen or other cells/ tissues for screening and for selection pressure
Lack of recombination	Irradiation, tissue culture, genetic manipulation of chromosome pairing systems
Stabilization of recombinants	Induction of pollen embryos

In addition to its application in the introgression of useful genes from wild species to the cultivars, wide hybridization is also important for the following objectives:

(i)   Production of haploids through chromosome elimination, as in barley and wheat (Hermsen and Ramanna 1981)

(ii) Synthesis of alloploid crops, such as *Triticale* and *Raphanobrassica* (Sareen et al. 1992)

(iii) Induction of cytoplasmic male sterility (Figure 1.2) through production of alloplasmic lines (having the cytoplasm of the wild species and nuclear genome of the cultivated species), as in wheat, tobacco, sunflower, and *Brassica* (Kaul 1988; Banga 1992)

Many recent studies have amply shown the potential of pollen for achieving genetic transformation. At present, cell and protoplast cultures are routinely used for genetic transformation. However, plant regeneration from transformed cells and protoplasts has been a major constraint of this approach. The use of pollen for the introduction of exogenous DNA could overcome this

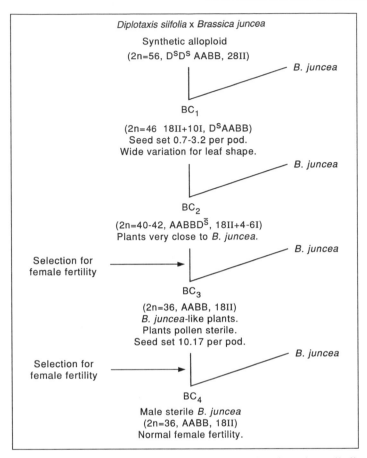

Figure 1.2 Protocol for the development of a new cytoplasmic male-sterile line through wide hybridization. *Brassica juncea* is used as an example.

major limitation since transformed pollen can be used to achieve fertilization and seed development. Significant progress has been made in this area in recent years (see Chapter 21).

One of the major difficulties for the integration of foreign DNA into pollen is the presence of a thick wall around the pollen. An ideal approach is to introduce foreign DNA into isolated sperm cells and/or the embryo sac/egg cell, and to achieve in vitro fertilization and subsequent development of the embryo. Because of the tremendous importance of this approach, studies on these aspects of biotechnology have been initiated in some laboratories. Recently, in vitro fertilization of isolated sperm and egg, and the development of an embryo have been realized in maize (see Chapter 19). The topics in "Pollen Biotechnology and Plant Breeding," Part IV, cover different aspects of pollen biotechnology that can be integrated into traditional breeding programs, as well as into direct pollen transformation.

Production of other economic products

Pollen grains have a direct role in honey production. Although honey is made up largely of nectar from flowers, pollen forms a critical component of the honey. Pollen is an important source of food for honeybees; protein, mineral, and vitamin requirements for the growth and development of the larvae and young bees are derived exclusively from pollen. A bee colony cannot exist without adequate pollen supply. Studies on pollen in relation to honey, termed mellittopalynology or mellissopalynology, play an important role in the honey industry. Pollen characters can also be used to identify the source and purity of honey (Louveaux et al. 1970).

Probably the most important commercial use of pollen is in the pharmaceutical industry. During the last three decades, the use of pollen as a health food and in pharmaceutical products has increased dramatically. An excellent account of these developments is available in Schmidt and Buchmann (1992). Pollen grains are rich in proteins, minerals, and vitamins, especially B vitamins, and low in fats, sodium, and fat-soluble vitamins (D, K, and E). The nutritional composition of pollen surpasses that of any food routinely used and, therefore, pollen forms an excellent human food supplement. There are a large number of reports on the beneficial effects of pollen on the physical stamina of individuals, particularly athletes (Steben and Boudreaux 1978). Pollen is also reported to be effective in the treatment of chronic prostatitis (Ask-Upmark 1967), in giving protection against the adverse effects of X-rays (Wang et al. 1984, cited in Schmidt and Buchmann 1992), and in reducing the symptoms of hay fever (Feinberg et al. 1940). There are also many reports on the use of pollen in animal diets, particularly those of racehorses. Although some of these claims are based on clinical trials, the majority of them are in nonscientific pamphlets published by major producers of pollen products. Clinical studies of the use of pollen in human and animal diets are, therefore, required.

A large number of pollen products in the form of capsules, granules, creams, candy bars, and oral liquids are available in the marketplace, and there is a great demand for such products. Information on the amount of pollen collected and processed in the pharmaceutical industry is difficult to obtain. The United States, China, Russia, and Europe are the main producers of pollen products; China alone is reported to harvest annually 3,000–5,000 metric tons of pollen (Wang 1980). Methods for the collection, cleaning, and processing of pollen are well described (Waller 1980; Benson 1984; Shaparew 1985). However, despite the tremendous market potential of pollen products, there are limited scientific data available on such products.

References

Allard, R.W. (1960). *Principles of Plant Breeding*. New York, London: John Wiley and Sons.

Anderson, D.C. (1938). The relation between single and double cross yields of corn. *J. Amer. Soc. Agron.* 30:209–211.

Ask-Upmark, E. (1967). Prostatitis and its treatment. *Acta Med. Scand.* 181:355–357.
Banga, S.S. (1992). Heterosis and its utilization. In *Breeding Oilseed Brassicas,* eds.
    K.S. Labana, S.S. Banga, and S.K. Banga, pp. 20–43. New Delhi: Narosa
    Publishing House.
Benson, K. (1984). Cleaning and handling pollen. *Amer. Bee J.* 124:301–305.
Boyer, J.S. (1982). Plant productivity and environment. *Science* 218:443–448.
Chopra, V.L. & Sharma, R.P. (1989). Innovative approaches for crop improvement.
    In *Plant Breeding: Theory and Practice,* ed. V.L. Chopra, pp. 453–466. New
    Delhi: Oxford & IBH Publ.
Cresti, M., Blackmore, S. & van Went, J.L. (1992). *Atlas of Sexual Reproduction in
    Flowering Plants.* Berlin: Springer-Verlag.
Dale, P.J. (1992). Spread of engineered genes to wild relatives. *Plant Physiol.*
    100:13–15.
Ducker, S.C., Pettitt, J.M. & Knox, R.B. (1978). Biology of Australian seagrasses:
    Pollen development and submarine pollination in *Amphibolis antartica* and
    Thalassodendron ciliata (Cymodoceaceae). *Aust. J. Bot.* 26:265–285.
Ellstrand, N.C. (1988). Pollen as a vehicle for the escape of engineered genes?
    Tibtech. 6:S30–S32.
Erdtman, G. (1966). *Pollen Morphology and Plant Taxonomy, Angiosperms.* New
    York: Hefner.
Faegri, K. & Iversen, J. (1989). *Textbook of Pollen Analysis,* IV ed., eds. K. Faegri,
    P.E. Kaland, and K. Krzywinski. New York: John Wiley and Sons.
Feinberg, S.M., Roran, F.L., Lichtenstein, M.R., Padnos, E., Rappaport, B.Z.,
    Sheldon, J. & Zeller, M. (1940). Oral pollen therapy in ragweed pollinosis. *J.
    Amer. Med. Assoc.* 115:23–29.
Free, J.B. (1970). *Insect Pollination of Crop Plants.* London: Academic Press.
Gardner, E.J. (1968). *Principles of Genetics,* 3rd ed. New York: John Wiley and Sons.
Goodman, R.M., Hauptli, H., Crossway, A. & Knauf, V.C. (1987). Gene transfer in
    crop improvement. *Science* 236:48–54.
Hawkes, J.G. (1977). The importance of wild germplasm in plant breeding.
    Euphytica 26:615–621.
Hermsen, J.G. Th. & Ramanna, M.S. (1981). Haploidy and plant breeding. *Phils.
    Trans. R. Soc. Lond. Ser. B* 292:499–507.
Iwanami, Y., Sasakuma, T. & Yamada, Y. (1988). *Pollen: Illustrations and Scanning
    Electron Micrographs.* Tokyo: Kodansha and Springer.
Jay, S.C. (1986). Spatial management of honeybees in crops. *Ann. Rev. Entomol.*
    31:49–65.
Kalloo, G. (1992). Utilization of wild species. In *Distant Hybridization of Crop
    Plants,* eds. G. Kalloo and J.B. Chowdhury, pp. 149–167. Berlin: Springer-
    Verlag.
Kappler, R. & Kristen, U. (1987). Photometric quantification of in vitro pollen tube
    growth: a new method suited to determine the cytotoxicity of various environ-
    mental substances. *Environ. Exp. Bot.* 27:305–309.
Kaul, M.L.H. (1988). *Male Sterility in Higher Plants.* Berlin: Springer-Verlag.
Kearns, C.A. & Inouye, D.V. (1993). *Techniques for Pollination Biologists.*
    Colorado: University of Colorado Press.
Knox, R.B. (1979). *Pollen and Allergy. Studies in Biology,* Vol. 107. London: Arnold.
Louveaux, J., Maurizio, A. & Verwohl, G. (1970). Methods of mellissopalynology.
    Bee World 51:125–138.
McGregor, S.E.M. (1976). *Insect Pollination of Cultivated Plants.* USDA Handbook
    No. 436. Washington, DC: U.S. Govt. Printing Office.
Mohapatra, S.S. & Knox, R.B., eds (1996). *Pollen Biotechnology: Gene Expression
    and Allergen Characterization.* New York: Chapman and Hall.

Moore, P.D. & Webb, J.A. (1978). *An Illustrated Guide to Pollen Analysis*. London: Hodder and Stoughton.

Pfahler, P.L. (1992). Analysis of ecotoxic agents using pollen tests. In *Modern Methods of Plant Analysis* Vol. 13, eds. H.F. Linskens and J.F. Jackson, pp. 317–331. Berlin: Springer Verlag.

Raynolds, A.F. & Gray, A.J. (1993). Genetically modified crops and hybridization with wild relatives: a U.K. perspective. *J. Applied Ecol.* 30:199–219.

Real, L., ed. (1983). *Pollination Biology*. Orlando, FL: Academic Press.

Richards, A.J. (1986). *Plant Breeding Systems*. London: George Allen & Unwin.

Sareen, P.K., Chowdhury, J.B. & Chowdhury, V.K. (1992). Amphidiploids/synthetic crop species. In *Distant Hybridization of Crop Plants,* eds. G. Kalloo and J.B. Chowdhury, pp. 62–81. Berlin: Springer-Verlag.

Schmidt, J.O. & Buchmann, S.L. (1992). Other products of the hive. In *The Hive and the Honey Bee,* ed. J.M. Graham, pp. 927–82. Hamilton, IL: Dadant & Sons.

Shaparew, V. (1985). Pollen trap – design optimization. *Amer. Bee J.* 125:173–175.

Shivanna, K.R. & Johri, B.M. (1985). *The Angiosperm Pollen: Structure and Function*. New Delhi: Wiley Eastern.

Shull, G.H. (1909). A pure line method of corn breeding. *Amer. Breed. Assoc. Rept.* 5:51–59.

Singh, M.B., Hough, T., Theerakulpist, P., Avjioglu, A., Davis, S., Smith, P.M., Taylor, P., Sampson, R.J., Ward, L.D., McClusky, J., Puy, R. & Knox, R.B. (1991). Isolation of cDNA encoding a newly identified major allergenic protein of rye-grass pollen: intracellular targeting to the amyloplast. *Proc. Natl. Acad. Sci.* USA 88:1384–1388.

Stalker, H.T. (1980). Utilization of wild species for crop improvement. *Adv. Agron.* 33:111–147.

Stanley, R.G. & Linskens, H.F. (1974). *Pollen: Biology, Biochemistry and Management*. Berlin: Springer-Verlag.

Steben, R.E. & Boudreaux, P. (1978). The effects of pollen and protein extracts on selected blood factors and performance of athletes. *J. Sports Med. Phys. Fitness* 18:221–226.

Strube, K., Janke, D., Kappler, R. & Kristen, U. (1991). Toxicity of some herbicides to in vitro growing tobacco pollen tubes (the pollen test). *Environ. Exp. Bot.* 31:217–222.

Torchio, P.F. (1990). Diversification of pollination strategies for U.S. crops. *Environ. Entomol.* 19:1649–1656.

Waller, G.D. (1980). A modification of O.A.C. pollen trap. *Amer. Bee J.* 120:119–121.

Wang, W. (1980). The development and utilization of resources of bee-pollen in China. *Proc. Intl. Congr. Apiculture (Apimondia)* 32:239.

Wolters, J.H.B. & Martens, J.M. (1987). Effects of air pollutants on pollen. *Bot. Rev.* 53:372–414.

Part I
Pollen biology: an overview

2

Pollen development and pollen–pistil interaction

K. R. SHIVANNA, M. CRESTI, and F. CIAMPOLINI

Summary   15
Introduction   16
Pollen development   16
    Syncytium and isolation   17
    Cytoplasmic reorganization   18
    Microspore development   18
    The tapetum   20
Structure of the pistil   22
    The stigma   22
    The style   24
    Ovary and ovule   25
Pollen–pistil interaction   28
    Pollen viability and stigma receptivity   28
    Pollination stimulus   29
    Pollen adhesion and hydration   30
    Pollen germination   31
    Pollen tube entry into the stigma and growth through the style   31
    Pollen tube entry into the ovule   32
    Double fertilization   33
    Pollen tube guidance   34
    Significance of pollen–pistil interaction   34
References   35

Summary

Structural details of pollen development are quite uniform in most of the species studied. The main structural events associated with pollen development are (i) the formation of a syncytium of microspore mother cells (MMCs), also referred to as pollen mother cells (PMCs) or meiocytes, in each anther locule, followed by the isolation of each MMC and the resulting microspores encased in a callose wall; (ii) cytoplasmic reorganization resulting in the breakdown of most of the RNA and ribosomes of MMCs, and dedifferentiation of plastids and mitochondria; (iii) release of microspores by the activation of callase; (iv) development of microspores accompanied by the synthesis and buildup of RNA, ribosomes, and proteins, and redifferentiation of plastids and mitochondria; (v) asymmetric division of the microspore; and (vi) desiccation and dispersal of pollen grains. The tapetum undergoes several changes and plays a crucial role in pollen development. Although the pistil shows tremendous morphological diversity, the surface of the stigma and the path of pollen tube growth in the pistil invariably contain extracellular components that come into contact with the pollen grain and the pollen tube. Pollination initiates a series of events leading to the discharge of sperm cells in the embryo sac and double fertilization. There is a close interaction between pollen and pistil throughout the postpollination period, and so far only a beginning has been made in understanding the details of these interactions.

Introduction

The development of normal, viable pollen grains, their transfer to the stigma, and pollen germination and successful completion of pollen–pistil interaction are prerequisites for fruit and seed development. Since the beginning of this century, extensive studies have been carried out on various aspects of pollen biology. A comprehensive discussion of all the areas of research is beyond the scope of this chapter. The intent here is to provide an overview of some of the fundamental aspects of pollen biology, that is, pollen development, structural features of the pistil, and pollen–pistil interaction. The references cited are subjective and selective.

Pollen development

Structural details of pollen development (Figures 2.1–2.8) have been investigated in a range of systems both at the light and electron microscopic level (Maheshwari 1950, 1963; Heslop-Harrison 1971; Bhandari 1984; Knox 1984; Shivanna and Johri 1985; Blackmore and Knox 1990; Bedinger 1992; Cresti et al. 1992; Goldberg et al. 1993). The young anther consists of a mass of homogeneous cells surrounded by the epidermis. As the anther develops, it becomes four lobed; four groups of archesporial cells (one in each lobe, corresponding to each microsporangium) differentiate in the hypodermal region. Archesporial cells divide periclinally to form an outer, primary parietal layer and an inner, sporogenous layer. The primary parietal layer undergoes periclinal and anticlinal divisions and gives rise to three to five concentric layers that differentiate into the endothecium (just below the epidermis), one to three middle layers, and the innermost tapetum. The primary sporogenous layer gives rise to MMCs directly or after some mitotic divisions (Figure 2.1).

Considerable information is available on the structural and biochemical details of meiosis and subsequent development of the microspore (Stern and Hotta 1974; Stern 1986; Dickinson 1987). Although most of the DNA in MMCs is synthesized during premeiotic S-phase, a small amount (0.3%) is also synthesized during the zygotene and pachytene stages of meiotic prophase; this is essential for the normal progress of meiosis. MMCs show

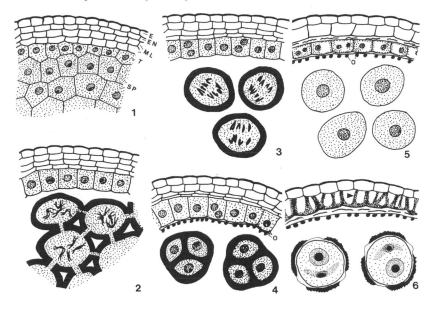

Figures 2.1–2.6 Diagrammatic representation of pollen development. Only a part of the microsporangium is shown. *2.1.* Differentiation of the wall layers (E = epidermis, EN = endothecial layer, ML = middle layers, T = tapetum, SP = sporogenous tissue). *2.2.* Formation of a syncytium by the development of cytoplasmic channels through incomplete callose wall (shown black) between neighboring microspore mother cells. *2.3.* Isolation of individual microspore mother cells by callose wall. *2.4.* Tetrads; individual microspores are enclosed by a callose wall. Middle layers have started degenerating and the orbicules (O) are laid down on the inner tangential wall of the tapetum. *2.5.* Release of microspores by the dissolution of callose wall. Middle layers have degenerated, cells of the endothecium have enlarged, and the tapetum has become vacuolated. *2.6.* Mature two-celled pollen. The tapetum is completely degenerated and the endothecial thickenings (arrow head) have developed.

active protein and RNA synthesis before the initiation of meiosis. However, there is a continuous reduction in the synthesis of proteins and RNA during meiotic prophase. The synthetic activity is initiated after meiosis.

Syncytium and isolation

Before the onset of meiosis, the wall layers, the tapetum, and the MMCs show plasmodesmatal connections with cells of the same layer as well as with those of adjacent layers. As the sporogenous cells enter meiosis, plasmodesmatal connections between wall layers and the tapetum, and between the tapetum and meiocytes, are progressively severed, although plasmodesmata are retained between cells of the same layer (Shivanna and Johri 1985). Initiation of meiosis is characterized by the deposition of a callose wall around individual MMCs, followed by the dissolution of the primary cell wall. Initially, the callose wall is incomplete, leaving many gaps (up to 2–3 μm) through which massive cytoplasmic channels between the neighboring MMCs are established (Bhandari 1984). The cytoplasm and organelles can readily pass through these cytomictic channels from one meiocyte to the other. Thus, all meiocytes of a sporangium form a single cytoplasmic entity, the syncytium (Figure 2.2).

Evidence from a number of studies indicates that syncytium formation is important for maintaining the synchrony for division of meiocytes (Heslop-Harrison 1972). At the end of metaphase I or II, the callose wall around MMCs becomes continuous. Following meiotic division, the callose wall is deposited around individual microspores (Figure 2.3). This results in the isolation of an individual microspore from its neighbors. The isolation of MMCs, and then the microspores, by a callose wall seems to be essential for normal development of pollen grains. Failure of callose wall development or its early breakdown results in pollen sterility (Chapters 7 and 8). A callose wall is also essential for the orderly deposition of the pollen wall, the exine.

Cytoplasmic reorganization

The cytoplasm of meiocytes undergoes profound changes during meiosis. Following the initiation of meiosis, there is a significant fall in the cytoplasmic RNA and elimination of a major portion of the ribosome population. Also, a part of the cytoplasm becomes enclosed with bi- or multimembrane units. The premeiotic cytoplasm enclosed by these membrane units is not subjected to RNA degradation and ribosome elimination, and is carried over to the microspore cytoplasm.

Conspicuous reorganization of the mitochondria and plastids of meiocytes is also observed. As prophase advances, both these organelles round off, become isodiametric, and lose most of their internal membranes. They are further simplified during zygotene and pachytene stages. Both the mitochondria and plastids initiate reorganization toward the end of meiosis, and by the time the microspores are released from tetrads, these organelles show normal profiles.

Isolation of meiocytes within the callose wall and cytoplasmic reorganization have been interpreted to be requirements for the transition from the sporophytic to the gametophytic phase and for the expression of the gametophytic genome (Heslop-Harrison 1971). The callose wall seems to give protection to meiocytes and microspores for the expression of the gametophytic genome without interference either from the parent sporophyte or from other microspores. Cytoplasmic reorganization leading to the elimination of diplophase information is believed to provide suitable conditions for the expression of the gametophytic genome by eliminating the information macromolecules of the sporophyte.

Microspore development

The blueprint of the exine – the primexine– is formed while the tetrads are still enclosed within the callose wall. The presumptive apertures are also demarcated during the formation of the primexine itself. Microspores are released by the digestion of callose by the enzyme callase secreted by the surrounding tapetum. The microspores enlarge and develop characteristic exine by the deposition of sporopollenin precursors on already laid down primexine and by polymerization of the precursors (Wiermann and Gubatz 1992). Pectocellulosic intine on the inner side of the exine is deposited simultaneously with the deposition of exine. Evidence from different sources has conclusively shown that the exine ornamentation is controlled by the sporophytic genome (Shivanna and Johri 1985). In the developing microspores, the cytoplasm that becomes enclosed by bi- or multilayered membrane units during early prophase, and thus escapes cytoplasmic reorganization, seems to carry the information moiety for the exine pattern.

At the time of release, the microspore contains a large nucleus in the center and normal cytoplasmic organelles. A large vacuole forms in the cytoplasm and results in the migration of the nucleus to the periphery. Most of the plastids and mitochondria in the cytoplasm are displaced to the region away from the nucleus, with the result that cytoplasm surrounding the nucleus contains very few mitochondria and plastids. The nucleus undergoes a mitotic division, giving rise to a large vegetative cell and a small generative cell. The generative cell initially remains in contact with the pollen wall, but it is soon detached from the intine and enclosed by the cytoplasm of the vegetative cell. By this time, the large vacuole that had developed earlier in the microspore is resorbed. Initially, the generative cell is bordered by a thin callosic wall. Later, the wall dissolves, and generative and vegetative cells are separated by the membranes of the respective cells. The generative cell is spheroidal to begin with but soon becomes elongated. The shape of the generative cell is maintained by the orientation of microtubules in the cytoplasm, just below the membrane and parallel to the long axis of the cell. The generative cell cytoplasm contains all the organelles; however, in many species the plastids are absent (Hagemann and Schroder 1989; Mogensen 1992).

Soon after the microspore mitosis, DNA in the generative nucleus increases to the 2C level, whereas the DNA of the vegetative nucleus remains at the 1C level. The nucleus of the generative cell is highly condensed, its DNA is associated with a high content of lysine-rich histones, and there is hardly any protein or RNA synthesis in the cytoplasm. In contrast, the nucleus of the vegetative cell is not condensed, lysine-rich histones are practically absent in its DNA, and the cell is active in RNA and protein synthesis.

In about 70% of angiosperm species, the pollen grains are dispersed at the two-celled stage (vegetative and generative cells). In the remaining 30% of species, the generative cell undergoes a mitotic division to give rise to two sperm cells (male gametes) before dispersal; such pollen grains are referred to as three-celled (a vegetative cell and two male gametes). In some species, the two sperm cells show differences in size, shape, and their organelles (Russell 1984; Knox and Singh 1987; Mogensen 1992). Often, the generative cell or the sperm cells are in association with the vegetative nucleus. Such a complex has been referred to as the male germ unit and its significance is not yet clearly established (Mogensen 1992; Yu and Russell 1994).

Both layers of the pollen wall, intine and exine, contain considerable amounts of proteins (Knox and Heslop-Harrison 1970). The intine proteins are present in the form of radially arranged tubules and are generally concentrated near the germpore(s). The exine proteins are present in the cavities between the baculae in tectate grains and in surface depressions in nontectate grains. Many enzymes have been identified in the pollen wall domain. Esterases are present predominantly in the exine, and acid phosphatases in the intine. In some systems, proteins are present in the poral region as well (Pacini et al. 1981). Intine proteins originate from the pollen cytoplasm, whereas exine proteins originate from the surrounding tapetum after the breakdown of tapetal cells. Pollen wall proteins play an important role in pollen–pistil interaction. The surface of pollen grains is also generally covered with lipids, flavonoids, and other degenerative products secreted by the tapetum, and this is referred to as pollenkitt or tryphine.

The tapetum

The tapetum is present as a distinct layer around the sporogenous tissue and plays a key role in pollen development (Figures 2.1–2.8). The tapetal cells are generally multinucleate, often with polyploid nuclei. Unlike the sporogenous cells, tapetal cells continue to remain metabolically active and exhibit protein and RNA synthesis, but do not undergo cytoplasmic reorganization. The tapetum can be one of two types – the secretory type (parietal or glandular) or the plasmodial type (amoeboid or invasive) (Echlin 1971; Pacini et al. 1985; Pacini and Franchi 1991).

In the secretory type (Figures 2.4 and 2.6), tapetal cells remain in the parietal position until their degeneration toward the end of pollen development. Following the in situ degeneration, the tapetal contents, especially proteins and lipids, are released into the thecal cavity and deposited in the cavities/depressions on the surface of the exine. In the plasmodial tapetum (Figures 2.7 and 2.8), the inner tangential and radial walls break down and the tapetal protoplasts intrude into the anther locule amidst the developing microspores. The tapetal protoplasts generally fuse in the anther locule to form a periplasmodium (Figure 2.8); in some species, however, the protoplasts maintain their identity through intact membrane (Figure 2.7). The plasmodium remains functional and comes into direct contact with individual microspores. Toward the end of pollen development, the periplasmodium loses its organized structure and its contents are deposited on the surface of pollen in the form of tryphine. Pollen coat substances play an important role in pollen transfer and in many postpollination functions.

The secretory tapetum generally shows the deposition of sporopollenin granules termed "Ubisch bodies," or orbicules, on an acetolysis-resistant

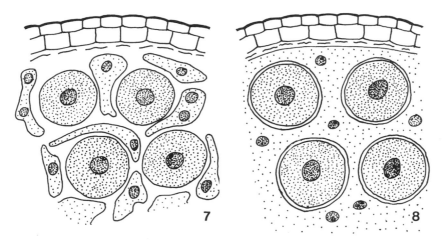

Figures 2.7–2.8 Diagrammatic representation of plasmodial tapetum at microspore stages comparable to that shown in Figure 2.5. The protoplasts of tapetal cells have entered the anther locule between the microspores. The identity of individual protoplasts is maintained in Figure 2.7, whereas the protoplasts have fused to form a periplasmodium in Figure 2.8.

membrane (tapetal membrane) that develops on the inner faces of tapetal cells (Figures 2.4–2.6). In the plasmodial tapetum, the orbicules are absent, but the tapetal membrane usually differentiates on the outer face of the tapetum.

Normal development of the tapetum is crucial for pollen development. Abnormalities in the structure and/or function of the tapetum invariably result in pollen sterility (Chapters 7 and 8). The following functions have been attributed to the tapetum: (i) supply of sporopollenin and a role in exine formation; (ii) breakdown of callose wall around microspore tetrads; (iii) supply of nutrients to developing pollen and reserve metabolites of mature pollen; and (iv) supply of exine proteins and surface coat substances (pollenkitt/tryphine).

In addition to the developmental changes that take place in the meiocytes and tapetum, wall layers also undergo changes. The middle layers degenerate (Figures 2.3–2.5), and the endothecium, situated just below the epidermis, elongates radially and develops radial thickenings along the wall (Figure 2.6). The endothecium is involved in anther dehiscence. At maturity, pollen grains and the anther lose most of the water and become dormant. The details of dehydration of pollen and anther, and of anther dehiscence, are not clearly understood. Dehydration seems to occur either by active resorption of water through anther filament (*Petunia*, *Tradescantia*) or by passive transpiration through anther surface (Pacini 1994). The anther eventually dehisces and the pollen grains are dispersed.

Table 2.1. *General classification of stigma types (with some examples for each group given in parentheses)*

Dry stigma (without apparent fluid secretion)
   Group I. Plumose (receptive surface dispersed on multiseriate branches,
             Gramineae)
   Group II. Receptive surface confined to the stigma
             A. Surface nonpapillate (Acanthaceae)
             B. Surface distinctly papillate
                1. Papillae unicellular (Cruciferae, Compositae)
                2. Papillae multicellular
                      a. Papillae uniseriate (Amaranthaceae)
                      b. Papillae multiseriate (Bromiliaceae, Oxalidaceae)

Wet stigma (surface secretion present during receptive period)
   Group III. Receptive surface papillate (some Rosaceae, some Liliaceae)
   Group IV. Receptive surface nonpapillate (Umbelliferae)
   Group V. Receptive surface covered with copius exudate in which detached
             secretory cells of the stigma are suspended (some Orchidaceae)

Structure of the pistil

The pistil is the female partner in the sexual reproduction of flowering plants. The structural features of the pistil are adapted to receive the pollen and to facilitate its germination, tube growth, and entry into the embryo sac for fertilization. The pistil consists of the stigma, style, and ovary.

The stigma

The stigma is the recipient of pollen. There is great variation in the morphology of the stigma (Heslop-Harrison and Shivanna 1977; Heslop-Harrison 1981; Cresti et al. 1992). The stigma can be one of two types: the wet stigma or the dry stigma, based on the presence or absence of a stigmatic exudate on its surface at the time of pollination (Figures 2.9 and 2.10). Each of these types is further divided on the basis of the presence or absence of papillae on the receptive surface and anatomical details of the papillae (Table 2.1).

Irrespective of the morphology, the stigma invariably contains extracellular components of the receptive surface (Heslop-Harrison and Shivanna 1977; Heslop-Harrison and Heslop-Harrison 1985; Knox et al. 1986). These components are highly heterogeneous and include lipids, proteins, and glycoproteins, different carbohydrates, amino acids, and phenols. A range of enzymes have also been localized; the nonspecific esterases are the predominant ones. Cytochemical demonstration of nonspecific esterases (Shivanna and Rangaswamy 1992) has become a standard method of localization of the receptive surface of the stigma.

In the dry stigma, extracellular components are present in the form of a thin extracellular membrane, the pellicle. The pellicle components originate from the epidermal cells of stigma and/or stigmatic papillae and are extruded onto the surface through discontinuities in the cuticle. In some systems, the pellicle shows ATPase and carbonic anhydrase activity (in addition to esterases and phosphatases); the pellicle also binds to lectins and contains arabinogalactans, a group of carbohydrates with adhesive properties (Clarke et al. 1979). Labeled studies have shown that proteinaceous components of the pellicle have a fast turnover rate.

In the wet stigma, extracellular components are present in the exudate. During the early stages of flower development, the wet stigma is comparable to the dry type with a cuticle–pellicle layer. At later stages, secretions pro-

Figures 2.9–2.12 Structure of the stigma and style. *2.9, 2.10*. Scanning electron micrographs of the stigma of *Ipomoea purpurea* (dry papillate type) and *Fragaria vesca* (wet papillate type), respectively. *2.11, 2.12*. Transmission electron micrographs of transections of the style of *Lycopersicon peruvianum* (Figure 2.11, solid style, im = intercellular matrix) and *Sternbergia lutea* (Figure 2.12, hollow style, c = disrupted parts of the cuticle, cc = canal cells, sc = stylar canal). Only a part of the transmitting tissue in Figure 2.11 and of the stylar canal with surrounding canal cells in Figure 2.12 is shown. The bars represent 10 μm. (*Figures 2.9–2.11 courtesy of M. Cresti; Figure 2.12 after Ciampolini et al. 1990, reproduced with permission of the* Annals of Botany Co.)

duced from cells of the stigma accumulate below the cuticle–pellicle layer; eventually, this layer is disrupted and the exudate spreads on the surface (Shivanna and Sastri 1981). The amount of exudate that accumulates on the stigma is highly variable; it may be confined to the interstices of papillae or may flood the entire surface. The exudate may be lipoidal, as in *Petunia* and *Oenothera*, or aqueous, as in *Lilium*. Proteinase inhibitors have been reported in the stigma of *Nicotiana* (Atkinson et al. 1993).

In some members of Leguminosae (Kenrick and Knox 1989; Owens 1989) and Orchidaceae (Calder and Slater 1985; Slater and Calder 1989), the stigma is cup or funnel shaped. The inner receptive surface is nonpapillate and lined with many layers of secretory cells covered with copious fluid secretion. The cells of the secretory zone are loosely arranged, and often detached cells are found free-floating in the stigma secretion.

Apart from their role in pollen–pistil interaction, different components of the stigma secretion have been implicated in other important functions. The lipoidal component is considered to prevent excessive evaporation and wetting by acting as a liquid cuticle. The phenolics and proteinase inhibitors have been suggested to give protection against insects and pathogens. The stigmatic exudate has also been reported to serve as a nutrient source for pollinating insects.

The style

The style can basically be one of two types – the solid or the hollow (Shivanna and Johri 1985). In the solid style (which is found in the majority of dicotyledons), a core of transmitting tissue, starting from the secretory tissue of the stigma, traverses the whole length of the style. The transmitting tissue is made up of elongated cells connected end to end through the plasmodesmata. In transection, these cells appear circular with conspicuous intercellular spaces filled with secretion products (Figure 2.11) (Knox 1984; Sanders and Lord 1992). The intercellular substance is composed predominantly of pectin but it also contains proteins, glycoproteins, and often lipids; it also responds to many enzymes, such as esterases, acid phosphatases, and peroxidases. A number of transmitting tissue-specific, proline-rich proteins have been localized in the intercellular matrix (Cheung et al. 1993; Gasser and Robinson-Beers 1993; Wang et al. 1993). The cells of the transmitting tissue exhibit normal ultrastructural profiles with numerous mitochondria, active dictyosomes, rough endoplasmic reticulum (RER), plastids, and ribosomes. Endoplasmic reticulum (ER) and Golgi vesicles have been implicated in the secretion of the intercellular matrix (Kristen et al. 1979). Pollen tubes grow down the style through the intercellular matrix of the transmitting tissue.

In the hollow style, a canal, termed the stylar canal, originating from the stigma surface traverses the whole length of the style and joins the ovarian cavity. The stylar canal is bordered by one layer or a few layers of glandular cells, the canal cells (Figure 2.12). The canal cells in the young bud are lined by a layer of cuticle. The secretion product from the canal cells accumulates below the cuticle. In many systems, such as *Gladiolus* and *Crocus*, the cuticle remains intact but is pushed toward the center of the canal (Figure 2.12) (Heslop-Harrison 1977). In others, such as *Lilium*, the cuticle is disrupted during the development of the pistil, and the stylar canal becomes filled with the secretion (Dickinson et al. 1982). In some hollow-styled systems, the inner tangential wall of the canal cells shows wall ingrowths characteristic of transfer cells. As in solid-styled systems, the stylar secretion is rich in carbohydrates and proteins, and shows esterase and acid phosphatase activity (Tilton and Horner 1980). In some species, lipids have also been reported in the stylar secretions.

In members of Leguminosae (Papilionoideae), although the style is hollow, it is not continuous with the stigmatic surface (Shivanna and Owens 1989). The stigma is solid; the upper part of the hollow style is derived secondarily by the dissolution of cells of the transmitting tissue, whereas the lower part is an extension of the ovary.

Thus, irrespective of the structural diversity of the stigma and style, the surface of the stigma and the path of pollen tube growth in the pistil invariably contain extracellular components, which come into direct contact with the pollen grain and pollen tube. In some systems, the proteins present on the surface of the stigma and in the style show qualitative differences, which may be related to their function (Heslop-Harrison and Heslop-Harrison 1982; Miki-Hirosige et al. 1987).

Ovary and ovule

The transmitting tissue/canal cells of the style continue into the ovary as the placenta. The ovule, the seat of the female gametophyte (embryo sac), develops on the placenta. Extensive studies have been carried out on the structural details of the ovule and embryo sac (Maheshwari 1950, 1963; Tilton 1981a, b; Tilton and Lersten 1981; Johri 1984; Cresti et al. 1992; Gasser and Robinson-Beers 1993; Russell 1993). Only the basic aspects are summarized here.

The ovule essentially consists of one or two outer coverings (the integuments), the nucellus, and the embryo sac (Figure 2.13). The integuments do not cover the nucellus completely but leave a narrow passage, the micropyle, at the tip. The ovule is attached to the placenta through a short stalk, the funicle. The ovules generally become curved to different degrees through unequal growth of the funicle. In the anatropous condition, which is the most common, the ovule is completely inverted and thus the micropyle is positioned close to the placenta (Figure 2.13). The nucellar tissue may be massive (crassinucellate) or confined to a few layers (tenuinucellate). In the tenuinucellate type, the nucellar cells surrounding the embryo sac degenerate during development and the mature embryo sac comes into contact with the inner layer of the integument, which differentiates into a specialized layer, the endothelium or integumentary tapetum. The cells of the endothelium are radially elongated and rich in cytoplasm with a prominent nucleus. There is great variation in the development and structure of the mature embryo sac (Reiser and Fischer 1993). A brief description of the most common type, the *Polygonum* type, follows.

A megaspore mother cell differentiates just below the nucellar epidermis at the micropylar region. It is distinguishable from other cells by its large size, dense cytoplasm, and a prominent nucleus. The megaspore mother cell undergoes meiotic divisions and gives rise to a linear row of four megaspores. Three of the megaspores, situated toward the micropylar region, degenerate. The chalazal megaspore persists and gives rise to the mature eight-nucleate, seven-celled embryo sac by undergoing three mitotic divisions of the nucleus and subsequent cellularization. During megasporogenesis, callose is deposited along the wall of the megaspore mother cell and along the transverse walls of the resulting megaspores, similar to the callose wall around microspores. After the completion of megasporogenesis, the callose wall disappears around the functional megaspore, although it may persist for a longer time around the degenerating megaspores.

The mature embryo sac essentially consists of the central cell, the egg apparatus at the micropylar end, and three antipodal cells at the chalazal end (Figures 2.13 and 2.14). The central cell is the largest cell of the embryo sac and contains two nuclei, the polar nuclei, toward the micropylar end. The central cell is highly vacuolated, and the cytoplasm is confined to a thin layer along the wall and around the polar nuclei. The polar nuclei eventually fuse to give rise to the secondary nucleus. At the micropylar pole where the central cell borders the egg apparatus, the central cell lacks the cell wall (Figure 2.14). In a few systems, wall ingrowths, characteristic of transfer cells, develop on the lateral wall of the central cell at the micropylar and/or chalazal region. The central cell does not show high metabolic activity, as revealed by ultrastructural and cytochemical studies.

The antipodal cells are ephemeral and degenerate by the time fertilization is effected. In many other types of embryo sac, there is great variation in the number and organization of antipodal cells. In members of Poaceae, the antipodals persist even after fertilization and proliferate into a mass of up to one hundred cells.

The egg apparatus located at the micropylar pole consists of three pear-shaped cells (two synergids and one egg) attached to the embryo sac wall near the micropyle (Figures 2.14–2.16). The egg cell is distinctly polar, containing a large vacuole at the micropylar region and most of the cytoplasm and nucleus at the chalazal end. The cell wall of the egg is usually absent or incomplete at the chalazal end. In general, the ultrastructural profile of the egg cell indicates that it is metabolically less active when compared to other cells of the embryo sac.

The synergids are also polarized cells; the micropylar half is densely cyto-

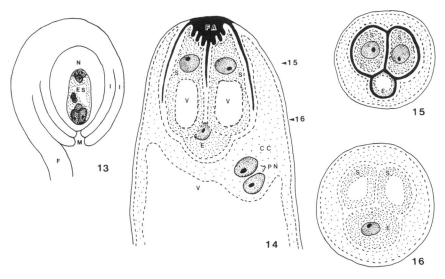

Figures 2.13–2.16 Diagrammatic representation of ovule and embryo sac. Cellulosic wall is shown in black and plasmamembrane in broken line. *2.13.* Longisection of an anatropous ovule with eight-nucleate, seven-celled embryo sac (ES = embryo sac, F = funiculus, I = integuments, M = micropyle, N = nucellus). *2.14.* The details of the micropylar part of the embryo sac in longisection (CC = central cell, E = egg cell, FA = filiform apparatus, PN = polar nuclei, S = synergid, V = vacuole). *2.15, 2.16.* Transections of embryo sac at levels 15 and 16 shown in Figure 2.14 to show the arrangement of cells of the egg apparatus. Cellulosic wall is present around the egg and the synergids in Figure 2.15, whereas in Figure 2.16, the egg, synergids and the central cell are separated through their respective membranes.

plasmic, whereas the chalazal half is highly vacuolated. The synergids at the micropylar region develop extensive wall ingrowths into the cytoplasm, resulting in an enormous increase in the surface area of the plasmalemma. This thickened wall region is called the filiform apparatus (Figure 2.14). The synergids are metabolically very active. They contain a large number of mitochondria and dictyosomes, abundant ribosomes, and extensive endoplasmic reticulum. The filiform apparatus is considered to be involved in the secretion of metabolic products of the synergids into the micropyle, and in the entry of pollen tubes. The cell wall of the synergids is attenuated toward the lower region and partially or completely absent in the chalazal region. Thus, in the lower part of the egg apparatus, the egg cell, synergids, and central cell are separated only by their respective plasma membranes (Figures 2.14 and 2.16). Frequently, one of the two synergids degenerates prior to the arrival of the pollen tube into the embryo sac. Together the egg, two synergids, and central cell have been referred to as the "female germ unit" (Huang and Russell 1992), because this unit plays a direct role in pollen tube entry, the discharge of sperm cells, and double fertilization.

Pollen–pistil interaction

Pollen grains are deposited on the stigma either by close proximity of the anthers or by biotic/abiotic agents. Successful pollination initiates a number of sequential events that culminate in the discharge of male gametes into the embryo sac (Figure 2.17). All these events, from pollination to the release of male gametes in the embryo sac, are included in pollen–pistil interaction.

Pollen viability and stigma receptivity

Pollen viability and stigma receptivity are critical for the effective initiation of pollen–pistil interaction. Pollen viability refers to the ability of pollen to successfully complete postpollination events on a receptive, compatible pistil and to deliver functional male gametes to the embryo sac. The period for which pollen grains remain viable after they are shed varies greatly between species. The details of pollen viability are discussed in Chapter 14.

Stigma receptivity refers to the ability of the stigma to support germination and tube growth of viable, compatible pollen. In general, the stigma becomes receptive by the time the flower opens and the pollen is shed. However, in many self-pollinated and protogynous species the receptivity of the stigma is advanced – and in protandrous systems is delayed – by one day or a few days in relation to flower opening (Lloyd and Webb 1986; Williams et al. 1991). The duration for which the stigma remains receptive is also variable (one day to many days). One of the standard methods to study stigma receptivity is to carry out controlled pollinations on the stigmas at different stages and to study pollen germination and pollen tube growth (Shivanna and Rangaswamy 1992).

For the successful initiation of pollen–pistil interaction, pollen grains have to come into contact with the receptive region of the stigma. In some species, the whole surface of the stigma may not be receptive. In chickpea, for example, the receptivity is confined to a small group of papillae at the extreme tip; the remaining part of the stigma is not receptive and cannot support pollen germination (Malti and Shivanna 1983; Turano et al. 1983). In some members of Boraginaceae, the stigma papillae are capitate with heavily cutinized, nonreceptive heads (Heslop-Harrison 1981). The receptive surface is confined to the junction of the bases of the papillae. Pollen grains do not germinate unless they reach the receptive area between the papillae. In some other legumes, the stigmatic exudate, which is required for pollen germination, remains below the cuticle at the time of pollination. Disruption of the cuticle and release of the exudate are necessary for pollen germination; this generally happens during tripping by visiting insects (Lord and Heslop-Harrison 1984; Shivanna and Owens 1989). In yet other cases, the receptive stigma is covered by a whorl of nonreceptive hairs that may prevent the pollen from coming into contact with the receptive surface; tripping is also needed for effective pollination in such systems (Shivanna and Owens 1989).

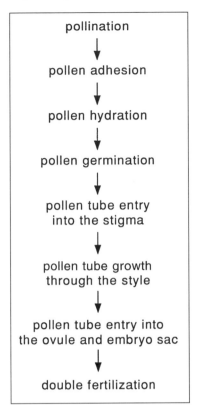

pollination

↓

pollen adhesion

↓

pollen hydration

↓

pollen germination

↓

pollen tube entry
into the stigma

↓

pollen tube growth
through the style

↓

pollen tube entry into
the ovule and embryo sac

↓

double fertilization

Figure 2.17 Sequential events during pollen–pistil interaction.

Pollination stimulus

Pollination initiates many changes in the pistil and ovary. In some species, such as watermelon (Sedgley and Scholefield 1980) and *Acacia* (Kenrick and Knox 1981), pollination stimulates another phase of stigmatic secretion. This may facilitate full hydration and germination of a large number of pollen grains. Pollination induces wilting of the corolla (Gilissen and Hoekstra 1984), which is preceded by a spurt of ethylene production in the pistil and corolla (Peach et al. 1987). Pollen-borne ethylene precursor, 1-aminocyclo-propane-1-carboxylic acid (ACC), is largely responsible for the early synthesis of ethylene in the pistil (Singh et al. 1992). Pollination induces the accumulation of flavonols, especially kaempferol, in the stigmatic exudate and outer cell layers of the stigma (Vogt et al. 1994). Flavonols either in the pollen or in the stigma are essential for pollen germination and pollen tube growth.

Many of the changes induced by pollination are observed in the lower part of the style and ovary. Postpollination changes in the pattern of RNA and protein synthesis and in the activity of several enzymes occur in the lower part of the style and ovary (van der Donk 1974; Bredemeijer 1982) before the arrival of the pollen tubes. These changes are specific to self- and cross-pollination in self-incompatible taxa. In pear (Herrero and Gascon 1987), pollination delays embryo sac degeneration, thus extending the period over which fertilization can be effected. As mentioned earlier, in many species pollination induces degeneration of one of the synergids. Studies clearly indicate that the stimulus of pollination is transferred to the ovary before pollen tubes reach the ovary. The nature of the stimulus is not yet clear (Spanjers 1978). Degeneration of synergid in cotton (Jensen and Ashton 1981) and prolongation of embryo sac viability in pear seem to be mediated through the release of gibberellic acid; the exogenous supply of gibberellic acid induces these responses. In general, flowers pollinated with compatible pollen senesce rapidly, whereas unpollinated flowers, as well as those pollinated with incompatible pollen, last much longer. This may be an adaptation to prolong the life of the flower in the absence of compatible pollination. In many parthenogenetic species (in which an embryo develops without fertilization), pollination stimulates the egg to develop parthenogenetically (Koltunow 1993).

In most species, ovules are fully developed by the time pollination takes place. In many orchids, however, ovules are not fully formed at the time of pollination, and they appear in the form of nucellar filaments along the placental ridges containing a quiescent archesporial cell. Further development of the ovule resulting in the differentiation of mature embryo sac takes place only after pollination. Thus, the time between pollination and fertilization is a long one and ranges from 2 weeks to 10 months in different species (Wirth and Withner 1959; Clifford and Owens 1988).

Pollen adhesion and hydration

Following pollination, pollen adhesion is the first postpollination event. Pollen adhesion largely depends on the nature and extent of pollen and stigma surface components. In species with wet stigma, pollen adhesion is not critical. Any pollen coming into contact with the exudate is held because of the stickiness and surface tension of the exudate. In dry stigma, pollen adhesion is more critical and depends on the components of the pellicle and pollen coat substances. In many systems with dry stigma, adhesive carbohydrates, particularly arabinogalactans, are present on the surface of the stigma and are likely to play an important role in pollen adhesion.

Effective pollen adhesion is followed by pollen hydration as a result of the passage of water from the stigma into the pollen, driven by the osmotic potential differences (Heslop-Harrison 1979). The rate of hydration depends on the nature of the stigma. In wet stigma, hydration is generally rapid, whereas in the dry type it is gradual and may last for 30–60 minutes. The pollen of such systems requires controlled hydration for satisfactory in vitro germination. The pollen of wet stigma generally does not require controlled hydration for in vitro germination.

Following hydration, pollen wall proteins are released onto the surface of the stigma and they come into contact with the components of the pellicle or the exudate. This is the first chemical interaction of the two partners and its details are not yet fully understood. In several systems, this triggers many other responses in the pollen and/or the pistil, indicating that this initial interaction may establish the identity of pollen.

Pollen germination

Although pollen hydration does not necessarily result in germination, it is an essential requirement for successful germination. The stigmatic surface provides ideal conditions for the germination of compatible pollen. In wet stigma, the role of stigmatic exudate in pollen germination is variable. In some taxa, such as *Amaryllis* and *Crinum*, the exudate seems to be necessary for pollen germination; stigmas free from the exudate do not support pollen germination (Shivanna and Sastri 1981). However, in others, such as *Petunia* and *Nicotiana*, the exudate does not seem to play an important role in pollen germination; even the younger stigmas free from the exudate support satisfactory pollen germination.

In dry stigma, the pellicle is involved in pollen germination. Removal of the pellicle components by enzymes or detergents inhibits pollen germination or pollen tube entry into the stigma in many systems (Knox et al. 1976; Shivanna et al. 1978; Clarke et al. 1979). The stigma surface also provides boron and calcium, which are required for pollen germination but are generally deficient in pollen. In *Vitis vinifera,* only those stigmas that contain 2–5 ppm of boron in the stigmatic secretion permit pollen germination (Gartel 1974). The accumulation of a high concentration of calcium in the pellicle and its transport to the pollen have been demonstrated in a few systems (Tirlapur and Shiggaon 1988; Bednarska 1989, 1991).

Pollen tube entry into the stigma and growth through the style

In wet stigma, in which the cuticle of the stigma surface/papillae is disrupted during secretion of the exudate, there is no physical barrier for pollen tube entry into the intercellular spaces of the transmitting tissue of the stigma. In dry stigma, the cuticle provides a physical barrier for pollen tube entry. The pollen tube has to erode the cuticle at the region of contact by the activation of cutinases. Many studies indicate that cutinases of the pollen require activators from the stigma. In some systems, digestion of pellicle proteins by pronase does not affect pollen germination but totally inhibits the entry of pollen tubes into the papillae (Knox et al. 1976; Shivanna et al. 1978).

In solid-styled systems, pollen tubes enter the intercellular spaces of the transmitting tissue and continue to grow through the intercellular matrix of the stigma and style. In hollow-styled systems, pollen tubes enter the stylar canal and grow on the surface of the canal cells. Pollen tubes invariably come into contact with the extracellular components secreted by cells of the transmitting tissue or the canal cells. As the glandular transmitting tissue or canal cells extend into the ovary as placental ridges, pollen tubes continue their growth into the ovary along the placentae.

The nutrients present in the pollen are not sufficient to sustain pollen tube growth to the ovules. In maize, for example, it has been calculated that the pollen reserve can sustain growth for no more than 2 cm, whereas the pollen tube may have to grow more than 10 cm (Heslop-Harrison et al. 1984). Pollen tubes take up water and nutrients from the pistil during growth (Shivanna and Johri 1985).

In two-celled pollen systems, pollen tube growth occurs in two phases: a period of slow growth in which pollen reserves are utilized, and a period of rapid growth during which the tube utilizes nutrients from the style (Mulcahy and Mulcahy 1983). Three-celled systems, on the other hand, seem to have only the period of rapid growth. In general, three-celled systems show a short lag phase (the period between hydration and germination) and a faster rate of pollen tube growth, as compared to two-celled pollen systems.

Pollen tube entry into the ovule

After entering the ovary, pollen tubes grow along the surface of the placenta toward the ovules and enter the micropyle. In some species, additional structures, such as an obturator (a placental outgrowth in the micropylar region of the ovule), are present and they seem to play an important role in regulating pollen tube entry. In peach, for example, pollen tubes come into contact with the obturator after entering the ovary and cease further growth (Herrero and Arbeloa 1989). Pollen tube growth is resumed only after 4–5 days, by which time starch present in the obturator is hydrolyzed with the concomitant production of a secretion that stains for carbohydrates and pro-teins. This secretion seems to be necessary for the resumption of pollen tube growth.

A pollen tube grows through the micropyle and enters one of the synergids through the filiform apparatus. In many species, such as *Gossypium*, *Hordeum*, and *Linum*, one of the synergids degenerates before the arrival of the pollen tube (Russell 1992). The cytoplasm of the degenerated synergid becomes electron-opaque and the plasma membrane loses its integrity. In such species, the pollen tube invariably enters the degenerated synergid (Jensen 1973). In species in which both the synergids remain intact, the pollen tube enters one of the two synergids and the synergid that receives the pollen tube starts degenerating; the other synergid remains intact for some time.

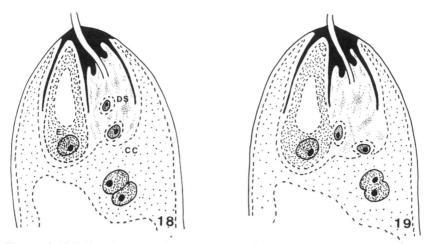

Figures 2.18–2.19 Diagrammatic representation of pollen tube entry into the embryo sac and fertilization. Longisection of the micropylar part of the embryo sac in a plane passing through one of the synergids and the egg. The synergid that receives the pollen tube is degenerated (DS) and lacks intact plasma membrane and, thus, the sperms are able to come into direct contact with the membrane of the egg (E) and central cell (CC).

Double fertilization

The pollen tube discharges its contents (two sperm cells, vegetative nucleus, and cytoplasm) into the synergid either through a pore or by rupture at the tip (Figure 2.18). One of the sperms enters the egg, and the other enters the central cell. The details of the movement of the sperm cells are not yet understood (Knox and Singh 1987). According to the classical model of double fertilization, the two sperms are identical, and fertilization of the egg and the central cell by the two sperms is random. Following the demonstration of sperm dimorphism in many systems, attempts have been made to investigate if there is preferential fertilization of the egg and the central cell by specific sperms. Studies of *Plumbago* (Russell 1985, 1993) indicate that the sperm cell rich in plastids preferentially fuses with the egg. If such preferential fertilization exists, there has to be a mechanism of recognition between one of the sperms and the egg cell, and the other sperm and the central cell. So far, no information is available on these aspects (Knox and Singh 1987). Also, there is no evidence of preferential fertilization in any other systems.

Very few studies have been carried out on fertilization per se. It has been shown that fertilization takes place through the contact and fusion of plasma membranes of the egg and the central cell with the respective sperm cells (Figure 2.19), which form a bridge through which the sperm nucleus enters (Jensen 1973). Nuclear fusion is also mediated through the fusion of nuclear membranes. In a majority of species, the cytoplasmic organelles of the sperm cells are excluded from fertilization (Hagemann and Schroder 1989). Electron microscopic studies on barley (Mogensen 1988) have also shown the exclusion of sperm cytoplasm from the synergid.

Pollen tube guidance

Pollen tubes follow a predetermined path in the pistil from the stigma through the style into the micropyle and eventually into one of the two synergids. It was thought that this directional growth of pollen tubes was due to the presence of a gradient of chemotropic substance in the pistil (Shivanna and Johri 1985). However, recent evidence indicates that the architecture of cells of the transmitting tissue from the stigma to the ovary provides a path of least resistance for the growth of pollen tubes. Once the pollen tubes orient in the correct direction in the transmitting tract, they continue to grow along the defined route, responding to the structural cues of the transmitting tract toward the ovary (Jensen and Fischer 1969; Heslop-Harrison and Heslop-Harrison 1986). However, chemotropic guidance seems to be necessary for changing the direction of pollen tube growth from the placenta into the micropyle. The details of the chemotropic substance are not yet clear. A great deal of evidence indicates that the nucellar cells along the micropyle and/or the synergids are the source of the chemotropic substance (Chao 1977). The release of calcium (accumulated in the synergids) following the degeneration of one of the synergids has been suggested to form a chemotropic gradient (Jensen and Ashton 1981). Recently, a high level of calcium accumulation has been reported in the synergids of wheat and *Pennisetum* (Chaubal and Reger 1990, 1992) and is believed to play a role in regulating the entry of pollen tubes into the synergid and subsequent fertilization (Chaubal and Reger 1992).

Significance of pollen–pistil interaction

Pollen–pistil interaction is unique to flowering plants and plays a significant role in sexual reproduction. The screening and selection of the male partner, which are essential requirements for sexual reproduction, take place during pollen–pistil interaction (Shivanna and Johri 1985). The screening is for the quality and compatibility of pollen. Generally, the number of pollen grains deposited on the stigma under adequate pollination conditions is much more than the number of ovules available for fertilization. This is particularly so in systems with one or fewer ovules in the ovary. Pollen grains are, therefore, subjected to competition during pollen–pistil interaction (Mulcahy 1979). Only those pollen that are more vigorous (that is, germinate early and show faster tube growth) are able to effect fertilization. As outcrossing takes place to a limited or greater extent in a majority of flowering plants, there is enormous genetic variability in the pollen population in which pollen competition and selection can operate.

The competition among pollen grains during pollen–pistil interaction results in the increased vigor of the progeny. Pollen competition can be experimentally manipulated by altering the density of pollen on the stigma, or the distance through which the pollen tube has to grow in the style (Schlichting et al. 1987; Hormaza and Herrero 1992). The progeny derived from intense pollen competition have been shown to be more vigorous and uniform when compared to the progeny derived from lesser competition (Chapter 16). Pollen competition during pollen–pistil interaction has, therefore, been considered an important causative factor in the evolutionary success of flowering plants.

Pollen grains are also screened for compatibility and incongruity during pollen–pistil interaction. Only compatible pollen grains are able to complete pollen–pistil interaction, whereas incompatible pollen (that is, self-pollen in self-incompatible species, and pollen from other species/genera in incongruity) are inhibited at some stage before entering the embryo sac (Chapters 12 and 13). Thus, pollen–pistil interaction offers enormous potential for the manipulation of pollen screening.

Acknowledgments

The help of Dr. M. N. B. Nair in the preparation of line diagrams is gratefully acknowledged.

References

Atkinson, A.H., Heath, R.L., Simpson, R.J. & Anderson, M.A. (1993). Proteinase inhibitors in *Nicotiana alata* stigmas are derived from a precursor protein which is processed into five homologous inhibitors. *Plant Cell* 5:203–213.
Bedinger, P. (1992). The remarkable biology of pollen. *Plant Cell* 4:879–887.
Bednarska, E. (1989). Localization of calcium on the surface of stigma in *Ruscus aculeatus* L. studies by the chlorotetracyline and X-ray microanalysis. *Planta* 179:11–16.
Bednarska, E. (1991). Calcium uptake from the stigma by germinating pollen in *Primula officinalis* L. and *Ruscus aculeatus* L. *Sex. Plant Reprod.* 4:36–38.
Bhandari, N.N. (1984). The microsporangium. In *Embryology of Angiosperms*, ed. B.M. Johri, pp. 53–121. Berlin: Springer-Verlag.
Blackmore, S. & Knox, R.B., eds. (1990). *Microspores: Evolution and Ontogeny.* London: Academic Press.
Bredemeijer, G.M.M. (1982). Mechanism of peroxidase isozyme induction in pollinated *Nicotiana alata* styles. *Theor. Appl. Genet.* 62:305–309.
Calder, D.M. & Slater, A.T. (1985). The stigma of *Dendrobium speciosum* Sm (Orchidacea): a new stigma type comprising detached cells within mucilaginous matrix. *Ann. Bot.* 55:297–307.
Chao, C.H. (1977). Further cytological studies of a periodic acid-Schiff's substance in the ovules of *Paspalum orbiculare*. *Amer. J. Bot.* 64:920–930.
Cheung, A.Y., May, B., Kawata, E.E., Gu, Q. & Wu, H.M. (1993). Characterization of cDNAs for stylar transmitting tissue-specific proline-rich proteins in tobacco. *Plant J.* 3:151–160.

Chaubal, R. & Reger, B.J. (1990). Relatively high calcium is localized in synergid cells of wheat ovaries. *Sex. Plant Reprod.* 3:98–102.

Chaubal, R. & Reger, B.J. (1992). The dynamics of calcium distribution in the synergid cells of wheat after pollination. *Sex. Plant Reprod.* 5:206–213.

Clarke, A.E., Gleeson, P., Harrison, S. & Knox, R.B. (1979). Pollen–stigma interactions: identification and characterization of surface components with recognition potential. *Proc. Natl. Acad. Sci. USA* 76:3358–3362.

Clifford, S.C. & Owens, S. (1988). Post-pollination phenomena and embryo development in the Orchidaceae. In *Sexual Reproduction in Higher Plants*, eds. M. Cresti, P. Gori, and E. Pacini, Heidelberg: Springer-Verlag.

Cresti, M., Blackmore, S. & van Went, J.L. (1992). *Atlas of Sexual Reproduction in Flowering Plants*. Berlin: Springer-Verlag.

Dickinson, H.G. (1987). The physiology and biochemistry of meiosis in the anther. *Intl. Rev. Cytol.* 107:79–109.

Dickinson, H.G., Moriarty, J. & Lawson, J. (1982). Pollen–pistil interaction in *Lilium longiflorum*: The role of pistil in controlling pollen tube growth following cross- and self-pollinations. *Proc. R. Soc. Lond. Ser. B* 215:45–62.

Echlin, P. (1971). The role of tapetum during microsporogenesis of angiosperms. In *Pollen: Development and Physiology*, ed. J. Heslop-Harrison, pp. 41–61. London: Butterworths.

Gartel, W. (1974). Micronutrients: their significance in vein nutrition with special regard to boron deficiency and toxicity. *Weinberg and Keller* 21:435–508.

Gasser, C.S. & Robinson-Beers, K. (1993). Pistil development. *Plant Cell* 5:1231–1239.

Gilissen, L.J.W. & Hoekstra, F.A. (1984). Pollination induced corolla wilting in *Petunia hybrida*. Rapid transfer through the style of a wilting-inducing substance. *Plant Physiol.* 75:496–498.

Goldberg, R.B., Beals, T.P. & Sanders, P.M. (1993). Anther development: basic principles and practical applications. *Plant Cell* 5:1217–1229.

Hagemann, R. & Schroder, M.B. (1989). The cytological basis of the plastid inheritance in angiosperms. *Protoplasma* 152:57–64.

Herrero, M. & Arbeloa, A. (1989). Influence of the pistil on pollen tube kinetics in peach (*Prunus persica*). *Amer. J. Bot.* 76:1441–1467.

Herrero, M. & Gascon, M. (1987). Prolongation of embryo sac viability in pear (*Pyrus communis*) following pollination or treatment with gibberellic acid. *Ann. Bot.* 60:287–293.

Heslop-Harrison, J., ed. (1971). *Pollen: Development and Physiology*. London: Butterworths.

Heslop-Harrison, J. (1972). Sexuality in angiosperms. In *Plant Physiology: A Treatise*, Vol. VI C, ed. F.C. Steward, pp. 134–289. New York: Academic Press.

Heslop-Harrison, J. (1979). An interpretation of the hydrodynamics of pollen. *Amer. J. Bot.* 66:737–743.

Heslop-Harrison, J. & Heslop-Harrison, Y. (1982). Pollen–pistil interaction in the Leguminosae. Constituents of the stylar fluid and stigma secretion of *Trifolium pratense*. L. *Ann. Bot.* 49:729–735.

Heslop-Harrison, J. & Heslop-Harrison, Y. (1985). Surfaces and secretion in the pollen–stigma interaction: a brief review. *J. Cell Sci. Suppl.* 2:287–300.

Heslop-Harrison, J. & Heslop-Harrison, Y. (1986). Pollen-tube chemotropism: Fact or delusion? In *Biology of Reproduction and Cell Motility in Plants and Animals*, eds. M. Cresti and R. Dallai, pp. 169–174. Siena: Univ. of Siena Press.

Heslop-Harrison, Y. (1977). The pollen–stigma interaction. Pollen tube penetration in *Crocus*. *Ann. Bot.* 41:913–922.

Heslop-Harrison, Y. (1981). Stigma characteristics and angiosperm taxonomy. *Nordic J. Bot.* 1:401–420.

Heslop-Harrison, Y., Reger, B.J. & Heslop-Harrison, J. (1984). The pollen–stigma interaction in the grasses. 6. The stigma (silk) of *Zea mays* L. as host to the pollen of *Sorghum bicolor* (L.) Moench and *Pennisetum americanum* (L.) Leeke. *Acta Bot. Neerl.* 33:205–227.

Heslop-Harrison, Y., and Shivanna, K.R. (1977). The receptive surface of angiosperm stigma. *Ann. Bot.* 401:1233–1258.

Hormaza, J.I. & Herrero, M. (1992). Pollen selection. *Theor. Appl. Genet.* 83:663–672.

Huang, B.Q. & Russell, S.C. (1992). Female germ unit: organization, isolation and function. *Intl. Rev. Cytol.* 140:233–293.

Jensen, W.A. (1973). Fertilization in flowering plants. *Bioscience* 23:21–27.

Jensen, W.A. & Ashton, M.F. (1981). Synergid pollen tube interaction in cotton. In *XIII International Botanical Congress*, p. 61. Sydney: Abstract.

Jensen, W.A. & Fischer, D.B. (1969). Cotton embryogenesis: the tissues of the stigma and style and their relation to the pollen tube. *Planta* 84:97–121.

Johri, B.M., ed. (1984). *Embryology of Angiosperms*. Berlin, Heidelberg, New York: Springer-Verlag.

Kenrick, J. & Knox, R.B. (1981). Post-pollination exudate from stigmas of *Acacia* (Mimosaceae). *Ann. Bot.* 48:103–106.

Kenrick, J. & Knox, R.B. (1989). Pollen–pistil interactions in Leguminosae (Mimosoideae). In *Advances in Legume Biology*, eds. C.H. Stirton and J.L. Zarucchi. Monographs Systematic Botany from the Missouri Botanical Garden, Vol. 29, pp. 127–156. St. Louis: Missouri Botanical Garden.

Knox, R.B. (1984). Pollen–pistil interaction. In *Encyclopedia of Plant Physiology*, Vol. 17, *Cellular Interactions*, eds. H.F. Linskens and J. Heslop-Harrison, pp. 508–608. Berlin, Heidelberg, New York: Springer-Verlag.

Knox, R.B., Clarke, A.E., Harrison, S., Smith, P. & Marchalonis, J.J. (1976). Cell recognition in plants: determinants of the stigma surface and their pollen interactions. *Proc. Natl. Acad. Sci. USA* 73:2788–792.

Knox, R.B. & Heslop-Harrison, J. (1970). Pollen-wall proteins: localization and enzyme activity. *J. Cell Sci.* 6:1–27.

Knox. R.B. & Singh, M.B. (1987). New prospectives in pollen biology and fertilization. *Ann. Bot. Suppl.* 4:15–37.

Knox, R.B., Williams, E.G. & Dumas, C. (1986). Pollen, pistil and reproductive function in crop plants. *Plant Breed. Rev.* 4:9–79.

Koltunow, A.M. (1993). Apomixis: embryo sacs and embryos formed without meiosis or fertilization in ovules. *Plant Cell* 5:1425–1437.

Kristen, U., Biedermann M., Liebezit,G. & Dawson, R. (1979). The composition of stigmatic exudate and the ultrastructure of the stigma papillae in *Aptenia cordifolia*. *Eur. J. Cell Biol.* 19:281–287.

Lloyd, D.G. & Webb, C.J. (1986). The avoidance of interference between presentation of pollen and stigma in angiosperms: dichogamy. *New Zeal. J. Bot.* 24:135–162.

Lord, E.M. & Heslop-Harrison, Y. (1984). Pollen–stigma interaction in the Leguminoseae: stigma organization and the breeding system in *Vicia faba* L. *Ann. Bot.* 54:827–836.

Maheshwari, P. (1950). *An Introduction to the Embryology of Angiosperms*. New York: McGraw-Hill.

Maheshwari, P., ed. (1963). *Recent Advances in the Embryology of Angiosperms*. Delhi: International Society of Plant Morphologists.

Malti & Shivanna, K.R. (1983). Pollen–pistil interaction in chickpea. *Intl. Chickpea Newslett.* 9:10–11.

Miki-Hirosige, H., Hoek, I.H.S. & Nakamura, S. (1987). Secretions from the pistil of *Lilium longiflorum*. *Amer. J. Bot.* 74:1709–1715.

Mogensen, H.L. (1988). Exclusion of male mitochondria and plastids during syngamy in barley as a basis of maternal inheritance. *Proc. Natl. Acad. Sci. USA* 85:2594–2597.

Mogensen, H.L. (1992). The male germ unit. *Intl. Rev. Cytol.* 140:129–148.

Mulcahy, D.L. (1979). Rise in angiosperms: a genecological factor. *Science* 206:20–23.

Mulcahy, G.B. & Mulcahy, D.L. (1983). A comparison of pollen tube growth in bi- and tri-nucleate pollen. In *Pollen Biology: Implications for Plant Breeding*, eds. D.L. Mulcahy and E. Ottaviano, pp. 29–33. New York: Elsevier Biomedical.

Owens, S.J. (1989). Stigma, pollen, and pollen–stigma interaction in Caesalpinioideae. In *Advances in Legume Biology*, eds. C.H. Stirton and J.L. Zarrucchi, Monographs Systematic Botany from the Missouri Botanical Garden, Vol. 29, pp. 113–126. St. Louis: Missouri Botanical Garden.

Pacini, E. (1994). Cell biology of anther and pollen development. In *Genetic Control of Self-Incompatibility and Reproductive Development in Flowering Plants*, eds. E.G. Williams et al., pp. 289–308. The Netherlands: Kluwer Academic Publications.

Pacini, E. & Franchi, G.G. (1991). Diversification and evolution of tapetum. In *Pollen and Spores*, eds. S. Blackmore and S.H. Barnes, pp. 301–316. Oxford: Clarendron Press.

Pacini, E., Franchi, G.G. & Hesse, M. (1985). The tapetum: its form, function and possible phylogeny in embryophyta. *Plant Syst. Evol.* 149:155–185.

Pacini, E., Franchi, G. & Sarfatti, G. (1981). On widespread occurrence of poral sporophytic proteins in pollen of dicotyledons. *Ann. Bot.* 47:405–408.

Peach, J.C., Latche. A., Larriguadiere, C. & Reid, M.S. (1987). Control of early ethylene synthesis in pollinated *Petunia* flowers. *Plant Physiol. Biochem.* 25:431–437.

Reiser, L. & Fischer, R.L. (1993). The ovule and the embryo sac. *Plant Cell* 5:1291–1301.

Russell, S.D. (1984). Ultrastructure of the sperm of *Plumbago zeylanica*: 2. Qualitative cytology and three-dimensional reconstruction. *Planta* 162:385–391.

Russell, S.D. (1985). Preferential fertilization in *Plumbago*: ultrastructural evidence for gamete-level recognition in an angiosperm. *Proc. Natl. Acad. Sci. USA* 82:6129–6132.

Russell, S.D. (1992). Double fertilization. *Intl. Rev. Cytol.* 140:357–390.

Russell, S.D. (1993). The egg cell: development and role in fertilization and early embryogenesis. *Plant Cell* 5:1350–1360.

Sanders, L.C. & Lord, E.M. (1992). A dynamic role for the stylar matrix in pollen tube extension. *Intl. Rev. Cytol.* 140:297–318.

Schlichting, C.D., Stephenson, A.G., Davis, L.E. & Winsor, J.A. (1987). Pollen competition and offspring variance. *Evol. Trends Plants* 1:35–39.

Sedgley, M. & Scholefield, P.B. (1980). Stigma secretion in the watermelon before and after pollination. *Bot. Gaz.* 141:428–434.

Shivanna, K.R., Heslop-Harrison, J. & Heslop-Harrison, Y. (1978). Pollen–stigma interaction: bud pollination in the Cruciferae. *Acta Bot. Neerl.* 27:107–119.

Shivanna, K.R. & Johri, B.M. (1985). *The Angiosperm Pollen: Structure and Function*. New Delhi: Wiley Eastern.

Shivanna, K.R. & Owens, S.J. (1989). Pollen–stigma interactions (Papilionoideae). In *Advances in Legume Biology*, eds. C.H. Stirton and J.L. Zarucchi, Monographs Systematic Botany from the Missouri Botanical Garden, Vol. 29, pp. 157–182. St. Louis: Missouri Botanical Garden.

Shivanna, K.R. & Rangaswamy, N.S. (1992). *Pollen Biology: A Laboratory Manual*. Berlin: Springer-Verlag.

Shivanna, K.R. & Sastri, D.C. (1981). Stigma-surface esterases and stigma receptivity in some taxa characterized by wet stigma. *Ann. Bot.* 47:53–64.

Singh, A., Evensen, K.B. & Kao, T. (1992). Ethylene synthesis and floral senescence following compatible and incompatible pollinations in *Petunia inflata*. *Plant Physiol.* 49:38–45.

Slater, A.T. & Calder, D.M. (1989). Fine structure of the wet, detached cell stigma of the orchid *Dendrobium speciosum* Sm. *Sex. Plant Reprod.* 3:61–69.

Spangers, A.W. (1978). Voltage variation in *Lilium longiflorum* pistils induced by pollination. *Experientia* 34:36–37.

Stern, H. (1986). Meiosis: some considerations. *J. Cell Sci. Suppl.* 4:29–43.

Stern, H. & Hotta, Y. (1974). Biochemical controls of meiosis. *Ann. Rev. Genet.* 7:37–66.

Tilton, V.R. (1981a). Ovule development in *Ornithogalum caudatum* (Liliaceae) with a review of selected papers on angiosperm reproduction. II. Megasporogenesis. *New Phytol.* 88:459–476.

Tilton, V.R. (1981b). Ovule development in *Ornithogalum caudatum* (Liliaceae) with a review of selected papers on angiosperm reproduction. IV. Egg apparatus structure and function. *New Phytol.* 88:505–532.

Tilton, V.R. & Horner, H.J., Jr. (1980). Stigma, style, and obturator of *Ornithogalum caudatum* (Liliaceae) and their function in the reproductive process. *Amer. J. Bot.* 67:1113–1131.

Tilton, V.R. & Lersten, N.R. (1981). Ovule development in *Ornithogalum caudatum* (Liliaceae) with a review of selected papers on angiosperm reproduction. III. Nucellus and megagametophyte. *New Phytol.* 88:477–504.

Tirlapur, U.K. & Shiggaon, S.V. (1988). Distribution of calcium and calmodulin in the papillar cells of stigma surface visualized using chlorotetracycline and fluorescing calmodulin binding phenothiazines. *Ann. Biol.* 4:49–53.

Turano, M.J., Baird, L.M. & Webster, B.D. (1983). Characteristic of the stigma of chickpea. *Crop Sci.* 23:1033–1036.

van der Donk, J.A.W.M. (1974). Differential synthesis of RNA in self- and cross-pollinated styles of *Petunia hybrida* L. *Mol. Gen. Genet.* 131:1–8.

Vogt, T., Pollak, P., Tarlyn, N. & Taylor, L.P. (1994). Pollination- or wound-induced kaempferol accumulation in *Petunia* stigmas enhances seed production. *Plant Cell* 6:11–23.

Wang, H., Wu, H.M. & Cheung, A.Y. (1993). Development and pollination regulated accumulation and glycosylation of a stylar transmitting tissue-specific proline-rich protein. *Plant Cell* 5:1639–1650.

Wiermann, R. & Gubatz, S. (1992). Pollen wall and sporopollenin. *Intl. Rev. Cytol.* 140:35–72.

Williams, E.G., Rouse, J.L., Kaul, V. & Knox, R.B. (1991). Reproductive timetable for the tropical Vireya rhododendron, *R. macregoriae*. *Sex. Plant Reprod.* 4: 155–165.

Wirth, M. & Withner, C.L. (1959). Embryology and development in the Orchidaceae. In *The Orchids: A Scientific Survey*, ed. C.L. Withner, pp. 155–188. New York: The Ronald Press Co.

Yu, H-S & Russell, S.D. (1994). Male reproductive cell development in *Nicotiana tabacum*: male germ unit associations and quantitative cytology during sperm maturation. *Sex. Plant Reprod.* 7: 324–332.

3

Gene expression during pollen development

DOUGLAS A. HAMILTON and
JOSEPH P. MASCARENHAS

Summary   40
Introduction   41
General pollen development   41
Numbers of pollen-expressed and pollen-specific genes   42
Timing of pollen gene expression   43
Pollen-specific genes   46
Expression of genes with homology to wall-degrading enzymes   46
Expression of genes with homology to cytoskeletal proteins   47
Pollen allergens   47
Other pollen-specific genes   48
Promoter sequence elements responsible for pollen specificity   50
Separable regulation of gametophytic and sporophytic expression   51
Trans-species expression of pollen-specific genes   53
Conclusions   53
References   54

Summary

The angiosperm microgametophyte displays a complex genetic program during its development. There appear to be two main classes of genes expressed during pollen development, with the transition occurring roughly at microspore mitosis. Although the earlier class of genes is presumed to be associated with immature microspore formation, the later genes probably represent those associated with pollen maturation, germination, and tube growth. This is supported by the observation that several pollen-specific clones of the late class have been identified whose sequences are similar to enzymes associated with cell wall metabolism, as well as many cytoskeleton genes, the products of which are clearly necessary for pollen tube growth.

   The genetic promoter elements required for a pollen response continue to elude our understanding, but the seemingly unrestricted interchangeability of pollen promoters among a wide range of host plants argues for some kind of universal pollen element within them. Numerous pollen-specific genes have been isolated recently, and when their promoters are studied in detail, it might be easier to identify common elements in the promoters. Eventually, characterization of genes and promoters involved in key regulatory processes in pollen should provide insights into the nature of haploid gene expression and its relation to that of the diploid plant. Findings of this nature should point the way toward practical applications of pollen molecular biology in the areas of plant breeding, biotechnology, and basic science.

Introduction

A pollen grain represents the male portion of the gametophytic stage in the angiosperm life cycle. Following the completion of microsporogenesis and microgametogenesis, the pollen grain is only a three-celled organism. This is certainly a very reduced system in a morphological sense, yet the pollen grain clearly contains all the genetic information necessary for the formation of a complete plant, since morphologically normal haploid plants can be generated by changing the normal developmental pathway of immature pollen (Raghavan 1976; see also Chapter 20). During the past few years, there has been an increasing body of literature on the genetic program operative in the male gametophyte in comparison to that in the sporophyte. Work from a number of laboratories is beginning to provide information regarding the genes expressed in pollen development of several plant species. This chapter presents some of what has been learned concerning male gametophyte–specific gene expression, the possible functions of these genes in pollen development, and current information about pollen-specific promoters.

General pollen development

In the discussion that follows, we will often use maize as an example. For recent general descriptions of pollen development, see Chapter 2 and the reviews by Bedinger (1992) and McCormick (1993). In the normal course of *Zea mays* pollen development (reviewed in Kiesselbach 1949; Mascarenhas 1989), sporogenous cells in the anther begin to differentiate within the nutritive tissue, the tapetum. That the tapetal tissue is important in normal pollen development is supported by the observation that cases of male sterility have been associated with abnormal development of the tapetum itself (Laughnan and Gabay-Laughnan 1983; Lonsdale 1987; also see Chapters 7 and 8). This is further supported by the finding that experimental disruption of anther function by the expression or inhibition of specific genes under the control of tapetum-specific promoters can result in male sterility (Mariani et al. 1990; van der Meer et al. 1992; Worrall et al. 1992; also see Chapter 11).

Following their initial differentiation, the pollen mother cells (microsporocytes) divide meiotically to produce a tetrad of four haploid cells (microspores). After meiosis there is a long interphase stage during which the microspores separate, enlarge greatly in size, and become highly vacuolate. Subsequent to this, there is an unequal mitotic division (microspore mitosis) that forms two cells, the vegetative and generative cells. These cells have a unique relationship in that the generative cell lies entirely within the cytoplasm of the vegetative cell. These young pollen grains now begin to accumulate starch. In most plants the pollen is shed at this stage, with the completion of the final mitotic division of the generative cell into the two sperm cells occurring during pollen germination. In maize, the mitotic division of the generative cell occurs prior to anthesis. Thus, the mature male gametophyte of maize consists of three cells. At the time of anthesis, pollen grains are relatively dehydrated and metabolically inactive.

The next phase in the life of the pollen grain occurs when the pollen is deposited by a vector such as an insect, bird, small mammal, or wind, upon a receptive stigma of a female flower. Barring some type of self-incompatibility reaction, the pollen germinates and extrudes a pollen tube that grows rapidly down the style; growth rates of as much as 240 μm/minute have been recorded in maize (Heslop-Harrison and Heslop-Harrison 1982). The two sperm cells move down the growing pollen tube as a unit (McConchie et al. 1987) and are deposited in the vicinity of the female gametophyte, where one sperm fuses with the egg and the other with the diploid central cell to form the zygote and the primary endosperm cell, respectively. The major biosynthetic activities within the growing pollen tube are those associated with cell membrane and cell wall synthesis (Mascarenhas 1975).

Clearly, pollen germination and tube growth are very different from the type of development seen prior to anthesis, and it might be reasonable to expect that two different sets of genes might be required for these two different phases. Research aimed at examining the genes expressed during pollen development has shed some light on the numbers, types, and timing of these genes, as well as the structure and expression of individual pollen-specific genes. This review will attempt to present some of these data and will focus on known pollen-specific genes and the *cis* elements involved in pollen expression.

Numbers of pollen-expressed and pollen-specific genes

In order to get an idea of the activity of genes expressed in pollen, the numbers of genes active in the pollen of several species have been directly estimated. In maize, it has been calculated that there are some 24,000 different mRNAs present in the mature pollen grain, compared with about 31,000 different mRNAs present in cells of shoot tissues (Willing et al. 1988). These values for the complexity of mRNAs from pollen and vegetative tissues in maize are roughly similar to those obtained by similar studies with *Tradescantia* (Willing and Mascarenhas 1984). The number of genes active in whole tobacco anthers has been estimated to be about 26,000 (Kamalay and Goldberg 1980). Thus, it is clear that there are a large number of genes active in the morphologically simple male gametophyte, especially when compared with the vegetative tissue that is composed of complex cell and tissue types.

Many of the genes expressed in pollen are cellular "housekeeping genes" that are also expressed in the sporophyte, but a proportion of them are specific to pollen. A study directed at estimating the extent of gametophytic and sporophytic gene expression in maize suggested that roughly 65% of genes expressed in mature maize pollen are also expressed in vegetative tissues (Willing et al. 1988). This estimate is in line with that obtained for *Tradescantia* (at least 64%; Willing and Mascarenhas 1984). By comparisons of the sporophytic versus gametophytic expression of isozymes, an estimate of 72% haplo-diploid expression was made for maize pollen (Sari-Gorla et al. 1986). Although all these estimates are roughly similar, they also have a rather large degree of uncertainty (Willing et al. 1988).

The percentage of genes expressed specifically in pollen can also be evaluated by screening for pollen-specific clones in a pollen cDNA library. In a maize cDNA library made to RNA from mature pollen, it was found that approximately 10% of the clones represented mRNAs expressed in pollen and not in vegetative tissues (Stinson et al. 1987). This is a smaller fraction than was predicted by the hybridization experiments, but the cloning procedure probably selects for the more abundant mRNAs, and it is possible that the pollen sequences might constitute a larger fraction of the rarer mRNAs. Indeed, Hodge et al. (1992) suggest that limitations in obtaining "positive" plaques or colonies during differential screening may preclude the isolation of low-abundance mRNAs. When one hundred randomly selected "cold" (that is, apparently nonhybridizing) plaques from a *Brassica napus* anther cDNA library were isolated and the inserts used as probes for Northern analysis, 34% of the isolated inserts hybridized specifically to pollen or anther RNA (Hodge et al. 1992). These results suggest that any estimates of the numbers of tissue-specific genes derived from standard screening in pollen or anther tissue might yield low values.

Timing of pollen gene expression

Many of the pollen genes characterized to date have expression patterns that have been characterized broadly as "early" or "late" (Mascarenhas 1990). Actin has been used as an example of an early gene (Stinson et al. 1987). In maize, mRNA hybridizing to an actin probe was found to be first detectable prior to microspore mitosis, reaching a maximum at late pollen interphase and decreasing thereafter. Alcohol dehydrogenase expression in maize pollen (Stinson and Mascarenhas 1985) and ß-galactosidase expression in *Brassica campestris* (Singh et al. 1985) are similar to those of actin and are thus early (Stinson et al. 1987). In contrast, genes that show expression starting at or after microspore mitosis and usually reaching maximum abundance of the mRNA at the mature pollen stage have been termed late genes. Genes of the late expression pattern make up the bulk of pollen-specific genes identified.

Early genes have been surmised to be involved with pollen developmental processes, whereas late genes probably encode for proteins with functions associated with pollen maturation and/or germination and pollen tube growth

Table 3.1. *Pollen-specific genes*

Gene	Species	Putative function (homology to)	References
Zm58	Maize	Pectate lyase (*lat56, lat59, G10, Amb a* I, II)*	Turcich et al. (1993)
G10	Tobacco, tomato	Pectate lyase (Zm58, *lat56, lat59, Amb a* I, II)*	Rogers et al. (1992)
Amb a I1–4, *Amb a* II	Ragweed	Allergens, pectate lyases (Zm58, *lat56, lat59, G10*)*	Rafnar et al. (1991); Rogers et al. (1991)
Lol p I, *Lol p* Ib	Rye-grass	Allergens (*zea m* I)	Singh et al. (1991)
Zea m I	Maize	Allergen (*Lol p* I)	Broadwater et al. (1993)
KGB family (*Poa p* IX)	Kentucky bluegrass	Allergens	Silvanovich et al. (1991)
Ole e I	Olive	Allergen (Zm13, *lat* 52, *PS1*)	Villalba et al. (1994)
Zm13	Maize	? (*Ole e* I, *lat52, PS1*)	Stinson et al. (1987); Hanson et al. (1989)
lat52	Tomato	? (*Ole e* I, Zm13, *PS1*, Kunitz trypsin inhibitor)	Twell et al. (1989)
PS1	Rice	? (*Ole e* I, Zm13, *lat52*)	Zou et al. (1994)
P2 family	*Oenothera*	Polygalacturonase	Brown and Crouch (1990)
PG1; W2247 (and others)	Maize	Polygalacturonase	Niogret et al. (1991); Allen and Lonsdale (1993); Barakate et al. (1993)
Npg1	Tobacco	Polygalacturonase	Tebbutt et al. (1994)
Sta 44-4	*Brassica napus*	Polygalacturonase	Robert et al. (1993)
Bp19	*Brassica napus*	Pectin esterase	Albani et al. (1991)
Bp10	*Brassica napus*	Ascorbate oxidase	Albani et al. (1992)
NTP303	Tobacco	Ascorbate oxidase	Weterings et al. (1992)
*chi*A	*Petunia*	CHI genes	van Tunen et al. (1990)
TUA1	*Arabidopsis*	α-tubulin	Carpenter et al. (1992)
ZmPRO 1, 2, 3	Maize	Profilin	Staiger et al. (1993)

Gene	Species	Putative function (homology to)	References
Tac25	Tobacco	Actin	Thangavelu et al. (1993)
SF3	Sunflower	DNA-binding protein (zinc finger)	Baltz et al. (1992a, b)
I3	*Brassica napus*	Oleosin	Roberts et al. (1991, 1993)
Tpc44, Tpc70	*Tradescantia*	?	Stinson et al. (1987); Turcich et al. (1994)
Bp4 (A, B, C)	*Brassica napus*	?	Albani et al. (1990)
Various	*Brassica napus*	?	Hodge et al. (1992)

Note: Genes marked with * are also homologous to the 9612 gene that is expressed in the styles of tobacco and tomato (Budelier et al. 1990).

(Stinson et al. 1987; Mascarenhas 1990). Additional experimental support for a biphasic (early/late) pattern of gene expression during microsporogenesis has been provided by examination of the proteins produced by developmentally staged maize microspores (Bedinger and Edgerton 1990). Changes in protein bands after gel electrophoresis during different stages of pollen development had been observed earlier in lily by Linskens (1966) and more recently in wheat (Vergne and Dumas 1988). One-dimensional SDS-polyacrylamide gel electrophoresis of proteins extracted from staged developing maize pollen showed a transition in the positions and intensities of numerous bands at around the time of the first microspore mitosis. In vitro translation of RNA isolated from staged microspores also reflected this change (Bedinger and Edgerton 1990). Eady et al. (1994) found evidence for the expression of the *lat52* (late anther tomato) gene just at the time of microspore mitosis in transgenic tobacco, but just prior to it in transgenic *Arabidopsis*, suggesting that passage through mitosis per se was not necessary for the transcriptional initiation of this gene. In general, these results indicate that there does appear to be a genuine developmental switch occurring in microspore gene expression around the time of microspore mitosis. Nonetheless, it must be noted that the division of gene activity into early and late categories is somewhat artificial and cannot describe all microspore-expressed genes, especially since several pollen-specific genes show intermediate or continuous expression during pollen development (Albani et al. 1991, 1992). These genes are obviously required during all of pollen development and might be housekeeping in function.

Experiments with inhibitors of protein and RNA synthesis have provided evidence that the mRNAs required for germination and early pollen tube growth are already present in the pollen grain at the time it is released from the anther (reviewed in Mascarenhas 1993). These experiments also provide

supporting evidence that the major functions of the late mRNAs are to produce proteins required for germination and tube growth.

Pollen-specific genes

Table 3.1 lists all the genes published to date that have been shown to be
pollen specific or nearly so, and that have been cloned and at least partially
characterized. This list is arbitrarily restricted and does not include any genes
that have been shown to be expressed in anthers but that are not yet defini-
tively localized to the pollen (for example, Kim et al. 1993). There are too
many pollen-specific genes described at this point to discuss each in detail, so
an overview of groups of related genes will be presented.

Many of the genes listed in Table 3.1 have been isolated via standard differential screening techniques, which select for RNAs that are present at much
higher levels in pollen than in other tissues. As a result, most of these are presumably among the most highly expressed and therefore the easiest to
recover. It must be mentioned that most of the putative proteins that have a
function or enzymatic activity cited have generally not been tested for that
activity; rather, the function has been assigned based on protein sequence
homologies to known proteins. What is striking from Table 3.1 is the number
of genes that bear sequence homology to wall-degrading enzymes and proteins involved in the cytoskeleton.

Expression of genes with homology to wall-degrading enzymes

The maize gene designated Zm58 (Turcich et al. 1993), the tobacco gene *G10*
(Rogers et al. 1992), and two tomato clones expressed in tomato anthers and
pollen, *lat56* and *lat59* (McCormick et al. 1989), show a similarity in their
amino acid sequences to pectate lyases, especially those known from the bacterial plant pathogen *Erwinia* (McCormick et al. 1989; Wing et al. 1989). This
commonality with a plant pathogenic bacterium is interesting in light of comparisons of the invasive growth of pollen tubes within the female tissue to a
pathogen–host interaction (Clark and Gleeson 1981; Wing et al. 1989). When
a pollen grain germinates on a receptive stigma, a pollen tube emerges and
then grows through the tissues of the style. These processes would require
both wall degradation and synthesis. Exactly how the cell-wall–degrading
enzymes participate in this haustorium-like growth is not clear, but the presence of large quantities of enzymes associated with wall synthesis and degradation would certainly be expected in pollen. In regard to the necessity of cell
wall breakdown during the growth of the pollen tube through the style, it is
interesting that a style-expressed gene also has homology to pectate lyase
(*9612*; Budelier et al. 1990). And of additional, but somewhat tangential,
interest is a report that the three-dimensional structure of the *Erwinia* pectate
lyase C has been resolved using X-ray crystallography, revealing that the protein folds with a newly identified configuration designated a parallel ß-helix
(Yoder et al. 1993). It is not clear whether the plant-expressed genes fold in

this manner, but they do contain conserved regions corresponding to the putative Ca^{2+}-binding region found in the bacterial enzymes.

In addition to pectate lyase, several cDNA clones corresponding to polygalacturonase have been isolated from the pollen of maize (Niogret et al. 1991; Allen and Lonsdale 1993; Barakate et al. 1993), *Oenothera organensis* (Brown and Crouch 1990), tobacco (Tebbutt et al. 1994), and *Brassica napus* (Robert et al. 1993). A cDNA has also been isolated that shows sequence similarity to pectin esterase in *Brassica napus* (Albani et al. 1991). The fact that enzymes of this type have been isolated from at least five pollen cDNA libraries suggests that the mRNAs coding for pectin-degrading enzymes are abundantly expressed in all pollens.

Expression of genes with homology to cytoskeletal proteins

Another large group of the pollen-expressed proteins characterized to date are cytoskeleton-related proteins such as actin (Thangavelu et al. 1993), α-tubulin (Carpenter et al. 1992), ß-tubulin (Rogers et al. 1993), profilin (Valenta et al. 1991b; Staiger et al. 1993), and actin depolymerizing factor (Kim ct al. 1993). In tobacco, there appear to be at least 25–30 actin genes that hybridize to a soybean actin gcnc clone (Thangavelu et al. 1993). RNA blot analysis of two hybridizing genomic clones showed that one was pollen specific (Tac25), whereas the other showed pollen-preferential expression. Likewise, a nearly pollen-specific member of the α-tubulin family has been isolated in *Arabidopsis* (*TUA1*; Carpenter et al. 1992). The *TUA1* promoter fused to the ß-glucuronidase gene (GUS) shows accumulation of blue color mainly in postmitotic pollen grains, with very much lower levels detectable in early anthers, flower receptacles, trichomes, and leaf axils. The ß-tubulin genes *tub3*, *tub4*, and *tub5* are all expressed in several tissues but are most abundant in pollen (Rogers et al. 1993). Staiger et al. (1993) have examined three profilin genes coding for three isoforms that are specifically expressed late in developing maize pollen. Anther-preferential cDNA clones with sequence homology to animal actin-depolymerizing factor (ADF) have been isolated from both lily and *Brassica napus* (Kim et al. 1993). The actin microfilament contractile system is responsible for the very active cytoplasmic streaming in the growing pollen tube and the control of growth of the tube (see the review in Mascarenhas 1993). It is thus not unexpected to find that genes for cytoskeletal proteins are actively expressed in pollen.

Pollen allergens

It is becoming clear that several of the highly expressed proteins in pollen are also human allergens. Sequence homology searches indicate that there is substantial homology between the previously mentioned pectate lyase proteins and proteins identified from the pollen of short ragweed, *Ambrosia artemisiifolia* (McCormick 1991; Rogers et al. 1991). Of the seasonal allergens that may cause hay fever, asthma, and hives, pollen from the short ragweed has been implicated as being among the most clinically important (King 1976; Rafnar et al. 1991). Several of the prominent allergenic proteins from this pollen have been immunochemically characterized, revealing that the *Amb a* I (formerly Allergen E) protein can account for at least 90% of the allergenicity of ragweed pollen (Rafnar et al. 1991). This protein is one of a multigene family of very similar proteins, including *Amb a* I.1, I.2, I.3, I.4, and II. The *Amb a* II protein is interesting in that it shows only approximately 65% homology with members of the *Amb a* I family yet demonstrates similar allergenic potential (Rogers et al. 1991). All the pollen-specific pectate lyase genes compared in Table 3.1 have homologies to *Amb a* I on the order of 45–48% amino acid identity and may share similar epitopes. There is also homology between these genes and the partial protein sequence of the major allergen of Japanese cedar pollen, *Cry j* I (Taniai et al. 1988; McCormick 1991).

Similarly, several pollen-expressed proteins responsible for IgE binding in allergic patients have been shown to be profilins. The first protein recognized as such was identified as one of the major allergens from birch pollen (Valenta et al. 1991b). It has since been shown that pollen profilins represent a common class of immuno-reactive agents and have been found in the pollen of several species, including timothy grass, *Phleum pratense* (Valenta et al. 1994). Another allergen that has been shown to exhibit high levels of expression in birch pollen is *Bet v* I (Breiteneder et al. 1989). Proteins homologous to *Bet v* I have been shown to be allergens in pollens from many trees in the order Fagales, for example, birch, alder, hazel, and hornbeam (Valenta et al. 1991a; Ebner et al. 1993). No known plant function has yet been assigned to *Bet v* I, but it does appear to have homology to plant disease resistance response genes (Breiteneder et al. 1989). There are several other pollen-expressed proteins that have been identified as human allergens, but whose function in pollen is not yet known (see Table 3.1).

Several of the allergenic proteins present in pollen appear to have important functions in pollen tube growth. Attempts to engineer plants with lowered levels of these allergens might thus be problematic from an environmental point of view, because male sterility might result.

Other pollen-specific genes

The pollen-specific maize gene Zm13 (Hamilton et al. 1989; Hanson et al. 1989) shows significant homology with a pollen-expressed gene (*lat52*) isolated from tomato pollen (Twell et al. 1989). The predicted amino acid sequences of these two clones show a 32% identity, including the presence of six conserved cysteine residues. In both cases, Southern analysis appears to indicate that the genes are present in only one or a few copies in the genome. Recently, two other clones have been described – one from rice pollen, *PS1* (Zou et al. 1994), and the other from olive tree pollen, *Ole e* I (Villalba et al.

1994) – that show significant homology to Zm13. It is interesting that the *Ole e* I protein is the major allergen from olive tree pollen (Villalba et al. 1994). The Zm13 class of pollen proteins might thus be potent human allergens. None of the clones show significant homology with any protein on record with a known function, in any protein or DNA sequence databank searched to date. Although it has been reported that *lat52* and Zm13 show partial sequence homology to several proteinase inhibitors (McCormick 1991), the pollen proteins do not show conservation of the amino acids at the active site of the inhibitors. Recently, a clone has been isolated from a sorghum cDNA library using a fragment of Zm13 as the probe (E. Pe, personal communication). This is a partial cDNA, but it shows a 90% sequence homology with Zm13, indicating significant conservation of this gene between these two monocots.

In situ hybridization studies using RNA probes have localized the mRNA for Zm13 to the cytoplasm of the vegetative cell (Hanson et al. 1989). Although there does not appear to be any association with the sperm cells, the localization is not precise enough to state this for certain. Hybridization studies show that the Zm13 mRNA is also present within the cytoplasm of the germinating pollen tube. A similar localization was determined for the *lat52* protein in an elegant experiment utilizing a fusion of the *lat52* promoter to the GUS gene containing a nuclear-targeting signal (Twell 1992). As a result of this fusion, GUS will be detected only in the nucleus of any cell in which it is expressed. The *lat52* promoter was active only in the vegetative cell; no activity was detected in the generative cell during maturation, germination, or pollen tube growth (Twell 1992). Since the generative and vegetative cells are products of a single mitosis, the basis for the onset of differential gene expression would be interesting to determine. Twell (1992) provides two hypotheses for this switch: The first proposes that an unequal partitioning of transcription factors associated with *lat52* expression occurs during cell division, and the second contends that the state of the chromatin may affect expression since that of the generative cell is highly condensed compared to that of the vegetative cell. Recent work by Eady et al. (1994) using carefully staged transgenic *Arabidopsis* microspores seems to show expression of a *lat52*/GUS construct slightly before microspore mitosis. If this evidence is confirmed, then any model of cell-specific pollen gene expression will require expression to be actively turned off in the generative cell, possibly by one of the aforementioned mechanisms. Regardless of the exact means, *lat52* and other late pollen genes respond to events that must occur around microspore mitosis, and they are expressed in a cell-specific as well as temporally regulated manner.

The potential function of the *lat52* protein has been further examined by antisense expression of the *lat52* mRNA in transgenic tomatoes (Muschietti et al. 1994). Reduction in the level of *lat52* protein resulted in pollen grains that were alive, but that were small and collapsed and whose pollen tubes grew abnormally. Manipulation of the water potential of the germination medium suggested that decreased levels of the *lat52* protein resulted in abnormal hydration characteristics of the pollen and a subsequent inability to effect fertilization. Although the exact role of this protein is still undetermined, it is clear that it plays a critical role in pollen maturation and/or germination.

Another group of genes that have been identified as pollen specific show homology to ascorbate oxidases. Genes with sequence homology to these oxidases have been isolated from *Brassica* (Bp10; Albani et al. 1992) and tobacco (NTP303; Weterings et al. 1992). Despite this homology, the functional identity of these proteins is still unclear, because the identity between them and the oxidases is low in the highly conserved copper-binding regions (Weterings et al. 1992).

One particularly interesting pollen-specific gene has been isolated from a cDNA library made to mature sunflower pollen (Baltz et al. 1992a). This gene, designated SF3, has zinc-finger domains that correspond to a so-called LIM motif present in several metal-binding cysteine-rich proteins of animals (Baltz et al. 1992b). Several proteins with the LIM motif have been identified as regulatory proteins, raising the possibility that SF3 might be involved in the regulation of late pollen genes.

Promoter sequence elements responsible for pollen specificity

It has been an important goal to find pollen-expressed genes that are pollen specific and to examine their promoter regions to determine the molecular basis for their specificity. The promoters of quite a number of pollen-specific genes have been isolated and characterized to some degree, usually by expression of a marker gene (normally GUS) in homo- or heterologous transformation systems and/or by deletion analysis.

In order to identify the region(s) of the 5' flanking region of the maize Zm13 gene that produces its pollen-specific expression, various deletion fragments of the promoter were transcriptionally fused to GUS (Guerrero et al. 1990). These constructs were introduced into tobacco via *Agrobacterium*-mediated transformation and showed that a 314-bp region was competent to induce pollen-specific expression. Similar results have recently been obtained in *Arabidopsis* (J. Rueda and J. P. Mascarenhas, unpublished results). It also shows that a monocot pollen promoter can be recognized by the transcriptional apparatus of a dicot (Guerrero et al. 1990).

To analyze the Zm13 promoter in greater detail, a transient expression system was employed (Hamilton et al. 1992), using high-velocity particle bombardment similar to that used with tomato pollen (Twell et al. 1989). Histochemical staining and fluorescent detection of GUS activity controlled by the Zm13 promoter indicated that full activity was displayed by a promoter fragment beginning at −260, whereas one beginning at −100 showed only 6% relative activity, and one beginning at −54 showed none. The −100 fragment showed a marked decrease in GUS activity, yet retained its pollen specificity of expression. These results indicate that quantitative element(s) responsible for efficient expression in pollen reside in the −260 to −100 region, whereas pollen-specific element(s) are present in the −100 to −54 region (Hamilton et al. 1992). Recent results have shown that 100% activity is retained in a fragment beginning at −118, indicating that element(s) necessary for high expression in pollen are located in the −118 to −100 region (D. A. Hamilton and J. P. Mascarenhas, unpublished results).

One of the most in-depth studies of the *cis*-elements necessary for pollen expression was described for the *lat* genes (Twell et al. 1990, 1991). Using deletion analysis, several groups of sequences in the promoters of the coordinately expressed *lat52*, *lat56*, and *lat59* genes were found to be in common in two, but not all, promoters. Deletion of these regions (termed the 52/56 box and 56/59 box) produced decreases in levels of expression of the promoters but not necessarily the loss of their pollen specificity. This is similar to the findings with Zm13 and suggests that pollen-specificity elements are separable from elements necessary for high expression. A core "pollen box" sequence was presented that has been identified in several pollen promoters (for example, *TUA1*; Carpenter et al. 1992), but not in others.

The *PS1* gene isolated from rice (Zou et al. 1994) is homologous to the Zm13 and *lat52* genes, but analysis of the promoter region shows only a few limited regions of homology. Presumably, continued isolation of pollen-specific promoter sequences will enable us to understand the specific promoter sequences or structures that confer pollen specificity.

Separable regulation of gametophytic and sporophytic expression

Pollen-expression signals must be discrete entities that function independently of other signals in the upstream area. There are numerous examples of promoters that express in both pollen and sporophytic tissues and whose activities can be deleted or altered independently. For example, the anther- and pollen-specific regions of the *chi*A promoter could be separated and expressed independently and correctly (van Tunen et al. 1990), indicating that sporophytic- and gametophytic-expression elements may reside together but individually in plant promoters. The promoter of a soybean gene encoding a cytosolic glutamine synthetase gene was placed in the control of GUS and transformed into *Lotus corniculatus* (Marsolier et al. 1993). Histochemical analysis showed greatest expression of the construct in root nodules, roots, pulvini, and anthers. Deletion analysis from the 5' end of the 3.5-kb promoter showed sequential loss of activity in nodules, pulvini, roots, and finally anthers. The activity in anthers (anther theca and pollen) was retained with as little as 240 bases of the promoter remaining. A similar result was seen with a *TUA1*/GUS promoter construct (Carpenter et al. 1992). Using a promoter fragment starting at −1500 resulted in the expression of GUS in flower receptacles as well as pollen, whereas the use of any fragment starting between −533 and −97 showed expression only in pollen.

Other studies have shown that, although the pollen-expression element in some promoters is correctly interpreted in transgenic plants, other elements may not be. With the 3.5-kb glutamine synthetase promoter just mentioned, the normal expression of the promoter seen in its native soybean includes sepals, petals, ovaries, and anthers, whereas in the transgenic *Lotus corniculatus* all nonanther floral expression is lost (Marsolier et al. 1993). These authors note that a similar result was seen with the petunia 5-enolpyruvyl-shikimate-3-phosphate (EPSP) promoter (Benfey and Chua 1989). When the petunia EPSP/GUS construct was transformed into tobacco, expression was lost in all floral tissues except pollen. In a somewhat similar fashion, the promoter of the *Brassica* S locus glycoprotein (SLG) gene can direct GUS expression in several tissues as well as in pollen, and like the glutamine synthetase promoter, the regions responsible for pollen expression were clearly separable from those necessary for expression in other tissues. Although GUS expression was noted in the pistils, anthers, and pollen of *Brassica* plants, expression was seen only in the pistils and pollen of transgenic tobacco (Dzelzkalns et al. 1993). Likewise, when the promoter of the pollen- and anther-expressed *Brassica campestris Bgp1* gene was linked to GUS and transformed back into *B. campestris*, GUS expression was detected in the same temporal and spatial fashion as was indicated by in situ hybridization. However, when the same construct was transformed into tobacco, GUS was detectable only in pollen (Xu et al. 1993).

Conversely, it is possible for sporophytic expression to appear where it has not been seen before. When a pollen-specific maize polygalacturonase promoter (W2247; Allen and Lonsdale 1993) was linked to GUS and stably transformed into tobacco, there was previously unseen low but detectable sporophytic expression in addition to the expected expression in pollen.

In each of these cases, although the expression seen in sporophytic tissues sometimes varied, the pollen expression was conserved. It is interesting to note that, with the exception of the *chi*A promoter, in promoters containing sporophytic elements in addition to a pollen signal, the region responsible for pollen activity was always closest to the transcription start site. Perhaps the location of sporophytic elements distal to those of pollen makes them more susceptible to position effects that affect their expression.

Pollen-specific promoter elements may not necessarily lie upstream of the transcription start site. The maize alcohol dehydrogenase-1 (*Adh-1*) gene that is expressed in several tissues of the maize plant, including pollen (Freeling and Bennet 1985), is an example of this. Mutations within the promoter region of *Adh-1* normally have similar effects on both sporophytic and gametophytic expression, but Dawe et al. (1993) have recently examined transposon-generated excision mutations between +42 and +52 of the untranslated leader that display changes of expression in specific tissues, as detected by changes in the steady-state mRNA levels. These alterations within this region of *Adh-1* 5' UTR caused either increased or decreased expression in pollen but not in any sporophytic tissues, and were localized to a seven-base motif (GGACTGA). Motifs similar to this were reported to exist in the 5' UTRs of the maize *waxy* gene (Klosgen et al. 1986) and the *Brassica napus* gene *BP19* (Albani et al. 1991). Although the expression patterns of these genes are not identical, it may be that this motif works to enhance the action of another region. A similar finding was reported by Kyozuka et al. (1994). They examined the tissue-specific expression of the maize *Adh-1* gene by observing *Adh-1* promoter-driven GUS activity in transgenic rice and concluded that pollen specificity could be disrupted by replacement of a region between +54 and +106 of the 5' UTR.

Trans-species expression of pollen-specific genes

The studies just mentioned indicate that the signals for pollen expression are robust and generally recognizable among a wide variety of plants. For example, the maize Zm13 promoter has been stably transformed into tobacco (Guerrero et al. 1990) and *Arabidopsis* (Mascarenhas et al., unpublished results) with complete maintenance of its temporal and spatial expression. The *lat52* promoter has recently been shown to express correctly in stably transformed species with bi- and tricellular transgenic pollens (tobacco and *Arabidopsis*, respectively; Eady et al. 1994). Transient expression in divergent pollens has also been shown. The Zm13 promoter is extremely active in transiently transformed *Tradescantia* pollen (Hamilton et al. 1992). The *lat52* tomato promoter has shown transient expression in *Nicotiana tabacum* (Twell et al. 1991) and *N. glutinosa* (van der Leede-Plegt et al. 1992), but interestingly, not in *Lilium longiflorum* (van der Leede-Plegt et al. 1992). Indeed, there has not yet been a report of a dicot pollen promoter that is active in a transformed monocot, but this may be simply due to a lack of attempts.

Conclusions

The molecular mechanism for pollen specificity is clearly not yet known. Sequence and functional comparisons among pollen-specific promoters simply have not resulted in the identification of consistent motif(s) that may be assigned to pollen specificity. This implies either that any trans-acting factor(s) responsible for pollen expression recognize multiple sequences, or perhaps that multiple mechanisms are at work. The isolation of trans-acting factors involved in pollen gene expression will shed some light on the promoter elements required.

Understanding the nature of gene expression in pollen is necessary for the manipulation of pollen genes. The ability to control pollen-targeted genes is currently manifesting itself in the production of, for example, male-sterile plants used for breeding and ornamental purposes. Such goals as manipulation of temperature sensitivity, control of self-(in)compatibility, and pest resistance have all been suggested as useful outgrowths of pollen gene research. And the genetic complexity of pollen, combined with its morphological simplicity, will continue to recommend it as an important experimental system for basic research on differentiation.

References

Albani, D., Robert, L.S, Donaldson, P.A., Altosaar, I., Arnison, P.G. & Fabijanski, S.F. (1990). Characterization of a pollen-specific gene family from *Brassica napus* which is activated during early microspore development. *Plant Mol. Biol.* 15:605–622.

Albani, D., Altosaar, I., Arnison, P.G. & Fabijanski, S.F. (1991). A gene showing sequence similarity to pectin esterase is specifically expressed in developing pollen of *Brassica napus;* sequences in its 5' flanking region are conserved in other pollen-specific promoters. *Plant Mol. Biol.* 16:501–513.

Albani, D., Sardana, R., Robert, L.S., Altosaar, I., Arnison, P.G. & Fabijanski, S.F. (1992). A *Brassica napus* gene family which shows sequence similarity to ascorbate oxidase is expressed in developing pollen: molecular characterization and analysis of promoter activity in transgenic tobacco plants. *Plant J.* 2:331–342.

Allen, R.J. & Lonsdale, D.M. (1993). Molecular characterization of one of the maize polygalacturonase gene family members which are expressed during late pollen development. *Plant J.* 3:261–271.

Baltz, R., Domon, C., Pillay, D.T.N. & Steinmetz, A. (1992a). Characterization of a pollen-specific cDNA from sunflower encoding a zinc finger protein. *Plant J.* 2:713–721.

Baltz, R., Evrard, J.L., Domon, C. & Steinmetz, A. (1992b). A LIM motif is present in a pollen-specific protein. *Plant Cell* 4:1465–1466.

Barakate, A., Martin, W., Quigley, F. & Mache, R. (1993). Characterization of a multigene family encoding an exopolygalacturonase in maize. *J. Mol. Biol.* 229:797–801.

Bedinger, P.A. (1992). The remarkable biology of pollen. *Plant Cell* 4:879–887.

Bedinger, P.A. & Edgerton, M.D. (1990). Developmental staging of maize microspores reveals a transition in developing microspore proteins. *Plant Physiol.* 92:474–479.

Benfey, P.N. & Chua, N.H. (1989). Regulated plant genes in transgenic plants. *Science* 244:174–181.

Breiteneder, H., Pettenburger, K., Bito, A., Valenta, R., Kraft, D., Rumpold, H., Scheiner, O. & Breitenbach, M. (1989). The gene coding for the major birch pollen allergen *Bet v* I, is highly homologous to a pea disease resistance response gene. *EMBO J.* 8:1935–1938.

Broadwater, A.H., Rubinstein, A.L., Chay, C.H., Klapper, D.G. & Bedinger, P.A. (1993). Zea-mI, the maize homolog of the allergen-encoding Lol-pI gene of rye grass. *Gene* 1331:227–230.

Brown, S.M. & Crouch, M.L. (1990). Characterization of a gene family abundantly expressed in *Oenothera organensis* pollen that shows sequence similarity to polygalacturonase. *Plant Cell* 2:263–274.

Budelier, K.A., Smith, A.G. & Gasser, C.S. (1990). Regulation of a stylar transmitting tissue-specific gene in wild-type and transgenic tomato and tobacco. *Mol. Gen. Genet.* 224:183–192.

Carpenter, J.L., Ploense, S.E., Snustad, D.P. & Silflow, C. (1992). Preferential expression of an α-tubulin gene of *Arabidopsis* in pollen. *Plant Cell* 4:557–571.

Clark, A.E. & Gleeson, P.A. (1981). Molecular aspects of recognition and response in the pollen–stigma interaction. *Rec. Adv. Phytochem.* 15:161–211.

Dawe, R.K., Lachmansingh, A.R. & Freeling, M. (1993). Transposon-mediated mutations in the untranslated leader of maize *Adh1* that increase and decrease pollen-specific gene expression. *Plant Cell* 5:311–319.

Dzelzkalns, V.A., Thorsness, M.K., Dwyer, K.G., Baxter, J.S., Balent, M.A.,

Nasrallah, M.E. & Nasrallah, J.B. (1993). Distinct *cis*-acting elements direct pistil-specific and pollen-specific activity of the Brassica *S* Locus Glycoprotein gene promoter. *Plant Cell* 5:855–863.

Eady, C., Lindsey, K. & Twell, D. (1994). Differential activation and conserved vegetative cell-specific activity of a late pollen promoter in species with bicellular and tricellular pollen. *Plant J.* 5:543–550.

Ebner, C., Ferreira, F., Hoffmann, K., Hirschwehr, R., Schenk, S., Szepfalusi, Z., Breiteneder, H., Parronchi, P., Romagnani, S., Scheiner, O. & Kraft, D. (1993). T cell clones specific for Bet V I, the major birch pollen allergen, crossreact with the major allergens of hazel, Cor A I, and alder, Aln G I. *Mol. Immunol.* 30:1223–1229.

Freeling, M. & Bennet, D.C. (1985). Maize *Adh-1*. *Ann. Rev. Genet.* 19:297–323.

Guerrero, F.D., Crossland, L., Smutzer, G.S., Hamilton, D.A. & Mascarenhas, J.P. (1990). Promoter sequences from a maize pollen-specific gene direct tissue-specific transcription in tobacco. *Mol. Gen. Genet.* 224:161–168.

Hamilton, D.A., Bashe, D.M., Stinson, J.R. & Mascarenhas, J.P. (1989). Characterization of a pollen-specific genomic clone from maize. *Sex Plant Reprod.* 2:208–212.

Hamilton, D.A., Roy, M., Rueda, J., Sindhu, R.K., Sanford, J. & Mascarenhas, J.P. (1992). Dissection of a pollen-specific promoter from maize by transient transformation assays. *Plant Mol. Biol.* 18:211–218.

Hanson, D.D., Hamilton, D.A., Travis, J.L., Bashe, D.M. & Mascarenhas, J.P. (1989). Characterization of a pollen-specific cDNA clone from *Zea mays* and its expression. *Plant Cell* 1:173–179.

Heslop-Harrison, J. & Heslop-Harrison, Y. (1982). The growth of the grass pollen tube: 1. Characteristics of the polysaccharide particles ("P-particles") associated with apical growth. *Protoplasma* 112:71–80.

Hodge, R., Paul, W., Draper, J. & Scott, R. (1992). Cold-plaque screening: a simple technique for the isolation of low abundance, differentially expressed transcripts from conventional cDNA libraries. *Plant J.* 2:257–260.

Kamalay, J.C. & Goldberg, R.B. (1980). Regulation of structural gene expression in tobacco. *Cell* 19:934–946.

Kiesselbach, T.A. (1949). *The Structure and Reproduction of Corn.* Res. Bull. 161, Agric. Exp. Sta., Univ. of Nebraska College of Agriculture, pp. 37–50.

Kim, S.R., Kim, Y. & An, G. (1993). Molecular cloning and characterization of anther-preferential cDNA encoding a putative actin-depolymerizing factor. *Plant Mol. Biol.* 21:39–45.

King, T.P. (1976). Chemical and biological properties of some atopic allergens. *Adv. Immunol.* 23:77–105.

Klosgen, R.B., Gierl, A., Schwarz-Sommer, Z. & Saedler, H. (1986). Molecular analysis of the *waxy* locus of *Zea mays*. *Mol. Gen. Genet.* 103:237–244.

Kyozuka, J., Olive, M., Peacock, W.J., Dennis, E.S. & Shimamoto, K. (1994). Promoter elements required for developmental expression of the maize *Adh1* gene in transgenic rice. *Plant Cell* 6:799–810.

Laughnan, J.R. & Gabay-Laughnan, S. (1983). Cytoplasmic male sterility in maize. *Annu. Rev. Genet.* 117:27–48.

Linskens, H.F. (1966). Die anderung des protein- und enzym-musters wahrend der pollenmeiose und pollenentwicklung. *Planta* 69:79–91.

Lonsdale, D.M. (1987). Cytoplasmic male sterility: a molecular perspective. *Plant Physiol. Biochem.* 25:265–271.

Mariani, C., DeBeuckeleer, M., Truettner, J., Leemans, J. & Goldberg, R.B. (1990). Induction of male sterility in plants by a ribonuclease gene. *Nature* 347:737–741.

Marsolier, M.C., Carrayol, E. & Hirel, B. (1993). Multiple functions of promoter sequences involved in organ-specific expression and ammonia regulation of a cytosolic soybean glutamine synthetase gene in transgenic *Lotus corniculatus*. *Plant J.* 3:405–414.

Mascarenhas, J.P. (1975). The biochemistry of angiosperm pollen development. *Bot. Rev.* 41:259–314.

Mascarenhas, J.P. (1989). The male gametophyte of flowering plants. *Plant Cell* 1:657–664.

Mascarenhas, J.P. (1990). Gene activity during pollen development. *Ann. Rev. Plant Physiol. Plant Mol. Biol.* 41:317–338.

Mascarenhas, J.P. (1993). Molecular mechanisms of pollen tube growth and differentiation. *Plant Cell* 5:1303–1314.

McConchie, C.A., Hough, T. & Knox, R.B. (1987). Ultrastructural analysis of the mature sperm cells of mature pollen of maize, *Zea mays*. *Protoplasma* 139:9–19.

McCormick, S. (1991). Molecular analysis of male gametogenesis in plants. *Trends Genet.* 7:298–303.

McCormick, S. (1993). Male gametophyte development. *Plant Cell* 5:1265–1275.

McCormick, S., Twell, D., Wing, R., Ursin, V., Yamaguchi, J. & Larabell, S. (1989). Anther-specific genes: molecular characterization and promoter analysis in transgenic plants. In *Plant Reproduction: From Floral Induction to Pollination,* eds. E. Lord and G. Bernier, pp. 128–135. Rockville, MD: American Society of Plant Physiology.

Muschietti, J., Dircks, L., Vancanneyt, G. & McCormick, S. (1994). LAT52 protein is essential for tomato pollen development: Pollen expressing antisense LAT52 RNA hydrates and germinates abnormally and cannot achieve fertilization. *Plant J.* 6:321–338.

Niogret, M.F., Dubald, M., Mandaron, P. & Mache, R. (1991). Characterization of pollen polygalacturonase encoded by several cDNA clones in maize. *Plant Mol. Biol.* 17:1155–1164.

Rafnar, T., Griffith, I.J., Kuo, M., Bond, J.F., Rogers, B.L. & Klapper, D.G. (1991). Cloning of *Amb a* I (antigen E), the major allergen family of short ragweed pollen. *J. Biol. Chem.* 266:1229–1236.

Raghavan, V. (1976). *Experimental Embryogenesis in Vascular Plants.* New York: Academic Press.

Robert, L.S., Allard, S., Gerster, J.L., Cass, L. & Simmonds, J. (1993). Isolation and characterization of a polygalacturonase gene highly expressed in *Brassica napus* pollen. *Plant Mol. Biol.* 23:1273–1278.

Roberts, M.R., Hodge, R., Ross, J.H.E., Sorensen, A., Murphy, D.J., Draper, J. & Scott, R. (1993). Characterization of a new class of oleosins suggests a male gametophyte–specific lipid storage pathway. *Plant J.* 3:629–636.

Roberts, M.R., Robson, F., Foster, G.D., Draper, J. & Scott, R.C. (1991). A *Brassica napus* mRNA expressed specifically in developing microspores. *Plant Mol. Biol.* 17:295–299.

Rogers, B.L., Morgenstern, J.P., Griffith, I.J., Yu, X.B., Counsell, C.M., Brauer, A.W., King, T.P., Garman, R.D. & Kuo, M.C. (1991). Complete sequence of the allergen *Amb a* II. *J. Immunol.* 147:2547–2552.

Rogers, H.J., Greenland, A.J. & Hussey, P.J. (1993). Four members of the maize ß-tubulin gene family are expressed in the male gametophyte. *Plant J.* 4:875–882.

Rogers, H.J., Harvey, A. & Lonsdale, D.M. (1992). Isolation and characterization of a tobacco gene with homology to pectate lyase which is specifically expressed during microsporogenesis. *Plant Mol. Biol.* 20:493–502.

Sari-Gorla, M., Frova, C., Binelli, G. & Ottaviano, E. (1986). The extent of gametophytic-sporophytic gene expression in maize. *Theor. Appl. Genet.* 72:42–47.

Silvanovich, A., Astwood, J., Zhang, L., Olsen, E., Kisil, F., Sehon, A., Mohapatra, S. & Hill, R. (1991). Nucleotide sequence analysis of three cDNAs coding for *Poa p* IX isoallergens of Kentucky bluegrass pollen. *J. Biol. Chem.* 266:1204–1210.

Singh, M.B., O'Neil, P., & Knox, R.B. (1985). Initiation of post-meiotic ß-galactosidase synthesis during microsporogenesis in oilseed rape. *Plant Physiol* 77: 225–228.

Singh, M.B., Hough, T., Theerakulpisut, P., Avjioglu, A., Davies, S., Smith, P.M., Taylor, P., Simpson, R.J., Ward, L.D., McCluskey, J., Puy, R. & Knox, R.B. (1991). Isolation of cDNA encoding a newly identified major allergenic protein of rye-grass pollen: intracellular targeting to the amyloplast. *Proc. Natl. Acad. Sci. USA* 88:1384–1388.

Staiger, C.J., Goodbody, K.C., Hussey, P.J., Valenta, R., Drobak, B. & Lloyd, C.W. (1993). The profilin multigene family of maize: differential expression of three isoforms. *Plant J.* 4:631–641.

Stinson, J.R., Eisenberg, A.J., Willing, R.P., Pe, M.E., Hanson, D.D. & Mascarenhas, J.P. (1987). Genes expressed in the male gametophyte of flowering plants and their isolation. *Plant Physiol.* 83:442–447.

Stinson, J.R. & Mascarenhas, J.P. (1985). Onset of alcohol dehydrogenase synthesis during microsporogenesis in maize. *Plant Physiol.* 77:222–224.

Taniai, M., Ando, S., Usui, M., Kurimoto, M., Sakaguchi, M., Inouye, S. & Matuhasi, T. (1988). N-terminal amino acid sequence of a major pollen allergen of Japanese cedar pollen (*Cry j* I). *FEBS Lett.* 239:329–332.

Tebbutt, S.J., Rogers, H.J. & Lonsdale, D.M. (1994). Characterization of a tobacco gene encoding a pollen-specific polygalacturonase. *Plant Mol. Biol.* 25:283–297.

Thangavelu, M., Belostotsky, D., Bevan, M.W., Flavell, R.B., Rogers, H.J. & Lonsdale, D.M. (1993). Partial characterization of the *Nicotiana tabacum* actin gene family: evidence for pollen-specific expression of one of the gene family members. *Mol. Gen. Genet.* 240:290–295.

Turcich, M.P., Hamilton, D.A. & Mascarenhas, J.P. (1993). Isolation and characterization of pollen-specific maize genes with sequence homology to ragweed allergens and pectate lyases. *Plant Mol. Biol.* 23:1061–1065.

Turcich, M.P., Hamilton, D.A., Yu, X. & Mascarenhas, J.P. (1994). Characterization of a pollen-specific gene from *Tradescantia paludosa* with an unusual cysteine grouping. *Sex Plant Reprod.* 7:201–202.

Twell, D. (1992). Use of a nuclear-targeted beta-glucuronidase fusion protein to demonstrate vegetative cell-specific gene expression in developing pollen. *Plant J.* 2:887–892.

Twell, D., Wing, R., Yamaguchi, J. & McCormick, S. (1989). Isolation and expression of an anther-specific gene from tomato. *Mol. Gen. Genet.* 247:240–245.

Twell, D., Yamaguchi, J. & McCormick, S. (1990). Pollen-specific gene expression in transgenic plants: coordinate regulation of two different tomato gene promoters during microsporogenesis. *Development* 109:705–713.

Twell, D., Yamaguchi, J., Wing, R., Ushiba, J. & McCormick, S. (1991). Promoter analysis of genes that are coordinately expressed during pollen development reveals pollen-specific enhancer sequences and shared regulatory elements. *Gene Devel.* 5:496–507.

Valenta, R., Ball, T., Vrtala, S., Duchene, M., Kraft, D. & Scheiner, O. (1994). cDNA cloning and expression of timothy grass (*Phleum pratense*) pollen profilin in *Escherichia coli*: comparison with birch pollen profilin. *Biochem. Biophys. Res. Commun.* 199:106–118.

Valenta, R., Breiteneder, H., Pettenburger, K., Breitenbach, M., Rumpold, H., Kraft,

D. & Scheiner, O. (1991a). Homology of the major birch pollen allergen, *Bet v* I, with the major allergens of alder, hazel and hornbeam at the nucleic acid level as determined by cross-hybridization. *J. Allergy Clin. Immunol.* 87:677–682.

Valenta, R., Duchêne, M., Pettenburger, K., Sillaber, C., Valent, P., Bettelheim, P., Breitenbach, M., Rumpold, H., Kraft, D. & Scheiner, O. (1991b). Identification of profilin as a novel pollen allergen; IgE autoreactivity in sensitized individuals. *Science* 253:557–560.

van der Leede-Plegt, L.M., van de Ven, C.E., Bino, R.J., van der Salm, T.P.M. & van Tunen, A.J. (1992). Introduction and differential use of various promoters in pollen grains of *Nicotiana glutinosa* and *Lilium longiflorum*. *Plant Cell Rept.* 11:20–24.

van der Meer, I.M., Stam, M.E., van Tunen, A.J., Mol, J.N.M. & Stuitje, A.R. (1992). Antisense inhibition of flavonoid biosynthesis in petunia anthers results in male sterility. *Plant Cell* 4:253–262.

van Tunen, A.J., Mur, L.A., Brouns, G.S., Rienstra, J.D., Koes, R.E. & Mol J.N.M. (1990). Pollen- and anther-specific promoters from petunia: tandem promoter regulation of the *chi*A gene. *Plant Cell* 2:393–401.

Vergne, P. & Dumas, C. (1988). Isolation of viable wheat male gametophytes of different stages of development and variations in their protein patterns. *Plant Physiol.* 88:969–972.

Villalba, M., Batanero, E., Monsalve, R.I., Gonzalez de la Pena, M.A., Lahoz, C. & Rodrigues, R. (1994). Cloning and expression of *Ole e* I, the major allergen from olive tree pollen. *J. Biol. Chem.* 269:15217–15222.

Weterings, K., Reijnen, W., van Aarssen, R., Kortstee, A., Spijkers, J., van Herpen, M., Schrauwen, J. & Wullems, G. (1992). Characterization of a pollen-specific cDNA clone from *Nicotiana tabacum* expressed during microgametogenesis and germination. *Plant Mol. Biol.* 18:1101–1111.

Willing, R.P., Bashe, D. & Mascarenhas, J.P. (1988). An analysis of the quantity and diversity of messenger RNAs from pollen and shoots of *Zea mays*. *Theor. Appl. Genet.* 75:751–753.

Willing, R.P. & Mascarenhas, J.P. (1984). Analysis of the complexity and diversity of mRNAs from pollen and shoots of *Tradescantia*. *Plant Physiol.* 75:865–868.

Wing, R.A., Yamaguchi, J., Larabell, S.K., Ursin, V.M. & McCormick, S. (1989). Molecular and genetic characterization of two pollen-expressed genes that have sequence similarity to pectate lyases of the plant pathogen *Erwinia*. *Plant Mol. Biol.* 14:17–28.

Worrall, D., Hird, D.L., Hodge, R., Paul, W., Draper, J. & Scott, R. (1992). Premature dissolution of the microsporocyte callose wall causes male sterility in transgenic tobacco. *Plant Cell* 4:759–771.

Xu, H., Davies, S.P., Kwan, B.Y.H., O'Brien, A.P., Singh, M. & Knox, R.B. (1993). Haploid and diploid expression of a *Brassica campestris* anther-specific gene promoter in *Arabidopsis* and tobacco. *Mol. Gen. Genet.* 239:58–65.

Yoder, M.D., Keen, N.T. & Jurnak, F. (1993). New domain motif: the structure of pectate lyase C, a secreted plant virulence factor. *Science* 260:1503–1507.

Zou, J.T., Zhan, X.Y., Wu, H.M., Wang, H. & Cheung, A.Y. (1994). Characterization of a rice pollen-specific gene and its expression. *Amer. J. Bot.* 81:552–561.

4

Pollination biology and plant breeding systems

PETER G. KEVAN

Summary 59

Introduction 59

Advertisements and rewards 63

   Advertisements 64

   Rewards 67

Breeding or mating systems 70

Pollen movement and neighborhoods 72

Evolutionary dynamics 77

Conclusion 78

References 79

Summary

Understanding genetics and breeding requires knowledge of how the processes of genetic recombination are initiated, are encouraged, and come about. Pollination is the first step by which genes are exchanged between plants (cross-pollination) or recombined within plants (self-pollination). The means by which this comes about range from simple to complex in floral structures, differential timing of floral maturation, and physiological mechanisms in self-compatibility and incompatibility. Pollination may occur by physical means (wind and water) or biotic means (insects, birds, bats, and other). The features of flowers reflect the nature of pollen vectors. Thus, recognizable combinations of floral characters and pollinator adaptations are known in terms of floral advertisements, pollinator senses, floral rewards (nectar, pollen, oil), and pollinators' nutritional needs. The spacing and arrangements of plants, and of flowers on plants, have a profound effect on the movement of pollinators and dispersal of pollen and genes within and between populations of plants. The ancient and fundamentally coevolved relationships of pollinators and flowering plants are continually changing in nature and by human direction, especially in crops. Pollination is a keystone ecological process in agricultural and natural sustainability throughout the world.

Introduction

Pollination is simply the transfer of pollen from the anthers of a flower to a stigma (Figure 4.1). It is an important and early step in the sexual reproduc-

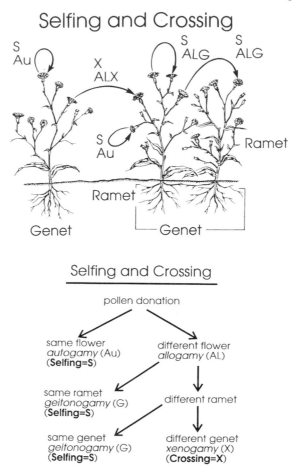

Figure 4.1 Pollen movement in self- and cross-pollination.

tion of higher plants from Cycadales to Angiospermae (the flowering plants). If successful, pollination gives rise to seeds, fruits, and the next generation of plants.

The simplicity of this idea was not realized scientifically until 1750, when Arthur Dobbs published an account of his observations from his estate in Ireland (Nelson 1990). Over the next 150 years, pollination biology grew into a refined and recognized science. It attracted many influential scientists to its complexities, including Charles Darwin, and culminated with Knuth's encyclopedic *Handbook* (1909–12). Nevertheless, in applied pollination biology, naiveté persists through pollinator biologists' misunderstandings of plant

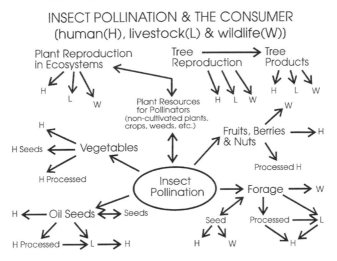

Figure 4.2 Insect pollination and the consumer.

breeding systems, and plant biologists' similar shortcomings in regard to pollinator biology and pollen movement.

The evolution and diversity of Angiospermae and advanced orders of Insecta (Lepidoptera, Diptera, Hymenoptera) are believed to be inextricably interrelated. Palaeoecologists indicate a history of over 140 million years of coevolution (Crepet 1983), but the roots of arthropod and spore interactions may stretch to the dawn of terrestrial life, some half a billion (10^9) years ago (Kevan et al. 1975). Pollination represents a mutualistic symbiosis (beneficial to both partners: plants and pollinators) that is fundamental to the sustainable productivity of all terrestrial ecosystems (except icecaps) (Figure 4.2), and should be taken most seriously in modern agriculture and conservation (Kevan et al. 1990; Kevan 1991).

As mentioned, insects are considered to have been the original pollinators of the early Angiospermae. There has been some debate about whether beetles (Coleoptera) or flies (Diptera) were the first pollinators, but today bees (Hymenoptera: Apoidea) are the most important in most ecosystems, including agricultural ones. Angiosperms that use vectors other than insects (for example, wind, birds, bats, and other mammals) for pollination are believed to be derived. Table 4.1 presents the array of pollinator types, the technical terms used to describe the system, and the characteristics of the system or syndrome (Faegri and van der Pijl 1979).

Although generalizations, or syndromes, can be recognized, they vary in tightness of definition. A mix of insect and wind pollination has been studied in *Plantago lanceolata* (Stelleman 1978) and probably operates in various species of grape (*Vitis* spp.) (Kevan et al. 1985). Fly pollination, or myophily, is such a broad category as to have little meaning. Some plants have flowers that can be pollinated by a wide diversity of vectors, from bats and birds to

Table 4.1. *Main floral syndromes, pollination characteristics, and some terminology*

Pollen vector	Terminology	Characteristics
Wind	Anemophily	Dry, smooth pollen, copious; expansive stigma
Water	Hydrophily	Hydrophobic pollen coating
Surface	Ephydrophily	Specialized female parts
Under	Hyphydrophily	Elongated, threadlike pollen and expansive stigma
Animals	Zoophily	Pollen often somewhat oily, ornamented; stigmata compact
Insects	Entomophily	
Beetles	Cantherophily	Flowers open bowls, often with characteristic scent, pollen all over body
Flies	Myophily	Flowers open bowls, pollen all over the body
Dung or carrion flies	Sapromyophily	Specialized flowers or inflorescences with mimetic scents, pollen all over the body
Moths	Phalaenophily	Flowers trumpet-shaped, nocturnally blooming, pale, and heavily scented, pollen on proboscis or head
Butterflies	Psychophily	Flowers tubular with horizontal, flat landing platform, often pinkish, scent faint, pollen on proboscis and head
Bees	Melittophily	Flowers variable, but often zygomorphic with hidden nectar and precise pollen placement on body of bee
Ants	Myrmecophily	Flowers small and open, plants intermixed, pollen generally on body
Birds	Ornithophily	Flowers mostly tubular, red, with copious nectar, pollen placed on beak, head, or precisely on body
Bats	Chiropterophily	Flowers mostly open, pale, heavily scented, with copious, often viscid nectar, anthers often numerous, pollen placed on the head
	Allophilic	Flowers without special adaptations to zoophily (as myophily)

Pollen vector	Terminology	Characteristics
	Hemiphilic	Flowers with intermediate degree of specialization to restrict pollinator diversity
	Euphilic	Flowers highly specialized
	Polyphilic	Flowers pollinated by many taxa
	Oligophilic	Flowers pollinated by some related taxa
	Monophilic	Flowers pollinated by one or few closely related taxa
	Allotropic	Pollinators not specialized for flower visitation/pollination
	Hemitropic*	Pollinators showing some specialization
	Eutropic*	Pollinators fully adapted for flower visitation/pollination
	Polylectic*	Pollinators visiting flowers of many plant taxa
	Oligolectic*	pollinators visiting flowers of few taxa, some times but not necessarily close related
	Monolectic*	Pollinators visiting flowers of one, or very few, taxa (more restricted than oligolectic)
	Dystropic	Pollinators may be destructive

* The suffixes *tropic* and *lectic* are sometimes used interchangeably.

insects. In such general systems, the relative importance of different pollinators may vary with place, time, and circumstances.

What is clear from the syndromes is that there are certain combinations of characters of flowers and pollen vectors that can be interpreted as having coevolved. Figure 4.3 presents the dimensions of pollination from the pollen vectors' (pollinators') perspective. This chapter allows for integration of those dimensions into plant breeding systems. We start with pollinators because pollination cannot start without a flower visit.

Advertisements and rewards

The interplay of attractants and rewards cannot be easily elucidated. The visual advertising by pear blossoms would seem as effective as that of other pome and stone fruits, yet pear blossoms are much less visited by insects, especially honeybees, than are flowers of others in the face of competition. This is presumed to be because of the inferior nectar (more watery) produced

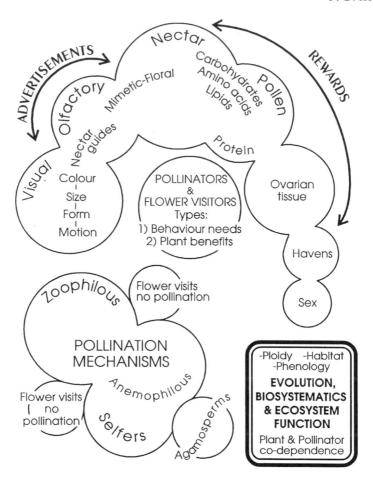

Figure 4.3 The dimensions of pollination showing the relationships of floral advertising and rewards on pollinators and other flower visitors and the roles of the former on plant reproduction and coevolutionary ecology.

by pears (McGregor 1976; Free 1993). Differences in visitation rates of honeybees and other bees to various cultivars of apples are known (Free 1993) but the reasons for them are not. Difficulties in producing hybrid seed in various crops come from the pollinators' fidelity to the flowers of one parent line or the other (Erikson 1983) for undiscovered reasons. The interplay between attractants and rewards involves learning and memory by pollinators and these are remarkably well developed (Papaj and Lewis 1993).

Advertisements

Visible floral displays have been the most studied, and, of these, colors are the best understood. In discussing floral colors in relation to pollination, ultraviolet reflections must be considered for insects and birds. Ultraviolet reflections from flowers have excited undue and biased interest. They cannot be treated alone in terms of color vision (Kevan 1979, 1983; Chittka et al. 1994). Recent advances in the mechanisms of insect color vision have shown how the colors of flowers should be examined in light of the trichromatic (ultraviolet, blue, and green) color vision of bees (Backhaus and Menzel 1987; Chittka et al. 1994).

Although it is well known that color differences between flowers allow pollinators to discriminate between species and varieties (for example, in *Phlox* by Levin 1978), appropriate colorimetry of crop plants to try to distinguish intervarietal differences has rarely been done. An examination of several commercial and hybrid-parental lines of canola (*Brassica napus*) detected no differences (Kevan, unpublished). Kay (1982) examined floral color polymorphism in *Raphanus raphanistrum* and found differences in the pollinator fauna (white with UV and yellow with UV). Other work on this species has also shown differential visitation rates by different pollinators and associated differences in seed set (Stanton 1987). Sometimes, though, color differences between cultivars of a species are ignored by pollinators (Kevan 1983).

Color changes in the perianth are used as cues by pollinators' decisions of which flowers to visit and which to ignore. The well-known change in the color spot in flowers of *Aesculus* spp. from yellow to red indicates to bumblebees (*Bombus*) that the flower is no longer rewarding and to bypass it (Kugler 1943). A similar effect has been shown for butterflies (Weiss 1991) and thrips (Mathur and Mohan Ram 1978) foraging at yellow flowers of *Lantana camara* but not at older, red flowers. Color changes in flowers are well known (Gori 1983; Weiss 1991) and some occur after pollination. These changes have rarely been measured with appropriate colorimetry (Kevan 1978, 1983).

Many flowers have color patterns within them. The importance of these as "nectar guides" that direct pollinators to the floral reward in such a way that pollination is assured was first systematically described by Sprengel (1793). Daumer (1958) and Kevan (1972, 1983) used the then most appropriate colorimetry to examine nectar guides in the insect visual spectrum. Similar research, but using the paradigms required by recent findings in vision research (Backhaus 1993; Chittka et al. 1994), is in progress at the Institut für Neurobiologie, Freie Universität Berlin (R. Menzel and team, personal communication, 1996). Variations in such patterns have rarely been investigated, but flowers of *Monarda fistulosa* that lack guides were ignored by pollinating bees while the normal flowers were visited (Scora 1964).

Flowers bloom against their backdrop of vegetation or ground with which their colors mostly contrast. Vegetation and ground are quite dull and reflect uniformly across the visible spectrum of most insects and so can be described as neutral or uncolored (grey) (Daumer 1958; Kevan 1972, 1983; Chittka et al. 1994). The color contrast of flowers against that background can thus be appreciated. Similarly, color patterns of nectar guides within flowers are contrasting and their importance to pollinator foraging has been shown (Lunau 1993).

Understanding the role of floral color in pollen movement is difficult, but when it has been examined, disassortative and assortative pollen flows have been found (Levin 1978; Stanton 1987). Similarity of colors between flowers of different plants but of the same species is probably extremely important in floral constancy by pollinators and intraspecific pollen flow (Christy 1884; Kevan 1983; Chittka and Kunze, in preparation), but this is not the case. Pollinators may be constant to species with flowers of variable colors, or may not be constant to species at all (Waser 1986; Chittka and Kunze, in preparation). The conditions under which a pollinator chooses to be flower constant or inconstant are probably complex and concern issues such as the distribution and abundance of reward, motivation, and type of pollinator. Nevertheless, floral colors allow pollinators to distinguish between flowers of different species (Kevan 1983), and it is to be expected that in contemporaneously and autochthonously flowering guilds of plants, the plants would isolate themselves reproductively and reduce interspecific competition for pollinators by having differently colored flowers (Kevan 1983). Another feature associated with the visibility of flowers is their size. It can be assumed that the larger a flower (or the group of flowers acting as a unit, as in an inflorescence), the greater the distance from which it can be detected (Dafni 1992). The actual measurement of floral size with respect to pollinator is complex and requires measurements of the receptance angle of the facets of the compound eye and understanding of the extent and nature of color contrast. Nevertheless, larger flowers or inflorescences generally receive more visits and are more successfully pollinated (Dafni et al. in press).

Associated with the size of flowers is the matter of form and presentation. Form classes are used frequently in studying pollination and can be incorporated into the context of syndromes (Table 4.1). Although insects (honeybees) can be trained to recognize and distinguish between certain forms and patterns, the means by which they do so is just starting to be understood (Dafni et al. in press). Within the problem of form is that of depth perception, which would seem poorly developed by stereopsis in many pollinating insects (the eyes are too close together and may have only small overlapping visual fields). Parallax as well as the relative angular speed of motion of distant (slow) versus nearby (fast) objects may be used (Wehner 1981). Coupled with all those factors are the parts of the visual system and brain used to detect size, shape, form, patterns, floral depth, and aspects of contrast. Some of these tasks in visual detection and discrimination may be color-blind, mediated by the green receptors of the compound eye and thus requiring contrast only in one waveband rather than difference in one or more (required for color discrimination).

Olfaction in many insects is highly developed, but it seems almost lacking in most birds. Some generalizations about floral scents can be made. They are volatile, oxygenated derivatives of alcohols, esters, aldehydes, and ketones; some are terpenoid or aminoid, and some are not. The heavy terpenoid and aminoid floral scents are associated with nocturnally pollinated flowers and presumably attract pollinators from a distance. Visual cues presumably come into play at close range. The lighter (or more delicate) floral scents are associated with diurnal bloomers in which visible cues attract pollinators from a distance and scent comes into play at close range. Some scents act solely as advertisements and have no counterpart outside flowers. Some are mimetic, such as the aminoid-, indole-, and scatole-containing blends that smell of carrion, dung, and mammal musk and deceive visitors into entering the flowers (for example, Araceae, *Stapelia,* Rafflesiaceae); others mimic insect pheromones and induce pseudocopulation of the male bee or wasp with the flower (for example, *Ophrys*), during which pollination occurs. In yet other orchids, scent is the reward collected by male bees (Englossini) to be used in their own elaborate mating procedures (Roubik 1989).

Some floral scents are distinctive, even to humans as well as to pollinators. Such distinctions can be made at the varietal level by, for example, rose fanciers and sniffers of *Narcissus'* scents (Richards 1986). Measures of discrimination and learning of scents by honeybees indicate that olfaction is an important component of floral recognition. Kauffeld and Sorensen (1971) noted that differences in honeybee pollination and visitation rates to flowers of various lines of *Medicago sativa* correlated with differences in scent. Galen and Kevan (1983) observed that the fauna of pollinators to *Polemonium viscosum* differed between flowers that smell "sweet" and those that smelled "skunky" to them. Kevan (1989b) found that honeybees visiting spathes of *Symplocarpus foetidus* did not discriminate among the kinds of scents emitted, but did not visit scentless spathes. The analysis of floral scents (Dobson 1991, 1994) has allowed for interspecific, intervarietal, and intersexual differences to be studied. Further, the detection of Duftmale or scent patterns within flowers (Lex 1954) has progressed (see Adey 1982, cited in Richards 1986).

Often overlooked as a means by which insects can discriminate between flowers of different species are structural features. Grooves and ridges in petals and sepals are used by insects as nectar guides. Remarkably, honeybees are able to discriminate between the arrangement and exposed wall forms of epidermal cells on petals that result in a microscopic form of "Braille" (Kevan and Lane 1985). Brookes and Small (1988) have noted that a reduction in these fine sculpturings is associated with the loss of oncrossing in *Medicago* spp. We looked for microsculptural differences in several varieties of canola (*Brassica napus*) but found none (unpublished results).

Rewards

The original reward that flowers offer their pollinators is probably pollen. Generally, pollen is highly nutritious, with average protein content at about 25%, amino acids at 10%, carbohydrates at 25%, and lipids at 5%. Also present in pollen are coenzymes, vitamins, minerals, and sterols (Schmidt and Buchmann 1992). The presence of starch or oil as the main energy storage compound is related to the size of the grains: smaller with oil, and larger with starch (Baker 1979).

Pollen has certain features that indicate its mode of dispersal. Pollen of anemophilous plants tends to be smooth, round, and dry (for example, Gramineae). Pollen of zoophilous plants tends to have an oily coating; such oily pollen is often ornamented and, in extreme cases, is held together in clumps by threads of viscin (as in Onagraceae), in polyads (as in Boraginaceae, Ericaceae, *Acacia*), or in complex pollinia (as in Asclepiadaceae, Orchidaceae). It is difficult to make precise and explanatory generalizations about the associations between the characteristics of pollen and the pollinators (Blackmore and Ferguson 1986).

Pollen is eaten by many kinds of flower visitors. In most insects, pollen is ingested whole; the protoplast is digested and removed by diffusion and the more or less empty exine defecated. In pollenivorous flies, females with developing ovaries consumes more than males do. Bees characteristically gather pollen in special parts of their bodies. The pollen baskets of honeybees and bumblebees are on the hind legs, but on leaf-cutting bees they are under the abdomen. The bees take the harvested pollen to their nests, where it is packed into cells for storage and later use (as in honeybees) or formed into balls or loaves onto which the bee lays its eggs. The completed cells (pollen loaf and egg) are sealed off separately in chambers, which may contain single or multiple cells.

Nectar is sought by most anthophiles. It is mostly sugars derived from phloem and dissolved in water. Minor constituents include amino acids, organic acids, minerals, and sometimes proteins, lipids, dextrins, vitamins, tannins, and alkaloids. Nectar is the main fuel for locomotion for anthophilous animals. Pure sucrose yields 16.53 joules/g of energy, which is quickly available through metabolism. The other sugars present in the nectar are mostly glucose and fructose, and a few others have been found in smaller amounts in some species (Baker and Baker 1983).

The importance of the main sugars and amino acids in nectar has been examined in detail by Baker and Baker (1983). They have pointed out the correlation of sugar concentrations and sugar ratios in nectar with the animals (Table 4.2). They have also noted the taxonomic correspondence of nectar sugar ratios (Table 4.2). Variability within species of crop plants may be important in applied pollination as Kevan et al. (1992) have discussed for canola, from which honeybees produce a honey with a high propensity for granulation, even in the beehive. Harvesting such honey is troublesome, and thus pollination services may cost more. The reasons for particular sugar ratios in nectar are not understood, either botanically or ecologically, but may relate to the energetics and metabolism of the respective pollinators.

A wide variety of amino acids have been detected in nectars. These also correlate with pollinator type and plant taxonomy (Table 4.2). However, the

Table 4.2. *Floral nectars and their characteristics according to floral type as described in Table 4.1*

Floral type	% Sugar	Sugar ratio	Volume	Amino acids micromoles/mL
Allophilic				
Cantherophilic		No trend	Little	–
Myophilic	10–80, solid	H	Little	0.56
Myrmecophilic	20–80, solid	H	Little	–
Sapromyophilic		H	Little	12.5
Melittophilic				
Short-tongued	20–80	H or S	Little more	1.02
Long-tongued	20–70	S	More	0.62
Psychophilic	15–50	S	Much	1.50
Sphyngophilic	15–30	S	Copious	0.54
Phalaenophilic	15–50	S	Much	1.06
Ornithophilic	20–40			
Hummingbirds		S	Copious	0.45
Other birds		H	Copious	0.26
Chiropterophilic	15–30	H	Copious	0.31

Note: % sugars are highly variable and depend on evaporation and floral form.
Volumes: *Little* represents mostly < 1 or 2 μL.
         More represents mostly 5–10 μL.
         Much represents mostly 10–15 μL.
         Copious represents tens of μL.
Sugar ratios: H is hexose-rich or dominant (glucose and fructose); S is sucrose-rich or dominant.
For more details, see Baker and Baker (1983).

roles of these and other minor constituents of nectar in pollination are not known. They affect the quality of honey, some of which is even poisonous to honeybees or people (Crane 1965). Oils are also secreted by the flowers of some plants (for examples, Barbados cherry, *Malpighia* spp., and nance, *Byrsonima*). The oil-producing glands (elaiophores) secrete free fatty acids that are energetically very rich and are gathered by specialist bees.

There are a number of insects, particularly beetles, that chew the ovarian tissue of flowers. It has been noted that flowers showing primitive character- istics and pollinated by beetles and other chewing insects have heavy and fleshy floral parts. The development of the closed conduplicate carpel (basic to Angiospermae) and the inferior ovary (in some angiosperm families) may have arisen in response to protection from chewing insects (Kevan 1984).

Some insects use flowers as havens. Kevan (1989a) has reviewed infor- mation concerning flower basking by insects during sunny weather in cold conditions. Some bees, especially males, sleep in flowers, as does the squash bee *Peponapis pruinosa* in the short-lived blossoms of *Cucurbita pepo* (Willis and Kevan 1995). These same males use the flowers as sites to find mates, and similar relations are thought to apply to other insects and flowers (Kevan and Baker 1983). Males of some hummingbirds and various insects, including bees, guard patches of flowers as part of their territories for sustenance and reproduction.

Breeding or mating systems

Breeding systems in most multicellular organisms are those combinations of processes by which the sexes come together, mating takes place, and the female gamete becomes activated to start development (that is, those processes involved in breeding). Wyatt (1983, pp. 55–56) defines a plant breeding system as "all aspects of sex expression in plants that affect the relative genetic contributions to the next generation of individuals within a species" and mating systems in plants can be defined as those factors that determine the pattern of gene inheritance between generations (after Kearns and Inouye 1993). For plants, those definitions include pollination mechanisms such as those just noted and pollen movement as next discussed. However, for ease of discussion, this section restricts the usage to means of sexual expression in plants and their consequences to sexual (and for agamospermy, asexual) reproduction.

The advertisements and rewards that flowers present to pollinators suggest how tightly the zoological phase of pollination is integrated with the botanical. However, the means by which pollination takes place for sexual reproduction in plants has thus far been examined only lightly (Figure 4.1). This part of the chapter involves the reproductive systems or the ways in which the sexual organs, gametes, and fertilization of plants are interconnected (Richards 1986). Vegetative reproduction (except some means of agamospermy) is not discussed.

Angiosperms are highly variable in the ways in which they express sexuality. Some plants are hermaphrodites, others are unisexual. Some species are characterized by hermaphroditic or unisexual individuals; in others, mixtures may occur. Overlying these conditions are issues of temporal and spatial separation of sexual function, extent of ability to self-pollinate and self-fertilize, and whether or not fertilization is needed for seeds to be set.

Hermaphroditic plants may have the male and female functions on separate flowers (monoecy) that may be separated in space on the plants (as in many conifers and in corn) or in time (as in many Palmae and Cucurbitaceae). When separation is temporally pronounced, self-pollination is not possible (for example, *Elaeis guineensis* – months between maturation of unisexual inflorescences), but when it is less pronounced (for example, *Cocos nucifera* – days in bisexual inflorescences), breeders have produced homogamous (both sexes blooming together) bisexual inflorescences (for example, Malaysian dwarf coconut). In Cucurbitaceae, plants often produce male flowers before females and may later again produce male ones. If homogamy on the plant is normal in monoecious species, they are often self-incompatible, as in conifers. Corn, *Zea mays*, is self-compatible. The open-pollinated varieties separate sexual function with tassels shedding pollen before the silks are receptive. Modern varieties are more homogamous. Nevertheless, self-pollination results in the decreased vigor of the next generation (Troyer 1993; Kannenberg, personal communication).

Self-incompatibility is also common in species with homogamous, perfect flowers. Breeders have selected, or bred, lines that are self-incompatible from self-compatible progenitors to produce hybrid seeds, as in canola (*Brassica napus*), or have selected for self-compatible lines from self-incompatible progenitors to circumvent the need for cross-pollination and pollinators and to ensure crop production, as in sunflower (*Helianthus annuus*). More details of self-incompatibility and self-compatibility are given in Chapter 9 and in Richards (1986). Self-incompatibility assures that self-pollination does not result in seed set and requires that the plant be cross-pollinated for sexual reproduction to be successful.

Some monoecious plants may show a range of variations with bisexual inflorescences or by having some flowers bisexual and others female (gynomonoecious), or others male (andromonoecious). These situations may arise through interspecific hybridizations (as in *Elaeis* spp. with bisexual inflorescences) or naturally.

Self-incompatibility is not the only means of avoiding autogamy in species with perfect flowers. The flowers may be formed so that the male and female parts are separated in space (herkogamy) (as in *Passiflora* and most Orchidaceae) or in time (dichogamy). Dichogamy is present in two forms: the more common is protandry, in which the anthers shed pollen before the stigma is receptive; protogyny is the reverse, and well exemplified by *Magnolia* and *Asimina*.

A special form of herkogamy is found in heteromorphy (heterostyly), in which flowers of more than one type are produced. Heterostyly (Figure 4.4) is also associated with self-incompatibility and has been studied extensively in both distylous and tristylous plants, and is a fertile area for more research (Richards 1986; Barrett 1990). Distyly is represented in a few crop plants, for example, Star fruit (*Averrhoa carambola*) and Mombin (*Spondias mombin*). Tristyly is well represented by the notorious adventive plant, purple loosetrife (*Lythrum salicaria*).

In some species, distyly seems to have given rise to dioecy in which the sexual roles of the plant are separated. Male plants seem to be thrum and female ones pin (Figure 4.4), as in *Sarcotheca celebica* in Indonesia (Lack and Kevan 1987) and sapodilla (*Spondias* sp.) in Mexico (Basarto Peña et al., unpublished results). There are various ways that dioecy is thought to have arisen (Thomson and Brunet 1990), but in crop plants the situation is uncommon. Examples are *Actinidia* spp.; Jojoba *Simondsia*; some Grapes, *Vitis* spp.; *Asparagus officinalis;* and Spinach, *Spinacia oleraceae*.

There are various intermediate arrangements of sexual organs on plants.

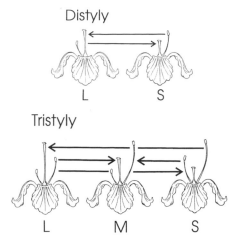

Figure 4.4 Heterostyly in which distyly is depicted with pin (L) and thrum (S) flowers, and tristyly with flowers containing long (L), medium (M), and short (S) styles and associated lengths of filaments. Arrows indicate pathways of appropriate pollen transfer.

With particular reference to dioecy is gynodioecy, in which some plants are male sterile. This situation is widely distributed in nature (Richards 1986) and is important in hybrid seed production because of the use of genic or cytoplasmic male sterility in artificially creating field populations of gynodioecious crop plants (Chapters 7 and 8). Androdioecy (that is, female-sterile and hermaphroditic plants in a population) seems very rare. Trioecy refers to plants that show three sexual forms – female, male, and hermaphrodite – but it is also a rarity.

Agamospermous plants produce seeds without the fertilization of ovules. There are numerous ways by which chromosomal numbers are maintained from parent to offspring. The phenomenon is quite common in flowering plants and mostly associated with polyploidy. Richards (1986) provides a detailed account of this system and notes that it is not an evolutionary dead end, as has been often suggested, because the opportunity for genetic recombination exists through pollen transfer. Dandelion (*Taraxacum officinale*) provides an excellent example.

All in all, there is a wide range of ways in which plants arrange their sexual organs from unisexual plants to bisexual plants with combinations of unisexual and bisexual flowers in various proportions.

Pollen movement and neighborhoods

At first glance, such an array of pollination mechanisms and reproductive systems is rather bewildering, but pollination biology must embrace it if the mode and consequences of pollen movement are to be understood. Similarly, plant scientists working with breeding systems must be aware of their diversity and recognize the importance of understanding how pollen moves within and between populations of plants.

Pollen of anemophilous plants blows in the air. It also has a thin, smooth exine and a high surface-to-volume ratio. It is subject to dehydration or hydration, depending on the prevailing environment. Anemophilous plants synchronously produce large clouds of pollen over very short durations of a season – for example, days in conifers, blackspruce, and jackpine (Di-Giovanni et al. 1996) and hours in pasture grasses in Canada (Tikhmenev and Kevan in press). The longevity of such pollen in nature is expected to be short, but has rarely been tested. In some steppe grasses of Russia, longevity is less than an hour (Ponomarev 1966).

The biophysical and biometeorological aspects of pollen dispersal are presented in Figures 4.3, 4.5, and 4.6, which show, in diagrammatic form, the way pollen can be thought to move through the air. The mathematical details of the models are beyond the scope of this chapter, but are considered in Di-Giovanni and Kevan (1991). These studies relate to the highly applied concerns of maintaining high levels of genetic superiority in seeds harvested from seed orchards for reforestation with improved trees.

The manner in which the pollen of anemophilous plants actually impacts on the stigma has been studied by Niklas (1985). Remarkable eddies swirl the airborne pollen into the receptive female reproductive parts (Figure 4.7), and the pollen capture efficiency of the micropylar arms of receptive female cones of jackpine is the same over a wide range of naturally occurring wind speeds (Roussy 1994).

In zoophilous pollination, pollen movement depends on the movements of the pollinators. In general, if a pollinator collects a load of pollen available for

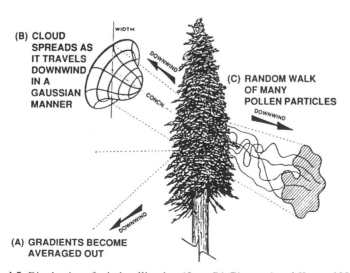

Figure 4.5 Biophysics of wind pollination (*from Di-Giovanni and Kevan 1991*).

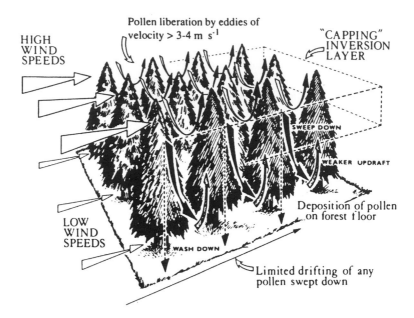

Figure 4.6 Biophysics of wind pollination (*from Di-Giovanni and Kevan 1991*).

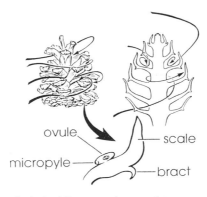

Figure 4.7 Patterns of wind eddies around an ovulate cone, showing how pollen is swept into the micropylar area of a receptive cone (*see Niklas 1984*).

pollination (that is, either ingested or packed), some proportion of that load will be removed by the stigmata of subsequently visited flowers. Pollen carry over by most animal pollinators seems to be limited to a few meters or a few flowers in herbaceous plants (Thomson and Plowright 1980; Handel 1983; Richards 1986) but depends on the dispersion of plants (Linhart 1973) and the kind of pollinator (Schmitt 1980). The usual form of carryover is an exponential decay function (Waser 1988) with flower visitation. The actual form depends on the value (ease of dislodgement and proportion dislodged) of the

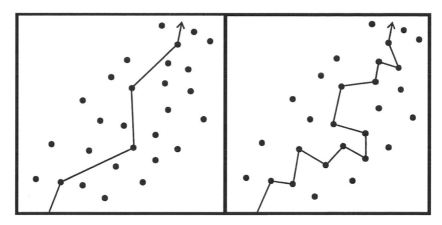

Figure 4.8 Path of pollinator foraging at a patch of poorly or nonrewarding flowers (left) and a patch of richly rewarding flowers (right). Note the overall "forwarding" motion, but shorter hops and sharper angles between flowers in the richly rewarding patch.

exponent and the variance (clumpiness of pollen grains) of the proportion dislodged after each subsequent visit. For honeybees and, presumably, other social pollinators, the situation is further complicated because pollen is transferred between individual bees in the hive and then redistributed on the bodies of different foragers (DeGrandi-Hoffmann et al. 1986) (see also Chapters 5 and 6). Nevertheless, cross-pollination is enhanced.

In general, the movements of pollinators between flowers is strongly influenced by the amount of reward present. The more reward, the shorter the distance the forager flies between flowers, inflorescences, or plants (stations), and the sharper are its turning angles as it progresses from one station to another. The result of this activity (Figure 4.8) is that on rewarding patches of blooming plants, the pollinator visits more flowers than it would on less rewarding patches (Waddington 1983). That influences pollen carryover and gene flow by distance. The shorter durations of visits on lesser rewarding flowers would also increase the pollen carryover according to flower sequence (Thomson and Plowright 1980; Waser 1988).

A genetic consequence of limited pollen carryover and gene flow within neighborhoods is optimal outcrossing distance by which the advantages of hybridity and matings with neighbors, presumably or actually related, are in the best balance (Waser and Price 1983, 1989). These ideas, and that of genetic neighborhood size (Crawford 1984), are important in the evolutionary ecology of natural stands of plants. Richards (1986) provides a discussion of the calculations of neighborhood sizes used in various studies, but the details of this are beyond the scope of this chapter. For most herbaceous plants, neighborhood size rarely exceeds a few hundred square meters, or 5,000 plants, for butterflies on subalpine *Senecio* spp. in Colorado (Schmitt 1980). For trees, the sizes are mostly 1,000–2,000 m^2 but may be much greater

pollinator movements from bottom to top -
both on a large inflorescence and on plants

Figure 4.9 Pollinator movements on vertical arrangement of flowers in and between inflorescences. Note that pollen is taken from younger flowers in male phase of one plant to older flowers in female phase of the next.

(Richards 1986). For crop plants, these concepts have rather little applicability because neighborhoods cannot develop. However, they do have relevance in situations where genetics are important, as in seed production or intervarietal isolation. Such situations have evoked special concern in regard to the movement of biotechnologically altered genomes between crops or to wild plants (Dale 1994; Schmitt and Linder 1994).

Once on a given plant, flower visitors forage according to certain rules (Pyke et al. 1977; Waddington 1983). Mostly, they tend to start low on inflorescences and work upward (Figure 4.9). Thus, they visit older flowers first, which are mostly in female phase because the flowers are protandrous, and then visit younger flowers in male phase. As they move from inflorescence to inflorescence, they transfer pollen from flowers in male phase to those in female phase. Variations on this theme include the pollinators' working centrifugally on horizontally or radially (as in sunflowers, Asteraceae) arranged inflorescences. The same pattern of movements also applies to flower visitors on trees, but in this situation the pollinators work upward between closely spaced inflorescences (apparently treating them as a unit), downward between more distantly spaced inflorescences, and tend to depart from the tree lower than the level at which they entered (Kevan 1990). Exceptions to these patterns are discussed by Corbet et al. (1981).

The patterns of pollinator movements between and on patches of flowers, and on inflorescences, have been interpreted in terms of optimal foraging theory (Pyke 1984), by which foragers should maximize their *net* rates of energy accumulation. Necessarily, the conditions under which they can accomplish that vary with circumstances and other demands for life placed on the forager (for example, avoiding predation, finding mates or oviposition sites, territory protection, rest, and so forth).

Evolutionary dynamics

The dynamics of pollination systems consist of complex interactions (Figure 4.10) by which not only do pollinators interact with plants and vice versa, but also pollinators interact with each other and plants do likewise. The ecological and evolutionary processes of plant–plant and pollinator–pollinator interactions are the same and involve competition and mutualism. Both processes are important in natural and agricultural environments.

In nature, plant–plant competition for pollinators has been suggested to be a weak force. However, such competition is difficult to measure and would vary in strength inversely with the strength of the pollinator force (population), the population size, and characteristics of the competing species, and other factors (Rathke 1983). The curious phenomenon of pollen allelopathy is noteworthy in that pollens of some species, when on the stigmata of others, inhibit the germination of that species' pollen (Murphy and Aarssen 1995). A consequence of suppressing one's neighbors' reproductive output is the reduction of competition. Other issues of pollen competition are presented in Chapters 5 and 6. Nevertheless, in agricultural settings such competition can be serious because flowers of weeds and other plants draw pollinators from the crop plants (Free 1993).

Evidence of competition between pollinators for floral rewards has been documented in various ways (see Kevan and Baker 1984) from the direct observation of species hierarchies at flowers, of aggressive encounters, and of more rapid and gainful foraging trips in the absence of competitors.

The outcomes of competition are change or extinction. Change comes about through character displacement by which the competing species reduce the amount of competition between themselves. Shifts in flowering time (floral phenology) has received the most attention, but changes in floral form, color, reward quality or quantity, and breeding system (for example, the poorer competitor shifts toward self-pollination and self-compatibility) are also important.

Through such character displacement, plants may become mutualists in supporting each other cooperationally by, for example, (a) supporting long-lived pollinators (hummingbirds and colonies of bees) through providing a continual source of sustenance, (b) providing continual availability of populations of short-lived pollinators (thrips for Dipterocarp pollination with the first blooming species' being self-compatible – Appanah 1981), and (c) pollinators' needing two kinds of resources, pollen from one species and nectar from another, or complementary amino acids (Dafni and Kevan 1994) or sugars (Kevan 1993) in nectars from dioecious or diclinous plants, or different species.

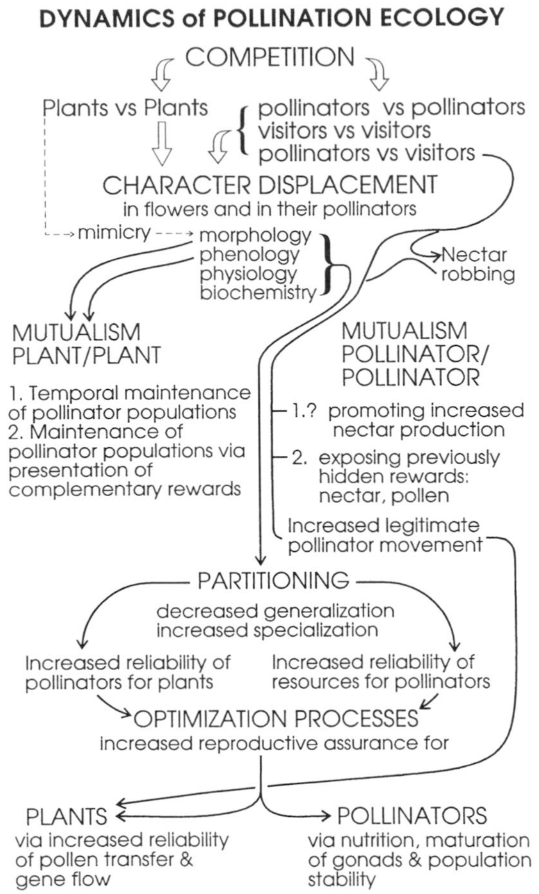

DYNAMICS of POLLINATION ECOLOGY

Figure 4.10 The ecological and evolutionary dynamics of pollination ecology (*redrawn from Kevan and Baker 1984*).

Conclusion

Agriculture mostly simplifies existing ecosystems. Oversimplification associated with high application rates of exogenous materials (fertilizers and biocides) and energy (tillage) has been questioned seriously regarding its sustainability and the extent to which the global environment has been adversely affected. Along with changes in agricultural practices, there has been some emphasis placed on alternative and diversified crops. At the same time, public awareness of the deplorable state of global environments has generated pressure for conservation in agricultural lands and wilderness.

As pointed out at the beginning of this chapter, pollination is a keystone ecological process upon which global terrestrial productivity and sustainability depend to an immeasurable extent. There are figures on the value of pollination by insects and, especially, honeybees to U.S. agriculture (Southwick and

Southwick 1992), but this is minuscule by comparison with the invaluable contribution of pollinators to the world's terrestrial ecosystems. These issues have been raised (Kevan et al. 1990; Kevan 1991), but they remain largely ignored.

In this chapter, the complexities and importance of pollination and plant reproductive systems have been introduced. Our knowledge of these intricacies at the levels of the ecological community and ecosystem is woefully inadequate for planning. In agriculture, the importance of alternative (non-honeybee) pollinators is well recognized by a few, yet all but ignored by policy makers with few exceptions (alfalfa and alfalfa leaf-cutting bees; bumblebees and tomatoes). Experiments on pollination requirements of new crops and new varieties of other crops all too frequently are not sufficiently interdisciplinary to answer the practical questions of pollination requirements for field production. Free's (1993) remarkable compendium directly and indirectly points to the need for species-by-species trials, and within-species variety-by-variety trials, to determine what the actual breeding systems involved are and what sorts of pollinators to use and how to manage them.

References

Appanah, S.A. (1981). Pollination in Malaysian primary forests. *Malays. For.* 44:37–42.

Backhaus, W. (1993). Color vision and color choice behavior of the honey bee. *Apidologie* 24:309–331.

Backhaus, W & Menzel, R. (1987). Color distance derived from a receptor model of color vision in the honey bee. *Biol. Cyber.* 55:321–331.

Baker, H.G. (1979). Starch in angiosperm pollen grains and its evolutionary significance. *Amer. J. Bot.* 66:591–600.

Baker, H.G. & Baker, I. (1983). A brief historical review of the chemistry of floral nectar. In *The Biology of Nectaries*, eds. B. Bentley and T. Elias, pp. 126–152. New York: Columbia University Press.

Barrett, S.C.H. (1990). The evolution and adaptive significance of heterostyly. *Trends Ecol. Evol.* 5:144–148.

Blackmore, S. & Ferguson, I.K., eds. (1986). Pollen and spores: form and function. *Linnean Society Symposium Series No. 12*. London: Academic Press.

Brookes, B. & Small, E. (1988). Enhanced floral analysis by low temperature scanning electron microscopy. *Scan. Micro.* 2:247–256.

Chittka, L., Shonida, A., Troje, N. & Menzel, R. (1994). Ultraviolet as a component of flower reflections, and colour perception of Hymenoptera. *Vision Res.* 34:1489–1508.

Christy, R.M. (1884). On the methodic habits of insects when visiting flowers. *J. Linn. Soc.* 7:186–194.

Corbet, S.A., Cuthill, I., Fallows, M., Harrison, T. & Hartley, G. (1981). Why do nectar-foraging bees and wasps work upwards on inflorescences? *Oecologia* 51:79–83.

Crane, E. (1965). *Honey: A Comprehensive Review*. New York: Crane, Russak & Co.

Crawford, T.J. (1984). The estimation of neighbourhood parameters for plant populations. *Heredity* 52:273–283.

Crepet, W.L. (1983). The role of insect pollination in the evolution of the angiosperms. In *Pollination Biology*, ed. L. Real, pp. 31–50. Orlando, FL: Academic Press.

Dafni, A. (1992). *Pollination Ecology, a Practical Approach*. Oxford: IRL Press.

Dafni, A. & Kevan, P.G. (1994). An hypothesis on complementary amino acids in nectar. *Evol. Theory* 10:259–260.

Dafni, A., Lehrer, M. & Kevan, P.G. (in press). Spatial flower parameters and insect spatial vision. *Biol. Reviews*.

Dale, P.J. (1994). The impact of hybrids between genetically modified crop plants and their related species: general considerations. *Mol. Ecol.* 3:31–36.

Daumer, K. (1958). Blumenfarben wie sie die Bienen sehen. *Zeitschrift für vergleichende Physiologie* 41:49–110.

DeGrandi-Hoffmann, G., Hoopingarner, R.A., Baker, K.K. & Klomparens, K. (1986). Influence of honey bee (Hymenoptera: Apidae) in-hive pollen transfer on cross-pollination and fruit set in apple. *Environ. Entomol.* 15:723–725.

Di-Giovanni, F. & Kevan, P.G. (1991). Factors affecting pollen dynamics and its importance to pollen contamination: a review. *Can. J. For. Res.* 21:1155–1170.

Di-Giovanni, F., Kevan, P.G. & Caron, G.E. (1996). Estimating the timing of maximum pollen release from jack pine (*Pinus banksiana* Lamb.) in northern Ontario. *Forestry Chronicle* 72:166–168.

Dobbs, A. (1750). A letter from Arthur Dobbs sent to Charles Stanhope Esq; F.R.S. concerning bees, and their method of gathering wax and honey, October 22nd, 1750. *Phil. Trans.* 16:536–549.

Dobson, H.E.M. (1991). Analysis of flower and pollen volatiles. In *Essential Oils and Waxes: Modern Methods of Plant Analysis*, eds. H.F. Linskens and J.F. Jackson, pp. 231–251. Berlin: Springer-Verlag.

Dobson, H.E.M. (1994). Floral volatiles in insect biology. In *Insect–Plant Interactions, Vol. V*, ed. E.A. Bernays, pp. 48–81. Boca Raton, FL: CRC Press.

Erikson, E.H. (1983). Pollination of entomophilous hybrid seed parents. In *Handbook of Experimental Pollination Biology*, eds. C.E. Jones and R.J. Little, pp. 493–535. New York: Van Nostrand Reinhold.

Faegri, K. & van der Pijl, L. (1979). *The Principles of Pollination Ecology*, 3rd ed. New York: Pergamon Press.

Free, J.B. (1993). *Insect Pollination of Crops*, 2nd ed. Orlando, FL: Academic Press.

Galen, C. and Kevan, P.G. (1983). Bumblebee foraging and floral scent dimorphism: *Bombus kirbyellus* Curtis (Hymenoptera: Apidae) and *Polymonium viscosum* Nutt. (Polemoniaceae). *Can. J. Zool.* 61:1207–1213.

Gori, D.F. (1983). Post pollination phenomena and adaptive floral changes. In *Handbook of Experimental Pollination Biology*, eds. C.E. Jones and R.J. Little, pp. 31–49. New York: Van Nostrand Reinhold.

Handel, S.N. (1983). Pollination ecology, plant population structure, and gene flow. In *Pollination Biology*, ed. L. Real, pp. 163–211. New York: Academic Press.

Kauffeld, N.M. & Sorensen, E.L. (1971). *Interactions of Honeybee Preference on Alfalfa Clones and Flower Colour, Aroma, Nectar Volume, and Sugar Concentration*. Publication No. 163. Manhattan, KS: Kansas Agricultural Experimental Station.

Kay, Q.D.N. (1982). Intraspecific discrimination by pollinators and its role in evolution. In *Pollination and Evolution*, eds. J.A. Armstrong, J.M. Powell, and A.J. Richards, pp. 9–28. Sydney: Royal Botanic Gardens.

Kearns, C.A. & Inouye, D.W. (1993). *Techniques for Pollination Biologists*. Niwot Ridge, CO: University Press of Colorado.

Kevan, P.G. (1972). Floral colors in the high arctic with reference to insect–flower relations and pollination. *Can. J. Bot.* 50:2289–2316.

Kevan, P.G. (1978). Floral coloration, its colorimetric analysis and significance in anthecology. In *Pollination of Flowers by Insects by Insects*, ed. A.J. Richards, pp. 51–78. Linnean Society Symposium Series No. 6. London: Academic Press.

Kevan, P.G. (1979). Vegetation and floral colours revealed by ultraviolet light: interpretational difficulties for functional significance. *Amer. J. Bot.* 66:749–751.

Kevan, P.G. (1983). Floral colors through the insect eye: what they are and what they mean. In *Handbook of Experimental Pollination Biology*, eds. C.E. Jones and R.J. Little, pp. 3–30. New York: Van Nostrand Reinhold.

Kevan, P.G. (1984). Pollination by animals and angiosperm biosystematics. In *Plant Biosystematics*, ed. W.F. Grant, pp. 271–292. Toronto: Academic Press.

Kevan, P.G. (1989a). Theroregulation in arctic insects and flowers: adaptation and co-adaptation in behaviour, anatomy, and physiology. In *Thermal Physiology*, ed. J.B. Mercer, pp. 747–754. Netherlands: Elsevier Science Publishers B.V.

Kevan, P.G. (1989b). How honey bees forage for pollen at skunk cabbage, *Symplocarpus foetidus* (L.) Nutt. (Araceae). *Apidologie* 20:485–490.

Kevan, P.G. (1990). How large bees, *Bombus* and *Xylocopa* (Apoidea: Hymenoptera), forage on trees. *Ethol., Ecol., and Evol.* 2:233–242.

Kevan, P.G. (1991). Pollination: keystone process in sustainable global productivity. *Acta Horticulturae* 288:103–110.

Kevan, P.G. (1993). Wind or insects: pollination of coconut (*Cocos nucifera*) in the Maldive Islands. In *Asian Apiculture*, eds. L.J. Connor, T. Rinderer, A. Sylvester, and S. Wongsiri, pp. 372–377. Cheshire, CT: Wicwas Press.

Kevan, P.G. & Baker, H.G. (1983). Insects as flower visitors and pollinators. *Ann. Rev. Entomol.* 28:407–453.

Kevan, P.G. & Baker, H.G. (1984). Insects on flowers: pollination and floral visitations. In *Insect Ecology*, eds. C.B. Huffaker and R.C. Rabb, pp. 607–631. New York: J. Wiley and Sons.

Kevan, P.G. & Lane, M.A. (1985). Floral petal microtexture is a tactile cue for bees. *Proc. Nat. Acad. Sci. USA* 82:4750–4752.

Kevan, P.G., Chaloner, W.G. & Savile, D.B.O. (1975). Inter-relationships of early terrestrial arthropods and plants. *Palaeontology* 18(2):391–417.

Kevan, P.G., Clark, E.A. & Thomas, V.G. (1990). Insect pollinators and sustainable agriculture. *Amer. J. Altern. Agric.* 5:13–22.

Kevan, P.G., Lee, H. & Shuel, R. (1992). Sugar ratios in nectar of varieties of canola (*Brassica napus*). *J. Apic. Res.* 30:99–102.

Kevan, P.G., Longair, R.A. & Gadawski, R.M. (1985). Dioecy and pollen dimorphism in *Vitis riparia* Michx. (Vitaceae). *Can. J. Bot.* 63:2263–2267.

Knuth, P. (1909–1912). *Handbook of Flower Pollination* (trans. J.R. Ainsworth-Davis). Oxford: Clarendon Press. 3 vols.

Kugler, M. (1943). Hummeln als Blütenbesucher. *Ergeb. Biol.* 9:143–323.

Lack, A.J. & Kevan, P.G. (1987). The reproductive biology of a distylous tree, *Sarcotheca celebica* (Oxalidaceae) in Sulawesi, Indonesia. *Biol. J. Linn. Soc.* 95:1–8.

Levin, D.A. (1978). Pollinator behaviour and the breeding structure of plant populations. In *The Pollination of Flowers by Insects*, ed. A.J. Richards, pp. 133–150. Linnean Society Symposium Series No. 6. London: Academic Press.

Lex, T. (1954). Duftmale und Blüten. *Z. vergl. Physiol.* 36:212–234.

Linhart, Y.B. (1973). Ecological and behavioral determinants of pollen dispersal in a hummingbird pollinated *Heliconia*. *Amer. Naturalist* 107:511–523.

Lunau, K. (1993). Interspecific diversity and uniformity of flower colour patterns as cues for learned discrimination and innate detection of flowers. *Experientia* 49:1002–1010.

Mathur, G. & Mohan Ram, H.Y. (1978). Significance of petal colour in thrips-pollinated *Lantana camara* L. *Ann. Bot.* 42:1473–1476.

McGregor, S.E. (1976). *Insect Pollination of Cultivated Crop Plants.* USDA Agriculture Handbook No. 496. Washington, DC: USDA.

Murphy, S.D. & Aarssen, L.W. (1995). Allelopathic pollen extracts from *Phleum pratense* L. (Poaceae) reduces seed set in sympatic species. *Internal. J. Plant Sci.* 156:435–444.

Nelson, E.C. (1990). Of bees and fly traps: the natural history of Arthur Dobbs. In *Curious in Everything. The Career of Arthur Dobbs of Carrickfergus 1689–1765*, eds. D.H. Rankin and E.C. Nelson, pp. 7–13. Carrickfergus, Ireland: Carrickfergus and District Historical Society.

Niklas, K. (1984). The motion of windborne pollen grains around conifer ovulate cones: implications on wind pollination. *Amer. J. Bot.* 71:356–374.

Niklas, K. (1985). The aerodynamics of wind pollination. *Bot. Rev.* 51: 323–386.

Papaj, D.R. & Lewis, A.C., eds. (1993). *Insect Learning: Ecological and Evolutionary Perspectives.* London and New York: Chapman & Hall.

Ponomarev, A.N. (1966). Nekotorye prisposobleniya zlakov k opyleniyu vetrom. *Bot. Zhurn.* 51:28–39.

Pyke, G.H. (1984). Optimal foraging theory: a critical review. *Ann. Rev. Ecol. Syst.* 15:523–575.

Pyke, G.H., Pulliam, H.R. & Charnov, E.L. (1977). Optimal foraging theory: a selective review of theory and tests. *Quart. Rev. Biol.* 52: 137–154.

Rathke, B. (1983). Competition and facilitation among plants for pollination. In *Pollination Biology,* ed. L. Real, pp. 305–330. Orlando, FL: Academic Press.

Richards, A.J. (1986). *Plant Breeding Systems.* London: Allen and Unwin.

Roubik, D.W. (1989). *Ecology and Natural History of Tropical Bees.* Cambridge, England: Cambridge Univ. Press.

Roussy, A.M. (1994). Alleles, clones and pollen: a discreet look into jack pine (*Pinus banksiana* Lamb.). M. Sc. dissertation, University of Guelph, Guelph, ON, Canada.

Schmidt, J.D. & Buchmann, S.L. (1992). Other products of the hive. In *The Hive and the Honey Bee*, ed. J.M. Graham, pp. 927–988. Hamilton, IL: Dadant & Sons.

Schmitt, J. (1980). Pollinator foraging behavior and gene dispersal in *Senecio* (Compositae). *Evolution* 34:934–943.

Schmitt, J. and Linder, C.R. (1994). Will escaped transgenes lead to ecological release? *Mol. Ecol.* 3:71–74.

Scora, R.W. (1964). Dependency of pollination patterns in *Monarda* (Labiatea). *Nature (Lond.)* 204:1011–1012.

Southwick, E.E. & Southwick, L. (1992). Estimating the economic value of honey bees (Hymenoptera: Apidae) as agricultural pollinators in the United States. *J. Econ. Ent.* 85:621–633.

Sprengel, C.K. (1793). *Das Entdecke Geheimnis in Natur im Bau und in der Befruchtung der Blumen.* Berlin: F. Vieweg.

Stanton, M.L. (1987). Reproductive biology of petal color variants in wild populations of *Raphanus sativus*: I. Pollinator response to color morphs. *Amer. J. Bot.* 74:178–187.

Stelleman, P. (1978). The possible role of insect visits in pollination of reputedly anemophilous plants, exemplified by *Plantago lanceolata*, and syrphid flies. In *The Pollination of Flowers by Insects*, ed. A.J. Richards, pp. 41–46.

Linnean Society Symposium Series No. 6. London: Academic Press.

Thomson, J.D. & Brunet, J. (1990). Hypotheses for the evolution of dioecy in seed plants. *Trends Ecol. Evol.* 5:11–16.

Thomson, J.D. & Plowright, R.C. (1980). Pollen carryover, nectar rewards and pollinator behaviour with special reference to *Diervilla lonicera. Oecologia* 46:68–74.

Tikmener, E.A. & Kevan, P.G. (in press). [Dynamics of anemophily, the ecology and species packing in some common grasses of eastern Canada.] (in Russian) *Ekologiya.*

Troyer, A.F. (1993). Silk delay, drought tolerance and evolution in corn. *Agron. Abstracts* 1993:104.

Waddington, K.D. (1983). Foraging behavior of pollinators. In *Pollination Biology,* ed. L. Real, pp. 213–239. Orlando, FL: Academic Press.

Waser, N.M. (1986). Flower constancy: definition, cause, and measurement. *Amer. Nat.* 127: 593–603.

Waser, N.M. (1988). Comparative pollen and dye transfer by pollinators of *Delphinium nelsoni. Funct. Ecol.* 2:41–48.

Waser N.M. & Price, M.V. (1983). Optimal and actual outcrossing in plants. In *Handbook of Experimental Pollination Biology,* eds. C.E. Jones and R.J. Little, pp. 314–359. New York: Van Nostrand Reinhold.

Waser, N.M. & Price, M.V. (1989). Optimal outcrossing in *Ipomopsis aggregata*: seed set and offspring fitness. *Evolution* 43:1097–1109.

Wehner, R. (1981). Spatial vision in arthropods. In *Handbook of Sensory Physiology, Vol. VII/6c,* ed. H. Autrum, pp. 287–616. Berlin and Heidelberg: Springer-Verlag.

Weiss, M.R.I. (1991). Floral colour changes as a cue for pollinators. *Nature (Lond.)* 354:227–229.

Willis, D.S. & Kevan, P.G. (1995). Foraging dynamics of *Peponapis pruinosa* on pumpkin (*Cucurbita pepo*) in southern Ontario. *Can. Entomol.* 127:167-175.

Wyatt, R. (1983). Pollinator–plant interactions and the evolution of breeding systems. In *Pollination Biology,* ed. L. Real, pp. 51–95. Orlando, FL: Academic Press.

Part II

Pollen biotechnology and optimization of crop yield

5

Pollination efficiency of insects

ARTHUR R. DAVIS

Summary 87
Introduction 87
Interactions between foraging insects 89
Comparative pollinating abilities of insect sexes and castes 90
Influence of insect tongue length on flower visitation pattern 91
Influence on pollen movement of pollen grain traits and insect body hairs 92
Pollination by bees foraging for nectar or pollen 94
Quantified removal of pollen by insects, and associated losses of pollen from
    flowers 96
How best to evaluate an insect's importance as a pollinator on the flower? 105
References 111

Summary

Insects visiting flowers to collect nectar, pollen, and other rewards often serve as important incidental agents of pollination. The efficiency of insect species as pollinators varies greatly. In this chapter, emphasis is placed on (a) the interactions between foraging insects, (b) comparison of pollinating abilities of insect sexes and castes, (c) comparisons of nectar and pollen foragers, (d) insect tongue length and its influence on flower visitation patterns, (e) characteristics of pollen grains and insect body hairs and their influence on pollen movement, and (f) pollen removal, and various other pollen losses, during insect visits. To date, most of our knowledge on insect pollination efficiency is derived from plant species growing in natural settings. Additional research is required for agricultural and horticultural crops. Some recommendations are given for future studies aimed at identifying the relative importance of different insect species as pollinators in agriculture. The importance of obtaining data within a plant species, for both insect removal and deposition of pollen, is emphasized.

Introduction

Insects are important vectors of pollen for many agricultural crops, and this chapter focuses on various characteristics that influence their efficiency as pollinators. Although some comparisons to non-insect agents of pollination are referred to herein, readers seeking information on pollen dynamics in other pollination syndromes may refer to the following: anemophily (wind pollination) – Di-Giovanni and Kevan (1991), McCartney (1994), and Niklas (1985); zoophily (mammal and bird pollination) – Fleming and Sosa (1994), and Chapter 4 of this volume.

Perhaps no other term in pollination biology has carried such ambiguity as "pollination efficiency." In their recently proposed lexicon for pollination biologists, Inouye et al. (1994) scrutinized pollination efficiency in 18 studies over the past three decades, and disclosed 12 separate meanings. Representations for the term range from the number of fruits produced per insect visit, or percent fruit set (Kendall and Smith 1975, 1976), to the proportion of stigmata touched per insect visit (Robinson 1979); to the number of pollen grains deposited per stigma as a fraction of the total number of grains produced per flower (Cruden et al. 1990) or number removed per flower (Snow and Roubik 1987); to the proportion of flowers visited positively, from the front, versus the fraction of flowers robbed of nectar from the flower base (Poulsen 1973); to the rate of insect foraging (K. W. Richards 1987; K. W. Richards and Edwards 1988). Another favored interpretation held pollination efficiency as the summation of both the costs (flower damage, pollen consumption, and so forth) and benefits (pollen deposited and transferred, seeds produced, and so forth) attributable to a solitary flower visit by an individual animal species (Neff and Simpson 1990; Kearns and Inouye 1993). This last usage was most applicable to those cases of mutualism where an animal not only pollinated a species, but also had immature life stages that inhabited and consumed parts of flowers of that same species (for example, figs – Janzen 1979; Wiebes 1979; *Heracleum lanatum* – Kearns 1992; yucca – Aker and Udovic 1981; Addicott 1986).

The recommendations of Inouye et al. (1994), to replace "pollination efficiency" with a series of more discrete descriptors, are followed throughout this chapter. Specifically, this chapter includes sections on interactions between foraging insects; comparisons of pollinating abilities of insect sexes and castes, and of nectar and pollen foragers; the influence of insect tongue length on flower visitation patterns; the influence of pollen characteristics and insect body hairs on pollen movement; various losses of pollen from flowers during insect visits; and methodological recommendations for future assessment of the relative importance of insect species as pollinators of agricultural crops. The literature on these topics is comprehensive, and it is impossible to describe every pertinent study. My apologies are extended for references not cited. Readers seeking information on other parameters that influence efficiency in entomophily should consult reviews by Levin and Kerster (1974), Levin (1981), and A. J. Richards (1986). Further, selected references on specific topics are: floral constancy (Free 1963, 1970; Waser 1986), plant density (Schmitt 1983; Burdon et al. 1988; Fenster 1991; Roumet and Magnier 1993), flight directionality (Zimmerman and Cook 1985; Ginsberg 1986; Thomson and Thomson 1989; Zietsman 1990), pollen carryover (Galen and Rotenberry 1988; Waser 1988; Cresswell 1994), pollination distance (Jackson and Clarke 1991; Ellstrand 1992; Raybould and Gray 1993; Scheffler et al. 1993; Broyles et al. 1994), and abiotic factors, such as weather (DeGrandi-Hoffman 1979; Corbet 1990; Corbet et al. 1993).

The material drawn upon for this chapter includes the rich literature of pollination ecology in nonagricultural settings. Indeed, a large proportion of our current knowledge of efficiency, as it pertains to insect pollination biology, derives from the research efforts of field pollination biologists (anthecologists). Many of the insights derived from studies of plants in natural settings have direct application to pollination in agricultural crops.

Interactions between foraging insects

Interactions exist between foraging insects and can have direct impact on pollination. For instance, encounters causing disturbance, both between and within insect species, can change the duration of flower visits (L. K. Johnson and Hubbell 1974; Morse 1982; Zietsman 1990) and alter "foraging rate" (Inouye et al. 1994). Interactions range from aggressive to subtle.

Insects foraging at the same floral resource may become physically competitive. Brian (1957) found that competition between *Bombus* spp. sometimes involved deliberate displacement of a floral occupant by a newcomer. In *Rosa carolina*, contact with other bees averaged once every 14–15 flowers (Morse 1978a), and smaller bees (*B. ternarius*, *B. vagans*) retreated when confronted by larger ones (*B. terricola*) (Morse 1982). The observations agree with Kikuchi's (1965) discovery of a strong positive correlation between dominance rank and insect-head width; dominant species of hover flies forced subordinate flies to visit old (relatively unrewarding) flowers, or they displaced subordinates to plant foliage. Size of the foraging force, however, can overcome differences in body size and governed successful displacement of the larger *Apis mellifera* by aggressive *Trigona* spp. (Roubik 1980). In a second study of honeybees and stingless bees, involving an artificial feeder, interspecific aggression was directed mainly to the next largest species (Koeniger and Vorwohl 1979). The smallest, and most aggressive, bees (*Trigona iridipennis*) had over twice the number of pollen species stored in their colonies than the largest bee (*Apis dorsata*), yet their respective flight ranges were 100 m and >5 km. Aggression is more common on artificial than on natural flowers (Brian 1957) and on smaller than larger inflorescences (Kikuchi 1965) and can affect related species very differently (Morse 1981).

Certainly far from all competitive interactions between species take the form of overt hostility. Individuals may avoid confrontation by changing their foraging patterns (Heinrich 1976b; Strickler 1979; Morse 1982; Thomson 1989). Kikuchi (1965) found that some occupant flies avoided newcomers to heads of Asteraceae by migrating to involucres and also that closely related flies visited the same flower species by foraging at different times of the day, thereby avoiding competition. Indeed, sometimes insect competition for floral resources can benefit agriculture; fully formed, symmetrical fruits were the result of the complementary foraging locations of solitary bees and honeybees on flowers of strawberry (Chagnon et al. 1993).

Interestingly, individuals of the same species seldom attack each other (Brian 1957; Kikuchi 1965; Morse 1978a; Koeniger and Vorwohl 1979). Still, competition between conspecifics occurs. For instance, feral honeybees influenced the foraging area of honeybees introduced to a *Macadamia* orchard (Gary et al. 1972a). And competition between workers of *Bombus vagans* reduced significantly the maximum sequences of consecutive-flower visits per trip (Heinrich 1979).

A subtle intraspecific interaction concerns scent-marking of flowers (Giurfa and Núñez 1992). Evidence suggests that many bees reject recently visited, nectar-depleted flowers on the basis of scents imparted by immediately previous visitors. Rejections by conspecific females of *Xylocopa virginica* occurred for up to 10 minutes (Frankie and Vinson 1977). Intraspecific messenger systems such as these raise the foraging efficiency of insects by increasing their visits to more rewarding (unmarked) flowers (Giurfa and Núñez 1992). At the same time, a temporary hiatus of visitors from scent-marked flowers could benefit the plant species not only by causing searches for recently unvisited flowers, but possibly by reducing any dislodgement from stigmata of pollen not yet anchored by germinating tubes (and see Osborne and Corbet 1993). Whether scent-marks of insects are discernible to other species remains to be investigated.

The final interaction considered is also the consequence of activity of a previous floral visitor. Certain "primary robbers" (Inouye 1980a, 1983) chew through the base of sympetalous corollae to steal nectar without contacting the anthers or stigmata at the flower front. "Secondary robbers," or opportunists (usually species lacking the mandibular dentition necessary to be primary robbers themselves), take advantage of these perforations at the flower base by collecting nectar from there. Most reports are from legumes, where holes chewed by *Bombus* spp. were utilized by honeybees; such diversionary foraging by the latter reduced their pollinating value (Fridén et al. 1962; Free 1965, 1968; Dennis and Haas 1967a, b; Poulsen 1973; Kendall and Smith 1976).

Comparative pollinating abilities of insect sexes and castes

The pollination capabilities of male versus female individuals of the same insect species have been studied. In bees, it is significant that males do not provision brood with pollen; instead, when foraging they seek only nectar and do not actively gather pollen, nor groom it from their bodies into pellets for return to the nest. Indeed, male bees often live outside the nesting area

(Ordway et al. 1987; Jennersten et al. 1991). On balance, the evidence to date marginally favors females as superior pollinators.

The foraging behavior of males and females can vary. Male bees worked faster, but made less extensive contact with the reproductive parts of *Claytonia* flowers (Motten et al. 1981), visited pistillate flowers of *Cucurbita pepo* only half as often (Tepedino 1981), and visited greater numbers of flowers per plant (Jennersten et al. 1991) than their female counterparts. Both male and female bees of the same species robbed nectar without causing pollination (Benedek 1973; Macior 1975a). Female bees exhibited higher fidelity than males to a particular plant species (Brittain and Newton 1933). Non-ovipositing, female cabbage-white butterflies were as constant to plant species as foraging males, although egg-laying females switched plants regularly (Lewis 1989).

Regarding body loads and viability of pollen, female bees (*Melissodes agilis*) on male-sterile heads of sunflower carried greater quantities than males (F. D. Parker 1981); the same occurred with checkerspot butterflies (Murphy 1984). Female (worker) bumblebees attending *Asclepias syriaca* did not transport more pollinaria than males, but carried a higher proportion on their legs than tongues; thus they were more effective pollinators (Jennersten et al. 1991). Viability of apple pollen was lower on male than female bodies of *Andrena wilkella* (Kendall 1973), but higher for male ants than females visiting an Australian orchid (Peakall et al. 1987).

Experiments with bees of different sex also have been conducted by allowing single visits to virgin flowers or capitula. Within a bee species, stigmata (lavender) received equal quantities of pollen (Herrera 1987), and no differences in seed set (red clover) or fruit set (apple) were evident (Free 1965; Kuhn and Ambrose 1984). Despite males of *Melissodes agilis* spending longer times (including overnight) on heads of male-sterile *Helianthus*, only two-thirds of their visits resulted in seed set; capitula visited by females averaged almost four times more seed (F. D. Parker 1981). The Parker report strongly supports the detected scarcity of body pollen on males of that species, because theoretically only a single pollen tube is required for seed set on sunflower heads.

Relatively less information is available on the importance of castes (queens and workers) as pollinators. The evidence to date comes from bumblebees, and, on a per bee basis, favors the queens. Although the abilities of queens at pollen removal were similar (Harder 1990), flower-handling times were the same (Galen and Blau 1988) or faster (Reader 1977; Inouye 1980b). Differences in tongue length, a significant factor of pollinator foraging behavior, partially explains this variation in foraging speed.

Influence of insect tongue length on flower visitation pattern

Numerous studies have focused on the question of the length of the proboscis of flower visitors and its influence on pollinator foraging behavior. Tongue length was always longer in queen bumblebees than in workers (Brian 1957; Macior 1975b; Heinrich 1976b; Ranta and Lundberg 1980; Inouye 1983), and in males it usually was intermediate (Ranta and Lundberg 1980; Inouye 1983). Thus, queens of "short-tongued" bumblebees (*Bombus lucorum* and *B. pratorum*) foraged on flowers with deeper corolla tubes than did workers. Workers may assort to flower species possessing corolla depths in accordance to their tongue lengths (Dennis and Haas 1967b; Heinrich 1976a), longer tongues allowing exploitation of both short- and long-tubed flowers (Ranta and Lundberg 1980). In the same manner, large workers within a single *Bombus* species may attend different flowering species than smaller individuals (Heinrich 1976b; Cumber, cited by Morse 1978b; Inouye 1980b). Interestingly, the same pattern of segregation extended to a single plant species, wherein large workers of *Bombus vagans* foraged on deeper florets of *Vicia cracca* than did small workers (Morse 1978b).

A relatively short tongue limits accessibility to nectar at the base of a tubular corolla. Bumblebees do not always yield to inaccessible flowers; instead, some chew through the perianth base, thereby robbing nectar without necessarily pollinating the flower or dispersing its pollen (Inouye 1983). Disparity in tongue length was apparently insufficient to always give queens the advantage, because nectar robbing by both queens and workers of the same species can occur (Brian 1957). In *B. vagans*, workers with longer tongues robbed nectar less often (Morse 1978b).

In studies of several insect species (mostly bees) that visited plants in common, individuals with longer tongues invariably were quicker foragers than short-tongued individuals (Holm 1966; Dennis and Haas 1967b; Benedek 1973; Inouye 1980b; Harder 1983). On *Lavandula latifolia*, flower-handling time decreased exponentially with increasing proboscis length, with bees (range 0.8–6.3 seconds per flower) working faster than flies (4.2–77.7) and butterflies (1.5–248.0) (Herrera 1989; see Figure 5.1).

Not only do insect tongues play a role in dictating foraging behavior, they are also important vehicles of pollen transfer on bees (Spencer-Booth 1965; Wolfe and Barrett 1989; Jennersten et al. 1991), butterflies (Levin and Berube 1972; Galen and Kevan 1980; Courtney et al. 1982; Jennersten 1984) and flies (Courtney et al. 1982; Zietsman 1990). On flies, pollen grains were found on tongues almost exclusively (Zietsman 1990).

Influence on pollen movement of pollen grain traits and insect body hairs

Size of pollen grains is variable and influences pollen removal and transfer. Even within a single species, pollen size depends on ploidy, flower position within the plant, heteroanthery, and environmental conditions (Stanley and Linskens 1974; Harder et al. 1985; Stanton and Preston 1986; Harder 1990). Pollen grains of species pollinated by wind are intermediate in size (17–58 μm), whereas those adhering to insects are much more variable, ranging from 4.5 μm

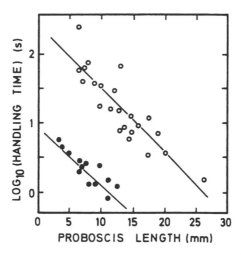

Figure 5.1 Relationship of flower-handling time and tongue length for hymenopteran (dark circles) and lepidopteran (light circles) visitors to the mint, *Lavandula latifolia*. *(Reproduced from C. M. Herrera 1989,* Oecologia *80:241-248, with permission from Springer-Verlag GmbH & Co.)*

(*Myosotis*) to >200 μm (*Cucurbita*) (Wodehouse 1935). Pollen grain size was negatively correlated with number of grains adhering to body hairs of honey- and bumblebees (Lukoschus 1957; Thomson 1986) and butterflies (Levin and Berube 1972). Harder (1990) discovered interesting, somewhat contradictory, relationships with pollen removal during bumblebee visits, but it is not certain whether pollen size was directly responsible for the results. For instance, queens of *B. flavifrons* not actively seeking pollen from *Mertensia paniculata* tended to remove more of it from flowers producing larger grains. On the other hand, workers departed *Pedicularis bracteosa* with more pollen from flowers bearing smaller grains; in *P. contorta*, an increase of just 1 μm (5%) in grain size accounted for a 20% reduction in pollen removal.

Besides size, adherence of pollen to insects also has been attributed to pollen morphology. Murphy (1984) found that grains of *Lasthenia* adhered to checkerspot butterflies better than to those of three other species visited more frequently, and he suggested that small grains with rough surfaces may be key parameters for pollen transfer by Lepidoptera. Pollen grains that are spherical but strongly echinate (for example, Malvaceae – cotton and okra) adhered poorly and were rarely carried as pollen loads by honeybees (Loper and DeGrandi-Hoffman 1994; Vaissière and Vinson 1994). A form of pollen "packaging," attributed to threads of viscin, may account for the clumping of pollen on visitors to *Epilobium* and *Oenothera* (Wolin et al. 1984; Galen and Plowright 1985b).

Adherence is not only a function of pollen characteristics, but also of the insect. The number, length, and complexity of body hairs differ markedly between insect species (Figure 5.2) and are inferred to have profound influence

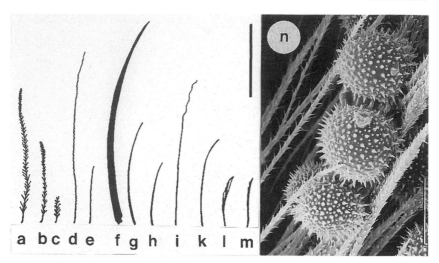

Figure 5.2 Illustrations of the size and morphology of various body hairs from flower-visiting bees (a–c, n), wasps (d, e), and flies (f–m). (**a–c**) *Apis mellifera*; (**d, e**) *Vespa germanica*; (**f, g**) *Calliphora erythrocephala*; (**h**) *Lucilia silvarum*; (**i**) *Syrphus pyrastri*; (**k, l**) *Helophilus floreus*; (**m**) *Eristalis apiformis*; (**n**) Branched hairs along mid-abdomen of *Bombus perplexus*, showing adhering pollen of hollyhock (*Althaea rosea*). Scale bar for (a–m) = 0.5 mm, and for (n) = 100 μm. *(Line drawings a–m reproduced from F. Lukoschus 1957, Zeitschrift für Bienenforschung 4:3–21, with permission from Elsevier Éditions Scientifiques.)*

on an insect's carrying capacity (Lukoschus 1957). Hair density is also important. Kearns (1992) suggested that the small quantities of *Linum* pollen on muscoid flies occurred because the grains (56 μm) did not lodge between their body bristles. In contrast, body loads of *Cardamine* and *Chrysopsis* pollen (23–30 μm) were much larger. Electrostatic charges on bees and flower parts may also cause pollen to adhere to bees or to jump from bee to stigma (Corbet et al. 1982; Erickson and Buchmann 1983; Buchmann et al. 1990). In the special case of social insects, microgametophytes might also be transferred among nestmates inside the colony (von Planta, cited by Betts 1920; Free and Williams 1972; DeGrandi-Hoffman et al. 1986) by electrostatic means.

Pollination by bees foraging for nectar or pollen

Unlike many other flower-visiting insects, adult bees are involved with the nutrition of their immatures. Although there are several types of "floral rewards" sought by insects (Simpson and Neff 1981, 1983; Buchmann 1987), it is chiefly the nectar and pollen that are collected as primary food sources by bees for return to their nests. The pertinent floral sources – nectaries and anthers – often are situated near opposite extremities of the floral structure (Davis 1995), and different floral postures are taken and different behaviors (associated with specific body movements) are invoked during a bee's procuration of each food type (Robinson 1979; Kuhn and Ambrose 1984; Bosch and Blas 1994). Also, easily recognizable are foragers with or without loads of pollen in the corbiculae of their hind legs (for example, honey- and bumble bees) or in the scopa on the underside of their abdomen (for example, leafcutter bees). Consequently, many researchers have been able to compare nectar collectors with pollen gatherers (which often take a small amount of sticky nectar as a medium to consolidate their loads of pollen) for various aspects of pollinator effectiveness.

For bees to collect loads of nectar and pollen, the number of flowers visited, and the time required, are highly variable. Honeybees required much more time to collect nectar loads (Park, cited by Gary et al. 1972b) and spent seven times longer on inflorescences of *Banksia* when foraging for nectar (Ramsay 1988). Bumblebees stayed almost three times longer on flowers of *Pedicularis* for nectar (Harder 1990). However, the collection of pollen from *Kalmia* took longer than that of nectar for honeybees, bumblebees and *Andrena* bees (Reader 1977). Furthermore, honeybees attending *Helianthus* visited more florets per capitulum as nectar than as pollen gatherers (Free 1964); the reverse was true for red clover (Free 1965). This lack of agreement between studies, on the effort required to collect either pollen or nectar, is not surprising and indicates inherent interspecific differences in nectar and pollen production per flower, and variability caused by prior depletion of these floral rewards (Thomson and Thomson 1989, 1992).

Many studies of honeybee foraging have demonstrated eager preferences for nectar over pollen. Four separate studies of onion production found pollen gatherers never to exceed 15% of the total foraging force (see Benedek 1977). In other plants, nectar gatherers outnumbered pollen seekers by two to one (de Oliveira et al. 1991; Zietsman 1991), but it was the reverse for carrots (Gary et al. 1972b). In hybrid cotton, the ratio of nectar gatherers to pollen seekers was as high as nine to one, as foragers strongly avoided pollen-donor (B) lines in lieu of the pollen-lacking (A) lines (Waller and Moffett 1981). In comparative studies on macadamia and cranberry, *Trigona carbonaria* and *Bombus* spp. sought pollen much more regularly than did *Apis mellifera* (Heard 1994; MacKenzie 1994). In nectar-producing flowers, bumblebees always attempted collection of nectar and never visited exclusively for pollen (Heinrich 1976a; but see Harder 1990). Interestingly, a comparison of nectar-gathering and pollen-gathering bumblebees showed no difference in the quantities of pollen removed from flowers of *Pedicularis* (Harder 1990).

The active removal from circulation of pollen as it is raked from bees' bodies and consolidated as loads in the corbiculae or scopae has raised the question of whether pollen gatherers are *inferior* pollinators than bees collecting nectar alone (Thomson and Plowright 1980; Davis 1992). Careful experimentation, wherein bees were allowed single visits to previously unattended (virgin) flowers, invariably has demonstrated to date that bees carrying pollen loads are, in fact, *superior* pollinating agents. For instance, the percentage of once-visited apple blossoms that set fruit ranged from 1.5 to 5 times higher for bees collecting pollen (Free 1966; Robinson and Fell 1981; Kuhn and Ambrose 1984), with twice as many ovules fertilized (Kendall 1973), because honeybees only occasionally contacted stigmata when collecting nectar alone (Free 1966; Robinson 1979). Results for almond were similar (Bosch and Blas 1994). *Apis* foragers seeking both nectar and pollen set significantly more drupelets per raspberry than those gathering just nectar (de Oliveira et al. 1991). Free (1965) found that both honey- and bumblebees carrying pollen loads gave significantly higher seed sets per head of *Trifolium* than bees lacking pollen in their corbiculae. And almost five times as many pollen tubes were detected at the style base of emasculated flowers of *Echium* receptive when visited by corbicular-laden honeybees than when visited by nectar collectors (Davis 1992). These findings and others suggest that grooming does not remove all body pollen (Lukoschus 1957; Davis 1991). Indeed, with only one exception, Free and Williams (1972) found *more* grains present on the bodies of pollen-seeking honeybees (despite their grooming activities) than on nectar gatherers. The question still requires investigation on plants (see next section) where nectar-collecting bees deliberately rake off adhering pollen ("pollen-discarding, active vectorial pollen loss"; Inouye et al. 1994).*

Purity of body pollen is also lower in nectar-gathering honeybees. Perhaps indicative of their smaller quantities of pollen acquired per flower, their rarer grooming actions, and the incidental in-house transfer of pollen grains among nestmates (DeGrandi-Hoffman et al. 1986), nectar-gathering bees bore significantly higher proportions of foreign pollen than pollen gatherers did for five different Rosacean species in orchards (Free and Williams 1972).

However, flower-visitation sequence may render pollen-gathering bees less important as pollinators. On the basis of the floral position first visited upon arrival at vertical inflorescences of fireweed, Galen and Plowright (1985a) determined that *Bombus* pollen collectors visited relatively few receptive (female-phase) flowers and, hence, may be less efficient pollinators than nectar gatherers. Moreover, pollen-gathering bees were regarded as antagonists to the reproductive effort of *Impatiens* by avoiding receptive flowers, because little of their removed pollen found stigmata (Wilson and Thomson 1991). The order of insect visitation to flowers is considered important; less pollen may remain in flowers when the pollination-efficient, nectar-gathering bee species finally arrive (Wilson and Thomson 1991; Thomson and Thomson 1992).

Quantified removal of pollen by insects, and associated losses of pollen from flowers

Several studies have quantitatively investigated various aspects of pollen removal by insects. The bulk of these studies concern Hymenoptera, likely a

---

* See *Note added in proof* p. 111.

reflection of the vital pollen-collecting activities of bees. Bee hairs acquire pollen readily (Free and Williams 1972) and bees usually bear higher quantities of body pollen than do other insect taxa, such as beetles, butterflies (Wolin et al. 1984), and flies (Lukoschus 1957; Westerbergh and Saura 1994; but see Kearns 1992). Investigations have dealt with the effects on pollen removal of floral-visit duration, number of visits a flower receives, and various pollen losses from flowers.

As bee visits to previously unvisited flowers lengthen, consistently more pollen is removed (Buchmann and Cane 1989; Galen and Stanton 1989; Harder and Thomson 1989; Harder 1990). The effect of duration on pollen removal is not linear across all insects (Primack and Silander 1975; Strickler 1979). For example, workers of *Bombus flavifrons* spent shorter times (11.9 seconds) during initial visits to virgin flowers of *Aconitum delphinifolium* than did workers of *B. mixtus* (19.0 seconds) yet removed significantly more of a flower's total pollen (61.4% vs. 37.7%) (Harder 1990).

The quantity of pollen removed during the first visit to virgin flowers is shown in Table 5.1. Despite these plant species' possessing a wide range of floral morphology, bees usually removed large quantities (30%–80%) of microgametophytes. Interspecific variation in pollen removal is related to contact of the pollinator with the anthers (Strickler 1979; Harder 1990; Lebuhn and Anderson 1994). Visits to the legumes *Lupinus* and *Vicia*, whose flowers dispense pollen in small quantities per offering (Harder 1990), are exceptional (Table 5.1; also see Haynes and Mesler 1984). Although there are fewer studies available, evidence to date suggests that butterflies (on *Ipomoea* and *Phlox*) and beetles (on *Oenothera*) remove less pollen than bees do (Table 5.1).

The trend that pollen removal increases with increased duration of floral visits also pertains to the number of visits received by the flower or inflorescence (Percival 1950; Snow and Roubik 1987; Galen and Stanton 1989; Harder 1990; Lebuhn and Anderson 1994; see Table 5.1). However, the proportion of pollen extracted per successive visit stayed constant or, more typically, declined (Galen and Stanton 1989; Harder and Thomson 1989; Wolfe and Barrett 1989; Harder 1990; Young and Stanton 1990; Lebuhn and Anderson 1994). In fact, in a number of examples, three or four further visits were necessary to remove a pollen fraction equal to that extracted initially (Table 5.1; for example, *Bombus mixtus* on *Aconitum*, *Centris inermis* on *Cassia*, *Bombus* spp. on *Lupinus* and *Pedicularis contorta*). In the last case, for example, the quantity of residual pollen removed on successive visits was 46% (visit 1), 38% (2), 28% (3), and 23% (4) (Harder 1990). This reduction in rate of pollen removal is partially attributable to shorter subsequent visits (Harder 1990). Thomson and Thomson (1992) questioned whether pollen remaining after each visit is relatively more resistant to removal.

Pollen removed from flowers during insect visits has a range of destinies. Grains removed, but not carried by the insect, represent a "passive loss" (Thomson and Plowright 1980; Harder and Thomson 1989). The simple recoiling of the proboscis postfeeding resulted in 15% (*Phlox pilosa*) and 52%

Table 5.1. *Proportion of pollen removed by various insects (mostly bees) during one or more visits to previously unvisited flowers of diverse plant taxa, and the fraction of pollen that was deposited on the stigma by the same insects. For many species, estimates have been calculated from the authors' data.*

Plant species (family)	Insect species	Percentage pollen removed after (n) visits	Percentage pollen reaching stigma	Reference
Aconitum delphinifolium (Ranunculaceae)	*Bombus flavifrons* B. mixtus (workers)	61.4 (1), 81.9 (2), 89.9 (3), 94.0 (4) 37.7 (1), 53.3 (2), 65.9 (3), 71.4 (4)	-----	Harder (1990)
Aralia hispida (Araliaceae)	*Bombus sandersoni*, B. ternarius, and B. vagans (workers)	79.6 (1)	-----	Harder (1990)
Berberis thunbergii (Berberidaceae)	*Bombus* spp. (queens)	36.3 (1)	-----	Lebuhn and Anderson (1994)
Cassia reticulata (Fabaceae)	*Centris inermis* C. longimana Eulaema nigrita Pseudaugo- chloropsis sp.	26.2 (1), 27.0 (3), 66.7 (5) 25.0 (1) 11.9 (1) 0 (1)	0.036[a] (0.0095[b]) 0.080[a] (0.020[b]) ----- -----	Snow and Roubik (1987)
Echium vulgare (Boraginaceae)	*Ceratina calcarata* Hoplitis anthocopoides H. producta Megachile relativa Osmia coerulescens	72 (1) 78–83 (1) 74–84 (1) 61–79 (1) 73–77 (1)	----- ----- ----- ----- -----	Strickler (1979)

Plant species (family)	Insect species	Percentage pollen removed after (n) visits	Percentage pollen reaching stigma	Reference
Erythronium americanum (Liliaceae)	*Andrena carlini* *Apis mellifera*	56.5 (1) 59.3 (1)	---- ----	Harder et al. (1985)
Erythronium americanum (Liliaceae)	*Apis mellifera* *Bombus ternarius*, *B. terricola* (queens)	63.8 (1) 42.2 (1)	---- ----	Harder and Thomson (1989)
Erythronium grandiflorum (Liliaceae)	*Bombus occidentalis* (queens)	----	0.6[ac]	Harder and Thomson (1989)
Erythronium grandiflorum (Liliaceae)	*Bombus bifarius* *B. californicus* *B. nevadensis* *B. occidentalis*	83.3 (1) 65.0 (1) 31.0 (1) 59.5 (1)	0.75[ad] (0.62[bd]) 0.50[ad] (0.32[bd]) 0.69[ad] (0.38[bd]) 0.67[ad] (0.37[bd])	Thomson and Thomson (1989)
Ipomoea trichocarpa (Convolvulaceae)	*Bombus pennsylvanicus* *Enyo lugubris*✦	523 grains$^\infty$ ----	52.2 grains$^\infty$ 27.2 grains$^\infty$	Murcia (1990)
Lupinus sericeus (Fabaceae)	*Bombus* spp. (esp. *B. bifarius*)	18.9 (1), 27 (2), 35 (3), 43 (4), 48 (5)	----	Harder (1990)
Mertensia paniculata (Boraginaceae)	*Bombus flavifrons* (queens, workers)	59.0 (1), 76.6 (2), 86.2 (3)	----	Harder (1990)
Oenothera fruticosa (Onagraceae)	*Apis mellifera* *Chauliognathus marginatus*◗	1,100 grains$^\infty$ 217 grains$^\infty$	193 grains$^\infty$ 114 grains$^\infty$	Primack and Silander (1975)

Table 5.1 (cont.)

Plant species (family)	Insect species	Percentage pollen removed after (n) visits	Percentage pollen reaching stigma	Reference
Pedicularis bracteosa (Scrophulariaceae)	Bombus spp. (workers)	43.0 (1), 55.0 (2), 65.8 (3)	-----	Harder (1990)
Pedicularis contorta (Scrophulariaceae)	Bombus mixtus, B. bifarius, B. melanopygus, and B. occidentalis (workers)	43.5 (1), 65.0 (2), 74.8 (3), 80.6 (4)	-----	Harder (1990)
Phlox glaberrima (Polemoniaceae)	Colias eurytheme✤	7.0 (1)	10[a] (0.5[b*])	Levin and Berube (1972)
Phlox pilosa (Polemoniaceae)	Colias eurytheme✤	12.9 (1)	17[a] (0.5[b*])	Levin and Berube (1972)
Polemonium viscosum (Polemoniaceae)	Bombus kirbyellus (queens)	43.8 (1), 55.5 (2), 60.2 (3)	2.9[a*]	Galen and Stanton (1989)
Pontederia cordata (Pontederiaceae)	Bombus griseocollis	32–73 (1)	0.53 (L)[c] 0.24 (M) 0.021(S)	Barrett and Wolfe (1986)
Pontederia cordata (Pontederiaceae)	Bombus griseocollis	45 (1)	-----	Wolfe and Barrett (1989)

Plant species (family)	Insect species	Percentage pollen removed after (n) visits	Percentage pollen reaching stigma	Reference
Raphanus sativus (Brassicaceae)	*Apis mellifera*	52 (1), 64 (3.7), 61 (5.7)	0.33[b]	Young and Stanton (1990)
Trifolium pratense (Fabaceae)	*Apis mellifera*	87.2 (1)	-----	Dunham (1939)
Vicia faba (Fabaceae)	*Bombus terrestris* (workers)	1.3[f]	0.23[bg]	Carré et al. (1994)

∞ Total quantity of pollen presented per flower is unknown
● Beetle
❖ Moth or butterfly
[a] % of pollen *removed* from flower that reaches stigmata
[a*] % of pollen *removed* from flower that reaches stigmata of compatible recipierts
[b] % of pollen *presented* by flower that reaches stigmata
[b*] % of pollen *presented* by flower that reaches stigmata of compatible recipients
[c] Total pollen deposited on stigmata, as a sum from all 20 recipients
[d] Total pollen deposited on stigmata, as a sum from all 40 recipients
[e] % of presented pollen finding compatible stigmata of long (L), medium(M), and short (S) gynoecia, in this tristylous species
[f] % of presented pollen adhering to the hair coat on the anterior portions of a bee
[g] Total pollen deposited on stigmata, over all 18 (mean) recipients

101

(*P. glaberrima*) loss of pollen from the tongues of *Colias* butterflies (Levin and Berube 1972). On average, one-seventh of *Erythronium* pollen fell to the ground during bee visits, a loss correlated positively with visit duration (Harder and Thomson 1989). Furthermore, several reports of pollen rejection exist whereby some individuals deliberately groomed off much of the pollen that attached to their body hairs during nectar feeding, for example, honeybees (Free 1964; Free and Nuttall 1968; Gary et al. 1977; F. D. Parker and Hatley 1978) and bumblebees (Thomson et al. 1986; Galen and Stanton 1989). Pollen discarders were dedicated to that behavior (Free and Nuttall 1968).

When pollen does leave the flower on the body of the departing insect, it can be lost to another plant species, that is, "misplaced pollen" (Inouye et al. 1994). A lack of floral constancy by the vector can result in heterospecific pollination (Levin 1972; Levin and Berube 1972; Campbell 1985; Campbell and Motten 1985; Galen and Gregory 1989; Kato et al. 1991; Arroyo and Dafni 1993; Kearns and Inouye 1994).

Pollen carried on the body surface of certain insects can also lose its viability. Viability of *Brassica* pollen was reduced 20% when placed on *Atta texana* and other ant species; this loss was attributed to antibiotics on the ants' bodies. In contrast, viability averaged over 98% for the thorax of bees and wasps (Harriss and Beattie 1991). Parker and Hatley (1978) found that viability of onion pollen brushed experimentally from the body hairs of *Apis mellifera* was only 11%, compared to 24%–40% for flies, wasps, and solitary bees. This loss was related to inhibitory enzymes in the liquid utilized by honeybees to pack pollen in their corbiculae, that is, "active pollen loss – germination inhibition" (Inouye et al. 1994). On the contrary, apple pollen collected directly from flowers or brushed from the bodies of a variety of insect visitors showed a consistent reduction in pollen viability for only one hover fly species (Kendall 1973). Pollen residing on the hair coat of honeybees confined overnight to their hive (where the brood-nest temperature is maintained around 34°C) lost its viability by next morning (Kraai 1962). However, in one test, apple pollen was still viable after a 21-hour exposure inside the hive (Dicklow et al. 1985).

The loss of pollen from circulation as a result of a bee's grooming actions, performed in order to gather microgametophytes into loads for transport back to its home, has been termed "active loss" (Harder and Thomson 1989). Various factors are known to influence grooming behavior. The frequency or vigor of grooming movements by bumblebees rose with increasing amounts of pollen removed per flower (Thomson and Plowright 1980; Buchmann and Cane 1989; Harder 1990), and also varied between different bee species visiting the same plant (Strickler 1979; Ordway et al. 1987). Grooming activities by bees generally occur in flight, between flowers (Hodges 1952; Thomson et al. 1986; Harder and Thomson 1989; but see Snow and Roubik 1987), such that pollen grooming is reduced when bees simply have to walk (instead of fly) to nearby florets on an inflorescence (Thomson and Plowright 1980;

Thomson 1986). Grooming actions disturb the original location of pollen grains on the insect's body; such rearrangement can have consequences for pollination (Thomson and Plowright 1980; Wolfe and Barrett 1989).

There is a relationship between weight of pollen loads carried and bee size. Clements and Long (1923) provided data for several bee species that gathered pollen from raspberry; collected quantities of pollen varied widely (Table 5.2). Only the two bumblebee species carried loads per foraging trip that approached or exceeded half their body weight. The smaller, solitary bees removed less than 20% their body weight in pollen, with average loads well below 10 mg.

By far, most studies of pollen loads have focused on the two corbicular pollen pellets carried on the hind legs of *Apis mellifera*. R. L. Parker (1926) found that the average pollen load of honeybees weighed 20% of the unladen bee, with significant variability in pellet size among pollen taxa (R. L. Parker 1926; Synge 1947; Maurizio 1953; Vaissière and Vinson 1994). Strong accordance in range of load weights exists for four separate surveys: 8.9–20.5 mg per load (or 4.5–10.3 mg per pellet) (Parker 1926), 8.4–21.4 mg per load (4.2–10.7 mg per pellet) (Maurizio 1953), 4.0–11.7 mg per pellet (Warakomska, cited by Vaissière and Vinson 1994), and 8.0–23.4 mg per load (4.0–11.7 mg per pellet) (Buchmann et al. 1990). Each pellet contains 250,000–4,000,000 pollen grains, depending on microgametophyte size (Buchmann et al. 1990). Quantity of pollen presented per plant species also influences pollen load size and effort in acquisition. Dunham (1939) estimated that each honeybee collecting an average load of red clover pollen must visit at least 346 florets, a matter of 29.3 minutes' worth of foraging. Percival (1950) cites a number of similar scenarios, ranging from 585 florets of *Trifolium repens* (on 106–166 heads) to obtain a full load (11 mg) of pollen, to >10 full loads of pollen being available *per single flower* of *Papaver orientale*.

This transfer by bees of pollen from the flower to the nest ("predeposition pollinivory"; Inouye et al. 1994) accounts for a major loss and represents pollen that is unlikely ever to reach stigmata (Harder et al. 1985; Harder and Thomson 1989). Generally, this loss to bee food is without consequence to a plant's reproductive effort, as the pollen-to-ovule ratios of many xenogamous, bee-visited species are large (Cruden 1977); the ratio is almost 50,000:1 in *Echium* (Davis 1992). In addition, the efficiency of a bee's grooming actions is not absolute (Lukoschus 1957; Free and Williams 1972; Davis 1991, 1992). There was no correlation between corbicular load size and the amount of residual pollen on the honeybee body, probably indicative of a standard, noncorbicular quantity of pollen remaining after constant grooming (Free and Williams 1972). Furthermore, Buchmann et al. (1990) found that for *Apis mellifera*, an average of 0.007% of pollen that adhered to the body was lodged in sites "safe" from removal by grooming.

Despite these various types of pollen losses, much of a flower's pollen may never leave the anthers. Two studies involving relatively large, echinate

Table 5.2. *Fresh weights of pollen loads, calculated from Clements and Long (1923; Table 87), for bees carrying loads that shared the same pollen species* (Rubus deliciosus)

Bee species	n (Bee loads)	Fresh weight (mg; Mean ± SE)	Pollen load as % of body weight	
			Mean	Range
Andrena crataegi	6	6.77 ± 0.65	12.7	9.9–14.4
Apis mellifera	10	18.71 ± 1.35	24.2	13.0–32.4
Bombus juxtus	13	19.49 ± 4.78	25.3	4.9–61.7
Bombus occidentalis	5	51.54 ± 12.53	34.6	7.5–46.9
Halictus medionitens	2	1.65 ± 1.35	18.3	9.4–27.3
Monumetha albifrons	1	3.50	2.0	------

pollen grains, those of cantaloupe (Mann 1953) and *Malvaviscus arboreus* (Webb and Bawa 1983), showed that even after 40–50 flower visits, up to 25% of the pollen remained in the microsporangia. And although most radish flowers received numerous visits, at least 20% of presented pollen was not withdrawn (Young and Stanton 1990). In contrast, the combined visits of hover flies and halictid bees removed 99.7% of pollen produced by *Jepsonia* (Ornduff 1971).

How best to evaluate an insect's importance as a pollinator on the flower?

The question posed here is important and has been periodically addressed in the anthecological literature, particularly as our knowledge of the pertinent processes involved in pollination, fertilization, and seed and fruit maturation became refined. The question as presented is restricted to events in the flower and so ignores many important parameters that affect insect pollination, such as carryover, flight distance, insect and plant population size, and plant density. The involvement of so many complex factors, some of which are highlighted in Chapters 4 and 6 and earlier in this chapter, makes this question extremely difficult to answer unconditionally. There are, however, some recommendations that could provide greater insights in future research.

A wide range of measurements have been taken during insect visits to flowers, to estimate an insect's contribution to pollination of a particular species. These efforts can be classified broadly as "direct" and "indirect" (Spears 1983). Motten et al. (1981, p. 1281) state, "The most direct means of evaluating visitor effectiveness is to examine seed set resulting from a single visit." Use of controlled, single visits to virgin flowers is most important. However, use of the terms "direct" and "indirect" in this chapter varies somewhat from that of these authors. Included as indirect methods are assessments of the quantity of pollen carried on the body, the duration of floral visits, the number of stigma touches, the number or percentage set of fruits and seeds, and so forth. Direct methods include measurement of "stigma pollen load per visit" (Inouye et al. 1994) and number of pollen tubes at the style base following single visits (Davis 1992). The latter measures a postpollination process (Inouye et al. 1994) and may be argued to be indirect. Yet it does represent the reproductive stage following pollination and provides a prefertilization datum preferable to postfertilization events such as fruit and seed set, which are more distant measures of pollination.

In numerous reports, ovules that developed from successful pollination and fertilization events were subsequently aborted by the plant due to scarcity of maternal resources (Stephenson 1981; Casper 1983; Motten 1983; Willson and Burley 1983; Pellmyr 1987; Schemske and Horvitz 1988). As a result, serious underestimations of true pollination can occur, if seed counts are assumed to serve as indicators of insect pollination (and see F. D. Parker

1981). Seed counts are better estimates of fertilization than pollination, especially if embryo abortions and incompatibility occur. In some exceptional species possessing very large numbers of ovules per flower, all compatible pollen grains on the stigma may theoretically be registered as fertilizations. Even then, this practice can be risky science. For example, after allowing single visits by honeybees to virgin perfect flowers (functionally pistillate) of *Actinidia deliciosa* cv. Hayward, the maximum seed count obtained from a mature fruit was 1,359 (Donovan and Read 1991). This total matches the maximum number of ovules per flower in that kiwifruit variety (Hopping 1976) derived from approximately 1,870–2,960 pollen tubes (Hopping and Jerram 1979), suggesting that (even in thousand-ovuled fruits!) the pollination contribution of just a solitary visitor might not always be fully registerable by this seed-count method. Indeed, Bertin (1990) reported that in *Campsis*, no further increases in seed mass or number occurred after a certain stigmatic pollen load was reached. In cases of pollen surplus, when there are more pollen tubes than ovules, resulting fertilization and seed quality can be a nonrandom sample of genotypes of the compatible pollen grains deposited on a stigma (Willson and Burley 1983; Snow and Roubik 1987; Richardson and Stephenson 1991; Kearns and Inouye 1993).

Aware of the problem of competition for limited maternal resources, several researchers basing their pollination conclusions on fruit and seed set have anticipated the suppression of those parameters in their experimental flowers. Practices such as the elimination or minimization of previous fruits prior to experimentation (Adlerz 1966; Kendall and Smith 1975), roguing competing plants (Galen 1985), clipping stigmata from nonexperimental flowers (Galen et al. 1985), and providing additional water and fertilizer to experimental plants (Dieringer 1992; Kearns and Inouye 1993) have all been performed to heighten the success of fruit and seed set in experimental flowers.

Destruction of maturing seeds and fruits by phytophagous animals and disease has, unfortunately, caused the loss of research data (Arnold 1982; Pellmyr 1987; Berger et al. 1988; de Oliveira et al. 1991; Dieringer 1991; R. A. Johnson 1992). For this reason, it has been advised to avoid these indirect measures of pollination (Levin 1981; Snow and Roubik 1987; Nadel and Peña 1994). A large delay in acquisition of one's experimental data is another disadvantage of fruit and seed set – for example, nine months in *Dieffenbachia* (Young 1988). In *Protea*, some seeds were lost to insect predation during the seven months' postflowering necessary to harvest mature infructescences (Coetzee and Giliomee 1985).

There are other problems associated with using fruit and seed set to estimate pollination. Deterioration of climatic conditions during fruit and seed maturation also has caused data loss (Handel 1982; Pellmyr 1987). In addition, the arbitrary nature of the analysis of fruit set is another disadvantage of the use of a fertilization parameter to judge pollination. For example, allowance for "June drop" of apples (Robinson and Fell 1981; DeGrandi-Hoffman et al. 1984; and several others), a common phenomenon in many orchard Rosaceae, is an admittance of underestimation of actual pollination. Only a quarter of apple flowers that set fruit actually yield mature fruits (Free and Spencer-Booth 1964; Brain and Landsberg 1981). Tepedino (1981) also showed that the percentage of mature *Cucurbita* fruits varied, depending on the time that had passed after floral visitation.

Fruit and/or seed set sometimes has provided useful preliminary data on the relative importance of floral visitors that differ in size, without insect visits necessarily having been observed. For example, it has been possible to selectively exclude larger visitors by placing wire cages or cloth bags of different mesh around flowers or inflorescences, while leaving others fully exposed (Kikuchi 1965; Coetzee and Giliomee 1985; Paton and Turner 1985; Mathur and Mohan Ram 1986; Ramsay 1988; Waser and Price 1990; Heard 1994; Westerbergh and Saura 1994). Resulting fruit and seed set from these treatments were then compared to estimate the relative importance of excluded visitor groups.

Observations of stigma touches by various insects on flowers (Free and Nuttall 1968; Robinson 1979; Mohr and Jay 1988; Dafni and Dukas 1986; Dafni et al. 1987; Heard 1994; MacKenzie 1994) are helpful preliminarily as indicators of pollination potential. However, these data are indefinitive because there are no guarantees that this documented stigmatic contact actually results in pollination. For example, such observations do not generate quantitative data of stigmatic pollen loads.

Because pollination is the deposition of pollen on the stigma (Inouye et al. 1994), measurements of pollen grains on stigmata, of course, yield the most direct measure of pollination. This measurement is, however, often difficult to make accurately (Bertin 1982; Cruden et al. 1990; Davis 1992), and Spears (1983, p. 196) states that "counting pollen grains transferred to a stigma tells us nothing of the quality of that pollen." Motten (1983, p. 355) advises that "a more appropriate measurement is the quantity of pollen actually involved in fertilization," and he and Davis (1992) have advocated quantification of the numbers of pollen tubes reaching the base of the style in self-incompatible *Erythronium umbilicatum* (wherein tubes from self-pollen rarely reached the style base) and in self-compatible *Echium plantagineum*. This technique is also preferable over seed counts when a species possesses relatively few ovules per flower, so that not all pollination events could get recorded as seeds.

A limitation on the absolute reliance on pollen-tube counts can be found in certain species like *Cochlospermum vitifolium* (Bawa and Webb 1984), where even self-incompatible grains send pollen tubes as far as the ovules and so cannot be distinguished from compatible grains. In those species, counts of all pollen tubes may be useful only as estimates of the stigmatic pollen load, and it might be possible only to distinguish cross- from self-pollen by genetic analysis of seeds. This can be done by electrophoresis.

Table 5.3 lists studies that have provided direct measurements of pollen deposition after one or more visits to flowers that had not been visited before.

Table 5.3. *Summary of selected studies that provide direct measurements of pollen deposition (stigmatic pollen loads, or pollen tube counts at the style base) allowing discrimination of the contribution of various insect visitors to pollination, following single (or multiple) visits to previously-unvisited (virgin) flowers*

Plant species (family)	Insect order(s)	Reference
Agalinis strictifolia (Scrophulariaceae)	Hymenoptera	Dieringer (1992)
Brassica napus (Brassicaceae)	Diptera	Ohsawa and Namai (1988)
Campsis radicans (Bignoniaceae)	Hymenoptera	Bertin (1982)
Cassia reticulata (Fabaceae)	Hymenoptera	Snow and Roubik (1987)
Clintonia borealis (Liliaceae)	Hymenoptera	Thomson and Plowright (1980)
Cnidoscolus urens (Euphorbiaceae)	Lepidoptera	Webb and Bawa (1983)
Cochlospermum vitifolium (Cochlospermaceae)	Hymenoptera	Snow and Roubik (1987)
Cucurbita pepo (Cucurbitaceae)	Hymenoptera	Tepedino (1981)
Delphinium nelsonii (Ranunculaceae)	Hymenoptera	Waser (1988); Waser and Price (1990)
Delphinium virescens (Ranunculaceae)	Hymenoptera	Waddington (1981)
Diervilla lonicera (Caprifoliaceae)	Hymenoptera	Thomson and Plowright (1980)
Echium plantagineum (Boraginaceae)	Hymenoptera	Davis (1992)
Epilobium angustifolium (Onagraceae)	Hymenoptera	Galen and Plowright (1985b)
Erythronium americanum (Liliaceae)	Hymenoptera	Thomson and Plowright (1980); Thomson (1986)
Erythronium grandiflorum (Liliaceae)	Hymenoptera	Thomson (1986); Thomson et al. (1986); Harder and Thomson (1989); Thomson and Thomson (1989)
Erythronium umbilicatum (Liliaceae)	Hymenoptera	Motten (1983)
Ipomoea trichocarpa (Convolvulaceae)	Hymenoptera Lepidoptera	Murcia (1990)

Plant species (family)	Insect order(s)	Reference
Lavandula latifolia (Lamiaceae)	Diptera Hymenoptera Lepidoptera	Herrera (1987)
Linaria vulgaris (Scrophulariaceae)	Hymenoptera	Arnold (1982); Thomson (1986)
Linum lewisii (Linaceae)	Diptera Hymenoptera	Kearns and Inouye (1994)
Mertensia ciliata (Boraginaceae)	Hymenoptera	Geber (1985)
Oenothera fruticosa (Onagraceae)	Coleoptera	Primack and Silander (1975)
	Hymenoptera	Silander and Primack (1978)
Oenothera speciosa (Onagraceae)	Hymenoptera	Wolin et al. (1984)
Phlox glaberrima (Polemoniaceae)	Lepidoptera	Levin and Berube (1972)
Phlox pilosa (Polemoniaceae)	Lepidoptera	Levin and Berube (1972)
Polemonium viscosum (Polemoniaceae)	Diptera Hymenoptera Lepidoptera	Galen and Kevan (1980)
Polemonium viscosum (Polemoniaceae)	Hymenoptera	Galen and Newport (1987); Galen and Stanton (1989); Cresswell and Galen (1991)
Pontederia cordata (Pontederiaceae)	Hymenoptera	Wolfe and Barrett (1989)
Raphanus sativus (Brassicaceae)	Hymenoptera	Young and Stanton (1990)
Stellaria pubera (Caryophyllaceae)	Diptera	Campbell and Motten (1985)
	Hymenoptera	Campbell (1985)
Vicia faba (Fabaceae)	Hymenoptera	Carré et al. (1994)

Strikingly evident is the small number of species of agricultural importance for which "stigma pollen load per visit" (Inouye et al. 1994) or pollen-tube counts at the style base have been determined. Obviously, much research lies ahead for crop plants. The relative importance of bees (Hymenoptera) as visitors to these diverse floral taxa is also clear (Table 5.3). In studies involving different insect orders, overall deposition of pollen by bees has exceeded that by butterflies and flies (Galen and Kevan 1980; Herrera 1987; Kearns and Inouye 1994).

Apparent from these studies is the highly stochastic nature of pollination by insects (Thomson and Plowright 1980; Thomson 1986; Wolfe and Barrett 1989). For example, during a bumblebee run on *Clintonia borealis*, a flower received 484 grains after the previous recipient flowers received 153 and 28 (Thomson and Plowright 1980). Similarly, pollen-deposition data from one run by *Bombus grisecollis* on nine consecutive recipients of *Pontederia cordata* were 20, 10, 2, 12, 0, 42, 0, 5, and 30 grains (Wolfe and Barrett 1989). Thus, not every visit by a "reliable" pollinator of that plant species results in pollination. For instance, 1% and 8% of bee visits to *Cassia reticulata* and *Cochlospermum vitifolium*, respectively, were nonpollinating (Snow and Roubik 1987). In *Erythronium umbilicatum*, 6.7% and 11.6% of virgin flowers visited once by *Apis* and *Andrena*, respectively, lacked pollen tubes at the style base (Motten 1983). In studies of consecutive visits to recipient flowers, 52% (average over several runs of 40 flowers each, *Bombus* spp. on *Erythronium grandiflorum*) and 67% (average over 10-flower runs, *Bombus terrestris* on *Vicia faba*) "misses" were recorded (Thomson and Thomson 1989; Carré et al. 1994). In *Pontederia*, all visits by bees removed pollen, but on average, 30% of consecutive visits did not result in deposition on stigmata (Barrett and Wolfe 1986; Wolfe and Barrett 1989). Stigma exertion and receptivity are important factors, even more important than visit duration (Kwak et al. 1985; Davis 1992), influencing pollen deposition by insects (Galen and Plowright 1985b; Thomson and Stratton 1985; Galen and Stanton 1989; Murcia 1990; Davis 1992; Carré et al. 1994), and they play a significant role in the "hit-or-miss," stochastic nature of pollen deposition. Interestingly, the number of flowers previously visited may have little influence on the quantity of compatible pollen deposited on succeeding flowers (Cresswell and Galen 1991).

In order to fully evaluate insect visitors as pollinators of a species, various researchers (Wilson and Thomson 1991; Thomson and Thomson 1992; and others) have stressed the need for measurement of not only pollen removal, but also the proportion of removed pollen that eventually is deposited on conspecific stigmata. These data are available for 10 plant species, shown in the penultimate column of Table 5.1. Although beetles and butterflies are represented in only a few studies, they deposited, on a per visit basis, 10% or more of that pollen which resided on their bodies upon departing conspecific flowers (Table 5.1). Bees, on the other hand, which are better represented, usually deposited only 0.5%–3% of *removed* pollen, occasionally 10%–17.5% (*Ipomoea, Oenothera*). Furthermore, bees deposited only 0.01%–0.5% of *presented* pollen on stigmata. That is, in those bee-visited species that have been investigated thoroughly, as few as 1 in 200–1,000 pollen grains that develops in an anther of a given species ever reaches a stigma of that same species. As well, as the quantity of pollen removed from *Erythronium* during single visits by *Bombus* queens increased, the proportion of removed pollen that landed on conspecific stigmata decreased (Harder and Thomson 1989).

It is recommended, therefore, that assessments of the efficiency of insects as pollinators include a combination of both pollen removal and deposition.

Determination of the amount of pollen withdrawn from a virgin flower during a single visit, as well as the proportion of removed pollen that ultimately is deposited on conspecific stigmata of recipient flowers, will provide a direct and effective measure of an insect's worth as a pollinator to the plant. In addition, paternity testing of resulting seed can provide important data about the insect's value for cross- versus self-pollination.

Note added in proof (see p. 96)
"Viability of pollen of *Cucumis melo* on the bodies of pollen foragers of *Apis mellifera* has been found to be significantly lower than that carried on the bodies of nectar-collecting bees." Vaissière, B. E., Malaboeuf, F. & Rodet, G. (1996). Viability of cantaloupe pollen carried by honeybees *Apis mellifera* varies with foraging behaviour. *Naturwissenschaften* 83:84–86.

References

Addicott, J.F. (1986). Variation in the costs and benefits of mutualism: the interaction between yuccas and yucca moths. *Oecologia* 70:486–494.

Adlerz, W.C. (1966). Honeybee visit numbers and watermelon pollination. *J. Econ. Ent.* 59:28–30.

Aker, C. & Udovic, D. (1981). Oviposition and pollination behavior of the yucca moth *Tegeticula maculata* (Lepidoptera: Proxidae), and its relation to the reproductive biology of *Yucca whipplei* (Agavaceae). *Oecologia* 49:96–101.

Arnold, R.M. (1982). Pollination, predation and seed set in *Linaria vulgaris* (Scrophulariaceae). *Amer. Midl. Nat.* 107:360 369.

Arroyo, J. & Dafni, A. (1993). Interspecific pollen transfer among co-occurring heteromorphic and homomorphic species. *Israel J. Bot.* 41:225–232.

Barrett, S.C.H. & Wolfe, L.M. (1986). Pollen heteromorphism as a tool in studies of the pollination process in *Pontederia cordata*. In *Biotechnology and Ecology of Pollen,* eds. D.L. Mulcahy, G. Bergamini Mulcahy, and E. Ottaviano, pp. 435–442. New York: Springer-Verlag.

Bawa, K.S. & Webb, C.J. (1984). Flower, fruit, and seed abortion in tropical forest trees: implications for the evolution of maternal and paternal reproductive patterns. *Amer. J. Bot.* 71:736–751.

Benedek, P. (1973). Relationship between the tripping rate and the tongue length of lucerne pollinating wild bees. *Z. angew. Entomol.* 73:113 116.

Benedek, P. (1977). Behaviour of honeybees (*Apis mellifera* L.) in relation to the pollination of onion (*Allium cepa* L.) inflorescences. *Z. angew. Entomol.* 82:414–420.

Berger, L.A., Vaissière, B.E., Moffett, J.O. & Merritt, S.J. (1988). *Bombus* spp. (Hymenoptera: Apidae) as pollinators of male-sterile upland cotton on the Texas High Plains. *Environ. Entomol.* 17:789–794.

Bertin, R.I. (1982). Floral biology, hummingbird pollination and fruit production of trumpet creeper (*Campsis radicans*, Bignoniaceae). *Amer. J. Bot.* 69:122–134.

Bertin, R.I. (1990). Effects of pollination intensity in *Campsis radicans*. *Amer. J. Bot.* 77:178–187.

Betts, A.D. (1920). The constancy of the pollen-collecting bee. *Bee World* 2:10.

Bosch, J. & Blas, M. (1994). Foraging behavior and pollinating efficiency of *Osmia cornuta* and *Apis mellifera* on almond (Hymenoptera, Megachilidae and Apidae). *Appl. Ent. Zool.* 29:1–9.

Brain, P., & Landsberg, J.J. (1981) Pollination, initial fruit set and fruit drop in apples: analysis using mathematical models. *J. Hort. Sci.* 56:41-54.

Brian, A.D. (1957). Differences in the flowers visited by four species of bumble-bees and their causes. *J. Anim. Ecol.* 26:71–98.

Brittain, W.H. & Newton, D.E. (1933). A study of the relative constancy of hive bees and wild bees in pollen gathering. *Can. J. Res.* 9:334–349.

Broyles, S.B., Schnabel, A. & Wyatt, R. (1994). Evidence for long-distance pollen dispersal in milkweeds (*Asclepias exaltata*). *Evolution* 48:1032–1040.

Buchmann, S.L. (1987). The ecology of oil flowers and their bees. *Ann. Rev. Ecol. Syst.* 18:343–369.

Buchmann, S.L. & Cane, J.H. (1989). Bees assess pollen returns while sonicating *Solanum* flowers. *Oecologia* 81:289–294.

Buchmann, S.L., Shipman, C.W. & Hansen, H.M. (1990). Pollen residing on safe sites on honey bee foragers. *Amer. Bee J.* 130(12):798–799.

Burdon, J.J., Jarosz, A.M. & Brown, A.H.D. (1988). Temporal patterns of reproduction and outcrossing in weedy populations of *Echium plantagineum*. *Biol. J. Linn. Soc.* 34:81–92.

Campbell, D.R. (1985). Pollen and gene dispersal: the influences of competition for pollination. *Evolution* 39:418–431.

Campbell, D.R. & Motten, A.F. (1985). The mechanism of competition for pollination between two forest herbs. *Ecology* 66:554–563.

Carré, S., Badenhausser, I., Taséi, J.N., LeGuen, J. & Mesquida, J. (1994). Pollen deposition by *Bombus terrestris* L, between male-fertile and male-sterile plants in *Vicia faba* L. *Apidologie* 25:338–349.

Casper, B.B. (1983). The efficiency of pollen transfer and rates of embryo initiation in *Cryptantha* (Boraginaceae). *Oecologia* 59:262–268.

Chagnon, M., Gingras, J. & de Oliveira, D. (1993). Complementary aspects of strawberry pollination by honey and indigenous bees (Hymenoptera). *J. Econ. Entomol.* 86:416–420.

Clements, F.E. & Long, F.L. (1923). *Experimental Pollination: An Outline of the Ecology of Flowers and Insects*. Washington: Carnegie Institution.

Coetzee, J.H. & Giliomee, J.H. (1985). Insects in association with the inflorescence of *Protea repens* (Proteaceae) and their role in pollination. *J. Ent. Soc. S. Africa* 48:303–314.

Corbet, S.A. (1990). Pollination and the weather. *Israel. J. Bot.* 39:13–30.

Corbet, S.A., Beament, J. & Eisikowitch, D. (1982). Are electrostatic forces involved in pollen transfer? *Plant, Cell and Environ.* 5:125–129.

Corbet, S.A., Fussell, M., Ake, R., Fraser, A., Gunson, C., Savage, A. & Smith, K. (1993). Temperature and the pollinating activity of social bees. *Ecol. Entomol.* 18:17–30.

Courtney, S.P., Hill, C.J. & Westerman, A. (1982). Pollen carried for long periods by butterflies. *Oikos* 38:260–263.

Cresswell, J.E. (1994). A method for quantifying the gene flow that results from a single bumblebee visit using transgenic oilseed rape, *Brassica napus* L. cv. Westar. *Transgenic Res.* 3:134–137.

Cresswell, J.E. & Galen, C. (1991). Frequency-dependent selection and adaptive surfaces for floral character combinations: the pollination of *Polemonium viscosum. Amer. Nat.* 138:1342–1353.

Cruden, R.W. (1977). Pollen–ovule ratios: a conservative indicator of breeding systems in flowering plants. *Evolution* 31:32–46.

Cruden, R.W., Baker, K.K., Cullinan, T.E., Disbrow, K.A., Douglas, K.L., Erb, J.D., Kirsten, K.J., Malik, M.L., Turner, E.A., Weier, J.A. & Wilmot, S.R. (1990). The mating system and pollination biology of three species of *Verbena* (Verbenaceae). *J. Iowa Acad. Sci.* 97:178–183.

Dafni, A., & Dukas, R. (1986). Insect and wind pollination in *Urginea maritima* (Liliaceae). *Plant Syst. Evol.* 154:1-10.

Dafni, A., Eisikowitch, D. & Ivri, Y. (1987). Nectar flow and pollinators' efficiency in two co-occurring species of *Capparis* (Capparaceae) in Israel. *Plant Syst. Evol.* 157:181–186.

Davis, A.R. (1991). Mixed loading of pollen from *Echium plantagineum* L. (Boraginaceae) and *Hirschfeldia incana* (L.) Lagrèze-Foss. (Brassicaceae) by an individual honey bee (*Apis mellifera* L.). *Amer. Bee J.* 131:649–655.

Davis, A.R. (1992). Evaluating honey bees as pollinators of virgin flowers of *Echium plantagineum* L. (Boraginaceae) by pollen tube fluorescence. *J. Apic. Res.* 31:83–95.

Davis, A.R. (1995). Floral biology and research techniques: nectar. In *Pollination of Cultivated Plants in the Tropics,* ed. D.W. Roubik, pp. 109–121. Agricultural Services Bulletin 118. Rome: Food and Agricultural Organization of the United Nations.

DeGrandi-Hoffman, G. (1979). The honey bee pollination component of horticultural crop production systems. *Hort. Rev.* 9:237–272.

DeGrandi-Hoffman, G., Hoopingarner, R. & Baker, K.K. (1984). Identification and distribution of cross-pollinating honey bees (Hymenoptera: Apidae) in apple orchards. *Environ. Entomol.* 13:757–764.

DeGrandi-Hoffman, G., Hoopingarner, R. & Klomparens, K. (1986). Influence of honey bee (Hymenoptera: Apidae) in-hive pollen transfer on cross-pollination and fruit set in apple. *Environ. Entomol.* 15:723–725.

de Oliveira, D., Gingras, J. & Chagnon, M. (1991). Honey bee visits and pollination of red raspberries. *Acta Hort.* 288:415–419.

Dennis, B.A. & Haas, H. (1967a). Pollination and seed-setting in diploid and tetraploid red clover (*Trifolium pratense* L.) under Danish conditions. I. Seed-setting in relation to the number and type of pollinating insects. *Kon. Vet. og Landboh. Aarsskrift (Denmark),* pp. 93-117.

Dennis, B.A. & Haas, H. (1967b). Pollination and seed-setting in diploid and tetraploid red clover (*Trifolium pratense* L.) under Danish conditions. II. Studies of floret morphology in relation to the working speed of honey and bumble bees (Hymenoptera:Apoidae). *Kon. Vet. og Landboh. Aarsskrift (Denmark),* pp. 118–133.

Dicklow, M.B., Firman, R.D., Rupert, D.B., Smith, K.L. & Ferrari, T.E. (1985). Controlled enpollination of honeybees (*Apis mellifera*): bee-to-bee and bee-to-tree pollen transfer. In *Biotechnology and Ecology of Pollen*, eds. D.L. Mulcahy, G. Bergamini Mulcahy, and E. Ottaviano, pp. 449–454. Berlin: Springer-Verlag.

Dieringer, G. (1991). Pollination ecology of *Streptanthus bracteatus* (Brassicaceae): A rare central Texas endemic. *Southwestern Naturalist* 36:341–343.

Dieringer, G. (1992). Pollinator effectiveness in *Agalinis strictifolia* (Scrophulariaceae). *Amer. J. Bot.* 79:1018–1023.

Di-Giovanni, F. & Kevan, P.G. (1991). Factors affecting pollen dynamics and its importance to pollen contamination: a review. *Can. J. Forestry Res.* 21:1155–1170.

Donovan, B.J. & Read, P.E.C. (1991). Efficacy of honey bees as pollinators of kiwifruit. *Acta Hort.* 288:220–224.

Dunham, W.E. (1939). Collection of red clover pollen by honeybees. *J. Econ. Entomol.* 32:668–670.

Ellstrand, N.C. (1992). Gene flow by pollen: implications for plant conservation genetics. *Oikos* 63:77–86.

Erickson. E. & Buchmann, S.L. (1983). Electrostatics and pollination. In *Handbook of Experimental Pollination Biology*, eds. C.E. Jones and R.J. Little, pp. 173–184. New York: Scientific and Academic Editions.

Fenster, C.B. (1991). Gene flow and dispersal. *Evolution* 45:398–409.

Fleming, T.H. & Sosa, V.J. (1994). Effects of nectarivorous and frugivorous mammals on reproductive success of plants. *J. Mammal.* 75:845–851.

Frankie, G.W. & Vinson, S.W. (1977). Scent marking of passion flowers in Texas by females of *Xylocopa virginica texana* (Hymenoptera: Anthophoridae). *J. Kansas Ent. Soc.* 50:613–625.

Free, J.B. (1963). The flower constancy of honeybees. *J. Anim. Ecol.* 32:119–131.

Free, J.B. (1964). The behaviour of honeybees on sunflowers (*Helianthus annuus* L.). *J. Appl. Ecol.* 1:19–27.

Free, J.B. (1965). The ability of bumblebees and honeybees to pollinate red clover. *J. Appl. Ecol.* 2:289–294.

Free, J.B. (1966). The pollinating efficiency of honey-bee visits to apple flowers. *J. Hort. Sci.* 41:91–94.

Free, J.B. (1968). The behavior of bees visiting runner beans (*Phaseolus multiflorus*). *J. Appl. Ecol.* 5:631–638.

Free, J.B. (1970). The flower constancy of bumblebees. *J. Anim. Ecol.* 39:395–402.

Free, J.B. & Nuttall, P.M. (1968). The pollination of oilseed rape *(Brassica napus)* and the behaviour of bees on the crop. *J. Agric. Sci., Cambridge* 71:91–94.

Free, J.B. & Spencer-Booth, Y. (1964). The effect of distance from pollinizer varieties on the fruit set of apple, pear and sweet-cherry trees. *J. Hort. Sci.* 39: 54–60.

Free, J.B. & Williams, I.H. (1972). The transport of pollen on the body hairs of honeybees (*Apis mellifera* L.) and bumblebees (*Bombus* spp. L.). *J. Appl. Ecol.* 9:609–615.

Fridén, F., Eskilsson, L. & Bingefors, S. (1962). Bumblebees and red clover pollination in central Sweden. *Sveiges Froodlareforbund. Medd.* 7:17–26.

Galen, C. (1985). Regulation of seed-set in *Polemonium viscosum*: floral scents, pollination, and resources. *Ecology* 66:792–797.

Galen, C. & Blau, S. (1988). Caste-specific patterns of flower visitation in bumblebees (*Bombus kirbyellus*) collecting nectar from alpine sky pilot, *Polemonium viscosum*. *Ecol. Entomol.* 13:11–17.

Galen, C. & Gregory, T. (1989). Interspecific pollen transfer as a mechanism of competition: consequences of foreign pollen contamination for seed set in the alpine wildflower, *Polemonium viscosum*. *Oecologia* 81:120–123.

Galen, C. & Kevan, P.G. (1980). Scent and color, floral polymorphisms and pollination biology in *Polemonium viscosum* Nutt. *Amer. Midl. Nat.* 104:281–289.

Galen, C. & Newport, M.E.A. (1987). Bumble bee behaviour and selection on flower size in the sky pilot, *Polemonium viscosum*. *Oecologia* 74:20–23.

Galen, C. & Plowright, R.C. (1985a). Contrasting movement patterns of nectar-collecting and pollen-collecting bumble bees (*Bombus terricola*) on fireweed (*Chamaenerion angustifolium*) inflorescences. *Ecol. Entomol.* 10:9–17.

Galen, C. & Plowright, R.C. (1985b). The effects of nectar level and flower development on pollen carry-over in inflorescences of fireweed (*Epilobium angustifolium*) (Onagraceae). *Can. J. Bot.* 63:488–491.

Galen, C. & Rotenberry, J.T. (1988). Variance in pollen carryover in animal-pollinated plants: implications for mate choice. *J. Theor. Biol.* 135:419–429.

Galen, C. & Stanton, M.L. (1989). Bumble bee pollination and floral morphology: factors influencing pollen dispersal in the alpine sky pilot, *Polemonium viscosum* (Polemoniaceae). *Amer. J. Bot.* 76:419–426.

Galen, C., Plowright, R.C. & Thomson, J.D. (1985). Floral biology and regulation of seed set and seed size in the lily, *Clintonia borealis*. *Amer. J. Bot.* 72:1544–1552.

Gary, N.E., Mau, R.F.L. & Mitchell, W.C. (1972a). A preliminary study of honey bee foraging range in macadamia (*Macadamia integrifolia*, Maiden and Betche). *Proc. Hawaii. Ent. Soc.* 21:205–212.

Gary, N.E., Witherell, P.C. & Marston, J. (1972b). Foraging range and distribution of honey bees used for carrot and onion pollination. *Environ. Entomol.* 1:71–78.

Gary, N.E., Witherell, P.C., Lorenzen, K. & Marston, J.M. (1977). Area fidelity and intra-field distribution of honey bees during the pollination of onions. *Environ. Entomol.* 6:303–310.

Geber, M.A. (1985). The relationship of plant size to self-pollination in *Mertensia ciliata. Ecology* 66:762–772.

Ginsberg, H. (1986). Honey bee orientation behaviour and the influence of flower distribution on foraging movements. *Ecol. Entomol.* 11:173–179.

Giurfa, M. & Núñez, J.A. (1992). Honeybees mark with scent and reject recently visited flowers. *Oecologia* 89:113–117.

Handel, S.H. (1982). Dynamics of gene flow in an experimental population of *Cucumis melo* (Cucurbitaceae). *Amer. J. Bot.* 69:1538–1546.

Harder, L.D. (1983). Flower handling efficiency of bumble bees: morphological aspects of handling time. *Oecologia* 57: 274–280.

Harder, L.D. (1990). Pollen removal by bumble bees and its implications for pollen dispersal. *Ecology* 71:1110–1125.

Harder, L.D. & Thomson, J.D. (1989). Evolutionary options for maximizing pollen dispersal of animal-pollinated plants. *Amer. Nat.* 133:323–344.

Harder, L.D., Thomson, J.D., Cruzan, M.B. & Unnasch, R.S. (1985). Sexual reproduction and variation in floral morphology in an ephemeral vernal lily, *Erythronium americanum. Oecologia* 67:286–291.

Harriss, F.C.L. & Beattie, A.J. (1991). Viability of pollen carried by *Apis mellifera* L., *Trigona carbonaria* Smith and *Vespula germanica* (F.) (Hymenoptera: Apidae, Vespidae). *J. Aust. Ent. Soc.* 30:40–47.

Haynes, J. & Mesler, M. (1984). Pollen foraging by bumblebees: foraging patterns and efficiency on *Lupinus polyphyllus. Oecologia* 61:249–253.

Heard, T.A. (1994). Behaviour and pollinator efficiency of stingless bees and honey bees on macadamia flowers. *J. Apic. Res.* 33:191–198.

Heinrich, B. (1976a). The foraging specializations of individual bumblebees. *Ecological Monographs* 46:105–128.

Heinrich, B. (1976b). Resource partitioning among some eusocial insects: bumblebees. *Ecology* 57:874–889.

Heinrich, B. (1979). "Majoring" and "minoring" by foraging bumblebees, *Bombus vagans*: an experimental analysis. *Ecology* 60:245–255.

Herrera, C.M. (1987). Components of pollinator "quality": comparative analysis of a diverse insect assemblage. *Oikos* 50:79–90.

Herrera, C.M. (1989). Pollinator abundance, morphology, and flower visitation rate: analysis of the "quantity" component in a plant–pollinator system. *Oecologia* 80:241–248.

Hodges, D. (1952). *The Pollen Loads of the Honeybee.* London: International Bee Research Association.

Holm, S.N. (1966). The utilization and management of bumble bees for red clover and alfalfa seed production. *Ann. Rev. Entomol.* 11:155–182.

Hopping, M.E. (1976). Effect of exogenous auxins, gibberellins, and cytokinins on fruit development in Chinese gooseberry (*Actinidia chinensis* Planch.). *N.Z. J. Bot.* 14:69–75.

Hopping, M.E. & Jerram, E.M. (1979). Pollination of kiwifruit (*Actinidia deliciosa* Planch.): stigma–style structure and pollen tube growth. *N.Z. J. Bot.* 17:233–240.

Inouye, D. (1980a). The terminology of floral larceny. *Ecology* 61:1251–1253.

Inouye, D. (1980b). The effect of proboscis and corolla tube lengths on patterns and rates of flower visitation by bumblebees. *Oecologia* 45:197–201.

Inouye, D.W. (1983). The ecology of nectar robbing. In *The Biology of Nectaries*, eds. B. Bentley and T. Elias, pp. 153–173. New York: Columbia University Press.

Inouye, D., Gill, D.E., Dudash, M.R. & Fenster, C.B. (1994). A model and lexicon for pollen fate. *Amer. J. Bot.* 81:1517–1530.

Jackson, J.F. & Clarke, G.R. (1991). Gene flow in an almond orchard. *Theor. Appl. Genet.* 82:169–173.

Janzen, D.H. (1979). How many babies do figs pay for babies? *Biotropica* 11:48–50.

Jennersten, O. (1984). Flower visitation and pollination efficiency of some North European butterflies. *Oecologia* 63:80–89.

Jennersten, O., Morse, D.H. & O'Neil, P. (1991). Movements of male and worker bumblebees on and between flowers. *Oikos* 62:319–324.

Johnson, L.K. & Hubbell, S.P. (1974). Aggression and competition among stingless bees: field studies. *Ecology* 55:120–127.

Johnson, R.A. (1992). Pollination and reproductive ecology of Acuña cactus, *Echinomastus erectrocentrus* var. *acunensis* (Cactaceae). *Int. J. Plant Sci.* 153:400–408.

Kato, M., Itino, T., Hotta, M. & Inoue, T. (1991). Pollination of four Sumatran *Impatiens* species by hawkmoths and bees. *Tropics* 1:59–73.

Kearns, C.A. (1992). Anthophilous fly distribution across an elevation gradient. *Amer. Midl. Nat.* 127:172–182.

Kearns, C.A. & Inouye, D.W. (1993). *Techniques for Pollination Biologists*. Niwot, CO: University Press of Colorado.

Kearns, C.A. & Inouye, D.W. (1994). Fly pollination of *Linum lewisii* (Linaceae). *Amer. J. Bot.* 81:1091–1095.

Kendall, D.A. (1973). The viability and compatibility of pollen on insects visiting apple blossom. *J. Appl. Ecol.* 10:847–853.

Kendall, D.A. & Smith, B.D. (1975). The pollinating efficiency of honeybee and bumblebee visits to field bean flowers (*Vicia faba* L.). *J. Appl. Ecol.* 12:709–717.

Kendall, D.A. & Smith, B.D. (1976). The pollinating efficiency of honeybee and bumblebee visits to flowers of the runner bean *Phaseolus coccineus* L. *J. Appl. Ecol.* 13:749–752.

Kikuchi, T. (1965). Role of interspecific dominance–subordination relationship on the appearance of flower-visiting insects. *Sci. Rep. Tôhoku Univ. Ser. IV (Biol.)* 31:275–296.

Koeniger, N. & Vorwohl, G. (1979). Competition for food among four sympatric species of Apini in Sri Lanka (*Apis dorsata, Apis cerana, Apis florea* and *Trigona iridipennis*). *J. Apic. Res.* 18:95–109.

Kraai, A. (1962). How long do honey-bees carry germinable pollen on them? *Euphytica* 11:53–56.

Kuhn, E.D. & Ambrose, J.T. (1984). Pollination of 'Delicious' apple by megachilid bees of the genus *Osmia* (Hymenoptera: Megachilidae). *J. Kansas Entomol. Soc.* 57:169–180.

Kwak, M.M., Holthuijzen, Y.A. & Prins, H.H.T. (1985). A comparison of nectar characteristics of the bumblebee-pollinated *Rhinanthus minor* and *R. serotinus*. *Oikos* 44:123–126.

Lebuhn, G. & Anderson, G.J. (1994). Anther tripping and pollen dispensing in *Berberis thunbergii*. *Amer. Midl. Nat.* 131:257–265.

Levin, D.A. (1972). Pollen exchange as a function of species proximity in *Phlox*. *Evolution* 26:251–258.

Levin, D.A. (1981). Dispersal versus gene flow in plants. *Ann. Missouri Bot. Gard.* 68:233–253.

Levin, D.A. & Berube, D.E. (1972). *Phlox* and *Colias*: the efficiency of a pollination system. *Evolution* 26:242–252.

Levin, D.A. & Kerster, H.W. (1974). Gene flow in seed plants. *Evol. Biology* 7:139–220.

Lewis, A.C. (1989). Flower visit consistency in *Pieris rapae*, the cabbage butterfly. *J. Anim. Ecol.* 58:1–13.

Loper, G.M. & DeGrandi-Hoffman, G. (1994). Does in-hive pollen transfer by honey bees contribute to cross-pollination and seed set in hybrid cotton? *Apidologie* 25:94–102.

Lukoschus, F. (1957). Quantitative Untersuchungen über den Pollen-Transport im Haarkleid der Honigbiene. *Z. Bienenforsch.* 4:3–21.

Macior, L.W. (1975a). The pollination ecology of *Delphinium tricorne* (Ranunculaceae). *Amer. J. Bot.* 62:1009-1016.

Macior, L.W. (1975b). The pollination ecology of *Pedicularis* (Scrophulariaceae) in the Yukon Territory. *Amer. J. Bot.* 62:1065–1072.

MacKenzie, K.E. (1994). The foraging behaviour of honey bees (*Apis mellifera* L.) and bumble bees (*Bombus* spp) on cranberry (*Vaccinium macrocarpon* Ait). *Apidologie* 25:375–383.

Mann, L.K. (1953). Honey bees on cantaloupes. *Amer. J. Bot.* 49:545–553.

Mathur, G. & Mohan Ram, H.Y. (1986). Floral biology and pollination of *Lantana camara*. *Phytomorph.* 36:79–100.

Maurizio, A. (1953). Weitere Untersuchungen an Pollenhöschen – Beitrag zur Erfassung der Pollen Trachtverhältnisse in verschiedenen Gegenden der Schweiz. *Beih. Schweiz. Bienen-Ztg.* 2:486–556.

McCartney, H.A. (1994). Dispersal of spores and pollen from crops. *Grana* 33:76–80.

Mohr, N.A. & Jay, S.C. (1988). Nectar- and pollen-collecting behaviour of honey-bees on canola (*Brussica campestris* L. and *Brassica napus* L.). *J. Apic. Res.* 27(2):131–136.

Morse, D.H. (1978a). Interactions among bumble bees on roses. *Insectes Sociaux* 25:365–371.

Morse, D.H. (1978b). Size-related foraging differences of bumble bee workers. *Ecol. Entomol.* 3:189–192.

Morse, D.H. (1981). Interactions among syrphid flies and bumblebees on flowers. *Ecology* 62:81–88

Morse, D.H. (1982). Foraging relationships within a guild of bumble bees. *Insectes Sociaux* 29:445–454.

Motten, A.F. (1983). Reproduction of *Erythronium umbilicatum* (Liliaceae): pollination success and pollinator effectiveness. *Oecologia* 59:351–359.

Motten, A.F., Campbell, D.R., Alexander, D.E. & Miller, H.L. (1981). Pollination effectiveness of specialist and generalist visitors to a North Carolina population of *Claytonia virginica*. *Ecology* 62:1278–1287.

Murcia, C. (1990). Effect of floral morphology and temperature on pollen receipt and removal in *Ipomoea trichocarpa*. *Ecology* 71:1098–1109.

Murphy, D.D. (1984). Butterflies and their nectar plants: the role of the checkerspot butterfly *Euphydryas editha* as a pollen vector. *Oikos* 43:113–117.

Nadel, H. & Peña, J.E. (1994). Identity, behavior, and efficacy of nitidulid beetles (Coleoptera: Nitidulidae) pollinating commercial *Annona* species in Florida. *Environ. Entomol.* 23:878–886.

Neff, J.L. & Simpson, B.B. (1990). The roles of phenology and reward structure in the pollination biology of wild sunflower (*Helianthus annuus* L., Asteraceae). *Israel J. Bot.* 39:197–216.

Niklas, K.J. (1985). The aerodynamics of wind pollination. *Bot. Rev.* 51:328–286.

Ohsawa, R. & Namai, H. (1988). Cross-pollination efficiency of insect pollinators in rapeseed, *Brassica napus* L. *Japan. J. Breed.* 38:91–102.

Ordway, E., Buchmann, S.L., Kuehl, R.O. & Shipman, C.W. (1987). Pollen dispersal in *Cucurbita foetidissima* (Cucurbitaceae) by bees of the genera *Apis*, *Peponapis* and *Xenoglossa* (Hymenoptera: Apidae, Anthophoridae). *J. Kansas Entomol. Soc.* 60:489–503.

Ornduff, R. (1971). The reproductive system of *Jepsonia heterandra*. *Evolution* 25:300–311.

Osborne, J.L. & Corbet, S.A. (1993). Evaluating pollinator effectiveness on crops: a step by step analysis. In *Pollination in Tropics,* eds. G.K. Veeresh, R. Uma Shaanker, and K.N. Ganeshaiah, pp. 179–183. Proceedings of the International Symposium on Pollination in Tropics, August 8–13, 1993, Bangalore, India.

Parker, F.D. (1981). How efficient are bees in pollinating sunflowers? *J. Kansas Entomol. Soc.* 54:61–67.

Parker, F.D. & Hatley, C.L. (1978). Onion pollination: viability of onion pollen and pollen diversity on insect body hairs. *Proc. IVth Int. Symp. Pollination*, Maryland Agricultural Experimental Station Special Miscellaneous Publication 1:201–206.

Parker, R.L. (1926). The collection and utilization of pollen by the honeybee. *Memoir, Cornell University Agricultural Experimental Station. Vol. 98.*

Paton, D.C. & Turner, V. (1985). Pollination of *Banksia ericifolia* Smith: birds, mammals and insects as pollen vectors. *Aust. J. Bot.* 33:271–286.

Peakall, R., Beattie, A.J. & James, S.H. (1987). Pseudocopulation of an orchid by male ants: a test of two hypotheses accounting for the rarity of ant pollination. *Oecologia* 73:522–524.

Pellmyr, O. (1987). Temporal patterns of ovule allocation, fruit set, and seed predation in *Anemonopsis macrophylla* (Ranunculaceae). *Bot. Mag. (Tokyo)* 100:175–183.

Percival, M.P. (1950). Pollen presentation and pollen collection. *New Phytol.* 49:40–63.

Poulsen, M.H. (1973). The frequency and foraging behaviour of honeybees and bumble bees on field beans in Denmark. *J. Apic. Res.* 12:75–80.

Primack, R.B. & Silander, J.A. (1975). Measuring the relative importance of different pollinators to plants. *Nature* 225:143–144.

Ramsay, M.W. (1988). Differences in pollinator effectiveness of birds and insects visiting *Banksia menziesii* (Proteaceae). *Oecologia* 76:119–124.

Ranta, E. & Lundberg, H. (1980). Resource partitioning in bumblebees: the significance of differences in proboscis length. *Oikos* 35:298–302.

Raybould, A.F. & Gray, A.J. (1993). Genetically modified crops and hybridization with wild relatives: a UK perspective. *J. Appl. Ecol.* 30:199–219.

Reader, R.J. (1977). Bog ericad flowers: self-compatibility and relative attractiveness to bees. *Can. J. Bot.* 55:2279–2287.

Richards, A.J. (1986). *Plant Breeding Systems.* London: George Allen & Unwin.

Richards, K.W. (1987). Diversity, density, efficiency, and effectiveness of pollinators of cicer milkvetch, *Astragalus cicer* L. *Can. J. Zool.* 65:2168–76.

Richards, K.W. & Edwards, P.D. (1988). Density, diversity, and efficiency of pollinators of sainfoin, *Onobrychis viciaefolia* Scop. *Can. Entomol.* 120:1085–1100.

Richardson, T.E. & Stephenson, A.G. (1991). Effects of parentage, prior fruit set and pollen load on fruit and seed production in *Campanula americana* L. *Oecologia* 87:80–85.

Robinson, W.S. (1979). Effect of apple cultivar on foraging behavior and pollen transfer by honey bees. *J. Amer. Soc. Hort. Sci.* 104:596–598.

Robinson, W.S. & Fell, R.D. (1981). Effect of honey bee foraging behaviors on 'Delicious' apple set. *HortScience* 16:326–328.

Roubik, D.W. (1980). Foraging behavior of competing Africanized honey bees and stingless bees. *Ecology* 61:836–845.

Roumet, P. & Magnier, I. (1993). Estimation of hybrid seed production and efficient pollen flow using insect pollination of male sterile soybeans in caged plots. *Euphytica* 70:61–67.

Scheffler, J.A., Parkinson, R. & Dale, P.J. (1993). Frequency and distance of pollen dispersal from transgenic oilseed rape (*Brassica napus*). *Transgenic Res.* 2:356–364.

Schemske, D.W. & Horvitz, C.C. (1988). Plant–animal interactions and fruit production in a neotropical herb: a path analysis. *Ecology* 69:1128–1137.

Schmitt, J. (1983). Flowering plant density and pollinator visitation in *Senecio*. *Oecologia* 60:97–102.

Silander, J.A. & Primack, R.B. (1978). Pollination intensity and seed set in the evening primrose (*Oenothera fruticosa*). *Amer. Midl. Nat.* 100:213–216.

Simpson, B.B. & Neff, J.L. (1981). Floral rewards: alternatives to pollen and nectar. *Ann. Miss. Bot. Gard.* 68:301–322.

Simpson, B.B. & Neff, J.L. (1983). Evolution and diversity of floral rewards. In *Handbook of Experimental Pollination Biology,* eds. C.E. Jones and R.J. Little, pp. 142–159. New York: Van Nostrand Reinhold.

Snow, A.A. & Roubik, D.W. (1987). Pollen deposition and removal by bees visiting two tree species in Panamá. *Biotropica* 19:57–63.

Spears, E.E. (1983). A direct measure of pollinator effectiveness. *Oecologia* 57:196–199.

Spencer-Booth, Y. (1965). The collection of pollen by bumble bees, and its transport in the corbiculae and the proboscidial fossa. *J. Apic. Res.* 4:185–190.

Stanley, R.G. & Linskens, H.F. (1974). *Pollen: Biology, Biochemistry, Management.* Berlin: Springer-Verlag.

Stanton, M.L. & Preston, R.E. (1986). Pollen allocation in wild radish: variation in pollen grain size and number. In *Biotechnology and Ecology of Pollen*, eds. D.L. Mulcahy, G. Bergamini Mulcahy, and E. Ottaviano, pp. 461–466. New York: Springer-Verlag.

Stephenson, A.G. (1981). Flowers and fruit abortion. *Ann. Rev. Ecol. Syst.* 12: 253–279.

Strickler, K. (1979). Specialization and foraging efficiency of solitary bees. *Ecology* 60:998–1009.

Synge, A.D. (1947). Pollen collection by honey bees. *J. Anim. Ecol.* 16:122–138.

Tepedino, V.J. (1981). The pollination efficiency of the squash bee (*Peponapis pruinosa*) and the honey bee (*Apis mellifera*) on summer squash. *J. Kansas Entomol. Soc.* 54:359–377.

Thomson, J.D. (1986). Pollen transport and deposition by bumble bees in *Erythronium*: influences of floral nectar and bee grooming. *J. Ecol.* 74:329–341.

Thomson, J.D. (1989). Reversal of apparent preferences of bumble bees by aggression from *Vespula* wasps. *Can. J. Zool.* 67:2588–2591.

Thomson, J.D. & Plowright, R.C. (1980). Pollen carryover, nectar rewards, and pollinator behavior with special reference to *Diervilla lonicera*. *Oecologia* 46:68–74.

Thomson, J.D. & Stratton, D.A. (1985). Floral morphology and cross-pollination in *Erythronium grandiflorum* (Liliaceae). *Amer. J. Bot.* 72:433–437.

Thomson, J.D. & Thomson, B.A. (1989). Dispersal of *Erythronium grandiflorum* pollen by bumblebees: implications for gene flow and reproductive success. *Evolution* 43:657–661.

Thomson, J.D. & Thomson, B.A. (1992). Pollen presentation and viability schedules and their consequences for reproductive success through animal pollination. In *Ecology and Evolution of Plant Reproduction: New Approaches*, ed. R. Wyatt, pp. 1–24. New York: Chapman and Hall.

Thomson, J.D., Price, M.V., Waser, N.M. & Stratton, D.W. (1986). Comparative studies of pollen and fluorescent dye transport by bumble bees visiting *Erythronium grandiflorum. Oecologia* 69:561–566.

Vaissière, B.E. & Vinson, S.B. (1994). Pollen morphology and its effect on pollen collection by honey bees, *Apis mellifera* L. (Hymenoptera: Apidae), with special reference to upland cotton, *Gossypium hirsutum* L. (Malvaceae). *Grana* 33:128–138.

Waddington, K.D. (1981). Factors influencing pollen flow in bumblebee-pollinated *Delphinium virescens. Oikos* 37:153–159.

Waller, G.D. & Moffett, J.O. (1981). Pollination of hybrid cotton on the Texas High Plains: nectar production and honey bee visits. *Proceedings of the Beltwide Cotton Production Research Conference, January 4–8, 1981, New Orleans, Louisiana.*

Waser (1986). Flower constancy: definition, cause and measurement. *Amer. Nat.* 127:593–603.

Waser, N.M. (1988). Comparative pollen and dye transfer by pollinators of *Delphinium nelsonii. Funct. Ecol.* 2:41–48.

Waser, N.M. & Price, M.V. (1990). Pollination efficiency and effectiveness of bumble bees and hummingbirds visiting *Delphinium nelsonii. Collectanea Botanica (Barcelona)* 19:9–20.

Webb, C.J. & Bawa, K.S. (1983). Pollen dispersal by hummingbirds and butterflies: a comparative study of two lowland tropical plants. *Evolution* 37:1258–1270.

Westerbergh, A. & Saura, A. (1994). Gene flow and pollinator behaviour in *Silene dioica* populations. *Oikos* 71:215–224.

Wiebes, J.T. (1979). Co-evolution of figs and their insect pollinators. *Ann. Rev. Ecol. System.* 10:1–12.

Willson, M.F. & Burley, N. (1983). *Mate Choice in Plants: Tactics, Mechanisms, and Consequences.* Princeton, NJ: Princeton University Press.

Wilson, P. & Thomson, J.D. (1991). Heterogeneity among floral visitors leads to discordance between removal and deposition of pollen. *Ecology* 72:1503–1507.

Wodehouse, R.P. (1935). *Pollen Grains: Their Structure, Identification and Significance in Science and Medicine.* New York: McGraw-Hill.

Wolfe, L.M. & Barrett, S.C.H. (1989). Patterns of pollen removal and deposition in tristylous *Pontederia cordata* L. (Pontederiaceae). *Biol. J. Linn. Soc.* 36:317–329.

Wolin, C.L., Galen, C. & Watkins, L. (1984). The breeding system and aspects of pollination effectiveness in *Oenothera speciosa* (Onagraceae). *Southwest. Nat.* 29:15–20.

Young, H.J. (1988). Differential importance of beetle species pollinating *Dieffenbachia longispatha* (Araceae). *Ecology* 69:832–844.

Young, H.J. & Stanton, M.L. (1990). Influences of floral variation on pollen removal and seed production in wild radish. *Ecology* 71:536–547.

Zietsman, P.C. (1990). Pollination of *Ziziphys mucronata* subsp. *mucronata* (Rhamnaceae). *S. Afr. Tydskr. Plantk.* 56:350–355.

Zietsman, P.C. (1991). Reproductive biology of *Grewia occidentalis* L. (Tiliaceae). *S. Afr. J. Bot.* 57:348–351.

Zimmerman, M. & Cook, S. (1985). Pollinator foraging, experimental nectar-robbing and plant fitness in *Impatiens capensis. Amer. Midl. Nat.* 113:84–91.

6

Pollination constraints and management of pollinating insects for crop production

R. W. CURRIE

Summary   121
Introduction   121
Increasing pollinator visitation rates   122
      Enhancing populations of native pollinators   122
      Importing commercially managed pollinators   123
      Maintaining pollinator populations on the target crops   125
      Spatial distribution of foragers on the crop   131
Resource limitation and managing overpollination   132
Increasing rate and quality of pollen transfer   133
      Agronomic practices   134
      Pollinator food preference   135
      Within-colony pollen transfer   137
Input costs   138
      Costs of increasing floral visitation rates   138
      Costs of increasing the rate and quality of pollen transfer   139
Conclusions   139
References   141

Summary

Enhancing insect pollination makes a significant contribution to the value of many commercially important crops by decreasing the time to crop maturity and by increasing crop uniformity, quantity, and quality. There are still many crops, however, that suffer yield losses in situations where adequate pollinator populations cannot be maintained. Plant and pollinator components of cropping systems can be managed to enhance pollinator visitation, increase the rate and quality of pollen transfer, or both. This chapter discusses the efficiency, costs, and benefits of different management strategies that have been developed to maintain and enhance pollinator populations on crops and to increase the rate and quality of pollen transfer.

Introduction

Insect pollinators are a crucial component in the production of many commercially important crops. Enhancing the amount or quality of insect pollination can lead to increases in crop value by decreasing the time to crop maturity and by increasing crop uniformity, quantity, and quality. The diverse group of insect species that play a role in crop pollination range from those that are incidental flower visitors to those that are superbly adapted for pollinating flowers on which they forage exclusively for their food.

Each species of insect has specific morphological, physiological, behavioral, and life history constraints that affect its abundance, effectiveness, and distribution on pollinated crops. These constraints have a direct bearing on the ways in which insects can be most effectively and economically manipulated for the purpose of enhancing crop pollination. Those species most suited to crop pollination forage under a wide range of environmental conditions, feed exclusively on flowers for pollen and nectar, visit flowers of the same species in succession, move frequently between flowers, and have morphological adaptations that facilitate pollen pick-up and its transfer to the flower's stigmatic surface (Free 1993).

Problems in crop pollination arise when pollinator and plant are not perfectly matched. This happens frequently because the few species of pollinators that are easily managed must be utilized on a diverse range of crops. There are two basic strategies to improve pollination: increase insect visitation rate or increase the rate and quality of pollen transfer to flowers. Each of these strategies must be considered within the context of their relative impacts on yield and input costs. This chapter discusses the range of management strategies that can be utilized to enhance and maintain populations of insect pollinators on agricultural crops and to enhance their rate and quality of pollen transfer.

Increasing pollinator visitation rates

When insect-pollinated crops are pollen limited, increasing the pollinator's visitation rate increases seed or fruit set up to the point when plant resources become limiting (see Figure 6.1A, B) (Plowright and Hartling 1981; Gori 1983). Beyond this point, there will be no further positive impact on crop yield. Management to increase pollinator visitation is warranted when crops are pollen limited and the costs associated with increasing pollinator visitation rates fall below the associated increase in crop value. The degree of impact that the increases in pollinator visitation rates will have on yield depends on the crop, its degree of self-incompatibility, and the need for outcrossing.

Pollinator visitation rates can be increased by either enhancing local populations of native pollinators or supplementing existing populations with commercially managed species.

Enhancing populations of native pollinators

One approach to increasing overall pollinator visitation on pollen-limited crops is to enhance local populations of native species through habitat management. There is considerable regional, annual, and seasonal variation in population abundance of the diverse pollinator guilds foraging on crops that reflect inherent differences in their morphology, sociality, foraging preference, number of generations per year, and adult phenologies (Cane and Payne 1993). Habitat management strategies are likely to be more effective if directed toward the members of a pollinator guild that are the most efficient but that also have low annual variation in abundance on the target crop.

Habitat management can be used to increase native pollinator populations when nest site availability is a limiting factor. For example, the practice of maintaining uncultivated strips along field margins or of providing artificial nesting sites increases populations of native *Megachile* spp. that forage on alfalfa, *Medicago sativa* L. (Peck and Bolten 1946). Provision of permanent nest boxes increases populations of bumblebees, *Bombus* spp., that forage on red clover, *Trifolium pratense* L. (Donovan and Wier 1978; Pomeroy 1981; Macfarlane et al. 1983).

The guilds of pollinators associated with a particular crop require diverse habitats to "fit" their behavioral and life history constraints. In addition to having nesting sites provided, a succession of forage sources must be available before and after the target crop blooms to maintain populations of long-lived solitary and social species. Pollinators must also be provided with a supply of appropriate nesting materials, and sites for overwintering and mating (Osborne et al. 1991). In some cases, it may be desirable to alter cropping practices to more closely suit the characteristics of the pollinator. Timing of planting dates, so that flowering periods of red clover coincide with peak populations of bumblebees for example, increases seed yield in red clover (Swart 1960, cited in Free 1993). Reintroducing native species without taking steps to improve habitat provides only a short-term solution. Hobbs (1967) showed that local populations of bumblebees can be increased by capturing queens in artificial nests and moving them to a target crop. Such introductions can triple local populations on red clover. However, without further management, populations must be reestablished each year (Clifford 1973).

To date, extensive habitat management programs have not been very successful. Torchio (1990) points out that in areas with intensive agronomic practices, high land values, extensive pesticide usage, and crops that are often grown on short rotation schedules in combination with nonflowering crops, the costs of implementing a program of habitat management may exceed the benefits derived by individual growers. Even where populations of native pollinators are increased through habitat management, many crops may still be pollen limited and require annual supplementation. For most crops that benefit from insect pollination, commercially managed pollinators (usually bees) are moved to crops to increase floral visitation (Torchio 1987; Free 1993; Richards 1993).

Importing commercially managed pollinators

For the purpose of commercial crop pollination, we currently can effectively and economically manage a number of social species of bees, including the honeybee, *Apis mellifera* L., and several species of bumblebees, *Bombus* spp. The honeybee's large foraging populations, transportability, year-round availability, and broad dietary preference at the colony level, coupled with floral constancy of individual foragers, make it the pollinator of choice for a wide range of crops. Honeybees are suitable for use in diverse cropping systems and are commonly used to pollinate oilseeds, legumes, fruits, vegetables, and nuts (McGregor 1976). Bumblebees have been commercially viable in applications where their large size, long tongue-length, ability to vibrate flowers requiring "buzz-pollination," and ability to fly at relatively low temperatures or ability to fly in enclosed spaces, make them more efficient pollinators than honeybees. To date, in part because of the technical difficulties and high costs associated with bumblebee management, use of bumblebees has been restricted chiefly to use in high-value greenhouse crops (Banda and Paxton 1991; Straver and Plowright 1991) or in crops, such as red clover, where their populations can be enhanced with little difficulty (Holm 1966; Macfarlane et al. 1983).

Several solitary species of bees have also been exploited successfully for crop pollination, including the alkali bee, *Nomia melanderi* Ckll., leaf-cutter bee, *Megachile rotundata* F., and orchard bee, *Osmia* spp. These species all live in close proximity to one another and reproduce in artificial nesting sites. These two aspects facilitate the manipulation and management of their populations. Alkali and leaf-cutter bees have both proven extremely valuable in commercial alfalfa seed pollination (Richards 1993; Torchio 1987). Leaf-cutter–bee management systems for alfalfa can also be easily adapted to increase seed production in other crops, such as clover, *Trifolium* spp.; polish rape, *Brassica campestris* (Fairey and Lefkovitch 1991); and hybrid canola, *Brassica napus* (Currie, unpublished data). The commercial use of the orchard bee has been more limited, but there is good potential to further develop these bees for use in pollinating a number of fruit and nut crops (Torchio, 1987; Yoshida and Maeto 1988).

Before bees are moved to a crop, they should be managed to maximize the number of foragers that are available when the target crop comes into bloom (Free 1993). Techniques for population management and for control of diseases, parasites, and predators are well established for a number of solitary and social insect pollinators and will not be covered here (see reviews by Dadant 1978; Torchio 1987; Richards 1989, 1993; Heemert et al. 1990; Free 1993).

In order to make meaningful pollination management decisions, it is essential to estimate accurately the number of bees to be imported, but there is often difficulty in determining the number of bees that will forage on the crop. With some solitary species, such as the leaf-cutter bee, the number of bees can be estimated precisely and the bees can be incubated and released into nests placed within the crop when required (Richards 1989). Because these pollinators do not fly far from their nests, the number and distribution of pollinators on the crop can be easily defined and controlled (Richards 1989,

1993). With social species, however, there can be considerable variation in the state of the colony contained within the hive that is moved to the crop. Hives or "pollination units" contain colonies of bumblebee and honeybee, *Apis mellifera* L., which vary with respect to the number of foragers, area of brood, amount and type of food stores, and presence of a queen. All of these factors influence the colony's foraging potential (S. C. Jay 1986; Free 1993). Thus when colonies for pollination are rented, it is often recommended that growers specify a minimum standard for a colony's worker population, and other attributes, in a written contractual agreement (for example, see Dadant 1978).

Recommendations for the populations of bumble- or honeybees required to pollinate a given area are available for a variety of crops (Holm 1966; McGregor 1976; DeGrandi-Hoffman 1987; Macfarlane et al. 1991; Straver and Plowright 1991; Free 1993). Where colony size is defined, desired levels of pollinator visits can be obtained by moving the appropriate number of hives to the target crop. However, worker foraging may not be directly proportional to the number of hives introduced, because species with large foraging areas are often attracted away from the target crop to competing sources of bloom (S. C. Jay 1986; Berger et al. 1988).

Maintaining pollinator populations on the target crops

Pollination of the target crop can be reduced if neighboring crops, native plants, or weeds provide greater nectar or pollen rewards for pollinators (S. C. Jay 1986; Free 1993). The negative impacts of floral competition on pollinator abundance on a target crop can be reduced by the following methods: altering the timing when bees are moved to the crop; replacing or rotating colonies on the crop; spatially or temporally isolating the crop; enhancing crop attractiveness through crop breeding and agronomic practices; using attractant sprays; training bees to crop odors; and reducing the attractiveness of competing plants. Each of these management techniques will be reviewed.

Timing of pollinator introduction

Floral visits are usually maximized when managed pollinators are moved to crops after the initiation of flowering (S. C. Jay 1986). Importing bees too late in the bloom period can be detrimental because flowers produced early in the season are not pollinated, which can reduce yields (Woodrow 1952; Shimanuki et al. 1967; Howell et al. 1972). Importing bees too early can cause other problems. For example, sources of nectar and pollen are required to support populations of species with limited flight ranges, such as the leaf-cutter bee, and to prevent their emigration from the field (Richards 1989). Their visitation rate is maximized when incubation is timed carefully, so that adult emergence is synchronized with floral density (Richards 1982). In contrast, emigration from the field is not a problem with bumble- and honeybees, as these social species are capable of surviving on nonblooming crops for extended periods. They maintain a store of food and, with larger flight ranges, are capable of exploiting alternative sources of forage beyond the confines of the crop border. Nevertheless, timing of colony introduction does have a major effect on the proportion of bees foraging on the target crop. Floral visitation rates are higher when bees are moved to crops after the initiation of bloom, because bees moved to crops before bloom quickly become trained to alternative forage sources (Karmo 1958; Free et al. 1960). The exact timing of colony placement is not strictly controlled by crop phenology because environmental conditions can alter bee foraging patterns and delay the "optimal" time to introduce colonies and maximize fruit set (Moeller 1973).

Colony replacement and rotation

In crops that bloom over long periods, sequential introduction of pollinators can be used to keep pace with expanding bloom (S. C. Jay 1986) or to compensate for bee losses (Currie et al. 1990). When honeybee colonies are moved to a new location, workers tend to visit the closest crop available and slowly radiate outward from their colony to forage (Free and Smith 1961; Lee 1961; M. D. Levin 1960). Thus, even if competing crops are nearby, many workers will be retained on the target crop for a few days until they discover more attractive sources of forage. Foragers can be maintained on crops by replacing colonies or sequentially moving in additional colonies throughout the bloom period, even though the bees from colonies introduced early in the bloom period have switched to competing crops (Al-Tikrity et al. 1972b). This technique has variable success, however, when honeybee colonies are sequentially increased on alfalfa fields. The number of nectar foragers increases but foragers trip fewer flowers (Palmer-Jones and Forster 1972). In this case the practice is not economically feasible, especially once the high costs of colony rental and transport are considered.

Spatial and temporal isolation

Spatial isolation can be used to maintain foragers on the target crop by increasing the distance to competing crops to the point where it is beyond the economical flight range of the pollinator. Isolating crops pollinated by species such as the leaf-cutter bee, which typically forages within several hundred meters of its nest, is much easier than isolating crops pollinated by honey- and bumblebees, which forage several kilometers from their colony (Eckert 1933; Gary et al. 1972; Heinrich 1983; Free 1993). Increasing the isolation distance from other crops can increase bee foraging but can also alter the type of forage collected and the quality of pollinator visits. Honeybees are ineffective pollinators of alfalfa in many geographical regions due to their tendency to avoid pollen collection and to "thieve" nectar from flowers. However, bees do trip flowers regularly in dry areas where alfalfa is grown under irrigation and little floral competition exists (Vansell and Todd 1946; McMahon 1954).

Temporal isolation may be effective in areas where, due to the acreage grown, spatial isolation is difficult to achieve. Plants that are highly attractive to pollinators can be grown in close proximity if overlap in bloom is avoided (Vansell 1942). In regions with long growing seasons or in crops with short flowering periods, it may be feasible to avoid overlap in bloom by altering planting dates to avoid competition with neighboring crops. In orchards where dandelion and white clover often attract pollinators away from the target crop (Free 1968a; Ford 1971; D. Jay and Jay 1984), spatial isolation to reduce competition can be achieved through mowing or herbicide applications (Karmo and Vickery 1954; Free 1968a; Ford 1971). Competing crops should not be removed from orchards, however, if they retain nectar-collecting bees that subsequently switch to foraging for pollen on the target crop (S. C. Jay 1986).

Plant breeding and agronomic practices

Polytrophic bees tend to forage on the most attractive source available. The attractiveness of crops to pollinators can be increased by long-term selective breeding programs (Shuel 1989; Free 1993). In cases where pollinators are poorly adapted to crops, it may be feasible for plant breeders to select for lines that contain one or more attractive characteristics. Most insect pollinators forage on flowers to collect nectar, pollen, or both. Thus, selection of plants to enhance the quantity, quality, or accessibility of nectar or pollen rewards should improve the ability of the plant to attract pollinators (Pedersen 1953; Murrell et al. 1982; S. C. Jay 1986; Davis and Gunning 1991). If rewards produced by flowers are too high, however, the number of flowers or plants visited by each pollinator may decrease, reducing the amount of cross-pollination.

Pollinators preferentially visit plants based on nonrewarding characters such as floral color, the numbers of flowers per inflorescence, and odor (Kevan 1978; Schemske 1980; Stanton et al. 1989; Dobson 1994). Thus breeding to enhance visual or olfactory cues may also be effective. Attractiveness of alfalfa to honeybees depends, in part, upon the relative amounts of attractive and repellent components in volatile floral odors. Selection of plants to optimize quantities of these compounds may also lead to increased pollinator visitation (Loper and Waller 1970; Henning et al. 1992).

Nectar production is influenced by external factors affecting photosynthesis and plant growth. Thus, crop attractiveness can be increased by implementing cultural practices to alter one or more aspects of soil moisture or fertility (Shuel 1967). Soil moisture can be regulated by irrigation. Watering increases nectar production in crops such as cotton, *Gossypium* spp. (Kaziev 1959), but attempts to use irrigation to regulate nectar secretion have had mixed success. Experiments on snapdragons (*Antirrhinum majus*) indicate that deviations from optimal soil moisture levels in either direction can lead to reductions in nectar yield (Shuel and Shivas 1953). Irrigation can also negatively affect bee visitation. Applications of water during evening periods significantly reduced nectar production in hybrid canola, *Brassica napus* L., fields, resulting in a 1.7-fold reduction in the abundance of foraging bees (Currie and Jay, unpublished data). Bee activity was reduced for up to a day after irrigation occurred.

Nectar production is enhanced through fertilization where soils are deficient in one or more nutrients or trace elements. In addition to enhancing floral nectar secretion, fertilizer applications can increase the attractiveness to pollinators by increasing the number of flowers per plant (Holdaway and Burson 1960). Fertilizing with nitrogen, phosphorous, potassium, and various trace elements such as boron, alone or in combination, increased nectar secretion and flower production (Shuel 1957; Kaziev 1967). In some cases, higher nectar production resulting from fertilizer treatment increases bee visitation and crop yields (Bobrzecka and Bobrzecki 1973; Novotna 1973). The effects of nutrients on plant growth and nectar secretion are interdependent and cannot be studied in isolation (Ryle 1954; Shuel 1957). In general, combinations of soil conditions that place no restrictions on growth but promote a balance between vegetative and reproductive growth are most likely to produce optimal conditions for nectar production (Shuel 1967). The approach of influencing nectar production through breeding or manipulation of soil fertility may have limited application, since climatic factors often have an overriding influence on nectar production (Shuel 1967; Free 1993).

Attractant sprays

Various substances have been tested as attractant sprays, some of which show promise (Waller 1970; van Praagh and von der Ohe 1983; S. C. Jay 1986; Mayer et al. 1989; Currie et al. 1992a,b). Although some attractants increase honeybee abundance on a crop, they do not always increase crop yield. Sugar syrup sprays (Roberts 1956; Free 1965a) and food supplement sprays (Burgett and Fisher 1979; Brakefield 1980; Belletti and Zani 1981; Margalith et al. 1984) often increase the abundance of foragers but rarely increase crop yields, because the foragers tend to collect the sugar syrup rather than visit the flowers for nectar or pollen (Free 1965a; Zyl and Strydom 1968).

Pheromones – chemicals used for communication between members of the same species – can increase activity of bees foraging on flowers when applied to crops as dilute sprays. Most of the research on directing pollinators to crops has focused on the honeybee because of its widespread use as a pollinator on many crops and our knowledge of its pheromonal systems (Free 1987). Because these chemical messages are species specific, pheromone sprays directed at one species do not enhance activity of alternative pollinators.

The worker honeybee's Nasonov gland produces pheromones used to attract other workers at the hive entrance, also during swarm clustering, when collecting water, and possibly when foraging at flowers (Free 1968b; Winston 1987). Sprays containing the Nasonov-gland components, citral and geraniol, increase honeybee foraging activity (Waller 1970; Woyke 1981), fruit set (Mayer et al. 1989) and yield (van Praagh and von der Ohe 1983). However, many attempts to use Nasonov-pheromone sprays have failed, and their economic benefit in commercial crop production has not been demonstrated (Free 1979a; Winston and Slessor 1993).

Queen honeybees produce a five-component pheromone blend secreted from their mandibular glands. This blend promotes retinue formation, attracts workers to swarm clusters, and suppresses queen rearing and swarming (Winston and Slessor 1993). When synthetic queen mandibular pheromone is sprayed on crops, it increases honeybee foraging under a wide range of environmental conditions, crop management systems, and geographical locations. The increased bee visitation can significantly increase crop quality, yield, or both (Currie et al. 1992a, b; Winston and Slessor 1993). Since the sprays are ineffective in attracting foragers after 18 hours (Currie et al. 1992a, b), they should be applied to crops in the early morning, with consideration given to environmental conditions suitable for bee flight activity, the availability of compatible pollen, and the period of optimal blossom quality (Wauchope 1968; Brain and Landsberg 1981; DeGrandi-Hoffman et al. 1987). Stickers are ineffective at increasing the longevity of the pheromone (Naumann 1993). Repeated applications would be required to maintain high levels of bee foraging on crops that bloom for extended periods.

Interestingly, the relative attractiveness of pheromone sprays is both dose and crop dependent (Currie et al. 1992a, b). Doses of pheromone that are too low or too high do not attract foraging workers (Currie et al. 1992a). The effective dose also varies with crop architecture. On a three-dimensional surface such as a tree or bush, the pheromone is more widely dispersed than on a more two-dimensional crop surface, such as a cranberry bog. Thus applications in orchards require a higher effective dose per unit of land area (Currie et al. 1992a, b). These factors should be carefully considered when assessing other attractant sprays.

It is important to note that although it is economically effective to apply queen mandibular pheromone sprays on some commercial crops, it is not effective in increasing bee activity, crop yield, or both, in all years or in all situations (Currie et al. 1992a, b; Winston and Slessor 1993). Applications of pheromone appear to be most effective at increasing honeybee foraging when bee numbers are low, either in crops relatively unattractive to bees or during poor weather conditions where pollinator activity falls below the threshold required for adequate crop pollination. Although application of pheromone attracts more honeybees to the crop, this increase in foraging activity results in increased yield only when crops are pollen limited (Figure 6.1A).

In addition to using pheromone sprays, it may be feasible to attract other pollinators using synthetic plant volatiles isolated from nectar, pollen, or foliage (Hopkins et al. 1969). A wide range of compounds are attractive to pollinators (reviewed by Dobson 1994). Sprays using combinations of synthetic plant volatile components might conceivably be directed at more than one species of pollinator, although no single compound has been identified that is attractive to a wide range of species (Dobson 1994).

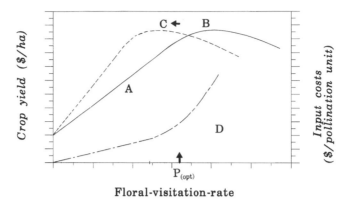

Figure 6.1 Model of the relationship between crop production, pollinator visitation, and pollination inputs. (A) Pollinator visitation rate where pollen limitation occurs (solid line). (B) Pollinator visitation rate above which plants are resource limited (solid line). (C) Impact on threshold curve of management practices to enhance pollen transfer (dashed line -----). (D) Input costs associated with increasing pollinator visitation rates (long-dashed line — — —). $P_{(opt)}$ indicates the optimal pollinator visitation rate required to maximize returns from management to increase visitation rate.

Plants offering limited or no rewards often "deceive" pollinators and attract them to their flowers by mimicking the pollinator's sex pheromones or aspects of an attractive flower's morphology, odor, or both (Faegri and van der Pijl 1979; Buchmann 1983; Dafni 1984; Barth 1985; Armstrong and Drummond 1990). This concept could be exploited in hybrid seed production cropping systems by developing sprays composed of pollen volatiles to attract pollen-foraging bees to male-sterile lines that contain no pollen or pollen odor.

Crop odors

Social insect pollinators, such as honeybees and bumblebees, can be directed to forage on target crops by feeding colonies with sugar or protein supplements laced with floral odors (von Rhein 1957; von Frisch 1967; Free 1969; Free 1968b; Free 1970). Increased pollinator visitation has been achieved by feeding colonies of honeybees from containers of sugar syrup covered with flowers of the target crop (Glushkov 1958; von Frisch 1967) or by feeding sugar syrup containing distillates of floral odors (Homann 1970; Johnson and Wenner 1970). In some crops, the increased recruitment of foragers increases crop yield (Skrebtsova 1957; Pritsch 1959; Kurennoi and Barabash 1966; Homann 1970), although results using this technique have been inconsistent and the impact on crop production has not been established (Free 1958; Free 1965b). Workers also become somewhat conditioned to their food sources and, when moved to a new site, tend to visit the same type of floral sources if available (Free 1959; M. D. Levin and Bohart 1959). Thus, rotating colonies between two locations several kilometers apart may help to maintain bees on crops where bees are moved from areas with abundant bloom to smaller plots near competing forage sources (M. D. Levin and Bohart 1959; S. C. Jay 1986). Further research is required before techniques to condition foragers to specific crops can be recommended to commercial growers (Free 1965b; Free 1993).

Reducing the attractiveness of competing crops

The relative attractiveness of competing crops can be reduced by flooding an area with large numbers of pollinators. Pollinators are unable to assess a plant's capacity to produce nectar but they do assess nectar availability (Heinrich 1976, 1979). When pollinators forage on the most attractive population of flowers in an area, they change the reward spectrum for other pollinators (Heinrich 1976). Thus, as the pollinator population increases, the available foraging niches "fill up" and, over time, the available nectar rewards of all the plants in an area tend to become equivalent (Heinrich 1976). In cropping systems, increasing pollinator abundance on more attractive competing crops decreases the "value" of their nectar and pollen rewards relative to the target crop. Introducing large numbers of honeybee colonies near a target crop has been used successfully to ameliorate the impact of floral competition (Thomas 1951; Free 1993).

It has been suggested that repellent compounds could be sprayed on competing crops to reduce their attractiveness to bees (Woodrow et al. 1965; S.C Jay 1986). Many compounds have been identified that could be used to reduce pollinator visitation when sprayed on competing crops (Woodrow et al. 1965; Atkins et al. 1975; Dobson 1994; Naumann et al. 1994). However, the use of repellent sprays to prevent floral competition is likely to be limited only to specialized situations involving highly valuable crops, such as breeder seed plots, because application of repellent sprays to other flowering crops might reduce their yields, permission to spray would have to be obtained from neighboring land owners, and costs in most situations would be prohibitive.

Spatial distribution of foragers on the crop

The optimal placement of hives in relation to crops varies with the species of bee and depends in part on the species' respective foraging areas. The abundance of pollinators decreases with increased distance from the hive or nesting site, but in order to maximize yields, it is important to maintain high visitation rates throughout the entire crop (Ribbands 1951; Torchio 1987; S. C. Jay 1986; Richards 1993). Honeybee colonies should be placed as close as possible to the target crop because bee visitation, amount of pollen collected, and crop yields all decrease with increasing distance to the crop (MacVicar et al. 1952; Braun et al. 1953; Peterson et al. 1960; Hammer 1961; Free and Williams 1974; Fries and Stark 1983; Alpatov 1984). Honeybees forage over large areas and their placement along the crop margin may be sufficient to achieve an even distribution of foragers in some crops (Phillips and Simpson 1989; see also Figure 6.2B). However, it is often desirable, and in some cases necessary, to distribute hives evenly within fields. In periods of inclement weather, fewer honeybees forage and foraging distances are shorter (Singh 1950; Free and Williams 1974; D. Jay and Jay 1984), so that more even pollination is obtained when bees are distributed throughout the crop. The foraging distance from honeybee colonies in some crops is limited enough to obtain significant benefit from distributing colonies evenly, in small groups of three to five, throughout the crop (Hutson 1925; Free 1962; Free and Williams 1974; D. Jay and Jay 1984). Spreading colonies singly throughout these crops may not increase yield enough to justify the higher cost (S. C. Jay 1986).

In species with more limited flight ranges than honeybees', an even distribution of nest sites throughout crops is more essential. Mason bees, *Osmia* spp., provide uniform pollination throughout orchards as long as their nesting material is evenly distributed throughout the crop (Richards 1993). Leaf-cutter bees' hives must also be distributed throughout crops; in alfalfa the bees do not fly far from their nests, and seed set decreases slightly with increasing distance from the shelter (Richards 1989). The distance leaf-cutter bees fly is also influenced by the direction of flight from the nest. Bees tend to fly twice as far to the east than to the west of shelters. Therefore, it is usually recommended that shelters on the western sides of fields be placed closer to the crop margin than shelters on the east (Richards 1989). Similar, but more dramatic, declines in leaf-cutter bee activity and yield occur in hybrid canola seed production fields (Figure 6.2A–E). Foraging distance within a crop is inversely related to crop "attractiveness" (Gary et al. 1977; Waddington 1983). Hives may have to be spaced closer together within crops that are more attractive, while constant bee populations are maintained in order to maintain consistent yields throughout the crop. More data are required on a variety of crops to determine both the optimal bee densities required to maximize yield and the relative benefits of a more uniform distribution of pollinators.

Resource limitation and managing overpollination

Increases in pollinator visits increase crop yields only up to the point that plants are resource limited (Figure 6.1B). In many crops, such as canola and strawberry (*Fragaria* x *ananassa* Duch.), increases in pollinator visitation increase the number of ovules that set seed to a point where further increases do not affect yield (I. H. Williams et al. 1987; Chagnon et al. 1989). In natural plant populations, yield reductions in resource-limited plants could result if high visitation levels cause pollinators to dislodge more pollen than they deposit, cause physical damage to plant reproductive structures, or increase improper pollen transfer (the transfer of foreign pollen to the stigmatic surface) (Gori 1983; Rathcke 1983). Improper pollen transfer can lower seed set through stigma clogging, exploitation, chemical or physical interference, or production of inviable or sterile hybrids (Rathcke 1983), although these effects do not occur commonly in agricultural crops (Free 1993). "Overpollination" commonly occurs in fruit crops, however, where high visitation rates set too many flowers and the resulting high levels of fruit set reduce profit (a) by increasing crop density, which causes limb breakage; (b) by affecting size, color, and quality of fruit; or (c) by reducing floral initiation, which, in turn, affects production of subsequent crops (Westwood, 1978). In such cases, the effects of overpollination can be compensated for by hand, mechanical, or chemical thinning of flowers or fruit. Because many of the chemical thinners are insecticides (Batjer and Westwood 1960; Westwood 1965), they should be used with caution where native or managed pollinators are present.

Improper pollen transfer can also reduce crop value in agroecosytems when bees import pollen from other cultivars, which subsequently contaminates certified, registered, or foundation seed (Free 1993). Contamination of seed production fields could be reduced by increasing isolation distance, field size, and border size, and by manipulating field shape (Pedersen et al. 1969). Contamination rates are also affected by the relative areas of main cultivar and contaminant, the extent to which self- and cross-fertilization occurs, and the number and type of pollinating insects and their respective foraging areas (Free 1993).

In geographical regions where sufficient land with adequate isolation distances is difficult to obtain, temporal isolation (discussed previously) could also be used to reduce contamination from external pollen sources. In hybrid seed production, lack of floral synchrony between lines is also a factor that increases the chance of improper pollen transfer and that lowers hybridity, if female lines continue to bloom after male lines cease flowering. Synchrony of bloom can be achieved by altering planting dates or by using fertilization or mowing to increase the period of flowering on male lines. Chemical thinners similar to those used on orchard crops (Batjer and Westwood 1960; Westwood 1965; Westwood 1978) or growth regulators (Phillips and Simpson 1989) might also reduce contamination by removing flowers from female lines after the male line is finished, although potential negative impacts of these compounds on seed yield and quality require investigation.

Increasing rate and quality of pollen transfer

Management to improve pollen transfer is directed at increasing "visit quality" so that fewer visits are required to obtain the same level of seed set. Reductions in the number of visits required would reduce the size of pollinator populations that are needed (Figure 6.1C). The quality of pollinator visits is related to the frequency of pollen deposition, the amount of pollen deposited on the stigmatic surface, and the "quality" of the pollen deposited (Herrera 1987). Pollen "quality" reflects factors such as pollen compatibility, the average distance from the pollen source, the purity of the pollen load, and the viability of the pollen deposited (Stanton and Preston 1988). Ecological studies have shown that pollen transfer (male fitness) can be enhanced by manipulation of both the plant and pollinator components through producing superior pollen grains, increasing pollen or flower production, or enhancing efficiency of pollen transfer by insect vectors (Stanton et al. 1986; DeGrandi-Hoffman 1987; Stanton and Preston 1988). In agroecosystems, visit quality is enhanced by breeding crops with flowers more closely adapted to their insect pollinators, choosing a species of pollinator more suited to the target crop, selectively breeding pollinators for specific crops, arranging crop plantings to exploit behavioral patterns of the chosen pollinator, and, to a limited extent, managing pollinator populations to enhance pollen transfer during foraging. Information relevant to the relative efficiency of alternative pollinators on specific crops (Torchio 1987, 1990; Cane and Payne 1993; Free 1993; Richards 1993) and information relevant to increasing the efficiency of pollination through plant breeding (E. H. Erickson, Jr. 1983; S. C. Jay 1986; Free 1993) are covered in several reviews. This review will focus on aspects of crop–pollinator systems that can be enhanced through crop or pollinator management.

Agronomic practices

Changes in plant density, floral density, or floral rewards can affect visit quality and the number and duration of visits. In natural populations of plants, pollinators tend to travel shorter distances in more rewarding patches and to increase their rate of turning, so that they cover less area (D. A. Levin 1978; Heinrich 1983; Waddington 1983). Thus distance of pollen flow is reduced. Manipulation of plant density might also be effective in increasing outcrossing in some agricultural crops, although changes in plant density do not affect the distance of pollen travel in hybrid onion, *Allium cepa* L., seed fields (Erickson and Gabelman 1956). Microclimate within the crop is also affected by plant density. In alfalfa, establishing thinner stands improves conditions for plant growth and increases air movement, light penetration, and bloom (Vansell and Todd 1947; Pedersen et al. 1959; Plews 1973). In addition to promoting plant growth, thinner stands make the flowers more attractive and accessible to pollinating honeybees, alkali bees (*Nomia melanderi* Ckll.), and leaf-cutter bees, resulting in more floral tripping. The improved pollination contributes to observed yield increases. Changes in floral density in meadow form, *Limnanthes alba* Hartw., affect pollination but interact with other factors. Seed yields are improved by maximizing the relationship between floral density, floral receptivity, and honeybee foraging activity (Norberg et al. 1993). The "pollinator component" of yield increases obtained by manipulating plant rewards, plant densities, or floral densities may result from increases in visitation rates, increases in visit quality, or both.

With many pollinators, pollen carryover is low and thus, in crops requiring cross-pollination, yield drops off rapidly as distance from the pollen source is increased (Free and Spencer-Booth 1964; Erickson 1983; Cresswell 1994). In orchard crops, pollen transfer between pollinizer and main cultivars can be enhanced by increasing the number of pollinizer trees or by grafting branches from pollinizers to those of the main cultivar. Similarly, increasing the proportion of male to female lines increases crossing in hybrid seed production fields (E. H. Erickson, Jr. 1983).

Plot size and shape are also important: Cross-pollination is predicted to be lower in larger-sized plots, and higher in rectangular plots than in square plots (Pedersen et al. 1969; D. A. Levin 1978). Thus, hybridity should be maximized and contamination minimized in large square fields containing smaller rectangular plots of male and female lines.

Although the degree of foraging fidelity varies between different species of pollinators, most species tend to remain constant to one cultivar and tend to forage along, rather than across, rows in both fruit and field crops (Stephen 1958; Free 1963; Lederhouse et al. 1972; Free and Williams 1973; Currie et al. 1990). Interspersing pollinizer trees at frequent intervals within rows enhances cross-pollination in orchards (Stephen 1958; Free 1993). In row-planting schemes, where interplanting is not feasible (E. H. Erickson, Jr. 1983), equal between-plant and row spacing encourages the movement of pollinators between rows (Free and Williams 1973) and thus enhances pollen transfer. Altering the placement of hives relative to rows of different cultivars may also be beneficial. When leaf-cutters fly over rows of male lines (oriented north–south) in hybrid seed production fields, seed set is greater than when they fly within the row containing their nest (Figure 6.2E).

Behavioral constraints on pollinators must be taken into consideration when selecting planting patterns. In apple orchards (*Malus domestica* Borkh), for example, the pollen carryover by honeybees is low; most bees restrict foraging to one or two trees and deposit most of their pollen on the first few flowers visited (Free 1993). As a result, a high ratio of pollinizer trees to main-cultivar trees and an even distribution of pollinizers are required. When *Andrena* spp., which exhibit lower floral constancy and have larger foraging areas, are present, pollinizer arrangement is less critical (Kendall 1973; Free 1993).

Pollinator food preference

Many nectar collectors remove nectar from the sides of flowers without touching the stigmatic surface (called "thieving" or "sideworking") (Robinson and Fell 1981; Free and Ferguson 1983; McVetty et al. 1989) or they bite holes in the corolla tube through which they and other pollinators feed (called "robbing") (Jany 1950). Because these visits do not contribute to pollination, methods to reduce the proportion of bees that thieve or rob would increase the efficiency of pollen transfer. In crops where pollen collectors are more effective pollinators, this can be done by increasing the ratio of pollen foragers to nectar foragers (Free 1960, 1966; Free and Williams 1973). There are a number of management strategies that can be used quite successfully to increase pollen foraging. However, their impact on crop yields still needs to be established.

In some solitary species, such as the leaf-cutter bee, the female forages actively for both nectar and pollen to provision the cells in which she lays eggs. Males collect nectar only and are less effective as pollinators of alfalfa, canola, and faba bean (S. C. Jay and Mohr 1987; Currie, unpublished data). Management to increase the ratio of female to male progeny in nesting tunnels is desirable as it leads to a higher proportion of the more efficient female bees the following season. Sex ratio is affected by tunnel diameter, and the relationship of sex ratio to space utilization can be optimized by providing bees with tunnels that are about 6.0 mm wide (Stephen and Osgood 1965).

Both honey- and bumblebee foragers regulate their pollen collection based on colony need and, thus, decreasing the colonies' pollen stores increases pollen collection (Rakhmankulov 1957; Plowright et al. 1993). Manipulating colonies to remove stored pollen is cost prohibitive, but in honeybees, pollen loads can be removed from returning foragers by forcing them to pass through a series of screens called a "pollen trap" (Stephen 1958). Pollen traps do cause an increase in pollen collection (Stephen 1958; Moriya 1966; van Laere and Martins 1971), but further research is required to evaluate their effectiveness. Pollen traps can also cause a shift from pollen to nectar collection, reduce brood rearing and colony growth, and increase the overall proportion of pollen foragers without affecting total colony pollen income (Free 1963; Todd and McGregor 1960; Free 1993). In the last case, their use in pollination applications would still be justified, provided total foraging by the colony does not decrease.

Feeding dilute sugar syrup to honeybee colonies also increases their proportion of pollen foragers (Free and Spencer-Booth 1961; Free 1965b; Barker 1971) and the amount of pollen collected from the target crop (Goodwin et al. 1991). Systems for feeding sugar syrup are well developed. Thus, this could be a simple and economically viable technique to enhance pollen collection in many crops.

Pollen foraging in honeybees is influenced by the presence of a queen, the presence of brood, and both queen- and brood-based pheromones (Free 1967; Jaycox 1970a, b; Al-Tikrity et al. 1972a). Colonies without queens have a smaller foraging force than colonies with queens and have proportionately fewer pollen foragers, even when brood (that is, eggs, larvae, pupae) are present (Jaycox 1970a, b). Queen pheromones stimulate foraging (Jaycox 1970a) and may provide useful management tools for manipulating worker foraging. Synthetic queen mandibular pheromone stimulates brood rearing and pollen collection in small colonies initiated from packages, but is ineffective in commercial-sized pollination units (Higo et al. 1992; Winston and Slessor 1993). In queenless pollination units containing brood, however, the pheromone maintains foraging at levels equivalent to queenright colonies (Currie et al. 1994). Brood may have a more significant effect on stimulating pollen foraging than the presence of a queen (Jaycox 1970b; Free 1979b), since pollen is an important component of the food fed to developing larvae. Adding brood to colonies increases pollen foraging (Rakhmankulov 1957; Free 1967) but is not economical, because it is labor intensive and requires additional colonies to supply brood (Free 1993). Brood pheromone might also be used to stimulate pollen foraging but to date the component(s) have not been identified (Free 1979b). If brood and additional queen pheromones (Free 1979b) can be identified and manufactured, they may prove to be effective management tools. Development of appropriate release devices may allow their use under a wider range of colony states than is currently feasible.

Within-colony pollen transfer

Even though individual foragers often remain constant to a single plant species or cultivar, some pollen is also transferred between nestmates within the colonies of honeybees, bumblebees, and possibly other social insect pollinators (Free and Durrant 1966; Free and Williams 1972; DeGrandi-Hoffman et al. 1986). Although pollen breaks down quite rapidly in the hive (Wilson 1989), the outgoing foragers do carry enough viable pollen (transferred from their nestmates) to fertilize flowers (Free and Durrant 1966). In-hive pollen transfer may be an important component of cross-pollination that occurs in self-incompatible tree fruit and in the male-sterile lines of hybrid seed production fields (Free 1993; DeGrandi-Hoffman et al. 1986; DeGrandi-Hoffman and Martin 1993). A more complete understanding of this process is critical for the development of management systems to increase outcrossing rates and to decrease the potential for contamination.

The quantity of pollen transferred to outgoing foragers in the hive can be enhanced by placing strips of soft nylon brushes at the hive entrance (Free et al. 1991). These bristles increase both the number and the diversity of pollen grains found on outgoing foragers. This technique may be useful to increase pollen transfer from a pollen source to self-incompatible tree fruit or to male-sterile lines in hybrid seed production fields. If effective, the method could reduce or possibly eliminate the need for complex planting schemes in orchards and hybrid seed production fields. This technique warrants further study, but in hybrid seed production areas without complete isolation it should not be used; the chance of contamination would be greatly increased, and reduced hybridity of seed may result.

Pollen transfer can also be enhanced through a similar process called "enpollination," in which foragers are forced to pass through a device mounted on the hive entrance that contains pollen of a compatible cultivar (Burrel and King 1932; Legge 1976; Dicklow et al. 1986; Ferrari 1990; Free 1993). This device, called a "pollen dispenser" or "hive insert," distributes pollen on incoming and outgoing bees and between workers in the hive. Some pollen is subsequently transferred to crops by foragers (R. R. Williams et al. 1979). Pollen dispensers have several drawbacks. The pollen must be replaced continually, which is labor intensive; the pollen itself is expensive to obtain; and, to date, success with the technique has been variable (Townsend et al. 1958; Griggs and Iwakiri 1960; Legge 1976; Dicklow et al. 1986; Ferrari 1990; Free 1993). Nevertheless, further research with enpollination is required, as it could prove to be especially valuable in hybrid seed production (Farkas and Frank 1982). In a cage study, Farkas and Frank found that colonies fitted with dispensers containing pollen will set seed in male-sterile lines even in the absence of male-fertile plants. Dispensers are more suited to commercial hybrid seed production than are entrance brushes because they are not as likely to redistribute incoming pollen and to increase contamination. Development of a method to utilize bee-collected pollen could provide a source of pollen for use in dispensers (Okada et al. 1982; Sasaki 1984; Free 1993).

Input costs

Costs associated with pollination – whether associated with management to increase pollinator visitation rates or associated with management to improve the efficiency of pollination — must be lower than the corresponding increase in returns in order for the practice to be economically viable (Figure 6.1D). Some growers contract out pollination services and rent bees, with payment based on a cash or crop share settlement; other growers take responsibility for their own pollination management. Whatever the case, costs associated with all components of pollination management systems must be accounted for in order to justify their value and ensure that profit is maximized.

Costs of increasing floral visitation rates

Pollination management costs should reflect costs associated with maintaining managed pollinators. In addition to operating costs (related to labor, feeding, disease and parasite control, transportation, insurance, and the repair and maintenance of buildings and machinery), fixed costs (for investment and depreciation of bees, hives and/or shelters, machinery, and specialized facilities required for incubation or winter storage of bees) must be accounted for (Blawat et al. 1992; Blawat and Fingler 1992).

The input costs associated with management to increase pollinator visitation rates are not likely to increase linearly (Figure 6.1D). As the density of pollinators on the target crop is increased, availability of its nectar and pollen rewards decrease with the addition of more bees. In social species such as honey- or bumblebees, addition of colonies lowers the proportion of the available work force that forages on the target crop. In solitary species such as leafcutter bees, higher proportions of introduced bees may emigrate from the field. In some cases, maintaining high densities of introduced bees may decrease floral visitation by other species but result in no net gain in yield (Figure 6.2A–E).

Input costs also increase on a "per capita" basis at higher densities if increased feeding costs result from a lack of forage availability, or if bee-keeper income from the sale of honey, hive products, or bees is reduced. In the case of honeybees, increased competition caused by placing large numbers of colonies on a crop may increase the need to provide supplemental feed and/or to reduce honey production. Thus, input costs will increase as a result of lost beekeeper income (Free 1993). Reproduction rates of leaf-cutter bees also decrease with increases in bee density, resulting in a subsequent loss of income from the sale of excess bees (Stephen 1973; Richards 1982, 1984). When pollinator densities are high, it becomes more costly to increase visitation rates and, due to the factors previously discussed, it may be possible to obtain only marginal increases in floral visitation rates. When this occurs, management to enhance visit quality may be more cost effective.

Pollination requirements can also place restrictions on practices associated with other aspects of crop production, such as pest-management and irrigation schedules, thus increasing management costs. In some crops, increasing the density of pollinators requires the provision for additional space for shelters or hives, as well as associated access roads within the field, and this reduces the area of productive land.

Costs of increasing the rate and quality of pollen transfer

Costs associated with increasing the efficiency of pollen transfer must be balanced against the alternative of increasing yield through increased abundance of native or managed pollinators. In many cropping systems, the plants that must be provided as a pollen source for the main cultivar are of lower or no economic value, as with pollinizer trees in orchards (R. R. Williams and Simms 1977; Free 1993), male kiwifruit, *Actinidia chinensis* Planch., vines (D. Jay and Jay 1984), or restorer lines in hybrid seed production fields. In such cases, a balance must be maintained between maximizing fruit or seed set and maximizing the ratio of highly productive to nonproductive plants in the field, with the objective of optimizing yield per unit land area. Brain and Landsberg (1981) showed that better returns will often be achieved by increasing insect visitation rates rather than by utilizing management techniques to enhance pollen transfer, as the latter is a "diminishing returns" relationship. More research is required to isolate the economic costs and benefits of the quantity and quality components of crop pollination for a wide variety of crops.

Conclusions

The tremendous benefits resulting from our ability to manage insect pollinator populations to enhance crop pollination are clear. On a large number of crops, improvements in pollination increase crop value by decreasing the time to crop maturity and by increasing crop uniformity, quantity, and quality. However, there are still numerous crops where pollen limitation resulting from inadequate insect pollination reduces yield.

In some crops, habitat management strategies can be implemented to enhance native populations of pollinators and ensure adequate pollination. In

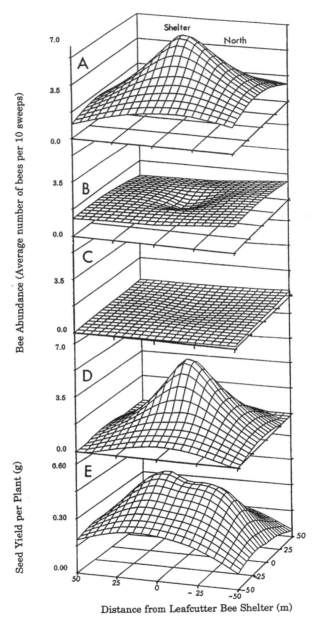

Figure 6.2 Relationship between distance from nest, pollinator activity, and crop yield of male-sterile lines in a hybrid canola, *Brassica napus* L., seed production field. Distribution of (A) total pollinators, (B) honeybees, *Apis mellifera* L., (C) native pollinators, (D) leaf-cutter bees, *Megachile rotundata* F., and (E) yield, radiating outwards from leaf-cutter bee shelters situated in hybrid canola. Shelter located at coordinates (0,0). (Currie, unpublished data.)

areas of intensive agriculture, where native species are neither abundant nor reliable, pollinator populations must be supplemented by importing managed pollinators. Our knowledge of the biology, ecology, life history, and behavior of insect pollinators is sufficiently developed to make management of their populations for crop pollination commercially viable. With basic research into the biology and ecology of insect pollinators, the number and effectiveness of commercially viable species will undoubtedly increase.

Progress has been made on methods to improve the quantitative and qualitative components of insect visitation through management of both plant and insect components of the pollination relationship. Improvements in our understanding of the evolution and ecology of plant–pollinator relationships should aid plant breeders in enhancing efficiency at the plant–pollinator interface. There is exciting potential to utilize current and future developments in the areas of insect and plant chemical-ecology to attract and maintain populations of pollinators on crops and, in social species, to manipulate forager behavior at the colony level, leading to improvements in colony efficiency and the quality of pollen transfer.

Our knowledge of the relative costs and benefits of management to enhance floral visitation or improve visit quality for most cropping systems is lacking. For many crops, accurate economic thresholds may be difficult to determine because of the complex interactions between pollinator activity, plant responses, the environment, and crop management practices. Nevertheless, further research should be directed toward establishing economic "pollination thresholds" or mathematical "threshold models" for crops and determining the exact levels of bee activity below which management to increase pollinator visitation is justified. Models should include the relative costs and benefits of enhancing pollinator visitation rates or visit quality, because this would help to focus research efforts in the areas of cropping systems where the greatest returns can be realized.

Acknowledgments

I thank Dr. D. Vanderwel for her valuable criticisms of the manuscript.

References

Al-Tikrity, W.S., Benton, A.W., Hillman, R.C. & Clarke, W.W. Jr. (1972a). The relationship between the amount of unsealed brood in honeybee colonies and their pollen collection. *J. Apic. Res.* 11:9–12.

Al-Tikrity, W.S., Benton, A.W., Risuis, M.L. & Clarke, W.W. Jr. (1972b). The effect of length of stay of a honeybee colony in a crown vetch field on its foraging behaviour. *J. Apic. Res.* 11:51–57.

Alpatov, V.V. (1984). Bee races and red clover pollination. *Bee World* 29:61–63.

Armstrong, J.E. & Drummond, B.A. (1990). Floral biology of *Myristica fragrans* Houtt. (Myristicaceae), the nutmeg of commerce. *Biotropica* 22:429.

Atkins, E.L., MacDonald, R.L. & Greywood-Hale, E.A. (1975). Repellent additives to reduce pesticide hazards to honeybees: field tests. *Environ. Entomol.* 4:207–210.

Banda, H.J. & Paxton, R.J. (1991). Pollination of greenhouse tomatoes by bees. The 6th International Symposium on Pollination, Tilburg, The Netherlands, August 1990. *Acta Horticulturae* 288:194–198.

Barker, R.J. (1971). The influence of food inside the hive on pollen collection by a honeybee colony. *J. Apic. Res.* 10:23–26.

Barth, F.G. (1985). *Insects & Flowers: The Biology of a Partnership*. Princeton: Princeton University Press.

Batjer, L.P. & Westwood, M.N. (1960). 1-Naphthyl N-methylcarbamate, a new chemical for thinning apples. *Proc. Amer. Soc. Hort. Sci.* 75:1–4.

Belletti, A. & Zani, A. (1981). [A bee attractant for carrots grown for seed.] (in Italian).*Sementi Elette*. 27:23–27.

Berger, L.A., Vaissière, B.E., Moffett, J.O. & Merritt, S.J. (1988). *Bombus* spp. (Hymenoptera: Apidae) as pollinators of male-sterile upland cotton on the Texas high plains. *Environ. Entomol.* 17:789–794.

Blawat, P. & Fingler, B. (1992). Guidelines for estimating alfalfa seed cost of production. *Manitoba Agriculture, Farm Business Management Information Update*, pp. 1–14.

Blawat, P., Fingler, B. & Dixon, D. (1992). Guidelines for estimating honey production costs. *Manitoba Agriculture, Farm Business Management Information Update*, pp. 1–13.

Bobrzecka, D. & Bobrzecki, J. (1973). Effect of different fertilization on the secretion of nectar, bee visiting, and seed yield of rapes. *Pszczelnicze Zeszyty Naukowe* 17:87–108.

Brain, P. & Landsberg, J.J. (1981). Pollination, initial fruit set and fruit drop in apples: analysis using mathematical models. *J. Hort. Sci.* 56:41–54.

Brakefield, J.J. (1980). Report of Beeline trials in top and soft fruit crops. *Report to Murphy Chemical Ltd*. Southampsted, St. Alvans. Herts, UK.

Braun, E., MacVicar, R.M., Gibson, D.A., Pankiw, P. & Guppy, J. (1953). Studies in red clover seed production. *Can. J. Agric. Sci.* 33:437-447.

Buchmann, S.L. (1983). Buzz pollination in angiosperms. In *Handbook of Experimental Pollination Biology*, eds. C.E. Jones and R.J. Little, pp. 77–113. New York: Van Nostrand Reinhold.

Burgett, M. & Fisher, G. (1979). An evaluation of Beeline as a pollinator attractant on red clover. *Amer. Bee J.* 119:356–357.

Burrel, A.B. & King, G.E. (1932). A device to facilitate pollen distribution by bees. *Proc. Amer. Soc. of Hort. Sci.* 28:85–86.

Cane, J.H. & Payne, J.A. (1993). Regional, annual, and seasonal variation in pollinator guilds: intrinsic traits of bees (Hymenoptera: Apoidea) underlie their patterns of abundance at *Vaccinium ashei* (Ericaceae). *Annals of the Entomol. Soc. of Amer.* 86:577–588.

Chagnon, M., Gingras, J. & de Oliveira, D. (1989). Effect of honey bee (Hymenoptera: Apidae) visits on the pollination rate of strawberries. *J. Econ. Entomol.* 82:1350–1353.

Clifford, T.P. (1973). Increasing bumble bee densities in red clover seed production areas. *N. Zeal. J. Exper. Agric.* 1:377–379.

Cresswell, J.E. (1994). A method for quantifying the gene flow that results from a single bumblebee visit using transgenic oilseed rape, *Brassica napus* L. cv. Westar. *Transgenic Research* 3:134–137.

Currie, R.W., Jay, S.C. & Wright, D.L. (1990). The effects of honeybees (*Apis mellifera* L.) and leafcutter bees (*Megachile rotundata* F.) on outcrossing between different cultivars of beans (*Vicia faba* L.) in caged plots. *J. Apic. Res.* 29:68-74.

Currie, R.W., Winston, M.L & Slessor, K.N. (1992a). Effect of synthetic queen mandibular pheromone sprays on honey bee (Hymenoptera: Apidae) pollination of berry crops. *J. Econ. Entomol.* 85:1300–1306.

Currie, R.W., Winston, M.L, Slessor, K.N. & Mayer, D.F. (1992b). Effect of synthetic queen mandibular compound sprays on pollination of fruit crops by honey bees (Hymenoptera: Apidae). *J. Econ. Entomol.* 85:1293–1299.

Currie, R.W., Winston, M.L. & Slessor, K.N. (1994). The effect of honey bee (*Apis mellifera* L.) synthetic queen mandibular compound on queenless "disposable" pollination units. *Amer. Bee J.* 143:200–202.

Dadant & Sons, Inc. (1978). *The Hive and the Honeybee.* Carthage,IL: Journal Printing Co.

Dafni, A. (1984). Mimicry and deception in pollination. *Ann. Rev. Ecol. and Systematics.* 15:259-278.

Davis, A.R. & Gunning, B.E.S. (1991). The modified stomata of the floral nectary of *Vicia faba* L. 2. Stomatal number and distribution as selection criteria for breeding for high nectar sugar production. *Acta Hort.* 288:329–334.

DeGrandi-Hoffman, G. (1987). The honey bee pollination component of horticultural crop production systems. *Hort. Rev.* 9:237–272.

DeGrandi-Hoffman, G. & Martin, J. H. (1993). The size and distribution of the honey bee (*Apis mellifera* L.) crosspollinating population on male-sterile sunflowers (*Helianthus annus* L.). *J. Apic. Res.* 32:135–42.

DeGrandi-Hoffman, G., Hoopingarner, R. & Klomparens, K. (1986). Influence of honey bee (Hymenoptera: Apidae) in-hive pollen transfer on cross-pollination and fruit set in apple. *Environ. Entomol.* 16:309–318.

DeGrandi-Hoffman, G., Hoopingarner, R. & Pulcer, R. (1987). REDAPOL: Pollination and fruit-set prediction model for 'Delicious' apples. *Environ. Entomol.* 16:309–318.

Dicklow, M.B., Firman, R.D., Rupert, D.B., Smith, K.L. & Ferrari, T.E. (1986). Controlled enpollination of honeybees (*Apis mellifera*): bee-to-bee and bee-to-tree pollen transfer. In *Biotechnology and Ecology of Pollen*, eds. D.L. Mulcahy, G. Bergamini-Mulchay, and E. Ottaviano, pp. 449–454. Berlin: Springer-Verlag.

Dobson, H.E.M. (1994). Floral volatiles in insect biology. In *Insect–Plant Interactions, Vol. V*, ed. E. Bernays, pp. 47–81. Boca Raton: CRC Press.

Donovan, B.J. & Wier, S.S. (1978). Development of hives for field population increase and studies on the life cycles of the four species of introduced bumble bees in New Zealand. *N. Zeal. J. Agric. Res.* 21:733–756.

Eckert, J.E. (1933). The flight range of the honeybee. *J. Agric. Res.* 47:257–285.

Erickson, E.H. & Gablemen, W.H. (1956). The effect of distance and direction on cross-pollination in onions. *Proc. Amer. Soc. Hort. Sci.* 68:351–357.

Erickson, E.H. Jr. (1983). Pollination of entomophilous hybrid seed parents. In *Handbook of Experimental Pollination Biology*, eds. C.E. Jones & R.J. Little, pp. 493-535. New York: Van Nostrand Reinhold.

Faegri, K. & van der Pijl, L. (1979). *The Principles of Pollination Ecology*. Oxford: Pergamon Press.

Fairey, D.T. & Lefkovitch, L.P. (1991). Reproduction of *Megachile rotundata* Fab., foraging on *Trifolium* Spp. and *Brassica campestris*. The 6th International Symposium on Pollination, Tilburg, The Netherlands, August 1990. *Acta Hort.* 288:185-189.

Farkas, J. & Frank, J. (1982). Experience gained when using honeybees for pollen dispersion in hybrid sunflower seed production. *Acta Agronomica Academiae Scientiarum Hungaricae* 31:267-270.

Ferrari, T.E. (1990). 'Enpollination' of honey bees with precollected pollen improves pollination of almond flowers. *Amer. Bee J.* 130:801.

Ford, I. (1971). Chinese gooseberry pollination. *N. Zeal. J. Agric.* 122:34-35.

Free, J.B. (1958). Attempts to condition bees to visit selected crops. *Bee World* 39:221-230.

Free, J.B. (1959). The effect of moving colonies of honey bees to new sites on subsequent foraging behaviour. *J. Agric. Sci.* 53:1-9.

Free, J.B. (1960). The behaviour of honey-bees visiting flowers of fruit trees. *J. Animal Ecol.* 29:385-395.

Free, J.B. (1962). The effect of distance from pollinizer varieties on the fruit set on trees in plum and apple orchards. *J. Hort. Sci.* 37:262-271.

Free, J.B. (1963). The flower constancy of honey bees. *J. Animal Ecol.* 32:119-131.

Free, J.B. (1965a). Attempts to increase pollination by spraying crops with sugar syrup. *J. Apic. Res.* 4:61-64.

Free, J.B. (1965b). The effect on pollen collection of feeding honeybee colonies with sugar syrup. *J. Agric. Sci.* 64:167-168.

Free, J.B. (1966). The pollinating efficiency of honey-bee visits to apple flowers. *J. Hort. Sci.* 41:91-94.

Free, J.B. (1967). Factors determining the collection of pollen by honeybee foragers. *Animal Behaviour* 15:134-144.

Free, J.B. (1968a). Dandelion as a competitor to fruit trees for bee visits. *J. Appl. Ecol.* 5:169-178.

Free, J.B. (1968b). The conditions under which foraging honeybees expose their Nasonov gland. *J. Apic. Res.* 7:139-145.

Free, J.B. (1969). Influence of the odour of a honey bee colony's food stores on the behaviour of its foragers. *Nature* 222:778.

Free, J.B. (1970). The flower constancy of bumble bees. *J. Animal Ecol.* 39:395-402.

Free, J.B. (1979a). Progress toward the use of pheromones to stimulate pollination by honey bees. *Proceedings of the 4th International Symposium on Pollination, College Park, Md.* 1:7-22.

Free, J.B. (1979b). Managing honeybee colonies to enhance the pollen-gathering stimulus from brood pheromones. *Applied Animal Ethology* 5:173-178.

Free, J.B. (1987). *Pheromones of Social Bees*. London: Chapman and Hall.

Free, J.B. (1993). *Insect Pollination of Crops*. San Diego: Academic Press.

Free, J.B. & Durrant, A.J. (1966). The transport of pollen by honey bees from one foraging trip to the next. *J. Hort. Sci.* 41:87-89.

Free, J.B. & Ferguson, A.W. (1983). Foraging behaviour of honey bees on oilseed rape. *Bee World* 64:22–24.

Free, J.B. & Smith, M.V. (1961). The foraging behaviour of honeybees from colonies moved into a pear orchard in full flower. *Bee World* 42:11–12.

Free, J.B. & Spencer-Booth, Y. (1961). The effect of feeding sugar syrup to honeybee colonies. *J. Agric. Sci., Cambridge* 57:147–151.

Free, J.B. & Spencer-Booth, Y. (1964). The effect of distance from pollinizer varieties on the fruit set of apple, pear and sweet cherry trees. *J. Hort. Sci.* 39:54–60.

Free, J.B. & Williams, I.H. (1972). The transport of pollen on the body hair of honeybees (*Apis mellifera* L.) and bumblebees (*Bombus* spp. L.). *J. Appl. Ecol.* 9:609–615.

Free, J.B. & Williams, I.H. (1973). The foraging behaviour of honeybees (*Apis mellifera* L.) on Brussels sprout (*Brassica oleracea* L.). *J. Appl. Ecol.* 10:489–499.

Free, J.B. & Williams, I.H. (1974). Influence of the location of honeybee colonies on their choice of pollen sources. *J. Appl. Ecol.* 11:925–935.

Free, J.B., Free, N.W. & Jay, S.C. (1960). The effect on foraging behaviour of moving honey bee colonies to crops before or after flowering has begun. *J. Econ. Entomol.* 53:564–566.

Free, J.B., Paxton, R.J. & Waghchoure, E.S. (1991). Increasing the amount of foreign pollen carried by honeybee foragers. *J. Apic. Res.* 30:132–136.

Fries, I. & Stark, J. (1983). Measuring the importance of honey bees in rape seed production. *J. Apic. Res.* 22:272–276.

Frisch, K. von (1967). *The Dance Language and Orientation of Bees*. Cambridge: Harvard Univ. Press.

Gary, N.E., Witherall, P.C. & Marston, J. (1972). Foraging range and distribution of honey bees used for carrot and onion pollination. *Environ. Entomol.* 1:71–78.

Gary, N.E., Witherall, P.C. Lorenzen, K. & Marston, J. (1977). The inter-field distribution of honey bees foraging on carrots, onions and safflower. *Environ. Entomol.* 6:637–640.

Glushkov, N.M. (1958). Problems of beekeeping in the U.S.S.R. in relation to pollination. *Bee World* 39:81–92.

Gori, D.F. (1983). Post-pollination phenomena and adaptive floral changes. In *Handbook of Experimental Pollination Biology*, eds. C.E. Jones & R.J. Little, pp. 31–49. New York: Van Nostrand Reinhold.

Goodwin, R.M., Houten, A. ten & Perry, J.H. (1991). Effect of variations in sugar presentation to honey bees (*Apis mellifera* L.) on their collection of kiwifruit (*Actinidia deliciosa*) pollen. *N. Zeal. J. Crop and Hort. Sci.* 19:259–262.

Griggs, W.H. & Iwakiri, B.T. (1960). Orchard tests of beehive pollen dispensers for cross pollination of almonds, sweet cherries and apples. *Proc. Amer. Soc. Hort. Sci.* 75:114–128.

Hammer, O. (1961). [Effect of distance on honey production.] (in Danish). *Nord Bitidskr.* 13:20–23.

Heemert, C. van, Ruijter, A. de, Eijnde, J. van den & Steen, J. van der (1990). Year-round production of bumble bee colonies for crop production. *Bee World* 71:54–57.

Heinrich, B. (1976). The foraging specializations of individual bumble bees. *Ecol. Monographs* 46:105–128.

Heinrich, B. (1979). "Majoring" and "minoring" by foraging bumblebees, *Bombus vagans:* an experimental analysis. *Ecology* 60:245–255.

Heinrich, B. (1983). Insect foraging energetics. In *Handbook of Experimental Pollination Biology*, eds. C.E. Jones & R.J. Little, pp. 187–214. New York: Van Nostrand Reinhold.

Henning, J.A., Peng, Y.S., Montague, M.A. & Teuber, L.R. (1992). Honey bee (Hymenoptera: Apidae) behavioral response to primary alfalfa (Rosales: Fabaceae) floral volatiles. *J. Econ. Entomol.* 85:233–239.

Herrera, C.M. (1987). Components of pollinator "quality": comparative analysis of a diverse insect assemblage. *Oikos* 50:79–90.

Higo, H A., Colley, S.J., Winston, M.L. & Slessor, K.N. (1992). Effects of honey bee (*Apis mellifera* L.) queen mandibular pheromone on foraging and brood rearing. *Can. Entom.* 124:409–418.

Hobbs, G.A. (1967). Obtaining and protecting red-clover pollinating species of *Bombus* (Hymenoptera: Apidae). *Can. Entom.* 99:943–951.

Holdaway, F.G. & Burson, P.M. (1960). Soil conditions and fertilizers in relation to pollination and seed production of some small-seeded forage legumes in Minnesota. *Scientific Journal Series, Minnesota Agricultural Experiment Statio*, pp. 38–46.

Holm, S.N. (1966). The utilization and management of bumble bees for red clover and alfalfa seed production. *Ann. Rev. Entomol.* 11:155–182.

Homann, H. (1970). [Effect of pollen extracts and scented oils on the foraging and recruiting activities of the honey bee.] (in German). *Apidologie* 1:157–158.

Hopkins, C.Y., Jevans, A.W. & Boch, R. (1969). Occurrence of octadeca-*trans*-2,*cis*-9-*cis*-12-trienoic acid in pollen attractive to the honey bee. *Can. J. Biochem.* 47:433–436.

Howell, G.S., Kilby, M.W. & Nelson, J.W. (1972). Influence of timing of hive introduction on production of highbush blueberries. *HortScience* 7:129–131.

Hutson, R. (1925). The honey bee as an agent in the pollination of pears, apples and cranberries. *J. Econ. Entomol.* 18:387–391.

Jany, E. (1950). Der "Einbruch" von Erdhummeln (*Bombus terrestris* L.) in die Bluten der Feuerbohne (*Phaseolus multilorus* Willd.). *Zeitschrift für Angew. Ent.* 32:172–183.

Jay, D. & Jay, S.C. (1984). Observations of honeybees on Chinese Gooseberries ("Kiwifruit") in New Zealand. *Bee World* 65:155–166.

Jay, S.C. (1986). Spatial management of honey bees on crops. *Ann. Rev. Entomol.* 31:49–65.

Jay, S.C. & Mohr, N. (1987). The effect of nest replacement on the production of females of the alfalfa leaf-cutter bee *Megachile rotundata* (F). *J. Apic. Res.* 26:69–72.

Jaycox, E.R. (1970a). Honey bee queen pheromones and worker foraging behaviour. *Annals Entomol. Soc. Amer.* 63:222–228.

Jaycox, E.R. (1970b). Honey bee foraging behaviour: Responses to queens, larvae, and extracts of larvae. *Annals Entomol. Soc. Amer.* 63:1689–1694.

Johnson, D.L. & Wenner, D.L. (1970). Recruitment efficiency in honeybees: studies on the role of olfaction. *J. Apic. Res.* 9:13–18.

Karmo, E.A. (1958). Honeybees as an aid in orchard and blueberry pollination in Nova Scotia. *10th Int. Cong. Entomol., 1956* 4:955–959

Karmo, E.A. & Vickery, V.R. (1954). The place of honeybees in orchard pollination. *Mimeo Circular, Nova Scotia Department of Agriculture* 67:1–7.

Kaziev, T.I. (1959). [The influence of fertilizers and watering on the nectar secretion of cotton.] (in Russian). *Pchelovodstvo* 36:25–28.

Kaziev, T.I. (1967). Some agrotechnical measures expediting increased nectar productivity of cotton. *Proceedings of the 21st International Apicultural Congress*, pp. 454–459. Bucharaest, Romania: Apimondia Publishing House.

Kendall, D.A. (1973). The viability and compatibility of pollen on insects visiting apple blossom. *J. Appl. Ecol.* 10:847–853.

Kevan, P.G. (1978). Floral coloration, its colorimetric analysis and significance in anthecology. In *The Pollination of Flowers by Insects*, ed. A.J. Richards, pp. 51–78. London: Academic Press.

Kurennoi, N.M. & Barabash, I.P. (1966). [The efficacy of methods of pollination and of increasing the grape crop.] (in Russian). *Sadovod. Vinograd. Vinodel. Mold.* 10:30–33.

Laere, O. van & Martins, N. (1971). [Artificial decreasing influence of the protein provision on collection activity of the bee colonies.] (in French). *Apidologie* 2:197–204.

Lederhouse, R.C., Caron, D.M. & Morse, R.A. (1972). Distribution and behavior of honey bees on onion. *Environ. Entomol.* 1:127–129.

Lee, W.R. (1961). The nonrandom distribution of foraging honeybees between apiaries. *J. Econ. Entomol.* 54:928–933.

Legge, A.P. (1976). Hive inserts and pollen dispensers for tree fruits. *Bee World* 57:159–167.

Levin, D.A. (1978). Pollinator behaviour and the breeding structure of plant populations. In *The Pollination of Flowers by Insects*, ed. A.J. Richards, pp. 133–150. London: Academic Press.

Levin, M.D. (1960). Do foraging bees retain their orientation to a certain distance from their hive, after the hive is moved? *Bee World* 41:281–282.

Levin, M.D. & Bohart, G.E. (1959). The effect of prior location on alfalfa foraging by honey bees. *J. Econ. Entomol.* 50:629–632.

Loper, G.M. & Waller, G.D. (1970). Alfalfa flower aroma and flower selection by honey bees. *Crop Sci.* 10:66–88.

Macfarlane, R.P., Griffin, R.P. & Read, P.E.C. (1983). Bumble bee management options to improve "Grasslands Pawera" red clover seed yield. *Proc. N. Zeal. Grasslands Assoc.* 44:47–53.

Macfarlane, R.P., Ende, H.J. van den & Griffin, R.P. (1991). Pollination needs of "Grasslands Pawera" red clover. The 6th International Symposium on Pollination, Tilburg, The Netherlands, August 1990. *Acta Horticulturae* 288:399–404.

MacVicar, R.M., Braun, D.R. & Jamieson, C.A. (1952). Studies in red clover seed production. *Scientific Agric.* 32:67–80.

Margalith, R., Lensky, Y. & Rabinowitch, H.D. (1984). An evaluation of Beeline as a honeybee attractant to cucumbers and its effect on hybrid seed production. *J. Apic. Res.* 23:50–54.

Mayer, D.F., Britt, R.L. & Lunden, J.D. (1989). Evaluation of beescent® as a honey bee attractant. *Amer. Bee J.* 129:41–42.

McGregor, S.E. (1976). *Insect Pollination of Cultivated Crop Plants*. Agriculture Handbook No. 496. Washington DC: USDA Agricultural Research Service.

McMahon, H.A. (1954). Pollination of alfalfa by honey bees. *Can. Bee J.* 62:4–6.

McVetty, P.B.E., Pinnisch, R. & Scarth, R. (1989). The significance of floral characteristics in seed production of four summer rape cultivar A-lines with *pol* cytoplasm. *Can. J. Plant Sci.* 69:915–918.

Moeller, F.E. (1973). Timing of placement of colonies of honey bees for pollination of cranberries. *J. Econ. Entomol.* 66:370–372.

Moriya, K. (1966). Effects of pollen trap on numbers of pollen foragers in honeybee colony. *Japanese J. Ecol.* 16:105–109.

Murrell, D.C., Tomes, D.T. & Shuel, R.W. (1982). Inheritance of nectar production in birdsfoot trefoil. *Can. J. Plant Sci.* 62:101–105.

Naumann, K. (1993). Stickers do not enhance longevity of honey bee queen mandibular gland pheromone. *Amer. Bee J.* 133:581–583.

Naumann, K., Currie, R.W. & Isman, M. (1994). Repellent effects of a NEEM insecticide on foraging honey bees and other pollinators. *Can. Entomol.* 126:225–230.

Norberg, O.S., Seddigh, M., Jolliff, G.D. & Fiez, T.E. (1993). Flower production and honey bee density effects on meadowform seed yield. *Crop Sci.* 33:108–112.

Novotna, H. (1973). Effect of B and Mo on the pollination of *Trifolium pratense* L. *Pol'Nohospodárstvo* 19:836–841.

Okada, I., Niijima, K. & Ashizawa, I. (1982). [Fruit set in apples hand-pollinated with bee-collected pollen stored for one year.] (in Japanese). *Honeybee Sci.* 2:63–64.

Osborne, J.L., Williams, I.H. & Corbet, S.A. (1991). Bees, pollination and habitat change in the European community. *Bee World* 72:99–115.

Palmer-Jones, T. & Forster, I.W. (1972). Measures to increase the pollination of lucerne (*Medicago sativa* Linn.). *N. Zeal. J. Agric. Res.* 15:186–193.

Peck, O. & Bolten, J.I. (1946). Alfalfa seed production in northern Saskatchewan as affected by bees, with a report on means of increasing the populations of native bees. *Sci. Agric.* 26:388–418.

Pedersen, M.W. (1953). Seed production in Alfalfa as related to nectar production and honey bee visitation. *Botany Gazette* 115:129–138.

Pedersen, M.W., Bohart, G.E., Levin, M.D., Nye, W.P., Taylor, S.A. & Haddock, H.L. (1959). Cultural practices for alfalfa seed production. *Bull. Utah Agric. Exper. Station* 408.

Pedersen, M.W., Hurst, R.L., Levin, M.D. & Stoker, G.L. (1969). Computer analysis of the genetic contamination of alfalfa seed. *Crop Sci.* 9:1–4.

Peterson, A.G., Furgala, B. & Holdaway, F.G. (1960). Pollination of red clover in Minnesota. *J. Econ. Entomol.* 53:546–550.

Phillips, S.A. Jr. & Simpson, J.L. (1989). Hybrid cotton pollination in relation to accumulated degree days. *Agron. J.* 81:975–980.

Plews, K.W. (1973). A study of the influence of management practices in alfalfa seed production in Manitoba. M. Sc. Thesis, Dept. of Plant Science, University of Manitoba.

Plowright, R.C. & Hartling, L.K. (1981). Red clover pollination by bumble bees: a study of the dynamics of a plant–pollinator relationship. *J. Appl. Ecol.* 18:639–647.

Plowright, R.C., Thomson, J.D., Lefkovitch, L.P. and Plowright, C.M.S. (1993). An experimental study of the effect of colony resource level manipulation on foraging for pollen by worker bumble bees (Hymenoptera: Apidae). *Can. J. Zool.* 71:1393–1396.

Pomeroy, N. (1981). Use of natural sites and field hives by a long-tongued bumble bee *Bombus ruderatus*. *N. Zeal. J. Agric. Res.* 24:409–414.

Praagh, J.P. van & Ohe, W. von der. (1983). The role of scents in pollination by the honeybee. *Acta Horticulturae* 139:65–67.

Pritsch, G. (1959). Investigations of scent-training of bees to red clover for increased seed production. *Arch. Geflugelz. Kleintierk.* 8:214–239.

Rakhmankulov, F.K. (1957). [How to increase the efficiency of the pollination of red clover by bees.] (in Russian). *Pchelovodstvo* 2:20.

Rathcke, B. (1983). Competition and facilitation among plants for pollination. In *Pollination Biology*, ed. L. Real, pp. 305–329. Orlando: Academic Press Inc.

Rhein, W. von (1957). [The influence of odours in the guidance of bees to increase seed production in red clover (*Trifolium pratense* L.).] (in German). *Z. Acker. Pflanzenbau* 103:273–314.

Ribbands, C.R. (1951). The flight range of the honey bee. *J. Anim. Ecol.* 20:220–226.

Richards, K.W. (1982). Inputs, expectation and management of the alfalfa leafcutter bee, *Megachile rotundata*. In *Proceedings of the First International Symposium on Alfalfa Leafcutting Bee Management*, ed. G.H. Rank, pp. 113–135. Saskatoon: University of Saskatoon Printing Services.

Richards, K.W. (1984). Relationship of leafcutter bee cell productivity to alfalfa seed yield. *Proceedings of the Vth International Symposium, Versailles, France*, pp. 449–453.

Richards, K.W. (1989). Alfalfa leafcutter bee management in western Canada. *Agric. Can. Publication 1495/E*.

Richards, K.W. (1993). Non-*Apis* bees as crop pollinators. *Revue Suisse de Zoologie* 100:807–822.

Roberts, D. (1956). Sugar sprays aid fertilization of plums by bees. *N. Zeal. J. Agric.* 93:206–207, 209, 211.

Robinson, W.S. & Fell, R.D. (1981). Effect of honey bee foraging behaviour on Delicious apple set. *HortScience* 16:326–328.

Ryle, M. (1954). The influence of nitrogen, phosphate and potash on the secretion of nectar. *J. Agric. Sci.* 44:400–419.

Sasaki, M. (1984). [A comparison of the honeybee (and *Osmia cornifrons* and *Eristalis cerealis*) as pollinators; and a proposal for the large-scale use of "revived" bee-collected pollen after long storage.] (in Japanese). *Honeybee Science* 5:55–62.

Schemske, D.W. (1980). Evolution of floral display in the orchid *Brassavola nodosa*. *Evolution* 34:489–493.

Shimanuki, H., Lehnert, T. & Stricker, M. (1967). Differential collection of cranberry pollen by honey bees. *J. Econ. Entomol.* 60:1031–1033.

Shuel, R.W. (1957). Some aspects of the relation between nectar secretion and nitrogen, phosphorous, and potassium nutrition. *Can. J. Plant Sci.* 37:220–236.

Shuel, R.W. (1967). The influence of external factors on nectar production. *Amer. Bee J.* 99:54–56.

Shuel, R.W. (1989). Improving honey production through plant breeding. *Bee World* 70:36–45.

Shuel, R.W. & Shivas, J.A. (1953). The influence of soil physical condition during the flowering period on nectar production in snapdragon. *Plant Physiology* 28:645–651.

Singh, S. (1950). Behavior studies of honey bees in gathering nectar and pollen. *New York Agric. Exper. Station, Ithaca, Memoirs* 288:1–57.

Skrebtsova, N.D. (1957). [Pollination of buckwheat flowers by bees.] (in Russian). *Pchelovodstvo* 34:48-50.

Stanton, M.L., & Preston, R.E. (1988). A qualitative model for evaluating the effects of flower attractiveness on male and female fitness in plants. *Amer. J. Bot.* 75:540–544.

Stanton, M.L., Snow, A.A. & Handel, S.N. (1986). Floral evolution: attractiveness to pollinators increases male fitness. *Science* 23:1626–1627.

Stanton, M.L., Snow, A.A., Handel, S.N. & Bereczky, J. (1989). The impact of a flower-colour polymorphism on mating patterns in experimental populations of wild radish (*Raphanus raphanistrum* L.). *Evolution* 43:335–346.

Stephen, W.P. (1958). Pear pollination studies in Oregon. *Technical Bulletin of the Oregon Agriculture Experimental Station* 43:1–43.

Stephen, W.P. (1973). Insects as natural resources and tools of management. *Ecol. Soc. of Australia Memoirs* 1:31–44.

Stephen, W.P. & Osgood, C.E. (1965). Influence of tunnel size and nesting medium on sex ratios in a leaf-cutter bee, *Megachile rotundata. J. Econ. Entomol.* 58:965–968.

Straver, W.A. & Plowright (1991). Pollination of greenhouse tomatoes by bumble-bees. *Greenhouse Canada* 11:10–12.

Thomas, W. (1951). Bees for pollinating red clover. *Gleanings in Bee Culture* 79:137–141.

Todd, F.E. & McGregor, S.E. (1960). The use of honey bees in the production of crops. *Ann. Rev. Entomol.* 5:265–278.

Torchio, P.F. (1987). Use of non-honeybee species as pollinators of crops. *Proc. Entomol. Soc. of Ontario* 118:111–124.

Torchio, P.F. (1990). Diversification of pollination strategies for U.S. crops. *Environ. Entomol.* 19:1649–1656.

Townsend, G.F., Riddell, R.T. & Smith, M.V. (1958). The use of pollen inserts for tree fruit pollination. *Can. J. Plant Sci.* 38:39–44.

Vansell, G. H. (1942). Factors affecting the usefulness of honeybees in pollination. *U.S. Dept. of Agriculture Circular No. 650.*

Vansell, G.H. & Todd, F.E. (1946). Alfalfa tripping by insects. *J. Amer. Soc. Agron.* 38:470–488.

Vansell, G.H. & Todd, F.E. (1947). Honeybees and other bees pollinate the alfalfa seed crop in Utah. *Gleanings in Bee Culture* 75:136–138.

Waddington, K.D. (1983). Foraging behaviour of pollinators. In *Pollination Biology*, ed. L. Real, pp. 213–239. Orlando: Academic Press Inc.

Waller, G.D. (1970). Attracting honeybees to alfalfa with citral, geraniol and anise. *J. Apic. Res.* 9:9–12.

Wauchope, D.G. (1968). Flowering and fruit set in the Packham's Triumph pear. *Austral. J. Exper. Agric. and Animal Husbandry* 8:456–460.

Westwood, M.N. (1965). A cyclic carbonate and three new carbamates as chemical thinners for apple. *Proc. Amer. Soc. Hort. Sci.* 86:37–40.

Westwood, M. N. (1978). *Temperate-Zone Pomology.* San Francisco: W.H. Freeman and Company.

Williams, I.H, Martin, A.P. & White, R.P. (1987). The effect of insect pollination on plant development and seed production in winter oil-seed rape (*Brassica napus* L.). *J. Agric. Sci.* 109:135–139.

Williams, R.R. & Sims, F.P. (1977). The importance of weather and variability in flower in time when deciding pollination schemes for Cox's Orange Pippin. *Exper. Hort.* 29:15–26.

Williams, R.R., Church, R.M., Wood, D.E.S. and Flook, V.A. (1979). Use of an anthocyanin progeny marker to determine the value of hive pollen dispensers in apple orchards. *J. Hort. Sci.* 54:75–78.

Wilson, R.L. (1989). Minimizing extraneous transfer of sunflower pollen by honey bees (Hymenoptera: Apidae) in field cages. *J. Kansas Entomol. Soc.* 62:387–391.

Winston, M.L. (1987). *The Biology of the Honey Bee.* Cambridge: Harvard University Press.

Winston, M.L. & Slessor, K.N. (1993). Applications of queen honey bee mandibular pheromone to beekeeping and crop pollination. *Bee World* 74:111–128.

Woodrow, A.W. (1952). Effect of time of pollination by honeybees on red clover seed yields. *J. Econ. Entomol.* 45:517–519.

Woodrow, A.W., Green, N., Tucker, H., Schonhorst, M.H. & Hamilton, K.C. (1965). Evaluation of chemicals as honey bee attractants and repellents. *J. Econ. Entomol.* 58:1094–1102.

Woyke, H.W. (1981). Some aspects of the role of the honeybee in onion seed production in Poland. *Acta Horticulturae* 111:91–98.

Yoshida, A. & Maeto, Y. (1988). Utilization of *Osmia cornifrons* (Radoszkowski) as a pollinator of apples in south-western Japan. *Honeybee Science* 9:1–6.

Zyl, H.J. van & Strydom, D.K. (1968). The problem of poor fruit set of Packham's Triumph pear trees. *Deciduous Fruit Grower* 18:121–123.

Part III

Pollen biotechnology and hybrid seed
production

7

Cytoplasmic male sterility

PETER B. E. McVETTY

Summary   155
Introduction   156
Cytoplasmic male sterility (CMS) origins   157
CMS characterization   159
CMS mechanism of action   159
Heterosis and other advantages of hybrids   167
CMS utilization in hybrid seed production   169
CMS limitations   172
CMS future developments   174
References   176

Summary

There has been long-term interest in cytoplasmic male sterility (CMS) in plants because it provides, at least in theory, a means to produce commercial quantities of hybrid seed for plants where this would otherwise be difficult or impossible. CMS has been observed in at least 150 different plant species. CMS systems have traditionally been characterized by the restorer genes required to overcome the CMS and to provide male-fertile progeny in the male-sterile cytoplasm. More recently, CMS systems have been characterized by DNA analysis techniques applied to cytoplasmic organelles that are able to distinguish genetic differences between normal and male-sterile cytoplasms. CMS systems of pollination control have been developed in several major field and horticultural crops. These CMS systems consist of a male-sterile female "A-line," a male-fertile maintainer "B-line," and a male-fertility restorer "R-line." There are a number of limitations associated with the use of CMS in the development of hybrid cultivars. Even with these limitations, CMS is the predominant method of pollination control for hybrid seed production for hybrid cultivars throughout the world. Current research is aimed at enhancing our understanding of the molecular basis of CMS and, through this, enhancing our understanding of improving and extending CMS as a pollination control mechanism. Research on future uses of CMS is focused on the development of cytoplasmic gene transformation systems using CMS components.

Introduction

Plants that do not produce viable, functional pollen grains are male sterile. If such male sterility is exclusively maternally inherited, it is described as cytoplasmic male sterility (CMS). Since the first reference to CMS (Bateson and Gairdner 1921), there have been more than seven hundred papers published on CMS in plants. Clearly, there has been a long-term, sustained interest in cytoplasmic male sterility.

The reproductive mechanisms of most higher plants make it difficult to produce hybrid seed derived from crossing two selected parents on a commercial scale. The stamens and carpels may be present within the same flower (as in wheat) or contained within separate flowers on the same plant (as in corn), and a high level of self-pollination can occur among plants of the female line or population. This selfing reduces the percentage of hybrid seed (hybridity) produced in the hybrid seed production process. To overcome this problem and ensure that crosses occur only between the selected female and male lines, several forms of pollination control have been used to produce hybrid seed in plants. These include (a) manual emasculation, (b) use of cytoplasmic male-sterility systems, (c) use of genic male-sterility systems (see Chapter 8), (d) use of self-incompatibility alleles (see Chapter 9), (e) use of male gametocides (see Chapter 10), and, more recently, (f) the use of genetically engineered "pollen killer" genetic systems (see Chapter 11). CMS has been extensively studied because it provides, at least in theory, a possible mechanism of pollination control in plants to permit the easy production of commercial quantities of hybrid seed.

Although CMS has been observed in more than 150 plant species (Laser and Lersten 1972; Kaul 1988), the mechanisms by which CMS interrupts normal pollen development are not well understood. Female fertility is generally not affected by CMS, so that male-sterile plants can and do set seed if viable pollen is provided.

The affected organs and tissues in CMS plants are the stamens (anther and filament) and pollen grains (microspores). The stamen plays an integral role in crop production because it is responsible for carrying out the male reproductive processes. Pollen grains develop within the anther in a precisely timed, high-metabolic-demand process, with the tapetal layer of the anther supplying enzymes, hormones, nutritive materials, nucleosides, and nucleotides for microsporogenesis to proceed (Vasil 1967; Pacini et al. 1985). It appears that CMS results because of a multitude of effects, all more or less related to insufficient or mistimed supply of necessary resources for developing microspores.

One of the major factors contributing to increased crop productivity in recent years has been the development of hybrid cultivars in a wide range of crops. In many crops, crosses between genetically distinct parental lines or populations give rise to progeny that exhibit heterosis or hybrid vigor (Feistritzer and Kelly 1987). Compared to their parents, hybrids with a heterotic phenotype are, in general, more resistant to disease and insects, less susceptible to environmental stress, and yield more seed (Feistritzer and Kelly 1987).

Although all CMS sources can theoretically be developed into functional CMS systems, only a few have actually been developed to the point where commercial quantities of hybrid seed are being produced. There are several reasons for this, including insufficient or unstable male sterility, difficulties with restoration systems, difficulties with seed production, and undesirable pleiotropic effects of the CMS pollination control system used. It is a long road from the discovery of CMS to the commercial use of CMS-based hybrids. Nevertheless, the majority of hybrids grown today are based on CMS as the pollination control mechanism.

Research on future uses of CMS appears focused on using CMS-related plasmids as vectors for cytoplasmic genes contained in both the mitochondria and chloroplasts. This research area is expected to develop capabilities similar to those already available in many crops for nuclear gene transformation.

This review will concentrate on nine agricultural or horticultural crops that are currently grown commercially as CMS-based hybrids.

Cytoplasmic male sterility (CMS) origins

CMS has several origins: intergeneric crosses, interspecific crosses, intraspecific crosses, and mutagen or antibiotic effects on cytoplasmic genes (Edwardson 1956, 1970; Frankel and Galun 1977; Kaul 1988); some origins are apparently spontaneous. To separate meiotic difficulties from CMS as a cause of male sterility in intergeneric and interspecific crosses, a backcross series to BC_2 or more is routinely completed to essentially introduce the nucleus of one species into the cytoplasm of another. In cases of true CMS, the degree of male sterility increases with each successive BC and exceeds 90% by BC_6 in most cases. In many cases, crosses have been intentionally done to try to discover new sources of CMS in plants, but only a few have been successful.

CMS occasionally arises from intergeneric crosses (Edwardson 1956, 1970). These cross types maximize the differences between nucleus and cytoplasm and may, therefore, increase the likelihood of obtaining a new CMS system due to incompatible nuclear–cytoplasmic interactions (Hanson and Conde 1985). These crosses are hard to make, however, and therefore few instances in the literature of these types of origins of CMS are noted. There appears to be a relationship between the width of crosses attempted, the likelihood of CMS's being produced, and the potential agronomic usefulness of the CMS produced. The wider the cross, the more likely the occurrence of incompatible nuclear–cytoplasmic interactions, and thus the more likely the discovery of CMS. However, such wide crosses result in many negative pleiotropic effects of the incompatible nuclear–cytoplasm interactions, and the agronomic usefulness of this type of CMS is very limited.

CMS frequently arises from interspecific crosses (Edwardson 1956, 1970). These cross types produce smaller differences between the nucleus and the cytoplasm and, therefore, somewhat decrease the likelihood of obtaining a new CMS system. These crosses are easier to make than intergeneric crosses, and many instances of these types of origins of CMS have been reported (Pearson 1981). There are few successful examples of CMS created by interspecific crosses, however. Pearson (1981) points out that for CMS created by interspecific crosses, in general, either the degree of male sterility was not economically satisfactory, or natural pollination mechanisms were not dependable, so that except for the marginal use in *Nicotiana* and *Gossypium*, these CMS systems have not become a factor in the production of hybrid seed.

CMS also arises from intraspecific crosses (Edwardson 1956, 1970). These are CMS types discovered in the wild or segregated out of cultivars. CMS arising from intraspecific crosses is quite common and is indicative of species existing in two or more cytoplasms, at least one of which is a male-sterile cytoplasm. CMS systems derived from these sources have been rapidly moved into seed production programs in onion, corn, sorghum, and rice (Pearson 1981).

CMS may also arise apparently spontaneously (Edwardson 1956, 1970). In this case, a male-sterile plant is discovered in a population of male-fertile plants. This has been quite common in occurrence, and many successful CMS systems, for example, in onions, corn, and pearl millet, originated spontaneously.

CMS may be induced through the use of mutagens (Edwardson 1956, 1970); a range of mutagens have been tried and occasional CMS has been uncovered. None of these sources of CMS have been found to be usable to date. Restricting our focus to the crops covered in this review, CMS has arisen by intergeneric crosses of *Raphanus sativus* × *Brassica napus* (Bannerot et al. 1977); and by interspecific crosses of *Helianthus petiolaris* × *H. annuus* (Leclercq 1969), *Petunia axillaris* × *P. hybrida* (Bianchi 1963), *Triticum timopheevi* × *T. aestivum* (Wilson and Ross 1962), and *Oryza spontanea* × *O. sativa* (Katsuo and Mizushima 1958). CMS from intraspecific crosses has originated in oilseed rape, where crosses of several spring- and winter-habit *B. napus* cultivars × *B. napus* cultivar Bronowski produced male-sterile (*nap* CMS) BC progeny (Shiga 1980); in onions, where the cross of *A. cepa* cultivars Scott Country Globe × [(Australian Brown × Persian White) F_4] produced CMS progeny (Peterson and Foskett 1953; Foskett 1954); and in sorghum, where crosses of *S. bicolor* types milo × kafir produced 99% CMS in the BC_2 (Stephens and Holland 1954).

The spontaneous origin of CMS has been reported in a single plant of the onion (*A. cepa*) variety Italian Red (H. A. Jones and Emsweller 1937). In corn (*Z. mays*), the CMS-T form (Rogers and Edwardson 1952), the CMS-S form (D. F. Jones et al. 1957), and the CMS-C form (Beckett 1971) all originated apparently spontaneously in plants grown in the field. In pearl millet, *P. americanum*, Burton (1958) described a male-sterile plant occurring naturally within inbred line 556. A stable male-sterile line, Tift 23A, was released in 1965 and formed the basis of a worldwide pearl millet hybrid cultivar development effort.

CMS has also been induced by mutation in corn (*Z. mays*) and sorghum (*S. bicolor*). Briggs (1971) reported that EMS treatment produced CMS-based, but unstable, male-sterile lines in corn. However, streptomycin produced stable CMS mutants in corn (Briggs 1973). Burton and Hanna (1976) reported that ethidium bromide produced some CMS mutants in pearl millet, but these were later found to be unstable. Mitomycin and streptomycin treatment did produce stable CMS mutants in pearl millet, some of which were identical to the A1 CMS already discovered (Burton and Hanna 1982). Finally, nuclear gene induction of CMS in corn has been reported by Lemke et al. (1985), who reported that F_2 progeny from crosses of *iojap*-gene–carrying males to male-fertile female lines were male sterile.

CMS characterization

CMS systems have traditionally been characterized by the restorer genes required to overcome the CMS and to provide male-fertile progeny in the male-sterile cytoplasm (Table 7.1). More recently, CMS systems have been characterized by DNA analysis techniques applied to cytoplasmic organelles that are able to distinguish genetic differences between normal and male-sterile cytoplasms (Hanson and Conde 1985). Current research in this area is primarily at the molecular level, where researchers are trying to elucidate the mechanisms underlying CMS and the relationship of nuclear-based restorer genes to cytoplasm-based male-sterility genes (Hanson 1991).

CMS restoration is by nuclear genes, frequently dominant in action, in many cases few in number (that is, one to four dominant genes, sometimes with minor male-fertility modifiers required to maintain complete restoration in some environments). The CMS restorer genes temporarily suppress the expression of CMS by mechanisms not well understood, permitting normal or near-normal pollen production. DNA analysis techniques are uncovering restriction map differences between CMS and fertile cytoplasms in all crops so far studied.

CMS mechanism of action

Abnormal behavior of the tapetum in the anthers has frequently been identified with CMS. Further, it has now been well established that the genetic determinants of CMS reside in the mitochondria and that nuclear genes control the expression of CMS (Newton 1988). The restorer genes described in the previous section are important components in the search for a CMS mechanism of action, since the restorer genes temporarily suppress male sterility even in the presence of the CMS cytoplasm.

Oilseed rape (Brassica napus)

Fan (1985) reported that failure of microsporogenesis for both the *nap* and *pol* CMS was associated with lack of differentiation of the first sporogenous cells

Table 7.1. *Characterization of CMS systems in selected crops*

Crop	CMS type	Characterization	
		Restorer genes	Organelle DNA
OILSEED RAPE			
B. napus Brown	*nap*	One dominant gene, *Rf1*, present in virtually all *B. napus* cultivars (Fan 1985)	Normal mitochondrial pattern used as the reference for all other CMS cytoplasm comparisons (M. Singh and 1991)
B. napus	*ogu*	Two dominant genes, *Rf1* and *Rf2*, from European radish (Heyn 1978)	A unique 2.5-kb NcoI mt DNA fragment is found (Erickson et al. 1986; Bonhomme et al. 1991)
B. napus	*pol*	One dominant gene, *Rf1*, from cv. Italy (Fang and McVetty 1988, 1989)	A novel unidentified reading frame upstream of the *atp6* gene is found (Erickson et al. 1986; Singh and Brown 1991; Handa and Nakajima 1992; Handa 1993)
SUNFLOWER			
H. annuus	*pet*	*Rf1* to *Rf4* dominant genes all from wild sunflower species, at least two dominant *Rf* alleles required (Vannozi 1987)	Two major rearrangements, a 12-kb inversion and a 5-kb insertion or deletion, distinguish the CMS and normal mt DNA (Crouzillat et al. 1987; Sicuella and Palmer 1988)
RICE			
O. sativa	WA	Two dominant genes, *Rf1* and *Rf2*, from many cultivated rice cultivars (Virmani et al. 1981; Govinda-Rej and Virmani 1988)	A 2,100-bp linear plasmid is found in WA mt DNA but not found in normal mt DNA (Mignouna et al. 1987)
PEARL MILLET			
P. americanum	A1	One dominant gene, *Rf1*, from most cultivated pearl millet cultivars (Andrews 1987; Patil 1987; Rai and Hash 1990)	No differences reported

Crop	CMS type	Restorer genes	Characterization
			Organelle DNA
SORGHUM			
S. bicolor	milo	One dominant gene, Rf1, from many milo cultivars (Chopra 1987)	Sorghum cytoplasms are distinguishable based on mt DNA restriction analyses (Pring et al. 1982; Conde et al. 1982; Bailey-Serres et al. 1986); a 65-kd variant polypeptide is produced that is not seen in normal cytoplasm (Dixon and Leaver 1982)
COMMON WHEAT			
T. aestivum	timopheevi	Two to four dominant genes, RF1 to Rf4, all from T. timopheevi (Maan 1985; Edwards 1987)	No differences reported
CORN			
Z. mays	S	One dominant gene, Rf3, gametophytic in action, from many commercial inbreds (Laughnan and Gabay 1978)	Two plasmidlike double-stranded DNA molecules, S1 and S2, are found that are not present in normal cytoplasm (Pring et al. 1977; Levings and Pring 1979)
Z. mays	T	Two dominant genes, Rf1 and Rf2, from commercial inbreds (Rf1 rare, Rf2 common) (Duvick 1956)	A unique gene, T-urf13, codes for a 13-kd polypeptide; URF13 produced by CMS-T mitochondria (Levings and Pring 1976, 1979; Dewey et al. 1986, 1987)
Z. mays	C	One dominant gene, Rf4, from many commercial inbreds (Khery-Pour et al. 1981)	Two circular DNA molecules are found that are not present in normal cytoplasm (Levings and Pring 1979; Kemble et al. 1980)

Table 7.1 (cont.)

Crop	CMS Type	Characterization	
		Restorer genes	Organelle DNA
ONIONS			
A. cepa	S	One dominant gene, *Rf1*, from some cultivars (Jones and Clarke 1943); *Rf* gene not required for hybrid onion cultivars	No differences reported
PETUNIA			
P. hybrida axillaris	S	One to three dominant genes, *Rf1* to *Rf3*, all from *P. axillaris* (Van Marrewijk 1969)	A unique mitochondrial locus, *pcf* (petunia CMS-associated fused), is present (Hanson 1991)

at the very beginning of pollen development, consistent with Shiga (1976) for the *nap* CMS. Ogura (1968), Bartkowiak-Broda et al. (1979), Fan (1985), and Polowick and Sawhney (1990, 1991) reported that abortion of microspores in *ogu* CMS is due to abnormal behavior of the tapetum. The failure of the tapetum to perform normal and necessary functions is a common presumed cause of CMS in many plant species (Edwardson 1956, 1970).

M. Singh and Brown (1991), Handa and Nakajima (1992), and Handa (1993) reported that the difference between normal and *pol* CMS mitochondria was that the latter contain a novel, unidentified reading frame (*pol-urf*) upstream of the *atp6* gene, which is cotranscribed with the *atp6* gene (M. Singh and Brown 1991). The transcription of *pol-urf* is affected by the *Rf1* gene for *pol* CMS restoration (Witt et al. 1991), which suggests that it plays a role in this CMS.

Sunflower (Helianthus annuus)

Sicuella and Palmer (1988) found a 5-kb insertion of unknown origin in the male-sterile–line mitochondria that was not present in the normal line. This 5-kb insertion results in the cotranscription of an *atpA* gene with urf522, an unidentified reading frame of 522 nucleotides (Kohler et al. 1991). Whether the sunflower region is truly correlated with CMS awaits critical testing of the transcription and translation of the aforementioned locus in isonuclear lines containing *Rf* alleles.

Rice (Oryza sativa)

Mignouna et al. (1987) reported that WA CMS lines of rice have a 2,100-bp linear plasmid present in the mt DNA not found in male-fertile rice lines. The relationship of this plasmid to WA CMS was not established.

Pearl millet (Pennisetum americanum)

The mechanism of action of CMS for pearl millet is presently unknown.

Sorghum (Sorghum bicolor)

A nonfunctional tapetum was the identified cause of milo CMS in sorghum (Brooks et al. 1966). The milo CMS cytoplasm produces a 65-kd variant polypeptide not present in the normal fertile cytoplasm, and the quantity of this polypeptide varied in response to the presence of nuclear restorer genes (levels were high in the absence of restorer genes, and low – less than 1% of the CMS line – in the presence of restorer genes) (Dixon and Leaver 1982). The mechanism of action of milo CMS gene and associated polypeptide remains unknown.

Common wheat (Triticum aestivum)

Degeneration of the tapetum and its subsequent lack of starch storage were reported to be a direct cause of microspore degeneration in *timopheevi* CMS common wheat (Joppa et al. 1966). Stamens of CMS common wheat had poorly developed vascular bundle. Joppa et al. (1966) concluded that decreased starch accumulation in the tapetum and the lack of starch storage in maturing microspores might be explained by reduced solute transport in stamens of sterile plants.

Corn (Zea mays)

Cytological studies of CMS-T anthers revealed that the mitochondria break down in the tapetal cells and later in the microspores (Warmke and Lee 1977). CMS-T mitochondria are distinguished by characteristic mt DNA restriction fragment patterns and produce a unique new 13-kd polypeptide (URF13) (Levings and Pring 1976). Synthesis of URF13 is greatly reduced in mitochondria of plants carrying the CMS-T restorer genes (Forde and Leaver 1980; Levings 1993). A mitochondrial gene designated *T-urf13* is responsible for the CMS and disease-susceptibility traits borne by CMS-T corn (Dewey et al. 1986, 1987). *T-urf13* encodes a 13-kd polypeptide (URF13) that is a component of the inner mitochondrial membrane (Dewey et al. 1987) and is uniquely associated with the CMS-T type of CMS.

The effect of restorer genes on *T-urf13* expression indicates an association between CMS and *T-urf13*. The dominant nuclear restorer genes *Rf1* and *Rf2* jointly suppress pollen sterility in CMS-T corn. *Rf1* decreases the abundance of URF13 by about 80% (Dewey et al. 1987). The reduction in the abundance of URF13 caused by *Rf1* is essential for restoration of male fertility; however, *Rf1* is not sufficient to restore male fertility without the activity of *Rf2*. What *Rf2* does is unknown, other than the fact that it does not alter the expression of *T-urf13* (Dewey et al. 1987).

Plant mitochondrial genes encode polypeptides that are components of the electron transport system (Complex I, III, IV), the F_0F_1 ATP synthase, and the mitochondrial protein synthesis system (Eckenrode and Levings 1986; Newton 1988). They also code for structural RNAs, such as the 26S and 45S rRNAs, and for tRNAs. Mitochondrial mutations among this restricted array of essential genes are predicted to be extremely deleterious or even lethal. However, CMS has been exploited in crop plants, in which anything other than normal growth, development, and productivity is unacceptable. This means that CMS is unlikely to result from a major lesion in an essential mitochondrial gene.

In CMS-T corn, the *T-urf13* gene is a unique chimeric sequence that codes for a nonessential polypeptide. The usual complement of essential mitochondrial genes appears to be normal in expression and function.

Almost any mitochondrial gene or its product could be involved in CMS if there is something unusual or different about the way mitochondrial genes are expressed or about the way they function during pollen development. In plants, a CMS gene manifests its adverse effect only in certain cells involved in pollen development, whereas other cells are relatively unaffected.

A possible explanation for the specific effects of CMS mutations on pollen development is that demands on the mitochondria of anthers are greatest at the time of pollen development. In corn, there is a 40-fold increase in the number of mitochondria per cell in the tapetum, and a 20-fold increase in the sporogenous cells (Lee and Warmke 1979). These mitochondria amplification events suggest an increased demand for energy, reducing equivalents or metabolites during pollen formation. This increase in the numbers of mitochondria occurs only in anthers, no where else. A mutant mitochondria gene product could be a serious impairment to development under conditions in which unusually heavy demands for energy exist. Under these conditions, proteins, such as URF13, that are located in the inner mitochondria membrane could interfere with the rapid amplification of mitochondria or, perhaps, with the electron transport or phosphorylation processes. Anther mitochondria may thus be uniquely sensitive to a mutant mitochondrial gene – a sensitivity that does not occur in other plant cells (Levings 1993).

A second mechanism of CMS involves the notion that anthers produce a specific substance that interacts with the URF13 protein to permeabilize the inner mitochondria membrane, resulting in mitochondria dysfunction and cell death. The precocious death of anther cells, particularly the tapetal cells, is expected to interfere with normal pollen development and cause male sterility. This model is attractive because it is already known that URF13 has the capacity to destroy mitochondria activity and cause cell death – events that could lead to pollen abortion if they occur in the anther (Levings 1993). There are no data to support this model at the present time, however.

URF13 may be overexpressed in anther cells, resulting in highly toxic levels that cause cellular dysfunction, death, and the cessation of pollen development. Overexpression of URF13 could be related to the increased levels of mitochondria biogenesis observed in the tapetum during pollen formation. Alternatively, anther cells, especially the tapetum, may be more sensitive to URF13 than are other cell types, because unique functions pertinent to pollen formation occur in these cells. It is possible that anther cells, with their specialized functions, are more vulnerable to the toxicity of URF13 than are other cell types. Even though the investigation of the mechanism of CMS-T has progressed this far, we still do not know the ultimate cause of CMS in CMS-T corn (Levings 1993).

The corn CMS-T locus is found near recombination repeats in the mitochondrial genome. The location of the CMS-correlated locus near recombination repeats may not be mere coincidence, but may reflect in part some of the recombination events involved in the CMS gene creation. Plant mitochondrial genomes are extremely fluid in their organization, though the gene sequences are highly conserved (Hanson and Folkerts 1992).

In CMS-T, restorer genes affect the expression of mitochondrial genes at the RNA level, though it is not known whether this is a direct effect on transcription or an indirect effect resulting from action at the translational level. A protein product unique to the CMS-T genotype is reduced in abundance by the restorer genes (Nivison and Hanson 1989). If the *T-urf13* gene polypeptide product (URF13) is the cause of male sterility, then an understanding of the mode of action of the polypeptide is required to understand the molecular basis of CMS. The corn URF13 protein is found in the mitochondrial membrane protein fraction (Dewey et al. 1987).

In CMS-S, the development of sterile CMS-S anthers is identical to that of normal cytoplasm anthers until very late in pollen development. Only the pollen grains themselves showed alterations in development compared to normal. Microspore disintegration started between the first and second divisions in the microspore and progressed rapidly thereafter (Lee et al. 1980).

Two plasmidlike strands are found in CMS-S (Pring et al. 1977); these two plasmids, designated S1 and S2, are approximately 6.2 kb and 5.2 kb in length, respectively, and are linear molecules with similar terminal inverted repeat sequences of about 200 base pairs at the termini of the duplex molecules (Levings and Pring 1979). These plasmids apparently do not respond to the *Rf3* (gametophytic restoration) gene for male-fertility restoration in CMS-S. The CMS-S mitochondria do produce unique polypeptides not found in CMS-T or CMS-C or normal cytoplasm, but these do not respond to the *Rf3* gene either (Forde and Leaver 1980). The mechanism for CMS-S thus remains a mystery.

For CMS-C, there are tapetal cell breakdowns that occur at the tetrad stage of pollen development. No mitochondrial structural changes, however, have been observed (Lee et al. 1979). In addition, in CMS-C there are two circular DNA species of about 1.57 kb and 1.42 kb that are not found in the other cytoplasmic types (Kemble et al. 1980). It is not known whether these plasmid species have a role in the determination of CMS-C (Kemble et al. 1980). The CMS-C mitochondria produce a unique 17,500-Mr polypeptide not seen in the other cytoplasms (Forde et al. 1978). This polypeptide does not respond to the *Rf4* gene required for male-fertility restoration and so a causal role for this polypeptide in CMS-C cannot be established.

Onion (Allium cepa)

CMS-S–related male sterility is caused by disturbance or delayed activity of the tapetum, which is believed to cause a disturbance in the hormone content of the anthers (Holford et al. 1991).

Petunia (Petunia hybrida)

In *axillaris* CMS, cells of the tapetal layer become vacuolated during early meiosis of the microspores, something that does not happen in the normal lines (Bino 1985). At the DNA level, the *axillaris*-CMS–encoding mitochondrial locus, termed *pcf* for "petunia CMS-associated fused," contains part of the coding region of *atp9*, part of each exon of *coxII*, and an unidentified reading frame termed *urf-S*. The locus most conclusively correlated with CMS in petunia contains one chimeric gene and one or more cotranscribed, standard mitochondrial genes with apparently normal coding regions. These chimeric genes are currently the most suspect as the cause of CMS. They encode protein products that are not present in fertile lines (Hanson 1991). The *axillaris* CMS locus is found near recombination repeats in the mitochondrial genome. The location of the CMS-correlated locus near recombination repeats may reflect in part some of the recombination events involved in the creation of the CMS genes. Plant mitochondrial genomes are extremely fluid in their organization, though the gene sequences are highly conserved (Hanson and Folkerts 1992). Synthetic peptide antibodies to peptides predicted by the *urf-S* portion of the *pcf* gene revealed the presence of a 25-kd protein (PCF protein) in CMS lines that was not present in fertile lines and that was much reduced in fertility-restored lines (Nivison and Hanson 1989).

In petunia, restorer genes affect the expression of mitochondrial genes at the RNA level, though it is not known whether this is a direct effect on transcription or an indirect effect resulting from action at the translational level.

The petunia PCF protein is found both in membrane and in soluble fractions (Nivison and Hanson 1989). Consistent differences in the partitioning of electron transport through the cytochrome oxidase and alternate pathway have also been detected in immature anthers of petunia (Connett and Hanson 1990). The immature anthers show a reduction of alternate pathway activity compared to normal fertile materials. A culture from a fertility-restored line was also restored in alternate pathway activity, a result providing a tentative correlation of the reduced activity with the sterile phenotype.

How an alteration in electron transport partitioning could result in disrupted pollen development is presently unknown, since in most plants the pathway's purpose is not known. However, the pathway is often found in tissues with high biosynthetic rates, implying some important function at such times – perhaps to permit oxidation of NADH when ATP levels are high (Siedow and Berthold 1986). Perhaps the tapetal layer, which is thought to synthesize many molecules important for pollen development, requires high levels of alternate pathway activity for biosynthesis.

Heterosis and other advantages of hybrids

Heterosis is defined as "hybrid vigour such that an F_1 hybrid falls outside the range of the parents with respect to some character or characters" (Allard 1960, p. 468). In a commercial context, heterosis for seed yield, fruit yield, or total dry matter yield is most important. Both high parent heterosis (performance of the F_1 that exceeds the better parent used in the cross) and standard heterosis (performance of the F_1 that exceeds the best nonhybrid cultivar of the day) are referred to in the hybrid advantage context.

Oilseed rape

High parent heterosis for seed yield in winter oilseed rape is approximately 50% (Shiga 1976); high parent heterosis for seed yield in summer rape is 30%–70% (Sernyk and Stefansson 1983; Grant and Beversdorf 1985; Brandle and McVetty 1989). High parent heterosis also occurs for total dry matter and harvest index in summer rape hybrids (McVetty et al. 1990).

Sunflower

High parent heterosis for seed yield is approximately 50% (Vranceanu 1987), and for oil content approximately 10%. High parent heterosis for plant height, head diameter, seed weight, and earlier flowering and maturity also occur in sunflower (Vranceanu 1987).

Rice

High parent heterosis for seed yield is approximately 20%–30% (Virmani 1987). High parent heterosis for total dry matter production, harvest index, plant height, earlier flowering, enhanced cold and salt tolerance, root growth, and seedling vigor also occur in rice (Virmani 1987).

Pearl Millet

High parent heterosis for seed yield is approximately 10%–15% (Andrews 1987). High parent heterosis for total dry matter and for improved yield stability also occur in pearl millet. Yield stability is the most important benefit of pearl millet hybrids in India, because they are traditionally grown under low-fertility and low-moisture conditions (Andrews 1987).

Sorghum

High parent heterosis for seed yield is approximately 20%–30% (Chopra 1987). High parent heterosis for earlier maturity, plant height, and greater responsiveness to added nutrients also occurs in sorghum. Hybrid sorghum cultivars are tolerant of lodging conditions and are easy to harvest (Chopra 1987).

Common wheat

High parent heterosis for seed yield in wheat is approximately 20%–30%, whereas standard heterosis is approximately 10%–20% (Edwards 1987). High parent heterosis is also displayed for plant height, for earlier heading and maturity (Virmani and Edwards 1984), and for total dry matter production and harvest index (Benson 1978). Other advantages of hybrid wheat include better disease and insect resistance, wide environmental adaptability, better milling and baking traits, seedling vigor, and improved root system development (Stroike 1987).

Corn

High parent heterosis for seed yield in maize is approximately 300%–400% over the best inbred lines parent, whereas standard heterosis is estimated to be approximately 50% (Hallauer et al. 1988). Corn hybrids also display high parent heterosis for lysine and oil content, forage digestibility, total dry matter production, disease and insect resistance, plant and ear height, ear length, leaf size, and earlier flowering date (Kim 1987).

Onion

High parent heterosis for marketable yield (not necessarily for actual yield) occurs in many hybrid vegetables, including onions (Riggs 1987). High parent heterosis for bulb size, earlier maturity, quality, and disease and insect resistance also occur in onion hybrids. In addition, hybrid onions are more uniform in phenology and form (Riggs 1987).

Petunia

Hybrid petunias display substantial growth vigor and overall excellent quality compared to inbred-line materials (Ewart 1984). The large (grandiflora) flower type and the double-flower-number characteristics in petunia are entirely dependent on hybrid production (Ewart 1984).

CMS utilization in hybrid seed production

The long-term sustained interest in CMS is related to the fact that it provides a possible mechanism of pollination control in plants to permit the easy production of commercial quantities of hybrid seed. A CMS system – consisting of a male-sterile line (the "A-line"), an isogenic maintainer line (the "B-line"), and, for crops where male-fertility–restored F_1's are required, a restorer (the "R-line") – must be developed (Allard 1960).

Traditionally, A-lines are developed by backcrossing selected B-lines to a CMS A-line for four to six times to generate a new A-line and B-line pair. R-lines are developed by similar backcross procedures using a CMS R-line as the female in the original cross and a new line as the recurrent parent in four to six backcrosses. Selfing the last backcross generation two successive times and selection of pure-breeding male-fertility–restored lines is required to complete the development of the new R-lines developed in the CMS cytoplasm.

Once the appropriate A-lines, B-lines, and R-lines are produced, seed production for commercial hybrid seed quantities is done in two stages: (1) the maintenance stage and (2) the crossing stage. The maintenance stage involves increasing the seed quantities of the A-line, the B-line, and the R-line; the crossing stage involves the production of hybrid seed using A-line and R-line parents. The maintenance stage of hybrid seed production is frequently controlled by the originator of the hybrid and is usually not problematical because relatively small quantities of seed of the A-line, B-line, and R-line are required for any one hybrid. The critical stage in hybrid seed production is the crossing stage, reviewed in Table 7.2.

The large number of similarities among hybrid seed production practices for a wide range of commercial hybrid crops grown throughout the world is striking (Table 7.2). Current commercial hybrid seed production relies entirely on the block method (alternating strips of female and male genotypes) of hybrid seed production. Research into the mixed method (A-line and R-line seed planted in a mixed stand and harvested in bulk) of hybrid seed production in oilseed rape is ongoing in Canada and Europe. Whether this approach to hybrid seed production will be successful is yet to be determined.

Table 7.2. *Hybrid seed production in selected crops: a summary of important information*

Crop	CMS type	Restor-ation	Seed prod. method	F:M ratios	F:M widths	Pollen vector	Borders	Isolation distance	Seed yield	References	Comments
OILSEED RAPE											
B. napus	*ogu*	*Rf1*	Block	3:1 7:1	9m:3m 14m:2m	Honey-bees	6m	?	1,000 to 3,000 kg/ha	Frauen (1987)	
B. napus	*ogu*	None	Block	3:1 7:1	9m:3m 4m:2m	Honey-bees	6m	?	1,000 to 3,000 kg/ha	Frauen (1987)	Use 20% pollinator in the commercial production fields
B. napus	*ogu*	*Rf1*	Mixed	90% F: 10% M	None	Honey-bees	6m	?	1,200 to 3,200 kg/ha	Frauen (1987)	80% or more hybridity expected
B. napus	*pol*	*Rf1*	Block	3:3	2m:2m	Leaf-cutter or honeybees	2m	800m	400 to 600 kg/ha	Pinnisch and McVetty (1990) McVetty (unpub.)	
B. napus	*pol*	*Rf1*	Mixed	90%F: 10%M	None	Leaf-cutter or honeybees	2m	800m	600 to 800 kg/ha	Hutchison (unpub.)	75% or more hybridity expected
SUNFLOWER											
H. annuus	*pet*	*Rf1* to *Rf4*	Block	2:1 7:1	1.8m:0.9m 6.3m:0.9m	Honey-bees	1.8m	3km	1,200 kg/ha	Vannozi (1987)	At least 2 dominant *Rf* alleles required
RICE											
O. sativa	WA	*Rf1*, *Rf2*	Block	5:1 6:1	1.0m:0.2m 1.2m:0.2m	Wind	100m	100m	200 kg/ha	Xizhi (1987) Virmani (1987)	Isolation area planted to male parent
PEARL MILLET											
P. americanum	A1	*Rf1*	Block	4:2 12:2	3.6m:1.8m	Wind	None	?	1,200 kg	Patil (1987)	

Crop	CMS type	Restor-ation	Seed prod. method	F:M ratios	F:M widths	Pollen vector	Borders	Isolation distance	Seed yield	References	Comments
SORGHUM											
S. bicolor	milo	*Rf1*	Block	6:2 24:8	5.4m:1.8m 21.6m:7.2m	Wind	7.2m	400m to 800m	1,500 to 1,800 kg/ha	Chopra (1987)	
COMMON WHEAT											
T. aestivum	*timo-pheevi*	*Rf1* to *Rf4*	Block	30:15 70:35	6m:3m 14m:7m	Wind	None	100m	1,000 to 3,000 kg/ha	Stroike (1987) Edwards (1987) Wilson (1984)	Two or more *Rf* genes plus male fertility modifiers required in some environments
CORN											
Z. mays	S	*Rf3* (gameto-phytic)	Block	4:1 6:2	4m:1m 6m:2m	Wind	5m	?	2,000 to 3,000 kg/ha	Wych (1988) Laughnan and Gabay (1978)	
Z. mays	C	*Rf4*	Block	4:1 6:2	4m:1m 6m:2m	Wind	5m	?	2,000 to 3,000 kg/ha	Wych (1988) Khery-Pour et al. (1981)	
ONIONS											
A. cepa	S	*Rf1*	Block	2:1	2m:1m	Honey-bees	None	1km?	400 to 600 kg/ha	Takahasi (1987)	*Rf* gene not required Seed production frequently done in tents or plastic tunnels
PETUNIA											
P. hybrida	*axill-aris*	*Rf*	Block	2:2	2m:2m	Hand-pollinated	None	None	0.5 to 5.0 g/plant	Ewart (1984) Izhar (1978, 1984)	All seed production done in greenhouses

CMS limitations

Many CMS systems have limitations that make it difficult or impossible to use them as pollination control systems in the production of commercial quantities of hybrid seed. These limitations are (1) pleiotropic negative effects of the CMS on agronomic quality performance of plants in the CMS cytoplasm, (2) enhanced disease susceptibility, (3) complex and environmentally unstable maintenance of male sterility and/or male-fertility restoration, and (4) inability to produce commercial quantities of hybrid seed economically because of poor floral characteristics for cross-pollination. In addition, the dangers of monoculture crop production in one CMS cytoplasm have been documented for corn (Levings 1993). Different crops have different problems or limitations, as indicated in the following.

Oilseed rape

The *nap* CMS system in oilseed rape is unstable at moderate to high temperatures and therefore not commercially usable (Fan and Stefansson 1986). Oilseed rape with the *ogu* CMS system growing at 12°C or less displays chlorosis and generally poor growth (Bannerot et al. 1977). In addition, female fertility is adversely affected by the *ogu* CMS *Rf* genes (Pellan-Delourme et al. 1988; Pellan-Delourme and Renard 1988). McVetty et al. (1990) reported that oilseed rape hybrids produced using the *pol* CMS system performed significantly poorer than hybrids in normal cytoplasm for seed yield, total dry matter, harvest index, and oil content. The frequency of good maintainers for oilseed rape in *pol* CMS is very low, due to minor modifying genes for male-fertility restoration (Fan et al. 1986; Burns et al. 1991). Polima CMS A-lines are also temperature sensitive, reverting to partial male fertility at temperatures over 30°C (Fan and Stefansson 1986).

Sunflower

The sunflower is both wind and insect cross-pollinated over great distances, requiring large isolation distances for commercial hybrid seed production. The hybrid seed production fields must be rogued three to four times to ensure genetic purity of hybrids, since even in the purest female line, 1%–2% of plants have broken male sterility (due to minor genes for male-fertility restoration). The large number of restorer genes (up to four) with at least two dominant alleles at a minimum of two different loci required for restoration adds to the complexity and cost of R-line development. Paired A-line and B-line plant crosses are made to ensure purity of genotype and maintenance of all agronomic and disease characteristics (Vannozi 1987).

Rice

For rice with the WA CMS, the primary problem with hybrid seed production is flower morphology related to necessary outcrossing to ensure economic yields of hybrid seed. The WA CMS A-lines have to be very carefully selected for outcrossing ability. Outcrossing percentages in China are 15%–45%, resulting in low hybrid seed production. The frequency of complete–male-sterility maintainers is low and many B-lines produce partial–male-sterile plants, even after seven to nine BC generations, due to linked minor genes for male-fertility restoration, similar to sunflower (Virmani 1987).

Pearl millet

For pearl millet with the A1 CMS system, there are a number of problems. The principal problem is the low frequency of maintainer genotypes and the presence of linked minor genes for male-fertility restoration, which results in partial male fertility in most A-lines produced in the A1 CMS. Most pearl millet A-lines are not perfectly male sterile and are therefore not usable in hybrid seed production. In addition, seed set is poor on many of the A-lines (Andrews 1987). To further complicate matters, the A-lines are quite susceptible to ergot (*Claviceps purpurea* Lovelass.), and the increased ergot susceptibility has been shown to be due to the A1 cytoplasm (Andrews 1987). The hybrids are more susceptible to downey mildew than are the conventional open-pollinated population varieties previously grown. This is not directly related to the A1 CMS cytoplasm, but due rather to the limited genetic potential of A-lines and B-lines presently available in pearl millet (Andrews 1987).

Sorghum

For sorghum with the milo CMS system, Kidd (1961) noted that higher temperatures tend to increase male fertility of the A-lines. This, in turn, results in much reduced hybridity of the hybrid seed lot.

Common wheat

For common wheat with the *timopheevi* CMS, the principal problems are high hybrid seed production costs, caused by a low seed multiplication rate (approximately 35:1) and by the limited extent of cross-pollination in seed production fields (Edwards 1987). In addition, male-fertility restoration for the *timopheevi* CMS system is complex, with up to four major restorer genes and one or more minor modifier genes required to effect full male-fertility restoration in some environments; even then, the restorer system is sensitive to environmental stresses (Stroike 1987).

Corn

For corn with the CMS-T, Duvick (1965) reported that the T cytoplasm resulted in plants that were 1%–3% shorter in height, lower in leaf number, and 1%–3% lower in yield than plants in the normal corn cytoplasm. The major and fatal limitation of CMS-T, however, was disease susceptibility. Southern corn leaf blight (*Bipolaris maydis* race T) is specifically virulent on CMS-T lines (Williams and Levings 1992). The susceptibility of CMS-T to *B. maydis* race T is caused by mitochondrial sensitivity to a host-specific pathotoxin (Levings 1993), which may be a pleiotropic effect of the CMS genes or due to different, but linked, genes in the CMS-T mitochondria (Pring and Lonsdale 1989). Yellow leaf blight (*Phyllosticta maydis*), another fungal pathogen, was also specifically virulent on corn CMS-T lines by the same mechanism as *Bipolaris maydis* race T, that is, a toxin (Levings 1993). In contrast, mitochondria from normal corn cytoplasm are resistant to both of these diseases. The use of CMS-T has been discontinued because of the disease susceptibility of this cytoplasm (Wych 1988).

For corn with CMS-S, the main limitation is the fact that male sterility is unstable, and reversions to male fertility occur, at varying frequencies, depending on nuclear genotype (A. Singh and Laughnan 1972)

For corn with CMS-C, the principal problem is instability of the male sterility, resulting in partial male-sterility breakdown, depending on the environment (Kim 1987; Tracy et al. 1991).

Onion

In onions with the CMS-S, the primary problem is that the male sterility of some A-lines is temperature unstable (Riggs 1987). The A-lines are completely male sterile at 14°C, but predominately male fertile at temperatures of 23°C or higher (Van Der Meer 1978). This reduces hybridity of the hybrid seed lot. Problems with male-fertility restoration are avoided; restoration is not required in hybrid onions because seed is not the economic component in the hybrid.

Petunia

For petunia with the *axillaris* CMS, hybrids produced using this CMS have been found almost always to be inferior to their fertile cytoplasm counterparts (Ewart 1984). Reduction in flower size, inferior flower color, and bud blasting are the problems most frequently encountered (Frankel and Galun 1977).

CMS future developments

In addition to a better understanding of CMS – and thus better control of CMS systems – and the extension of CMS to crops that presently have no CMS systems, other future uses of CMS can be suggested. These include new combinations of CMS mitochondria and normal or herbicide-resistant chloroplasts by cybrid formation. Pelletier (1983) produced *Brassica* cybrids with (1) *ogu* CMS mitochondria and *nap* cytoplasm chloroplasts and (2) *ogu* CMS mitochondria and atrazine-resistant chloroplasts using protoplast fusion and organelle sorting out. Similarly, Chuong et al. (1988) were able to combine *pol* CMS mitochondria and atrazine-resistant chloroplasts to produce oilseed rape cybrids.

In another approach, the use of CMS-related mt DNA plasmids such as S1 and S2 (of corn from CMS-S) – two plasmids that have many characteristics reminiscent of transposable elements of prokaryotes (Levings and Pring 1979) – as potential vehicles for genetic engineering of the mitochondria has been suggested. The S1 and S2 elements are linear duplexes with terminal repeat sequences. There is also good evidence that their disappearance as autonomous mt DNA species in association with CMS-S reversion is coupled with transposition of at least some of their sequences to the mitochondrial genome. In light of this endeavor, the vectoring of specific improved respiration-related genes or herbicide-resistance genes to mitochondria using S1 and S2 has been envisioned (Laughnan and Gabay-Laughnan 1983).

In yet another approach, the CMS trait carried in mitochondria is used as a genetic marker to permit the identification of mitochondria in protoplast-fusion–produced cybrids. Since mitochondria genome recombination also frequently takes place in cybrid cells, this approach allows for the recombination of mitochondria genes to create new and improved mitochondria, with combinations of traits not formerly available (Galun and Aviv 1983).

In a similar approach, Nilkamp et al. (1983) are developing genetic manipulation methods for plant-cell organelle genes. These authors are studying the molecular basis of CMS in petunia and the possibility of transfer of the trait to other members of the Solanaceae family by somatic cell cybridization. They are using autonomous replicating sequences (ARSs) to try to vector mitochondria and chloroplast genes to their respective organelles. They propose the insertion of such ARSs in plasmids with an appropriate marker gene that can function as a selection marker (for example, CMS) to produce a vector for plant organelles. The availability of such vectors and of techniques such as cybridization opens the way for introduction of organelle genes involved in economically important properties such as photosynthesis, CMS, resistance to herbicides, and resistance to plant pathogens into plant organelles (Nilkamp et al. 1983).

Future molecular genetic work will help to elucidate the molecular and functional basis of CMS. Techniques such as the isolation and study of organelle DNA, the development of restriction endonuclease, and cloning and sequencing techniques have permitted a closer look at CMS. In corn and petunia, normal and CMS strains were found to have significant mt DNA differences. In some crops, a specific mitochondrial gene has been directly implicated in CMS. For example, the *T-urf13* gene and CMS-T, and Petunia *pcf* and CMS-S are functionally related in corn and Petunia, respectively. Despite the identification of CMS-related loci in some crops, the molecular mechanism of CMS is still unknown (Hanson and Folkerts 1992). Thus, a cause-and-effect relationship for CMS has yet to be established for any crop. Once this relationship is established, the possibility of improving or creating CMS in any species desired, through the use of molecular genetic techniques, becomes closer to reality.

Acknowledgments

The assistance of Mr. L. Friesen in the preparation of this chapter is gratefully acknowledged.

References

Allard, R.W. (1960). *Principles of Plant Breeding*. New York: John Wiley and Sons.

Andrews, D.J. (1987). Breeding pearl millet grain hybrids. In *Hybrid Seed Production of Selected Cereal, Oil and Vegetable Crops*, eds. W.P. Feistritzer and A.F. Kelly, pp. 83–111. Rome: FAO.

Bailey-Serres, J., Dixon, L.K., Liddell, A.D. & Leaver, C.J. (1986). Nuclear–mitochondrial interactions in cytoplasmic male-sterile Sorghum. *Theor. Appl. Genet.* 73:252–260.

Bannerot, H., Boulidard, L. & Chupeau, Y. (1977). Unexpected difficulties met with the radish cytoplasm in *Brassica oleracea*. *Eucarpia Newsletter* 2:16.

Bartkowiak-Broda, I., Rouselle, P. & Renard, M. (1979). Investigations of two kinds of cytoplasmic male sterility in rapeseed (*Brassica napus* L.). *Genet. Polonica* 20:487–501.

Bateson, W. & Gairdner, A.E. (1921). Male sterility in flax subject to two kinds of segregation. *J. Genet.* 11:269–275.

Beckett, J.B. (1971). Classification of male-sterile cytoplasms in maize (*Zea maize* L.). *Crop Sci.* 11:721–726.

Benson, R.M. (1978). The performance of 61 hybrid hard red spring wheats. M.S. thesis, North Dakota State University, Fargo.

Bianchi, F. (1963). Transmission of male sterility in Petunia by grafting. *Genen et Phaenen* 8:36–43.

Bino, R.J. (1985). Ultrastructural aspects of cytoplasmic male sterility in *Petunia hybrida*. *Protoplasma* 127:230–240.

Bonhomme, S., Budar, F., Ferault, M. & Pelletier, G. (1991). A 2.5 kb NcoI fragment of Ogura radish mitochondrial DNA is correlated with cytoplasmic male-sterility in Brassica cybrids. *Curr. Genet.* 19:121–127.

Brandle, J.E. & McVetty, P.B.E. (1989). Heterosis and combining ability in hybrids derived from oilseed rape cultivars and inbred lines. *Crop Sci.* 29:1191–1195.

Briggs, R.W. (1971). Cytoplasmic male sterility research in corn. *Maize Genet. Coop. Newslett.* 45:13–16.

Briggs, R.W. (1973). Cytoplasmic male sterility research. *Maize Genet. Coop. Newslett.* 47:35–37.

Brooks, M.H., Brooks, J.S. & Chien, L. (1966). The anther tapetum in cytoplasmic-genetic male sterile sorghum. *Am. J. Bot.* 53:902–908.

Burns, D.R., Scarth, R. & McVetty, P.B.E. (1991). Temperature and genotypic effects on the expression of *pol* cytoplasmic male sterility in summer rape. *Can. J. Plant Sci.* 71:655–661.

Burton, G.W. (1958). Cytoplasmic male-sterility in pearl millet (*Pennisetum glaucum*) (L.) R. Br. *Agronomy J.* 50:230.

Burton, G.W. & Hanna, W.W. (1976). Ethidium bromide induced male sterility in pearl millet. *Crop Sci.* 16:731–732.

Burton, G.W. & Hanna, W.W. (1982). Stable cytoplasmic male sterility mutants induced in Tift 23 DB$_1$ pearl millet with mitomycin and streptomycin. *Crop Sci.* 22:651–652.

Chopra, K.R. (1987). Technical and economic aspects of seed production of hybrid varieties of sorghum. In *Hybrid Seed Production of Selected Cereal, Oil and Vegetable Crops*, eds. W.P. Feistritzer and A.F. Kelly, pp. 193–217. Rome: FAO.

Chuong, P.V., Beversdorf, W.D., Powell, A.D. & Pauls, K.P. (1988). The use of haploid protoplast fusion to combine cytoplasmic atrazine resistance and cytoplasmic male sterility in *Brassica napus*. *Curr. Plant Sci. Biotechnol. Agric.* 191–194.

Conde, M.F., Pring, D.R., Schertz, K.F. & Ross, W.M. (1982). Correlation of mitochondrial DNA restriction endonuclease patterns with sterility expression in six male-sterile sorghum cytoplasms. *Crop Sci.* 22:536–539.

Connett, M.B. & Hanson, M.R. (1990). Differential mitochondrial electron transport through the cyanide-sensitive and cyanide-insensitive pathways in isonuclear lines of cytoplasmic male sterile, male fertile, and restored Petunia. *Plant Physiol.* 93:1634–1640.

Crouzillat, D., Leroy, P., Perrault, A. & Ledoigt, G. (1987). Molecular analysis of the mitochondrial genome of *Helianthus annuus* in relation to cytoplasmic male sterility and phylogeny. *Theor. Appl. Genet.* 74:773–780.

Dewey, R.E., Levings, C.S. III & Timothy, D.H. (1986). Novel recombinations in the maize mitochondrial genome produce a unique transcriptional unit in the Texas male-sterile cytoplasm. *Cell.* 44:439–449.

Dewey, R.E., Timothy, D.H. & Levings, C.S.-III (1987). A mitochondrial protein associated with cytoplasmic male sterility in the T cytoplasm of maize. *Proc. Natl. Acad. Sci. (USA)* 84:5374–5378.

Dixon, L.K. & Leaver, C.J. (1982). Mitochondrial gene expression and CMS in Sorghum. *Plant Mol. Biol.* 1:89–102.

Duvick, D.N. (1956). Allelism and comparative genetics of fertility restoration of cytoplasmically pollen sterile maize. *Genet.* 41:544–565.

Duvick, D.N. (1965). Cytoplasmic pollen sterility in corn. *Advances in Genet.* 13:1–56.

Eckenrode, V.K. & Levings, C.S. III (1986). Maize mitochondrial genes. *In Vitro Cellular & Developmental Biol.* 22:169–176.

Edwards, I.B. (1987). Breeding of hybrid varieties of wheat. In *Hybrid Seed Production of Selected Cereal, Oil and Vegetable Crops*, eds. W.P. Feistritzer and A.F. Kelly, pp. 1–35. Rome: FAO.

Edwardson, J.R. (1956). Cytoplasmic male sterility. *Bot. Rev.* 22:796–838.

Edwardson, J.R. (1970). Cytoplasmic male sterility. *Bot. Rev.* 36:341 420.

Erickson, L., Grant, I. & Beversdorf, W.D. (1986). Cytoplasmic male sterility in rapeseed (*Brassica napus* L.). 1. Restriction patterns of chloroplast and mitochondrial DNA. *Theor. Appl. Genet.* 72:145–150.

Ewart, L. (1984). Plant breeding. In *Petunia,* ed. K.C. Sink, pp. 180–220. New York: Springer-Verlag.

Fan, Z. (1985). Cytoplasmic male sterility in rape (*Brassica napus* L.). Ph. D. thesis, University of Manitoba, Winnipeg.

Fan, Z. & Stefansson, B.R. (1986). Influence of temperature on sterility of two cytoplasmic male-sterility systems in rape (*Brassica napus* L.). *Can. J. Plant Sci.* 66:221–227.

Fan, Z., Stefansson, B.R. & Sernyk, J.L. (1986). Maintainers and restorers for three male-sterility–inducing cytoplasms in rape (*Brassica napus* L.). *Can. J. Plant Sci.* 66:229–234.

Fang, G.H. & McVetty, P.B.E. (1988). Inheritance of male fertility restoration for the Polima CMS system in *Brassica napus* L. *Proceedings of the 7th International Rapeseed Congress, Poznan, Poland*, pp. 73–78.

Fang, G.H. & McVetty, P.B.E. (1989). Inheritance of male fertility restoration and allelism of restorer genes for the Polima cytoplasmic male sterility system in oilseed rape. *Genome* 32:1044–1047.

Feistritzer, W.R. & Kelly, A.F., eds. (1987). *Hybrid Seed Production of Selected Cereal, Oil and Vegetable Crops*. Rome: FAO.

Forde, B.G. & Leaver, C.J. (1980). Nuclear and cytoplasmic genes controlling synthesis of variant mitochondrial polypeptides in male-sterile maize. *Proc. Natl. Acad. Sci. (USA)* 77:418•422.

Forde, B.G., Oliver, R.J. & Leaver, D.J. (1978). Variation in mitochondrial translation products associated with the male sterility cytoplasms in maize. *Proc. Natl. Acad. Sci. (USA)* 75:3841–3845.

Foskett, R.L. (1954). Nature and inheritance of male-sterility in the onion variety Scott Country Globe. *Iowa State Col. J.* 28:317.

Frankel, R. & Galun, E. (1977). Pollination mechanisms, reproduction and plant breeding. In *Monogr. Theor. Appl. Genet. No. 2*, pp. 235–268. New York: Springer-Verlag.

Frauen, M. (1987). Technical and economic aspects of seed production of hybrid varieties of rape. In *Hybrid Seed Production of Selected Cereal, Oil and Vegetable Crops*, eds. W.P. Feistritzer and A.F. Kelly, pp. 281–301. Rome: FAO.

Galun, E. & Aviv, D. (1983). Manipulation of protoplasts, organelles and genes: applicability to plant breeding. In *Efficiency of Plant Breeding, Proc. of the 10th Congress of the European Association for Research on Plant Breeding, Eucarpia*, eds. W. Lange, A.C. Zeven, and N.G. Hogenboom, pp. 228–237. Wageningen, The Netherlands.

Govinda-Rej, K. & Virmani, S.S. (1988). Genetics of fertility restoration of WA type cytoplasmic male sterility in rice. *Crop Sci.* 28:787–792.

Grant, I. & Beversdorf, W.D. (1985). Heterosis and combining ability estimates in spring oilseed rape (*Brassica napus* L.). *Can. J. Genet. Cytol.* 27:472–478.

Hallauer, A.R., Russell, W.A. & Lamkey, K.R. (1988). Corn breeding. In *Corn and Corn Improvement*, 3d ed., eds. G.F. Sprague and J.W. Dudley, pp. 463–564. Madison, WI: Am. Soc. Agronomy, Crop Sci. Soc. Am., Soil Sci. Soc. Am.

Handa, H. (1993). Molecular genetic studies of mitochondrial genome in rapeseed (*Brassica napus* L.) in relation to cytoplasmic male sterility. *Bull. Natl. Inst. Agrobiol. Resour.* 8:47–105.

Handa, H. & Nakajima, K. (1992). Different organization and altered transcription of the mitochondrial *atp6* gene in the male-sterile cytoplasm of rapeseed (*Brassica napus* L.). *Curr. Genet.* 21:153–159.

Hanson, M.R. (1991). Plant mitochondrial mutations and male sterility. *Ann. Rev. Genet.* 25:461–486.

Hanson, M.R. & Conde, M.F. (1985). Functioning and variation of cytoplasmic genomes: lessons from cytoplasmic–nuclear interactions affecting male fertility in plants. *Int. Rev. Cytol.* 94:213–267.

Hanson, M.R. & Folkerts, O. (1992). Structure and function of the higher plant mitochondrial genome. *Int. Rev. Cytol.* 141:129–172.

Heyn, F.W. (1978). Introgression of restorer genes from *Raphanus sativus* into male sterile *Brassica napus* and the genetics of fertility restoration. *Proceedings of the 5th International Rapeseed Conference Malmo, Sweden* 1:82–83.

Holford, P., Croft, J. & Newbury, H.J. (1991). Structural studies of microsporogenesis in fertile and male-sterile onions (*Allium cepa* L.) containing the cms-S cytoplasm. *Theor. Appl. Genet.* 82:745–755.

Izhar, S. (1978). Cytoplasmic male sterility in Petunia. III. Genetic control on microsporogenesis and male fertility restoration. *J. Hered.* 69:22–26.

Izhar, S. (1984). Male sterility in Petunia. In *Petunia*, ed. K.C. Sink, pp.77–91. New York: Springer-Verlag.

Jones, D.F., Stinton, H.T. Jr. & Khoo, U. (1957). Pollen restoring genes. *Connecticut Agric. Exp. Sta. Bull. Immed. Inform.*

Jones, H.A. & Clarke, A.E. (1943). Inheritance of male sterility in the onion and the production of hybrid seed. *Proc. Amer. Soc. Hort. Sci.* 43:189–194.

Jones, H.A. & Emsweller, S.L. (1937). A male-sterile onion. *Proc. Amer. Soc. Hort. Sci.* 34:583–585.

Joppa, L.R., McNeal, F.H. & Welsh, J.R. (1966). Pollen and anther development in cytoplasmic male sterile wheat (*Triticum aestivum* L.). *Crop Sci.* 6:296–297.

Katsuo, K. & Mizushima, U. (1958). Studies on cytoplasmic differences in rice. I.

On the fertility of direct and reciprocal crosses between cultivated and wild rices. *Jap. J. Breed.* 8:1–5.

Kaul, M.L.H. (1988). *Male Sterility in Higher Plants*. Berlin: Springer-Verlag.

Kemble, R.J., Gunn, R.E. & Flavell, R.B. (1980). Classification of normal and male sterile cytoplasms in maize. II. Electrophoretic analysis of DNA species in mitochondria. *Genet.* 95:451–458.

Khery-Pour, A., Gracen, V.E. & Everett, H.L. (1981). Genetics of fertility restoration in the C-group of cytoplasmic male sterility in maize. *Genet.* 98:379–388.

Kidd, H.J. (1961). The inheritance of restoration of fertility in cytoplasmic male-sterile sorghum: a preliminary report. *Sorghum Newslett.* 4:47–49.

Kim, S.K. (1987). Breeding of hybrid varieties of maize. In *Hybrid Seed Production of Selected Cereal, Oil and Vegetable Crops*, eds. W.P. Feistritzer and A.F. Kelly, pp. 55–83. Rome: FAO.

Kohler, R.H., Horn, R., Lossl, A. & Zetsche, K. (1991). Cytoplasmic male sterility in sunflower is correlated with the co-transcription of a new open reading frame with the *atpA* gene. *Mol. Gen. Genet.* 227:369–376.

Laser, K.D. & Lersten, N.R. (1972). Anatomy and cytology of microsporogenesis in cytoplasmic male sterile angiosperms. *Bot. Rev.* 38:425–454.

Laughnan, J.R. & Gabay, S. (1978). Nuclear and cytoplasmic mutations to fertility in S male sterile maize. In *Maize Breeding and Genetics*, ed. D.B. Walden, pp. 427–447.

Laughnan, J.R. & Gabay-Laughnan, S. (1983). Cytoplasmic male sterility in maize. *Ann. Rev. Genet.* 17:27–48.

Leclercq, P. (1969). Une stérilité male cytoplasmique chez le tournesol. *Ann. Amelior. Plantes* 19:99–106.

Lee, S.L.J. & Warmke, H.E. (1979). Organelle size and number in fertile and T-cytoplasmic male-sterile corn. *Amer. J. Bot.* 66:141–148.

Lee, S.L.J., Gracen, V.E. & Earle, E.D. (1979). The cytology of pollen abortion in C-cytoplasmic male sterile corn anthers. *Amer. J. Bot.* 66:656–667.

Lee, S.L.J., Earle, E.D. & Gracen, V.E. (1980). The cytology of pollen abortion in S cytoplasmic male sterile corn anthers. *Amer. J. Bot.* 67:237–245.

Lemke, C.A., Gracen, V.E. & Everett, H.L. (1985). A new source of cytoplasmic male sterility in maize induced by the nuclear gene, *iojap. Theor. Appl. Genet.* 71:481–485.

Levings, C.S. III (1993). Thoughts on cytoplasmic male sterility in CMS-T maize. *The Plant Cell* 5:1285–1290.

Levings, C.S. III & Pring, D.R. (1976). Restriction endonuclease analysis of mitochondrial DNA from normal and Texas cytoplasmic male-sterile maize. *Science* 193:158–160.

Levings, C.S.-III & Pring, D.R. (1979). Molecular basis of cytoplasmic male sterility in maize. In *Physiological Genetics*, ed. J.G. Scandalios, pp. 171–193. New York: Academic Press.

McVetty, P.B.E., Edie, S.A. & Scarth, R. (1990). Comparison of the effect of *nap* and *pol* cytoplasms on the performance of intercultivar summer oilseed rape hybrids. *Can. J. Plant Sci.* 70:117–126.

Mignouna, H., Virmani, S.S. & Briquet, M. (1987). Mitochondrial DNA modifications associated with cytoplasmic male sterility in rice. *Theor. Appl. Genet.* 74:666–669.

Newton, K.J. (1988). Plant mitochondrial genomes: organization, expression and variation. *Ann. Rev. Plant Physiol. Plant Mol. Biol.* 39:503–532.

Nilkamp, H.J.J., Overbeeke, N. & Kool, A.J. (1983). Genetic manipulation of cytoplasmic plant genes. *Efficiency in Plant Breeding*. Proceedings, 10th Congress, European Association for Research on Plant Breeding, EUCARPIA,

Wageningen, the Netherlands, 19–24 June 1983, eds. W. Lange, A.C. Zeven and N.G. Hogenboom, pp. 239–243.

Nivison, H.T. & Hanson, M.R. (1989). Identification of a mitochondrial protein associated with cytoplasmic male sterility in petunia. *Plant Cell* 1:1121–1130.

Ogura, H. (1968). Studies on the new male sterility in Japanese radish with special reference to the utilization of this sterility towards the practical raising of hybrid seed. *Mem. Fac. Agric. Kagoshima Univ.* 6:39–78.

Pacini, E., Franchi, G.G. and Hesse, M. (1985). The tapetum: its form, function, and possible phylogeny in Embryophyta. *Plant Syst. Evol.* 149:155–185.

Patil, T.T. (1987). Technical and economic aspects of seed production of hybrid varieties of millets. In *Hybrid Seed Production of Selected Cereal, Oil and Vegetable Crops*, eds. W.P. Feistritzer and A.F. Kelly, pp. 217–253. Rome: FAO.

Pearson, O.H. (1981). Nature and mechanisms of cytoplasmic male sterility in plants: a review. *Hort. Sci.* 16:482–487.

Pellan-Delourme, R., Eber, F. & Renard, M. (1988). Male fertility restoration in *Brassica napus* with radish cytoplasmic male sterility. *Proc. 7th International Rapeseed Congress, Poznan Poland,* pp. 199–203.

Pellan-Delourme, R. & Renard, M. (1988). Cytoplasmic male sterility in rapeseed (*Brassica napus* L.): female fertility of restored rapeseed with "Ogura" and cybrids cytoplasms. *Genome* 30:234-238.

Pelletier, G. (1983). Cytoplasmic hybridization in higher plants. *Efficiency in Plant Breeding.* Proceedings, 10th Congress, European Association for Research on Plant Breeding, EUCARPIA, Wageningen, The Netherlands, 19–24 June 1983, eds. W. Lange, A.C. Zeven and N.G. Hogenboom, pp. 244–247.

Peterson, C.E. & Foskett, R.L. (1953). Occurrence of pollen sterility in seed fields of Scott Country Globe onions. *Proc. Amer. Soc. Hort. Sci.* 62:443–448.

Pinnisch, R. & McVetty, P.B.E. (1990). Seed production of hybrid summer rape in the field using the *pol* cytoplasmic male sterility system: a first attempt. *Can. J. Plant Sci.* 70:611–618.

Polowick, P.L. & Sawhney, V.K. (1990). Microsporogenesis in a normal line and in the *ogu* cytoplasmic male-sterile line of *Brassica napus*. I. The influence of high temperatures. *Sex Plant Reprod.* 3:263–276.

Polowick, P.L. & Sawhney, V.K. (1991). Microsporogenesis in a normal line and in the *ogu* cytoplasmic male-sterile line of *Brassica napus*. II. The influence of intermediate and low temperatures. *Sex Plant Reprod.* 4:22–27.

Pring, D.R. & Lonsdale, D.M. (1989). Cytoplasmic male sterility and maternal inheritance of disease susceptibility in maize. *Ann. Rev. Phytopathol.* 27:483–502.

Pring, D.R., Conde, M.F. & Shertz, K.F. (1982). Organelle genome diversity in sorghum: male sterile cytoplasms. *Crop Sci.* 22:414–421.

Pring, D.R., Levings, C.S. III, Hu, W.W.L. & Timothy, D.H. (1977). Unique DNA associated with mitochondria in the "S"-type cytoplasm of male-sterile maize. *Proc. Nat. Acad. Sci. (USA)* 74:2904–2908.

Rai, K.N. & Hash, C.T. (1990). Fertility restoration in male sterile X maintainer hybrids of pearl millet. *Crop Sci.* 30:889–892.

Riggs, T.J. (1987). Breeding F_1 hybrid varieties in vegetables. In *Hybrid Seed Production of Selected Cereal, Oil and Vegetable Crops*, eds. W.P. Feistritzer and A.F. Kelly, pp. 149–175. Rome: FAO.

Rogers, J.S. & Edwardson, J.R. (1952). The utilization of cytoplasmic male-sterile inbreds in the production of corn hybrids. *Agron. J.* 44:8–13.

Sernyk, J.L. & Stefansson, B.R. (1983). Heterosis in summer rape (*Brassica napus* L.). *Can. J. Plant Sci.* 63:407–413.

Shiga, T. (1976). Studies on heterosis breeding using cytoplasmic male sterility in rapeseed, *Brassica napus* L. *Bull. Natl. Inst. Agric. Sci. Series D* 27:1–101.

Shiga, T. (1980). Male sterility and cytoplasmic differentiation. In *Brassica Crops and Wild Allies, Biology and Breeding*, ed. S. Tsunoda, pp. 205–211. Tokyo: Jap. Sci. Soc. Press.

Sicuella, L. & Palmer, J.D. (1988). Physical and gene organization of mitochondrial DNA in fertile and male sterile sunflower. CMS-associated alterations in structure and transcription of the *atpA* gene. *Nucleic Acids Res.* 16:3787–3799.

Siedow, J.N. & Berthold, D.A. (1986). The alternative oxidase: a cyanide resistant respiratory pathway in higher plants. *Physiol. Plant* 66:569–573.

Singh, A. & Laughnan, J.R. (1972). Instability of S male-sterile cytoplasm in maize. *Genet.* 71:607–620.

Singh, M. & Brown, G. (1991). Suppression of cytoplasmic male sterility by nuclear genes alters expression of a novel mitochondrial gene region. *Plant Cell* 3:1349–1362.

Stephens, J.C. & Holland, R.F. (1954). Cytoplasmic male-sterility for hybrid sorghum seed production. *Agron. J.* 46:20–23.

Stroike, J.E. (1987). Technical and economic aspects of hybrid wheat seed production. In *Hybrid Seed Production of Selected Cereal, Oil and Vegetable Crops*, eds. W.P. Feistritzer and A.F. Kelly, pp. 177–187. Rome: FAO.

Takahasi, 0. (1987). Utilization and economics of hybrid vegetable varieties in Japan. In *Hybrid Seed Production of Selected Cereal, Oil and Vegetable Crops*, eds. W.P. Feistritzer and A.F. Kelly, pp. 313–29. Rome: FAO.

Tracy, W.F., Everett, H.L. & Gracen, V.E. (1991). Inheritance, environmental effects, and partial male fertility in C-type CMS in a maize inbred. *J. Hered.* 82:343–346.

Van Der Meer, Q.P. (1978). Results of recent investigations of male-sterility in onion (*Allium cepa* L.). *Biul. Warzywniczy* 22:73–79.

Van Marrewijk, G.A.M. (1969). Cytoplasmic male sterility in *Petunia*. I. Restoration of fertility with special reference to the influence of environment. *Euphytica* 18:1–20.

Vannozi, G.P. (1987). Technical and economic aspects of seed production of hybrid varieties of sunflower. In *Hybrid Seed Production of Selected Cereal, Oil and Vegetable Crops*, eds. W.P. Feistritzer and A.F. Kelly, pp. 253–281. Rome: FAO.

Vasil, I.K. (1967). Physiology and cytology of anther development. *Biol. Rev. Cambridge Phil. Soc.* 42:327–373.

Virmani, S.S. (1987). Hybrid rice breeding. In *Hybrid Seed Production of Selected Cereal, Oil and Vegetable Crops*, eds. W.P. Feistritzer and A.F. Kelly, pp. 35–55. Rome: FAO.

Virmani, S.S., Chaudhary, R.C. & Khush, G.S. (1981). Current outlook on hybrid rice. *Oryza* 18:76–84.

Virmani, S.S. & Edwards, I.B. (1984). Current status and future prospects for breeding hybrid rice and wheat. *Adv. Agric.* 36:145–214.

Vranceanu, A.V. (1987). Breeding hybrid varieties of sunflower. In *Hybrid Seed Production of Selected Cereal, Oil and Vegetable Crops*, eds. W.P. Feistritzer and A.F. Kelly, pp. 111–133. Rome: FAO.

Warmke, H.E. & Lee, S.L. (1977). Mitochondrial degeneration in Texas cytoplasmic male-sterile corn anthers. *J. Hered.* 68:213–222.

Williams, M.E. & Levings, C.S. III (1992). Molecular biology of cytoplasmic male sterility. *Plant Breeding Rev.* 10:23–51.

Wilson, J.A. (1984). Hybrid wheat breeding and commercial seed development. *Plant Breeding Rev.* 2:303–319.

Wilson, J.A. & Ross, W.M. (1962). Male sterility interaction of the *Triticum aestivum* nucleus and *Triticum timopheevi* cytoplasm. *Wheat Info. Ser.* 14:29.

Witt, U., Hansen, S., Albaum, M. & Abel, W.O. (1991). Molecular analyses of the
    CMS-inducing "Polima" cytoplasm in *Brassica napus* L. *Curr. Genet.*
    19:323–327.
Wych, R.D. (1988). Production of hybrid seed corn. In *Corn and Corn Improvement.*
    Agronomy Monograph 18:565–607.
Xizhi, L. (1987). Technical aspects of seed production of hybrid varieties of rice in
    China. In *Hybrid Seed Production of Selected Cereal, Oil and Vegetable
    Crops*, eds. W.P. Feistritzer and A.F. Kelly, pp. 187–193. Rome: FAO.

8

Genic male sterility

V. K. SAWHNEY

Summary 183
Introduction 183
Genetics and morphology of GMS mutants 184
Environmental effects 186
    Temperature 186
    Photoperiod 187
Cytological changes 187
Biochemical changes 189
    Amino acids 189
    Soluble proteins 189
    Enzymes 190
Hormones and male sterility 191
Use of genic male sterility in hybrid programs 192
Conclusions 194
References 194

Summary

Nuclear encoded, genic male sterility (GMS) is a common occurrence in angiosperms and is reported in nearly every major crop species. GMS can result from mutations in any one of a number of genes controlling pollen and/or stamen development and, accordingly, the phenotypes of GMS mutants vary. There are a number of cytological, physiological, and biochemical processes affected in GMS mutants, but the causative mechanisms of GMS are not clearly understood. Although GMS is not commonly used in hybrid seed production, primarily because of the maintenance of pure male-sterile lines, there are several proposals put forward to circumvent these problems. One promising approach is to select for chemical- or environment-sensitive GMS lines in which fertility can be restored by appropriate treatments. Such systems are potentially useful in hybrid seed production.

Introduction

Normal development of the male reproductive organ (stamen) and male gametophyte (pollen grain) is essential for the successful completion of sex-

183

ual reproduction in angiosperms. Abnormalities at any stage of stamen and pollen development can result in male sterility.

Male sterility is a wide occurrence in flowering plants and has been of interest to various plant biologists and plant breeders. Male-sterile mutants serve as useful tools for investigations into the genetic, molecular, physiological, and developmental processes involved in stamen and pollen development. The sterility of the male reproductive organ is also considered a major mechanism by which gynodioecy is believed to have originated (Bawa 1980). For plant breeders, male-sterile plants are useful systems for interspecific hybridization and for performing backcrosses. Perhaps the most widely accepted use of male sterility is in the production of F_1 hybrid seed in monoecious and hermaphrodite crops (Frankel and Galun 1977; Kaul 1988). The use of male-sterile plants eliminates the labor-intensive process of flower emasculation and, therefore, significantly reduces the cost of hybrid seed production. The hybrid seed produced by the use of male-sterile systems also has high hybrid purity.

Male sterility can result from mutations in either the nuclear or the cytoplasmic genes. Nuclear or genic male sterility (GMS) is generally caused by mutations in genes that in the recessive condition affect stamen and pollen development. However, GMS regulated by dominant genes is also known (Kaul 1988). For discussions on cytoplasmic male sterility (CMS), see Chapter 7.

Genic male sterility is reported in nearly every major crop and in a large number of dicotyledon and monocotyledon species (Kaul 1988). Generally, GMS is of spontaneous origin, but it can also be induced by mutagens (that is, by chemicals) (Fujimaki et al. 1977; Cross and Ladyman 1991; see also Chapter 10), by radiations (Driscoll and Barlow 1976), and by genetic engineering (for example, Mariani et al. 1990; Worrall et al. 1992; Goldberg et al. 1993; see also Chapter 11). Novel male-sterile plants can also be generated by protoplast fusion (that is, by cybrid formation) (Kofer et al. 1990; Gourret et al. 1992), by t-DNA transposon tagging (Aarts et al. 1993), and by affecting the synthesis of flavonoids (Van Tunen et al. 1994).

In this chapter, cytological and biochemical processes associated with the breakdown of pollen development in GMS systems, the regulation of GMS by environmental and hormonal factors, and the practical aspects of the use of GMS in hybrid seed production are discussed. The focus is on major crop species, but information from some model systems is also included wherever appropriate.

Genetics and morphology of GMS mutants

In nearly every crop species a number of male-sterile (*ms*) loci have been identified. In corn, for example, 21 nonallelic *ms* genes have been characterized (Coe et al. 1987; Palmer et al. 1992) but more than 100 are known to

Figures 8.1–8.5 *8.1.* Mature flowers of a male-fertile (left) and a monogenic male-sterile line (right) of *Brassica napus*. Note the small size of male-sterile flowers and the reduced filament length. *8.2. Stamenless-1* (*sl-1*) mutant flower of tomato (*Lycopersicon esculentum*) at anthesis. Note the stamens are modified as carpels and fused to the ovary. *8.3. Stamenless-2* (*sl-2*) mutant flower of tomato at anthesis. The stamens contain an antherlike region at the distal end and bear ovules (ov) at the basal end on the adaxial surface. *8.4–8.5.* Flowers of *sl-2* mutant of tomato grown in high and low temperatures. In high temperatures (Figure 8.4), stamens are carpel-like, but in low temperatures (Figure 8.5) they are normal and resemble the wild-type stamens.

occur (Albertsen et al. 1993). In soybean, 16 (Palmer et al. 1992) and in tomato, 12 (Stevens and Rick 1986) *ms* genes have been reported. For some of the *ms* genes a number of alleles are known that exhibit varying degrees of male sterility. The linkage maps of *ms* genes are available for most of the major crops, for example, for maize (Coe et al. 1987), soybean (Palmer and Kiang 1990), and tomato (Mutschler et al. 1987).

The morphology of male-sterile mutants is variable. It ranges from the complete absence of male reproductive organs to the formation of normal stamens, with viable pollen, that fail to dehisce (Frankel and Galun 1977; Kaul 1988). In general, male-sterile plants are morphologically not distinguishable from the parent fertile plants with the exception of flower structure. Male-sterile flowers are commonly smaller in size in comparison to the fertile flowers, and the size of stamens is generally reduced (Figure 8.1). The anthers may or may not be collapsed and the filament length may be affected (Figure 8.1). In some cases, stamens may be modified as carpels, as in the *stamenless-1* (*sl-1*) mutant of tomato (Figure 8.2). In others, structures with both anther and carpel features are produced, as in the *stamenless-2* (*sl-2*) mutant of tomato (Figure 8.3).

Environmental effects

The expression of male sterility is known to be influenced by environmental factors. Indeed, the phenotypes of some of the male-sterile mutants are completely different from one another under different growth conditions. The two main physical factors that influence GMS are temperature and photoperiod.

Temperature

One of the earlier observations of the possible influence of temperature on GMS mutants was on the stamenless (*sl*) mutant of tomato. Flowers of *sl* plants grown in the summer completely lack stamens, whereas those grown in the winter produce abnormal stamens, often with viable pollen (Bishop 1954). This seasonal effect was in part related to the effect of temperature. Similarly, cooler temperatures restore fertility in *ms-15* and *ms-33* mutants of tomato (Schmidt and Schmidt 1981).

The effect of temperature on male sterility was examined in greater detail in the variable male-sterile (*vms*) mutant of tomato. *vms* plants exposed to temperatures of 30°C and above are male sterile, but those grown in the greenhouse are normal (Rick and Boynton 1967). Experiments done in temperature-controlled growth chambers on *sl-2* mutants of tomato showed more accurately the effects of temperature on male sterility. *sl-2* plants grown in high temperatures (28°C day/23°C night) produce flowers with carpel-like structures instead of stamens (Figure 8.4), whereas those grown in low temperatures (18°C day/15°C night) bear normal stamens with viable pollen (Figure 8.5; see also Sawhney 1983). In intermediate temperatures (23°C day/18°C night), *sl-2* flowers produce structures with half-stamens and half-carpels (Figure 8.3). Thus, a 5°C difference in temperature significantly modifies the expression of the *sl-2* allele.

In a GMS mutant in *Brassica oleracea*, temperature sensitivity is even more precise. At 10°–11°C, the male-sterile plants can be converted to fertile in 30 days, but at 12°C it takes over two months for partial fertility (Dickson 1970). In many other GMS mutants, however, male sterility is not affected by a change in temperature. For example, a GMS line in *Brassica napus* is insensitive to temperature fluctuations (A. Shukla, personal communication). Similarly, GMS mutants of other species are unaffected by a change in temperature (Kaul 1988). Nonetheless, as Kaul (1988) points out, the effect of temperature on GMS mutants needs to be examined in greater detail.

Photoperiod

A change in photoperiod can have a strong influence on the expression of male sterility. Ahokas and Hockett (1977) reported that *ms9* mutant of barley is completely sterile in Finland, but partially fertile in Bozeman, U.S.A. The authors suggested that the restoration of fertility in the United States is due to a different photoperiod. Similarly, a male-sterile mutant in cabbage is reportedly sterile in summer, but fertile in winter (Rundfeldt 1960). In this case, however, sterility may be related both to temperature and to photoperiod.

In rice, a spontaneous mutant, in sub sp. japonica cv. Nongken 58, has been isolated that is monogenic recessive and photoperiod sensitive. In long days (>14 hours), the line is completely sterile, but under short days (<13 hours) it is fertile (Shi 1986). Similarly, in EMS-induced male-sterile mutants of rice, fertility is increased when field-grown plants are transferred to a growth chamber set at 12h daylength (Oard et al. 1991).

We have recently isolated a photoperiod-sensitive male-sterile mutant (7B-1) in tomato. The mutant plants are completely male sterile in the summer field conditions in Saskatoon. However, at a daylength of 8–10 hours, 7B-1 plants produce many flowers with normal anthers that contain normal, viable pollen.

Cytological changes

Pollen development is a tightly programmed, sequential process that involves distinct changes in the sporogenous tissue (gametophytic) and in the tapetum and wall layers (sporophytic tissues) of the anther (Bedinger 1992; Cresti et al. 1992; see also Chapter 2).

In GMS mutants, the breakdown in microsporogenesis can occur at a number of pre- or postmeiotic stages. The abnormalities can involve aberrations in any one of the following stages: during the process of meiosis, in the formation of tetrads, during the release of tetrads (that is, the dissolution of callose), at the vacuolate microspore stage, or at mature or near-mature pollen stage (Gottschalk and Kaul 1974; Kaul 1988). Also, although there is presumed specificity of action of *ms* genes, there is, in some cases, inter- and intraplant variation. The following are selected examples documenting the cytological effects of *ms* genes.

In soybean, the cytological effects of *ms* genes on microsporogenesis have been examined at the light and electron microscope levels (see reviews; Graybosch and Palmer 1988; Palmer et al. 1992). In *ms1* mutant, meiosis occurs in the microspore mother cells (MMCs), but cytokinesis does not take place and the microspores produced contain four nuclei, that is, they are coenocytic (Albertsen and Palmer 1979). This abnormality is not associated with defects in tapetum development, and pollen-wall development is normal. In *ms4* mutant, also, coenocytic microspores are produced; however, the pollen wall is abnormal (Graybosch and Palmer 1985). In *ms2*, *ms3*, and *ms6* mutants, sterility occurs at specific postmeiotic stages and is always associated with abnormalities in the tapetum (Palmer et al. 1992). In contrast, in *msp* mutant, there is no specific stage of pollen abortion (Stelly and Palmer 1982).

In corn, there are a large number of homozygous recessive, and a couple of dominant, male-sterile mutants that affect a range of developmental stages in microsporogenesis (see review by Palmer et al. 1992). The different mutants exhibit abnormalities starting from the premeiosis stage, that is, before the formation of pollen mother cell (*ms22*), to a late stage in microsporogenesis, that is, after the formation of the generative cell (*ms24*). In other *ms* mutants, abnormalities can occur at the diad and tetrad stages, or at the early microspore or vacuolate microspore stages, or at the early or near-mature pollen stages (Palmer et al. 1992). In some mutants, for example, *ms10* and *ms13* microspore-wall development is specifically affected (Albertsen and Phillips 1981); in *ms10,* microspore degeneration is associated with abnormalities in the tapetum (Cheng et al. 1979). The work on soybean and maize male-sterile mutants indicates that a number of genes are involved in pollen development, and mutations in one or more of these genes can result in male sterility.

Abnormalities in pollen development in male-sterile mutants of tomato are commonly associated with aberrations in tapetum development (Rick 1948; Sawhney and Bhadula 1988). In the *sl-2* mutant, for example, tapetal cells enlarge prematurely, are often highly vacuolate, and their degeneration is delayed (Sawhney and Bhadula 1988). This mistiming of tapetum degeneration results in many microspores' being devoid of a wall. In the absence of the wall, microspores enlarge and eventually burst, leaving a mass of debris in the pollen sac. These observations on the *sl-2* mutant, and on other *ms* mutants in different crops, demonstrate the importance of the tapetum in microspore and pollen development (Pacini et al. 1985; Bedinger 1992). They also support the view that *ms* genes act on the tapetum, which in turn affects microspore development (Rick 1948; Novak 1971).

In male-sterile mutants induced by radiations (Theis and Röbbelen 1990) and by chemical mutagens (Chaudhury 1993; Dawson et al. 1993), pollen development can also be affected at pre- or postmeiotic stages. In these examples, the breakdown in microsporogenesis is also generally associated with abnormal tapetum development. The importance of the tapetum in pollen development has been used to develop strategies at the molecular level that selectively destroy the tapetum and thereby induce male sterility in crop plants (for further details, see Chapter 11).

Biochemical changes

Male sterility has been shown to be accompanied by qualitative and quantitative changes in amino acids, proteins, and enzymes in developing anthers. The general approach in such studies has been to analyze the levels of various amino acids, total proteins, and the activity and isozymes of specific enzymes believed to be associated with pollen development. The following is a summary of some of these studies. For additional information, the volume by Kaul (1988) may be consulted.

Amino acids

Analyses of amino acids in male-sterile anthers of a number of species have revealed both reduced and increased levels of amino acids. In general, the level of proline, leucine, isoleucine, phenylalanine, and valine (Kern and Atkins 1972; Kaul 1988) is reduced in male-sterile anthers, but that of asparagine, glycine, arginine, and aspartic acid is increased (reviewed in Kaul 1988).

In GMS mutants, the level of proline is particularly affected. In Japanese radish, for example, the mature male-sterile anthers contain one-eighth the amount of proline in comparison to the fertile anthers (Kakihara et al. 1988). The level of other amino acids is also affected, but not to that extent. Kakihara et al. (1988) showed that the filaments of male-sterile stamens accumulate proline, suggesting a problem in the translocation of this amino acid to developing anthers.

How proline is specifically required for pollen development is not understood. It is also not known whether the reduction in proline levels, or in other amino acids, is a primary or a secondary effect of mutations in *MS* genes.

Soluble proteins

It should not be surprising that mutations in *MS* genes result in qualitative and/or quantitative changes in soluble proteins in male-sterile anthers. Differences in total protein content and in polypeptide profiles, obtained by 1-D and 2-D gel electrophoresis, have been shown between male-sterile and fertile anthers. In general, the male-sterile anthers contain lower protein content and fewer polypeptide bands than the fertile anthers (Alam and Sandal 1969; Banga et al. 1984; Sawhney and Bhadula 1987). By feeding ^{35}S-methionine to fertile and male-sterile floral buds of tomato at different developmental stages, Bhadula and Sawhney (1991) showed that some polypeptides synthesized in normal stamens were absent in mutant stamens. As well, a 53-kd polypeptide was synthesized only in mutant stamens at the time when tapetum abnormalities appear in mutant anthers. Whether the 53-kd polypeptide, or the lack of a number of other polypeptides, is directly responsible for pollen degeneration in this tomato mutant is not known.

Enzymes

The sequential development of pollen grains must involve specific enzyme-mediated reactions at different stages. Aberrations in the activity of one or more of such enzymes can lead to abnormal pollen development and thus to male sterility.

The enzyme callase is required for the breakdown of callose (ß-1,3 glucans) that surrounds PMCs and, after meiosis, the tetrads. Microspores are released from the tetrads by the activity of callase. In some GMS and CMS systems, mistiming of callase activity (that is, early or delayed) results in premature or delayed release of meiocytes and microspores, resulting in male sterility (Izhar and Frankel 1971; Pritchard and Hutton 1972; Gottschalk and Kaul 1974). Thus, in transgenic tobacco plants in which callose dissolution is induced prematurely, male sterility results (Worrall et al. 1992).

Esterases have also been related to pollen development. They have been localized specifically in the tapetum at early stages of microsporogenesis and later at the surface of pollen grains. In both GMS and CMS systems, the activity of esterases is decreased (van Marrewijk et al. 1986; Bhadula and Sawhney 1987). In the *sl-2* mutant of tomato, the number and intensity of isozymes of esterases are reduced at stages when abnormalities are observed in the tapetum and when pollen-wall development is affected (Bhadula and Sawhney 1987). Esterases are believed to have a role in the hydrolysis of sporopollenin (Ahokas 1976), the polymer required for pollen-wall formation. Aberrations in the activity of esterases would, therefore, have an effect on pollen development.

The activities of other enzymes (for example, peroxidases, dehydrogenases, phosphotases, and cytochrome oxidase) are also known to be affected in male-sterile systems (for review, see Kaul 1988). Carbohydrates, particularly simple sugars, are essential components of pollen-wall development. It is, therefore, of interest to note that in a male-sterile tomato mutant the activity of amylases is decreased, and it corresponds with high starch content and reduced levels of soluble sugars (Bhadula and Sawhney 1989).

It is apparent from the biochemical studies that mutations in *MS* genes affect changes in protein and carbohydrate metabolism and in the activity of several enzymes. It is not clear, however, whether these changes are a consequence of microspore and pollen degeneration or whether they are part of the causative mechanisms in male sterility. In a study done on adenine phosphoribosyltransferase (APRT)–deficient mutants in *Arabidopsis*, the lack of APRT was directly related to the cause of male sterility (Moffatt and Somerville 1988). APRT is an enzyme involved in the purine salvage pathway and it converts adenine to AMP. Although the mechanism(s) by which APRT deficiency affect pollen development is not clearly understood, the accumulation of adenine itself may be toxic to developing microspores (Regan and Moffatt 1990).

Hormones and male sterility

Experimental studies on hermaphrodite, monoecious, and dioecious species have shown that hormones, that is, plant growth substances (PGSs), play an important role in stamen and pollen development (for review, see Kinet et al. 1985; Greyson 1994; Sawhney and Shukla 1994). Thus, in male-sterile plants, aberrant stamen and pollen development is known to be accompanied by changes in endogenous PGSs. In a few systems examined, male sterility has been associated with changes in one or more PGSs.

In a GMS line of rice, male sterility was related to a deficiency in gibberellins (GAs); the male-sterile anthers contain one-sixth the level of GA_1 and GA_4 in comparison to fertile anthers (Nakajima et al. 1991). In male-sterile *Mercurialis annua*, the concentration of indole acetic acid (IAA) is lower than in partial-sterile and fertile lines (Hamdi et al. 1987), but the male-sterile lines contain higher levels of active cytokinins (CKs) than the fertile lines (Louis et al. 1990). Similarly, in the apetalous male-sterile mutant in soybean, the concentrations of IAA and abscisic acid (ABA) are lower in comparison to the wild type, and the ratio of these substances changes from anthesis to the pod-fill stage (Skorupska et al. 1994). These studies indicate that male sterility cannot be related to changes in any one PGS.

In a comprehensive study done on two GMS systems, it was shown that male sterility is associated with changes in not one PGS, but several PGSs. In the *sl-2* mutant of tomato, the total level of GAs and GA-like substances is lower (Sawhney 1974; Sawhney and Shukla 1994), and that of IAA higher (Singh et al. 1992), than in the wild type, in both the vegetative and reproductive organs. Similarly, the concentration of CKs, particularly zeatin- and dihydrozeatin-nucleotide, is lower, and that of ABA is higher, in the *sl-2* mutant in comparison to the wild type (Singh and Sawhney, unpublished data). Also, stamens showed the maximum difference in all of the PGSs between mutant and wild-type flowers.

Analysis of PGSs in a GMS line of *Brassica napus* showed reduced content of active CKs (Shukla and Sawhney 1992, 1993), but increased levels of ABA (Shukla and Sawhney 1994) and IAA (Sawhney and Shukla 1994). The low level of CKs was again mainly in the nucleotides of zeatin and dihydrozeatin. The metabolism of [^{3}H]-dihydrozeatin in male-sterile and fertile *B. napus* floral buds confirmed that it is the nucleotide form that is most affected in GMS flowers (Shukla and Sawhney 1993). These studies on different systems show that male sterility is accompanied by changes in the metabolism of a number of PGSs, indicating that it is the altered balance of endogenous PGSs, rather than any one substance, that affects stamen and pollen development. This hypothesis is consistent with the view that it is the ratio of PGSs that is critical in plant developmental processes (Klee and Estelle 1991).

Additional support for this hypothesis comes from the analysis of a CMS system in *B. napus*. The *ogu* CMS plants contain low levels of CKs (Singh and Sawhney 1992) and IAA, but high levels of ABA (Singh, unpublished data). Thus, male sterility, regardless of nuclear or cytoplasmic origin, seems to involve some common metabolic pathways.

If PGSs are affected in male-sterile systems, it should, theoretically, be possible to restore fertility in GMS lines by exogenous application of PGSs. This has been shown in some male-sterile systems in tomato. Exogenous applications of GAs restore fertility in *sl-1*, *sl-2*, *ms-15*, and *ms-33* mutants in tomato, and in GMS mutants in barley and *Cosmos* (reviewed in Sawhney and Shukla 1994). However, this has not been possible in other male-sterile systems. The problem may have its basis in the difficulty of monitoring the level of several PGSs, for instance, when there are reduced levels of some, and increased levels of other, PGSs.

Use of genic male sterility in hybrid programs

Male-sterile plants of monoecious or hermaphrodite crops are potentially useful in hybrid programs because they eliminate the labor-intensive process of flower emasculation. In crops (for example, maize, soybean, tomato, and pepper), manual emasculation significantly increases the cost of hybrid seed production. A functional male-sterile system – genic, cytoplasmic, chemically induced, or genetically engineered – can overcome this problem, although each system has its limitations.

The major problem with GMS is the maintenance of the male-sterile line. Normally, a GMS line (A-line) is maintained by backcrossing with the heterozygote (B-line), but the progeny produced are 50% fertile and 50% male sterile (Figure 8.6). In the field, this creates the problem of the removal (roguing) of fertile plants (Frankel 1973; Frankel and Galun 1977).

Several proposals have been put forward to alleviate this problem. One suggestion is to identify marker genes that are closely linked to *ms* genes and affect some vegetative characters, such as seed color and shape, or leaf or

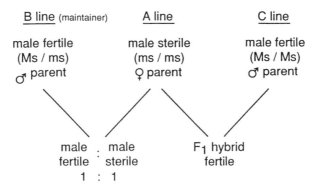

Figure 8.6 Schematic diagram for the normal maintenance and use of genic male sterility in plants *(adapted from Frankel and Galun 1977).*

stem pigmentation (Frankel and Galun 1977; Driscoll 1986; Rao et al. 1990). Such marker genes would help identify the *ms* plants at an early stage. Thus, either only selected *ms* plants are planted in the field, or prospective fertile plants are removed at the seedling stage. A follow-up of this suggestion is a proposal to introduce marker genes, by genetic engineering, in male-sterile plants (Jorgensen 1987). An alternative approach is to identify some visible pleiotropic effects of *ms* genes that would help sort out the male-sterile plants at an early stage (Driscoll 1986; Rao et al. 1990).

The ability to manipulate male sterility in GMS lines by environmental or chemical methods is another desirable approach in hybrid seed production. One of the simplest proposals by Smith (1947) was to select for GMS lines that are fertile in one environment and sterile in a different environment. Subsequently, Greyson and Walden (1976) argued for identifying male-sterile lines in corn that are hormone sensitive, that is, male-sterile plants in which fertility can be restored by hormones. Hockett et al. (1978) also stressed the need for increased research on inducing fertility in GMS lines in barley by chemical treatments. Similarly, Lapushner and Frankel (1967) pointed out the usefulness of male-sterile lines in tomato in which male fertility can be restored. The restoration of fertility in GMS plants by environmental or chemical treatments can lead to the production of 100% male-sterile seed (Figure 8.7), which can be used as female parents in hybrid seed production, without the need of roguing fertile plants (Sawhney 1984).

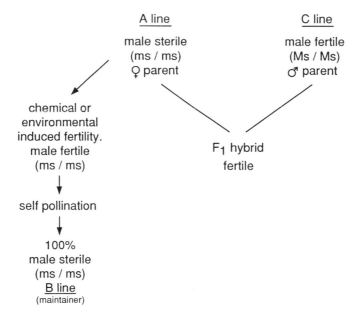

Figure 8.7 Schematic diagram for the maintenance of pure genic male-sterile lines by the restoration of fertility via environmental or chemical treatments (*adapted from Sawhney 1984*).

Perhaps the best example of regulation of male sterility in GMS lines is that of photoperiod-sensitive rice (previously discussed). This system has been used successfully in hybrid seed production. Similarly, in tomato a promising system is the photoperiod-sensitive, male-sterile 7B-1 line (see section under "Environmental effects"). Preliminary tests have shown that this line can be readily maintained by growing plants under appropriate photoperiod. The potential use of the 7B-1 line in tomato hybrid seed production is currently under investigation.

Conclusions

The nuclear-encoded male sterility is manifested in different forms. A number of genes are involved in normal stamen and pollen development, and mutations in any one of them can potentially lead to microspore and pollen abortion. Research on a number of GMS systems in different species has shown that the breakdown in microsporogenesis can occur at any stage and is accompanied by protein, enzymatic, and hormonal changes. However, further research is required to clarify whether these changes are some of the causative mechanisms or the pleiotropic effects of mutations in male-sterility genes.

Genic male-sterile systems – like CMS, chemically induced, genetically engineered, and self-incompatibility systems – are potentially useful in hybrid seed production and crop improvement. There are several approaches that may be employed for the effective use of GMS in hybrid programs. Future research on the exploitation of GMS systems at the commercial scale should be rewarding.

References

Aarts, M.G.M., Dirkse, W.G., Steikema, W.J. & Pereira, A. (1993). Transposon tagging of a male sterility gene in *Arabidopsis*. *Nature* 363:715–717.

Ahokas, H. (1976). Evidence of a pollen esterase capable of hydrolyzing sporopollenin. *Experientia* 32:175–177.

Ahokas, H. & Hockett, E.A. (1977). Male sterile mutants of barley. IV. Different fertility levels of *msg9ci* (cv. Vantage) an ecoclinal response. *Barley Genet. Newslett.* 7:10–11.

Alam, S. & Sandal, P.C. (1969). Electrophoretic analysis of anther proteins from male fertile and male sterile sudangrass, *Sorghum vulgare* var. *sudanese* (Piper). *Crop Sci.* 9:157–159.

Albertsen, M.C. & Palmer, R.G. (1979). A comparative light and electron-microscope study of microsporogenesis in male sterile *(ms1)* and male-fertile soybeans. *Amer. J. Bot.* 66:253–265.

Albertsen, M.C. & Phillips, R.L. (1981). Developmental cytology of 13 genetic male-sterile loci in maize. *Can. J. Genet. Cytol.* 23:195–208.

Albertsen, M.C., Fox, T.W. & Trimell, M.R. (1993). Cloning and utilizing a maize nuclear male sterile gene. *Proc. Ann. Corn Sorghum Res. Conf.* 48: 224–233.

Banga, S.S., Labana, K.S. & Banga, S.K. (1984). Male sterility in Indian mustard

(*Brassica juncea* L.) coss: a biochemical characterization. *Theor. Appl. Genet.* 68:515–519.

Bawa, K.S. (1980). Evolution of dioecy in flowering plants. *Ann. Rev. Ecol. Syst.* 11:15–39.

Bedinger, P. (1992). The remarkable biology of pollen. *Plant Cell* 4:879–887.

Bhadula, S.K. & Sawhney, V.K. (1987). Esterase activity and isozymes during the ontogeny of stamens of male fertile *Lycopersicon esculentum* Mill., a male sterile *stamenless-2* mutant and the low temperature reverted mutant. *Plant Sci.* 52:187–194.

Bhadula, S.K. & Sawhney, V.K. (1989). Amylolytic activity and carbohydrate levels during the stamen ontogeny of a male fertile and a "gibberellin-sensitive" male sterile mutant of tomato (*Lycopersicon esculentum*). *J. Exp. Bot.* 40:789–794.

Bhadula, S.K. & Sawhney, V.K. (1991). Protein analysis during the ontogeny of normal and male sterile *stamenless-2* mutant stamens of tomato (*Lycopersicon esculentum* Mill.). *Biochem. Genet.* 29:29–41.

Bishop, C.J. (1954). A stamenless male sterile tomato. *Amer. J. Bot.* 41:540–542.

Chaudhury, A.M. (1993). Nuclear genes controlling male fertility. *Plant Cell* 5:1277–1283.

Cheng, P.C., Greyson, R.I. & Walden, D.B. (1979). Comparison of anther development in genic male-sterile *(ms10)* and in male-fertile corn (*Zea mays*) from light microscopy and scanning electron microscopy. *Can. J. Bot.* 57:578–596.

Coe, E.H., Hoisington, D.A. & Neuffer, M.G. (1987). Linkage map of corn (maize) (*Zea mays* L.) $2n = 20$. *Maize Genetics Coop. Newslett.* 61:116–149.

Cresti, M., Blackmore, S. & van Went, J.L. (1992). *Atlas of Sexual Reproduction in Flowering Plants*. Berlin: Springer-Verlag.

Cross, J.W. & Ladyman, J.A.R. (1991). Chemical agents that inhibit pollen development: tools for research. *Sexual Plant Reprod.* 4:235–243.

Dawson, J., Wilson, Z.A., Aarts, M.G.M., Braithwaite, A.F., Briarty, L.G. & Mulligan, B.J. (1993). Microsporogenesis and pollen development in six male-sterile mutants of *Arabidopsis thaliana*. *Can. J. Bot.* 71:629–638.

Dickson, M.H. (1970). A temperature sensitive male sterile gene in broccoli, *Brassica oleracea* L. var. Italica. *J. Amer. Soc. Hort. Sci.* 95:13–14.

Driscoll, C.J. (1986). Nuclear male sterility systems in seed production of hybrid varieties. *CRC Critical Rev. Plant Sci.* 3:227–256.

Driscoll, C.J. & Barlow, K.K. (1976). Male sterility in plants: induction, isolation and utilization. In *Induced Mutations in Cross Breed*, pp. 123–131. Vienna: FAO/IAEA Publ.

Frankel, R. (1973). The use of male sterility in hybrid seed production. In *Agriculture Genetics*, ed. R. Moav, pp. 85–94. New York: J. Wiley & Sons.

Frankel, R. & E. Galun. (1977). *Pollination Mechanisms, Reproduction and Plant Breeding*. Berlin: Springer-Verlag.

Fujimaki, H., Hiraiwa, S., Kushibuchi, K. & Tanaka, S. (1977). Artificially induced male sterile mutants and their usages in rice breeding. *Jap. J. Breed.* 27:70–77.

Goldberg, R.B., Beals, T.P. & Sanders, P.M. (1993). Anther development: basic principles and practical applications. *Plant Cell* 5:1217–1229.

Gottschalk, W. & Kaul, M.L.H. (1974). The genetic control of microsporogenesis in higher plants. *Nucleus* 17:133–166.

Gourret, J.P., Delourme, R. & Renard, M. (1992). Expression of *ogu* cytoplasmic male sterility in cybrids of *Brassica napus*. *Theor. Appl. Genet.* 83:549–556.

Graybosch, R.A. & Palmer, R.G. (1985). Male sterility in soybean (*Glycine max*). II. Phenotypic expression of the *ms4* mutant. *Amer. J. Bot.* 72:1751–1764.

Graybosch, R.A. & Palmer, R.G. (1988). Male sterility in soybean: a review. *Amer. J. Bot.* 75:144–156.

Greyson, R.I. (1994). *The Development of Flowers.* New York: Oxford Univ. Press.

Greyson, R.I. & Walden, D.B. (1976). Possibilities for gibberellin–male sterile relationships in corn: a proposal. *Maize Genet. Coop. Newslett.* 50:116–117.

Hamdi, S., Teller, G. & Louis, J.P. (1987). Master regulatory genes, auxin level and sexual organogenesis in the dioecious plant *Mercurialis annua. Plant Physiol.* 85:393–399.

Hockett, E.A., Baenziger, P.S. & Steffens, G.L. (1978). A proposal for increased research on chemical induction of fertility in genetic male-sterile barley. *Euphytica* 27:109–111.

Izhar, S. & Frankel, R. (1971). Mechanisms of male sterility in *Petunia*: the relationship between pH, callase activity in the anthers and the breakdown of microsporogenesis. *Theor. Appl. Genet.* 41:104–108.

Jorgensen, R.A. (1987). Synthetic linkages and genetic stability in transgenic plants: implications for breeding. In *Tomato Biotechnology*, eds. D.J. Nevins and R.A. Jones, pp. 179–188. New York: Alan R. Liss Inc.

Kakihara, F., Masahiro, K. & Tokumasu, S. (1988). Relationship between pollen degeneration and amino acids, especially proline, in male-sterile japanese radish (*Raphanus sativus* L. var. *longipinnatus* Bailey). *Sci. Hort.* 36:17–23.

Kaul, M.L.H. (1988). *Male Sterility in Higher Plants.* Berlin: Springer-Verlag.

Kern, J.J. & Atkins, R.E. (1972). Free amino acid content of the anthers of male sterile and fertile lines of grain *Sorghum bicolor* (L.) Moench. *Crop Sci.* 12:835–838.

Kinet, J.M., Sachs, R.M. & Bernier, G. (1985). *The Physiology of Flowering. Vol. 3. The Development of Flowers.* Boca Raton, FL: CRC Press.

Klee, H. & Estelle, M. (1991). Molecular genetic approaches to plant hormone biology. *Ann. Rev. Plant Physiol. & Mol. Biol.* 42:529–551.

Kofer, W., Glimelius, K. & Bonnett, H.T. (1990). Modification of floral development in tobacco induced by fusion of protoplasts of different male-sterile cultivars. *Theor. Appl. Genet.* 79:97–102.

Lapushner, D. & Frankel, R. (1967). Practical aspects, and the use of male sterility in the production of hybrid tomato seed. *Euphytica* 16:300–310.

Louis, J.P., Augur, C. & Teller, G. (1990). Cytokinins and differentiation process in *Mercurialis annua* L. (2n = 16). Genetic regulation with auxin, indoleacetic acid oxidases, and sexual expression patterns. *Plant Physiol.* 94:1535–1541.

Mariani, C., De Beauckeleer, M., Truettner, J., Leemans, J. & Goldberg, R.B. (1990). Induction of male sterility in plants by a chimaeric ribonuclease gene. *Nature* 347:737–741.

Marrewijk, G.A.M. van, Bino, R.J. & Suurs, L.C.J.M. (1986). Characterization of cytoplasmic male sterility in *Petunia hybrida*. I. Localization, composition and activity of esterases. *Euphytica* 35:77–88.

Moffatt, B.A & Somerville, C. (1988). Positive selection for male-sterile mutants of *Arabidopsis* lacking adenine phosphoribosyl transferase activity. *Plant Physiol.* 86:1150–1154.

Mutschler, M.A., Tanksley, S.D. & Rick, C.M. (1987). 1987 Linkage maps of the tomato (*Lycopersicon esculentum*). *Tomato Genet. Coop. Rep.* 37:5–34.

Nakajima, M., Yamaguchi, I., Kizawa, S., Murofushi, N. & Takahashi, N. (1991). Semiquantification of GA_1 and GA_4 in male sterile anthers of rice by radioimmunoassay. *Plant & Cell Physiol.* 32:511–513.

Novak, F. (1971). Cytoplasmatisch männliche Sterilität bei Weidelgras (*Lolium* sp.). *Z. Pflanzenzücht* 65:206–220.

Oard, J.H., Hu, J. & Rutger, J.N. (1991). Genetic analysis of male sterility in rice mutants with environmentally influenced levels of fertility. *Euphytica* 55:179–186.

Pacini, E., Franchi, G.G. & Hesse, M. (1985). The tapetum: its form, function and possible phylogeny in embryophyta. *Plant Syst. Evol.* 149:155–185.

Palmer, R.G., Albertsen, M.C., Horner, H.T. & Skorupska, H. (1992). Male sterility in soybean and maize: developmental comparisons. *The Nucleus* 35:1–18.

Palmer, R.G. & Kiang, Y.T. (1990). Linkage map of soybean (*Glycine max* L. Merr.). In *Genetic Maps*, ed. S.J. O'Brien. New York: Cold Spring Harbor Laboratory Publisher.

Pritchard, A.J. & Hutton, E.M. (1972). Anther and pollen development in male-sterile *Phaseolus atropurpureus. J. Hered.* 63:280–282.

Rao, M.K., Uma Devi, K. & and Arundhati, A. (1990). Applications of genic male sterility in plant breeding. *Plant Breeding* 105:1–25.

Regan, S.M. & Moffatt, B.A. (1990). Cytochemical analysis of pollen development in wild-type *Arabidopsis* and a male-sterile mutant. *Plant Cell.* 2:877–889.

Rick, C.M. (1948). Genetics and development of nine male-sterile tomato mutants. *Hilgardia* 18:599–633.

Rick, C.M. & Boynton, J.E. (1967). A temperature-sensitive male-sterile mutant of the tomato. *Amer. J. Bot.* 54:601–611.

Rundfeldt, H. (1960). Untersuchungen zur Züchtung des Kopfkohls (*B. oleracea* L. var. *capitata*). *Z. Pflanzenzücht* 44:30–62.

Sawhney, V.K. (1974). Morphogenesis of the *stamenless-2* mutant in tomato. III. Relative levels of gibberellins in the normal and mutant plants. *J. Exp. Bot.* 25:1004–1009.

Sawhney, V.K. (1983). Temperature control of male sterility in a tomato mutant. *J. Hered.* 74:51–54.

Sawhney, V.K. (1984). Hormonal and temperature control of male-sterility in a tomato mutant. *Proc. VIII Int. Symp. on Sexual Reproduction in Seed Plants, Ferns and Mosses*, pp. 36–38. Wageningen: Purdoc. Publ.

Sawhney, V.K. & Bhadula, S.K. (1987). Characterization and temperature regulation of soluble proteins of a male sterile tomato mutant. *Biochem. Genet.* 25:717–728.

Sawhney, V.K. & Bhadula, S.K. (1988). Microsporogenesis in the normal and male sterile *stamenless-2* mutant of tomato (*Lycopersicon esculentum*). *Can. J. Bot.* 66:2013–2021.

Sawhney, V.K. & Shukla, A. (1994). Male sterility in flowering plants: are plant growth substances involved? *Amer. J. Bot.* 81:1640–1647.

Schmidt, H. von & Schmidt, V. (1981). Untersuchungen an pollensteriten, stamenless-ähnlichen Mutanten von *Lycopersicon esculentum* Mill. II. Normalisievung von *ms-15* und *ms-33* mit Gibberellinsäure (GA₃). *Biol. Zbl.* 100:691–696.

Shi, M.S. (1986). The discovery, determination and utilization of the Hubei photosensitive genic male-sterile rice (*Oryza sativa* subsp. *japonica*). *Acta Genet. Sinica* 13:107–112.

Shukla, A. & Sawhney, V.K. (1992). Cytokinins in a genic male sterile line in *Brassica napus. Physiol. Plant.* 85:23–29.

Shukla, A. & Sawhney, V.K. (1993). Metabolism of dihydrozeatin in floral buds of wild-type and a genic male sterile line of rapeseed (*Brassica napus* L.). *J. Exp. Bot.* 44:1497–1505.

Shukla, A. & Sawhney, V.K. (1994). Abscisic acid: one of the factors affecting male sterility in *Brassica napus. Physiol. Plant.* 91:522–528.

Singh, S. & Sawhney, V.K. (1992). Cytokinins in a normal and the Ogura (*ogu*) cytoplasmic male sterile line of rapeseed (*Brassica napus*). *Plant Sci.* 86:147–154.

Singh, S., Sawhney, V.K. & Pearce, D.W. (1992). Temperature effects on endogenous indole-3-acetic acid levels in leaves and stamens of the normal and male sterile *stamenless-2* mutant of tomato (*Lycopersicon esculentum*). *Plant, Cell and Environ.* 15:373–377.

Skorupska, H.T., Desamero, N.V. & Palmer, R.G. (1994). Development hormonal expression of *apetalous* male sterile mutations in soybean, *Glycine max* (L.) Merr. *Ann. Biol.* 10:152–164.

Smith, L. (1947). Possible practical method for producing hybrid seed of self-pollinated crops through the use of male sterility. *J. Amer. Soc. Agron.* 39:260–261.

Stelly, D.M. & Palmer, R.G. (1982). Variable development in anthers of partially male-sterile soybeans. *J. Hered.* 73:101–108.

Stevens, M.A. & Rick, C.M. (1986). Genetics and breeding. In *The Tomato Crop*, eds. J.G. Athenton and J. Rudich, pp. 35–109. London: Chapman and Hall Ltd.

Theis, R. & Röbbelen, G. (1990). Anther and microspore development in different male sterile lines of oilseed rape (*Brassica napus* L.). *Agen Bot.* 64:419–434.

Van Tunen, A.J., Van der Meer, I.M. & Moe, J.N.M. (1994). Flavonoids and genetic modification of male fertility. In *Genetic Control of Self-incompatibility and Reproductive Development in Flowering Plants*, eds. E.G. Williams, A.E. Clarke, and R.B. Knox, pp. 423–442. Dordrecht, Netherlands: Kluwer Academic Publ.

Worrall, D., Hird, D.L., Hodge, R., Wyatt, P., Draper, J. & Scott, R. (1992). Premature dissolution of the microsporocyte callose wall causes male sterility in transgenic tobacco. *Plant Cell* 4:759–771.

9

Self-incompatibility

ANDREW McCUBBIN and HUGH DICKINSON

Summary 199
Introduction 199
Gametophytic self-incompatibility 201
    Structural aspects 201
    The biochemical basis of GSI 201
    Molecular cloning of S-*gene products and* S-*locus structure* 202
    Mechanism of the SI response 204
Two-locus gametophytic SI systems 204
    Genetics 204
    The operation of bifactorial SI 205
Sporophytic self-incompatibility 206
    History and phenomenology 206
    Primary structure of S-*locus glycoproteins* 207
    Other families of glycoproteins homologous to SLSGs 208
    S-*gene action in* Brassica 209
Diallelic sporophytic SI and heteromorphy 210
    Features associated with heteromorphic SI 210
    Operation of diallelic SI 211
SI: The current position 212
References 213

Summary

The field of self-incompatibility research is at present highly dynamic, and an understanding of the mechanisms of operation of the two most widespread systems is almost within our grasp. This review brings together information on the main types of self-incompatibility systems, outlines the potential benefits of the ability to manipulate self-incompatibility systems to agriculture, and explains how recent breakthroughs made in this field may allow the plant breeders to have this capability within the foreseeable future.

Introduction

Self-incompatibility (SI) is a genetic system possessed by many flowering plants. Indeed, SI has been reported in 66 plant families, representing every major phylogenetic line of the angiosperms (Brewbaker 1957), and it can be defined as "the inability of a fertile hermaphrodite seed plant to produce zygotes after self-pollination" (Lundqvist 1964, p. 222). The significance of SI in the evolutionary context cannot be overstated, for its possession leads to obligate outbreeding and the maintenance of heterozygosity within a species (Stebbins 1950).

Among crop and ornamental plants, most of the perennial grasses (Gramineae), forage legumes (Fabaceae), and members of the Brassicaceae (cabbage, kale, and so forth), Asteraceae (sunflowers, cosmos), Rosaceae (apples, cherries, pears, and so forth) and Solanaceae (petunia, potato, tobacco, and tomato relatives) have SI mechanisms of varying kinds and degrees of effectiveness.

SI presents contrasting prospects to plant breeders. On one hand it will frustrate efforts to produce homozygous lines, but on the other it provides a way to hybridize two lines without emasculation, nuclear or cytoplasmic sterility, or resorting to gametocides. The extensive employment of F_1 hybrids in agricultural and horticultural practice to exploit the phenomenon of "heterosis," or hybrid vigor, that often results, has led to a requirement for intercrossing the parental lines *en masse* to provide adequate quantities of F_1 seed. All alternatives to SI in large-scale F_1 seed production have their flaws: the cost involved in emasculation of one parent is often prohibitive; the use of sterility can lead to difficulties in breeding of the parental lines and occasionally inadvertently introduces genetic weakness (for example, the Texas male cytoplasm in maize – Williams and Levings 1992). Further, although recently developed gametocides may be effective, growing trends toward reduction in the use of chemical sprays and the rising environmental concern render them a less-than-perfect solution. The employment of SI systems is thus valuable and, indeed, all of the new hybrid brussels sprouts, cabbages, and kales owe their origins to breeding programs employing SI. Unfortunately, SI systems rarely provide the perfect vehicle for hybrid seed production. Often the negation of self–seed set is not 100% and may be dependent upon environmental conditions. Further, not all species naturally possess SI.

Although SI systems are widespread among angiosperms, few have been subjected to genetic analysis. Studies carried out on a range of different plant families have demonstrated that the phenomenology of SI varies extensively with respect to genetic control and cytological manifestation. As a result, SI systems can be subdivided into three basic groups.

SI systems in which incompatibility is associated with a number of different floral morphs (heteromorphic systems) can be distinguished from homomorphic systems. Furthermore, within homomorphic systems two distinct classes are found: gametophytic, where the SI phenotype of the pollen is determined by the genotype of the gametophyte (that is, the pollen); and sporophytic, where the SI phenotype of the pollen is determined by the genotype of the sporophyte (that is, the pollen parent plant). For many species, genetic control appears to be mediated by a single locus, termed the *S*-locus, which exhibits extreme polymorphism. Systems employing two or more unlinked loci also exist, notably the two-locus gametophytic system in the grasses (Lundqvist 1964). This review focuses largely on the two model SI systems that have been most extensively studied: single-locus gametophytic SI, exemplified by members of Solanaceae; and single-locus sporophytic SI, epitomized by members of Brassicaceae.

Gametophytic self-incompatibility

Structural aspects

Gametophytic self-incompatibility (GSI) is a common outbreeding mechanism occurring in more than 60 families of the angiosperms. In its most common and genetically simplest form, GSI is governed by a single, highly polymorphic locus. Pollen carrying an *S*-allele identical to one of the two alleles carried by the pistil is prevented from effecting fertilization. The most-studied family displaying GSI is Solanaceae.

In Solanaceae, incompatible pollen germinates normally on the stigma and is able to penetrate the stigma surface. The pollen tubes then enter the stylar transmitting tract, which is composed of files of longitudinally interconnected cells. Initially, growth appears to be normal (de Nettancourt et al. 1973); however, shortly after entering the transmitting tract, incompatible tubes take on a characteristic appearance. Numerous authors have reported the thickening of the outer wall of these tubes at this juncture, giving them a bulbous appearance, combined with an increase in density of cytoplasmic organelles (de Nettancourt et al. 1973).

Another interesting cytological feature of incompatible pollen tubes is the organization of the endoplasmic reticulum (ER) into concentric whorls and its subsequent degradation throughout the cytoplasm of the pollen tube (Cresti et al. 1979). Whorls of this type are generally associated with the cessation of protein synthesis. In addition to being found in incompatible pollen tubes, these whorls have been found in dormant tissues, notably *Betula* buds (Dereuddre 1971) and dormant potato tubers (Shih and Rappaport 1971). However, the relationship between whorls of ER and the cessation of protein synthesis has not been established unequivocally. As yet, there is no clue as to whether the cessation of protein synthesis, if indeed it occurs, is a cause or an effect.

The biochemical basis of GSI

Over the last decade, much of the research in the field of GSI has stemmed from the observation that, within pistil extracts, certain proteins segregate with their respective *S*-alleles in genetic crosses. These S-proteins are present in high proportions in style tissues (generally 1%–10%) and are sufficiently diverse to be differentiated on SDS-polyacrylamide or isoelectric focusing gels. S-proteins have now been identified in the following solanaceous species: *Lycopersicum peruvianum* (Mau et al. 1986), *Nicotiana alata* (Bredemeijer and Blaas 1981), *Petunia inflata* (Ai et al. 1990), *Petunia hybrida* (Kamboj and Jackson 1986), *Solanum chacoense* (Xu et al. 1990a), and *Solanum tuberosum* (Kirch et al. 1989). The molecular weights of these S-proteins range from 23 to 34 kD; all possess high pI values – often higher than 8.0; and every S-protein of this type so far studied has been glycosylated.

In addition to segregating with S-alleles, the temporal and spatial distribution of these proteins correlates with the onset of incompatibility and the site of pollen tube rejection.

Molecular cloning of S-gene products and S-locus structure

The rise of molecular cloning techniques has considerably advanced our knowledge of the S-gene and its expression in female tissues. So far, the deduced amino acid sequences of 18 S-proteins from five solanaceous species have been reported. These include seven S-alleles of *Nicotiana alata* (Anderson et al. 1986, 1989; Kheyr-Pour et al. 1990), three of *Petunia inflata* (Ai et al. 1990), three of *P. hybrida* (Clark et al. 1990), two of *Solanum chacoense* (Xu et al. 1990b), and three of *S. tuberosum* (Kaufmann et al. 1991).

Two independent experiments have unequivocally established S-proteins as the pistillar component of the SI response. Both studies involved transformation with S-protein constructs to alter pistil phenotype. Lee et al. (1994) demonstrated both loss and gain of female S-function of the S_3 allele in *Petunia inflata* through transformation with antisense and sense constructs of the S_3 allele driven by its own promoter. Murfett et al. (1994) also demonstrated gain of female S-function in a *Nicotiana alata* × *langsdorffii* hybrid with a sense construct of the *N. alata* S_{a2} allele driven by a chitinase promoter (ChiP). The advent of these transformation experiments heralds a new era of opportunity in SI research, opening up the possibility of studying S-gene structure–function relationships.

Analysis of the aligned sequences reveals 34 perfectly conserved residues and 27 sites that are only conservatively replaced. Sixteen of the 34 conserved residues are also conserved in RNase T2 of *Aspergillus oryzae* and RNase Rh of *Rhizopus niveus*. These short stretches of homology would not normally be considered significant. However, because these stretches include the sequences surrounding the catalytic domains of the fungal ribonucleases – the histidine residues essential for catalysis (Kawata et al. 1990) and the four cysteine members of two disulfide bonds – this observation assumed greater significance. These findings led to the discovery that S-proteins do indeed exhibit ribonuclease (RNase) activity (McClure et al. 1989).

Amino acid sequence identity between the S-proteins so far published varies considerably, from 38.74% to 93.47% (Singh and Kao 1992). This tremendous diversity between alleles of the same species contrasts sharply to other multiallelic genes, which usually differ only in a few amino acids.

Conserved residues within S-proteins essentially fall into five regions, designated C1–C5. A window-averaged hydrophobicity plot shows that C1, C4, and C5 are hydrophobic, suggesting that they may form the core structure of S-proteins. C2 and C3 are hydrophilic and are also conserved in the fungal ribonucleases; hence, they are probably involved in catalytic activity. Two regions of extreme variability have also been identified – hypervariable (Hv) regions a and b (Ioerger et al. 1991); these are consistently the most hydrophilic regions in each S-protein and may constitute the domains involved in encoding *S*-allele specificity.

The role of the carbohydrate side chains of S-proteins in *S*-allele specificity has recently been investigated. This involved the transformation of *Petunia inflata* with an S_3 construct in which the single asparagine glycosylation site was converted to aspartic acid. It was demonstrated that transformants expressing this protein at levels comparable to wild-type S-proteins gained S_3 function (Karunanandaa et al. in press). Though some S-proteins possess more than one side chain, this result strongly suggests that the glycosyl groups of S-proteins play no role in *S*-specificity and thus this function must be encoded by the amino acid backbone.

Despite considerable advances in our understanding of the *S*-locus component expressed in the pistil, the identity of the pollen component remains unresolved. Dodds et al. (1993), through the use of polymerase chain reaction techniques, demonstrated detectable amounts of transcript identical to pistil S-RNase in postmeiotic microspores. Mutations leading to the loss of male function have – at least in some cases – been proposed to be related to a duplication event involving the *S*-locus (Pandey 1965; de Nettancourt 1977). This might be explained by the phenomenon of cosuppression, which may account for the frequently observed breakdown of SI in polyploids. The mutation of unlinked modifier loci has also been implicated in loss of male function (R. D. Thompson et al. 1991) and may also reconcile genetic data with a model involving identical *S*-locus products in the pollen and pistil. There is, however, a considerable amount of genetic data suggesting that pollen and pistil components of the SI interaction can mutate independently (Lewis 1949; Pandey 1956; van Gastel and de Nettancourt 1974). As a result, it appears unlikely that pollen and pistil components are identical.

It had been hoped that examination of genomic sequences flanking known *S*-genes might yield some information as to the organization of the locus. Genomic clones for two *S*-alleles from both *S. tuberosum* (Kaufmann et al. 1991) and *P. inflata* (Coleman and Kao 1992) have been isolated and characterized. Both *P. inflata* alleles were found to be embedded regions rich in repetitive sequences, but no open reading frames were identified (7 kb upstream and 13 kb downstream). This suggests that chromosome walking to linked sequences is not viable (at least in *P. inflata*) (Coleman and Kao 1992), and hence, novel approaches may have to be found to elucidate the components of the linkage group that participate in the interaction.

Mechanism of the SI response

The discovery that S-proteins exhibit RNase activity immediately raised the question of its relevance to biological function. A recent report demonstrated that RNase activity is essential for *S*-protein function (Huang et al. 1994). This demonstration was achieved through the transformation of *Petunia inflata* with an S_3 construct in which one of the histidine residues essential for catalytic activity was replaced with an asparagine residue. It was demonstrated that the modified protein did indeed lack RNase activity and that when expressed at levels comparable to wild-type S_3 protein, it did not lead to a gain of S_3 function.

Because RNase activity is an integral part of the SI reaction, the most obvious potential role would be in the degradation of pollen tube RNA. One hypothesis is that S-RNase enters incompatible pollen tubes as they grow through the extracellular matrix and degrades RNA, resulting in the demise of the pollen tubes. In view of the fact that rRNA genes are not transcribed in pollen (Mascarenhas 1990; McClure et al. 1990), this would provide an effective way of ensuring pollen tube death. This model necessitates a specificity in the interaction regulating the uptake of the S-RNase into the pollen tube.

Another model for the arrest of pollen tube growth envisages S-specific inhibitors that interfere with S-protein activity in an allele-specific manner. This model requires the male products of the *S*-locus to be highly specific inhibitors that interact with S-proteins over a wide range of sequence diversity whilst remaining ineffective against one molecule, the self–S-protein. Aside from the improbability of this, preliminary experiments show no evidence of any molecule from pollen tubes inhibiting S-protein RNase activity (McClure et al. 1990). It is not as yet possible to rule out the inhibitor model entirely, as it is possible that a complex interaction, involving one or more pistil components and an S-specific inhibitor present in the pollen tube cytoplasm, is required to inactivate S-RNases.

RNases have been implicated in the regulation of other key processes in plant growth and development (Farkas 1982). Certain plant lectins may provide an interesting parallel to S-proteins. Ricin A and phytoalaccin, for example, are able to inactivate ribosomes through an associated RNase activity (Obrig et al. 1985). The study of the role of RNase activity in SI is therefore not an isolated problem and is of potential significance to other areas of plant growth and development.

Two-locus gametophytic SI systems

Genetics

Two-locus, or bifactorial, GSI systems were first described in the grasses (Lundqvist 1956). Most grasses investigated possess SI of this form, although all cereals except rye are self-compatible. The two loci involved in the SI response are termed S and Z and each are polyallelic. There appears to be co-operation between S and Z in the pollen, but independent actions in the pistil. Each combination, therefore, gives rise to a distinct specificity in the haploid pollen, and rejection occurs when this specificity is matched by one of the four possible combinations of S- and Z-alleles in the diploid stigma. This results in the two-locus system's being more resistant to decay than that governed by a single locus, and the two-locus system is less likely to suffer from inbreeding and loss of alleles in small populations. Another consequence of employing two loci in an SI system is that it is less likely to acquire self-compatible mutants, for the S- and Z-loci act in both a complementary and an independent manner; thus, if one locus mutates, the other gives rise to incompatibility.

The operation of bifactorial SI

The bifactorial system differs from other gametophytic systems not only in having two-locus control, but also in exhibiting many cytological features that are much more akin to those of sporophytic systems (pollen is trinucleate, and inhibition occurs at the stigma surface). It has been suggested, therefore, that gametophytic SI in the grasses has arisen independently from self-compatible plants (Richards 1986). Thus it is not surprising that the operation of bifactorial SI differs fundamentally from that found in other gametophytic SI plants.

There are two major morphological differences between the operation of SI in grasses and in plants possessing single-locus gametophytic SI. First, although the pollen germinates well and the pollen tubes start to grow normally in grasses, tube growth ceases as the tube touches the stigma surface. Secondly, on contact with the papilla, nodules (probably of microfibrillar pectins) appear at the tip of the tube. After this initial response, there follows a rapid accumulation of pectin occluding the tube apex. As in single-locus systems, the cessation of growth is then accompanied by accumulation of an inner sheath of callose around the tube tip. The arrested tube can be shown to respire normally and it appears that the pectic occlusion of the tube tip is solely responsible for cessation of tube growth.

These observations led J. Heslop-Harrison and Heslop-Harrison (1982) to propose the following model of S-gene action in grasses. First, the female incompatibility factors are glycoproteins with binding properties, located on the papilla surface. Second, these glycoproteins bind specifically to sugar moieties present in the long-chain carbohydrates in the wall of the pollen tube apex. Binding of the glycoproteins to the tube tip mechanically prevents growth and extension of the tube tip, causing a build-up of wall precursor particles that include microfibrillar pectins. These particles, no longer dispersed into the tube wall, rapidly occlude the tube tip, terminating the chance of successful fertilization.

At present, there is little information as to the precise nature of products of the S- and Z-loci, one of the main reasons being the technical complications arising from the study of a two-locus system. Segregation analysis of S-phenotypes and attempts to determine segregating gene products require studying considerably larger populations than for one-locus systems. Despite the technical problems associated with studying this system, the potential benefits to agriculture are large. It might be expected that introduction of an SI system from closely related genera into cereal crops may be more straightforward than employing a system from completely unrelated plants, for genetic background effects have often thwarted attempts to introgress SI into self-compatible lines.

Sporophytic self-incompatibility

History and phenomenology

Sporophytic SI (SSI) was first observed in radish (*Raphanus sativus*) (Stout 1920); however, its genetic control remained unknown until analysis of the inheritance patterns of *Iberis amara* and *Raphanus* (Bateman 1955). In SSI, pollen phenotype is determined by the genotype of the mother plant, and dominance interactions occur that determine the phenotype of the pollen. The number of S-alleles at the S-locus is, as in GSI, unusually large, estimated at 22 in *Iberis* (Bateman 1955), 34 in *Raphanus* (Sampson 1957), and 60 in *B. oleracea* (Ockendon 1974).

The best-characterized sporophytic system is that which operates in members of Brassicaceae. The crucifer stigma is capped by a layer of papillate cells and, in a compatible pollination, there is a rapid series of events, including adhesion to the stigma, hydration and germination of the pollen grains, and subsequent invasion of the papillar cell wall by the emerging pollen tube (Elleman et al. 1992). The phenomena associated with incompatible pollinations vary according to the strength of the alleles involved. In the strictest form of incompatible pollination, adhesion to the dry stigma surface is poor and even hydration is absent, but with weaker alleles, hydration and germination may occur. Resultant tubes succeed in penetrating the stigmatic cuticle, but they fail to invade the stigma cell wall (Kanno and Hinata 1969). Generally, there is a deposition of the ß-1,3-glucan callose within the stigmatic papillae at the site of contact with incompatible pollen (J. Heslop-Harrison 1975); this can be a useful diagnostic phenomenon when determining the cross-compatibility. However, recent data suggest that this is related more to the leakage of gametophytic proteins from necrotic pollen grains or tubes than to playing a mechanistic role in the incompatibility reaction itself (Elleman and Dickinson 1994).

SSI is also developmentally regulated and comes into operation 1–2 days pre-anthesis. Because stigmas are receptive prior to this point, there is a window in development when it is possible to successfully self-pollinate immature flowers. This has been valuable in the generation and maintenance of S-locus–homozygous lines for both research and breeding purposes.

Early approaches to examining stigmatic components employed electrophoretic and immunological methods to identify polypeptides specific to *S*-alleles in *B. oleracea* (M. E. Nasrallah and Wallace 1967; M. E. Nasrallah et al. 1970). These *S*-locus–specific glycoproteins (SLSGs) are stigma specific and temporally associated with the onset of SI (J. B. Nasrallah et al. 1985a); hence they are strongly implicated in the SI response.

Primary structure of *S*-locus glycoproteins

The method of SLSG cDNA isolation, sequencing, and subsequent deduction of amino acid sequence has been extensively employed in *B. oleracea* and *B. campestris* (J. B. Nasrallah et al, 1985b, 1987; Lalonde et al. 1989; Trick and Flavell 1989; Chen and Nasrallah 1990; Dwyer et al. 1991; Goring et al. 1992; Scutt and Croy 1992). Analysis of sequence data provided by these studies has highlighted a number of points of interest. The various *S*-alleles exhibit high levels of sequence divergence – up to 30% (Chen and Nasrallah 1990; Stein et al. 1991), averaging around 20% at the protein level. Furthermore, the differences observed between *B. oleracea* *S*-alleles are of the same order as those observed between alleles from *B. oleracea* and *B. campestris*, indicating a common origin for the two species. Differences take the form of substitutions throughout the entire sequence and of deletions or additions of one or two residues at isolated points in the chain.

The possibility that the glycosyl groups of these SLSGs are involved in *S*-allele specificity has also been raised. Takayama et al. (1986) studied microheterogeneity within the glycosyl side chains. It was found that two oligosaccharide chains constitute the principal components of the SLSG side chains. Both chain types are common throughout plant glycoproteins; one is predominant and may correspond to the mature form, whereas the other is probably its precursor. The number and position of potential N-glycosylation sites have been identified as being characteristic for each allele (Dickinson et al. 1992). In a study of the 9 *S*-allele sequences available, 13 potential N-glycosylation sites were identified; 4 are conserved in all 9 sequences studied. An additional complication is that not all potential N-glycosylation sites are occupied, for in *B. campestris,* analysis of the S_8 allele suggests that only 7 of 9 potential sites are glycosylated (Takayama et al. 1986). At present, it is uncertain whether these carbohydrate moieties play a role in *S*-specificity.

Despite the extreme sequence variability between *S*-alleles, 11–12 cysteine residues are conserved in the carboxy-terminal regions of SLSGs. This conservation, along with the importance of cysteine residues in the formation of disulfide bridges and in the maintenance of three-dimensional structure, strongly suggests that these residues are vital to the function of *S*-allele products (J. B. Nasrallah et al. 1987).

A number of naturally occurring *S*-alleles have been arranged in a dominance series based on their genetic behavior relative to other alleles in heterozygous plants (K. F. Thompson and Taylor 1966). It has been possible to create two broad classes: high-activity (class I) alleles, which are placed high on the dominance scale and exhibit a strong SI phenotype, and low-activity (class II) alleles, which have a weak SI phenotype. J. B. Nasrallah et al. (1991) have correlated sequence data from SLSGs with their position in the dominance series. Within class I alleles, there was 90% DNA sequence homology and 80% at the peptide level. However, class I and II alleles share only 70% DNA homology and 65% amino acid homology. Further, it was found that the distinction could be made at the protein level via the antibody MAbH8, which binds to class I SLSGs but not to class II. This distinction extends beyond *B. oleracea* to *B. campestris*, and, importantly, this grouping is not restricted to *Brassica*, for it has been possible to distinguish two sets of *S*-alleles in *Raphanus* by employing MAbH8. This constitutes a useful method by which plant breeders may determine the strength of an unknown *S*-allele extremely rapidly and, significantly, provides a strong indication that the evolution and fixation of *S*-alleles may have preceded recent speciation in this group.

Other families of glycoproteins homologous to SLSGs

Southern analysis has demonstrated that the *S*-locus glycoprotein gene (*SLG*), as it is now termed, is a member of a multigene family comprising some 10–15 related sequences (J. B. Nasrallah et al. 1985b, 1988). Some family members have been sequenced and apparently are pseudogenes, having accumulated frameshift mutations (Nasrallah et al. 1988). Lalonde et al. (1989) demonstrated the stigma-specific expression of one of these sequences and, importantly, that it was not linked to the *S*-locus. Probing a stigma-specific cDNA library, two classes of clone were identified: one corresponding to the SLSG cDNA, and another hybridizing less strongly and representing a clone they termed *SLR1* (*S*-locus related 1). Nucleic acid sequences of *SLR1* and *SLSG* have been determined for a number of *S*-alleles. The predicted SLR1 protein shows several characteristics of SLSG molecules, including the 12 cysteine residues in the region of the carboxy-terminus of the protein. Translation of the *SLR1* transcripts in stigmas of *Brassica* has been confirmed with the aid of immunological and N-terminal protein sequence analyses (Umbach et al. 1990; Gaude et al. 1991).

The *SLR1* gene is not the only *S*-locus–related gene expressed in stigmas of *Brassica*; Scutt et al. (1990) have reported the presence of further *SLR*-class transcripts. These *SLR2s* have been shown to be present in the majority of lines of *B. oleracea* and, like *SLR1s*, are highly conserved (Umbach et al. 1990; Boyes et al. 1991). Conservation of *SLR* sequences extends beyond SI *Brassica* and has been detected in self-compatible *Brassica*, *Raphanus*, and *Arabidopsis*.

The most exciting recent development in this field arose from the observation that part of a putative receptor serine-threonine protein kinase from *Zea mays*, named ZmPK1, possessed 27% amino acid homology to the S_{13} SLSG from *B. oleracea* (Walker and Zhang 1990). An equivalent gene, termed *S*-related kinase (*SRK*), was rapidly shown to be expressed in stigmas and anthers of *B. oleracea* (Stein et al. 1991). The *SRK* gene is comprised of three domains: an *SLG*-like domain, a transmembrane domain, and a kinase domain. The *SRK* gene is linked genetically to the *SLG* and, hence, to the SI phenotype. Importantly, the *SLG* and *SLG*-like component of the *SRK* within an *S*-haplotype are highly homologous (around 90% amino acid identity) (Stein et al. 1991; Goring and Rothstein 1992).

Almost all *SLG* alleles reported to date lack introns (J. B. Nasrallah et al. 1988; Dwyer et al. 1989). In contrast, *SRKs* consist of seven exons separated by six introns (Stein et al. 1991). In common with many eukaryotic genes, the *SRK* gene appears to have originated by the shuffling of exons that represent distinct functional domains. Thus, exon 1 encodes the *S*-domain, exon 2 the transmembrane and juxtamembrane domain, and exons 3–7 encode the cytoplasmic kinase domain. The nature of exon 1 as a distinct structural unit is further emphasized by its physical separation from the other exons by the longest intron, which varies from 896 bp to over 5.5 kb (Stein et al. 1991; Tantikanjana et al. 1993).

A plausible hypothesis for the origin of the *SLG–SRK* gene pair emerged from the analysis of a class II haplotype (Tantikanjana et al. 1993). This haplotype is unusual because it possesses an *SLG* composed of two exons (the first exon encoding the *S*-domain of the gene) and produces two transcripts that differ at their 3' ends. One transcript contains sequences from the first exon and terminates within the intron, encoding a secreted glycoprotein. The second, larger transcript contains sequences from both exons and encodes a membrane-anchored form of SLSG. The sequence similarity shared between *SLG* and *SRK* in this haplotype not only includes the *S*-domain but extends into the intron and second exon of the genes. As a result, Tantikanjana et al. propose that *SLGs* are derived from *SRKs* by gene duplication.

S-*gene action in* Brassica

Present genetic and molecular data are consistent with the hypothesis that rejection of self-pollen is mediated by activation of the *SRK* on self-pollination. The operation of a signaling receptor is also consistent with the speed of inhibition of self-pollen and also with the nature of the cytological response. Phosphorylation of intracellular substrates by the SRK protein could couple the initial molecular recognition events at the stigma surface with a phosphorylation cascade leading ultimately to pollen rejection (Stein et al. 1991). The role of the SLSG is less obvious; an interaction between the transmembrane receptors and secreted forms of the SLSG's extracellular "recognition" domain may be a requirement for signaling across the cell wall (J. B. Nasrallah and Nasrallah 1993). In the stigma, SLSG and SRK might compete for the binding of some as-yet-unidentified, *S*-specific ligand borne by the pollen or might perhaps interact in a form of homophilic binding – again mediated through a pollen-borne S-specific ligand or modifier. This scenario is analogous to the mode of operation of ligand-activated tyrosine kinases of animal systems. The key question, as yet unanswered, is the nature of the male component of the interaction. The likely location of this component is in the sporophytically derived pollen coat, which through recent pollen coat modification experiments has been shown to be capable of altering the S-phenotype of the pollen grain (H. G. Dickinson, unpublished results).

One component of the pollen coat, termed PCP7, has been shown to interact with S-class molecules (Doughty et al. 1993). PCP7 is a highly charged molecule with a molecular weight of 7 kd and structural homology to thionins (J. Doughty, personal communication); as yet, it is uncertain whether it exhibits polymorphism between S-haplotypes. The PCP7 molecule appears to be a member of a family of 10–15 homologous peptides, only one of which is capable of interacting with S-class molecules. PCP7 interacts with both SLSG and SLR molecules (Doughty et al. 1993; S. J. Hiscock, unpublished results) and would be envisaged to interact with the SLSG domain of the SRK molecule. If S-polymorphism exists within the PCP7 class molecules, it is possible that the interaction product encodes a specificity in terms of a further binding to activate a phosphorylation event via SRKs, or perhaps a further as-yet-unknown component is required for specificity. Nevertheless, because this class of peptides is capable of interaction solely with S-class stigmatic molecules and is tapetally derived, it seems unlikely that it has no involvement in the incompatibility response.

Diallelic sporophytic SI and heteromorphy

Features associated with heteromorphic SI

Unlike gametophytic SI systems, sporophytic systems can exist with only two alleles, and in such cases the two alleles have been termed S (dominant) and s (recessive). This is possible because the diploid heterozygous male parent produces pollen of a single S-phenotype though the pollen genotypes are both S and s. The most-studied genus possessing this type of SI is *Primula*. Almost all diallelic SI systems display floral heteromorphism, usually in the form of di- or heterostyly. In distyly, one form has a short style and long anthers (the "thrum" form), and the other has the opposite (the "pin" form) (see Chapter 4). Various other floral characters often vary between morphs, including pollen size and sculpturing, cell shape in the stylar conducting tissue, and stigma morphology. Generally there is tight correspondence between the phenotype and the heteromorph, and it might be reasonably suggested that the incompatibility phenotype and the heteromorph are pleiotropic expressions of the same gene. However, recombinants between incompatibility phenotypes and the heteromorph, and between character states within the heteromorph, are known, the best studied being the *Primula* homostyles (Mather and De

Winton 1941). It is likely, therefore, that in most cases heteromorphy is controlled by two linkage groups – one comprising genes encoding morph-associated characters, and the other the *S/s*-incompatibility-locus – with the two groups themselves closely linked in a supergene. The rarity of recombinants within the linkage groups, either between the incompatibility phenotype and the heteromorph, or between the various characters involved in the heteromorphy, strongly suggests that the various features controlled by the loci of the supergene are co-adapted.

Operation of diallelic SI

A number of studies have been carried out with regard to the operation of diallelic SI; most have been in *Primula* (Pandey and Troughton 1974; Y. Heslop-Harrison et al. 1981; Stevens and Murray 1982; Shivanna et al. 1983), but others have been in *Linum* (Dickinson and Lewis 1974; Ghosh and Shivanna 1980). The various studies illustrate that, at least in *Primula*, within-morph incompatibility can occur at a number of stages in the fertilization process:

(a) lodgement, adhesion, and germination of pollen on the stigma surface
(b) penetration of the stigmatic papillae by the pollen tube
(c) growth of the pollen tube in the stigma
(d) growth of the pollen tube in the style

At any one stage, incompatibility is rarely total, and each stage seems to act in a quantitative rather than a qualitative manner. The cumulative effect of the quantitative action at each of these stages does, however, give rise to total within-morph incompatibility (Richards and Ibrahim 1982). The incompatibility mechanism, as a whole, appears to vary between species and usually between morphs (Richards 1986).

The assumption that heteromorphic features are involved in the incompatibility response is far from being proved certain. Several workers, notably Ornduff (1979), have examined "legitimate" (between-morph) and "illegitimate" (within-morph) stigmatic pollen loads to verify the assumption. There is evidence that heterostyly does promote legitimate pollen transfer, but the evidence is not always convincing.

The apparently co-adapted linkage between diallelic SI and heteromorphic features, the rarity of plants in which the linkage disequilibrium has been lost, and the uncanny correspondence between heteromorphic features in unrelated genera all strongly suggest that heteromorphic characters play an important role in the incompatibility reaction.

At present, there is an impressive amount of genetic information amassed concerning diallelic SI systems and linked heteromorphic characters; also, some physiological data have been gathered. However, no molecular data are available. The present sophistication of molecular methodology suggests that there is considerable potential for breaking down this complex syndrome into its various components and subjecting them to molecular analysis. Diallelic SI also provides the molecular biologist with a well-characterized system in which to study a supergene complex in plants.

SI: The current position

Recent years have seen considerable advances in our understanding of SI systems, made possible largely through the application of recombinant DNA technology. Despite this encouraging progress, a large gap remains between our present knowledge and an ability to comprehend the intricacies of the various systems. From an applied prospective, the vital question to address is which system or systems are likely to provide the most appropriate vehicle for the exploitation of SI in plant breeding. There is an ever-increasing amount of information available suggesting that SI systems are polyphyletic and have not arisen as modifications on a theme, and recent data suggest that even within one-locus gametophytic SI systems there is considerable diversity. For example, SI in *Papaver* does not appear to be mediated through S-RNases (Foote et al. 1994), and recent, as-yet-unpublished data from *Antirrhinum* (Scrophulariaceae) implicate S-RNases but indicate that these have arisen independently from those employed in Solanaceae.

The seemingly polyphyletic origins of SI systems, combined with the frequently reported problems associated with "modifier" or "suppression" loci found in compatible plants, strongly implies that, when introducing SI into a plant variety, it will be essential to employ a system naturally occurring in a closely related species. Consequently, it would be imperative to study a representative range of SI systems, thus developing a large information base.

One of the greatest dilemmas in the field is how to manipulate SI systems so that they are sufficiently stringent to use in the breeding of F_1 hybrids. This has already proven to be a problem with all types of SI systems, and there is no reason to assume that the same problem will not occur on introducing SI into other species or varieties. As yet, it is difficult to ascertain whether or not it will be possible to optimize any SI system so that it operates to the required levels of stringency. An alternative possibility, however, and one worth entertaining, is whether or not a combination of perhaps a one-locus gametophytic or sporophytic system and one or more heteromorphic traits might provide a solution. In terms of our present understanding, and also the likely size of a gene construct conferring all of the required components of such a system, this concept might seem rather implausible. However, as technology improves along with our knowledge of SI systems, this may not be beyond the realm of possibility and indeed may provide the simplest solution to a complex problem.

In conclusion, the field of SI research is at present extremely dynamic. Identification of the key specificity components of female *S*-gene products is not far over the horizon and is likely to be achieved within the near future. The great unanswered question in all systems concerns the nature of the male *S*-gene products.

Acknowledgments

The authors gratefully acknowledge the funding from the AFRC (U.K.) and EC BRIDGE.

References

Ai, Y., Singh, A., Coleman, C.E., Ioerger, T.R., Kheyr-Pour, A. & Kao, T.-h. (1990). Self-incompatibility in *Petunia inflata*: isolation and characterization of cDNAs encoding three S-allele associated proteins. *Sex. Plant Reprod.* 3:130–138.

Anderson, M.A., Cornish, E.C., Mau, S-L., Williams, E.G., Hoggart, R.M., Atkinson, R.J., Roche, P., Haley, J., Penschow, J., Niall, H., Tregear, G., Coghlan, J., Crawford, R. & Clarke, A.E. (1986). Molecular cloning of a cDNA of a stylar glycoprotein associated with the expression of self-incompatibility in *Nicotiana alata* Link and Otto. *Nature* 321:38–44.

Anderson, M.A., McFaddon, G.I., Bernatzsky, R., Atkinson, A., Orpin, T., Dedman, H., Tregear, G., Fernley, R. & Clarke, A.E. (1989). Sequence variability of three alleles of the self-incompatibility gene of *Nicotiana alata*. *Plant Cell* 1:483–489.

Bateman, A.J. (1955). Self-incompatibility systems in angiosperms. *Heredity* 9:52–68.

Boyes, D.C., Chen, C.H., Tantikanjana, T., Esch, J.J. & Nasrallah, J.B. (1991). Isolation of a second *S*-locus related cDNA from *Brassica oleracea*, genetic relationships between the two related loci. *Genetics* 127:221–228.

Bredemeijer, G.M.M. & Blaas, J. (1981). S-specific proteins in styles of self-incompatible *Nicotiana alata*. *Theor. Appl. Genet.* 59:185–190.

Brewbaker, J.L. (1957). Pollen cytology and incompatibility systems in plants. *J. Hered.* 48:217–277.

Chen, C. H. & Nasrallah, J.B. (1990). A new class of S-sequences defined by a pollen recessive self-incompatibility allele of *Brassica oleracea*. *Mol. Gen. Genet.* 222:241–248.

Clark, K.R., Okuley, J.J., Collins, P.D. & Simms, T. (1990). Sequence variability and developmental expression of S alleles in self-incompatibility and pseudocompatibility in *Petunia*. *Plant Cell* 2:815–826.

Coleman, C.E. & Kao, T.-h. (1992). The flanking regions of two *Petunia inflata* S-alleles are heterogeneous and contain repetitive sequences. *Plant Mol. Biol.* 18:725–737.

Cresti, M., Campolini, F., Pacini, E., Sarfatti, G., van Went, J.L. & Willemse, M.T.M. (1979). Ultrastructural differences between compatible and incompatible pollen tubes in the stylar transmitting tissue of *Petunia hybrida*. *J. Submicr. Cytol.* 11(2):209–219.

Dereuddre, J. (1971). Sur la présence de groupes de saccules appartenants au reticulum endoplasmique dans les cellules des ébauches foliares en vie ralentie de *Betula verrucosa* Ehrt. *C.R. Acad. Sci. Paris* 273(D):2239–2242.

Dickinson, H.G., Crabbe, M.J.C. & Gaude, T. (1992). Sporophytic self-incompatibility systems: *S* gene products. *Intl. Rev. Cytol.* 140:525–561.

Dickinson, H.G. & Lewis, D. (1974). Changes in the pollen grain wall of *Linum grandiflorum* following incompatible intraspecific pollination. *Ann. Bot.* 38:23–29.

Dodds, P.N., Bonig, I., Du, H., Rodin, J., Anderson, M.A., Newbiggin, E. & Clarke, A.E. (1993). S-RNase gene of *Nicotiana alata* is expressed in developing pollen. *Plant Cell* 5:1771–1782.

Doughty, J., Hedderson, F., M^{C}Cubbin, A. & Dickinson, H.G. (1993). Interaction between a coating borne peptide of *Brassica* pollen grains and stigmatic S (self-incompatibility)-locus-specific glycoproteins. *Proc. Natl. Acad. Sci. USA* 90:467–471.

Dwyer, K.G., Balent, M.A., Nasrallah, J.B. & Nasrallah, M.E. (1991). DNA sequences of self-incompatibility genes from *Brassica campestris* and *Brassica oleracea*: polymorphism predating speciation. *Plant Mol. Biol.* 16:481–486.

Dwyer, K.G., Chao, A., Cheng, B., Chen, C.H. & Nasrallah, J.B. (1989). The *Brassica* self-incompatibility multigene family. *Genome* 31:969–972.

Elleman, C.J. & Dickinson, H.G. (1994). Pollen stigma interaction during sporophytic self-incompatibility in *Brassica oleracea*. In *Genetic Control of Self-Incompatibility and Reproductive Development*, eds. E.G. Williams and R.B. Knox, pp. 67–87. Dordrecht: Kluwer Academic Publishers.

Elleman, C.J., Franklin-Tong, V. & Dickinson, H.G. (1992). Pollination in species with dry stigmas: the nature of the stigmatic response and the pathway taken by the pollen tubes. *New Phytol.* 121:413–424.

Farkas, G.L. (1982). Ribonuclease and ribonucleic acid breakdown. In *Nucleic Acids and Proteins in Plants,* eds. B. Parthier and D. Boulter, pp. 224–262. Berlin: Springer-Verlag.

Foote, H.C.C., Ride, J., Franklin-Tong, V.E., Walker, E.A, Lawrence, M.A. & Franklin, C.H. (1994). Cloning and expression of a distinctive class of self-incompatibility (S) gene from *Papaver rhoeas* L. *Proc. Natl. Acad. Sci. USA* 91:2265–2269.

Gastel van, A.J.G. & de Nettancourt, D. (1974). The effects of different mutagens on self-incompatibility in *Nicotiana alata* Link and Otto. I. Chronic gamma irradiation. *Radiation Bot.* 14:43–50.

Gaude, T., Denoroy, L. & Dumas, C. (1991). Use of a fast electrophoretic purification for N-terminal sequence analysis to identify S-locus related proteins in stigma of *Brassica oleracea*. *Electrophoresis* 12:646–653.

Ghosh, S. & Shivanna, K.R. (1980). Pollen–pistil interaction in *Linum grandiflorum*, scanning electron microscope observations and proteins of the stigma surface. *Planta* 149:257–261.

Goring, D.R., Banks, P., Falks, L., Baszczynski, C.L., Beversdorf, W.D. & Rothstein, S.J. (1992). Identification of an S-locus glycoprotein allele introgressed from *Brassica napus* ssp. *rapifera* to *B. napus* ssp. *oleifera*. *Plant J.* 2:983–989.

Goring, D.R. & Rothstein, S.J. (1992). The *S*-locus receptor kinase gene in a self-incompatible *Brassica napus* line encodes a functional serine/threonine kinase. *Plant Cell* 4:1273–1281.

Heslop-Harrison, J. (1975). Incompatibility and the pollen stigma interaction. *Ann. Rev. Plant Physiol.* 26:403–425.

Heslop-Harrison, Y., Helop-Harrison, J. & Shivanna, K.R. (1981). Heterostyly in *Primula*. I. Fine structure and cytochemical features of the stigma and style in *Primula vulgaris* Huds. *Protoplasma* 107:171–187.

Heslop-Harrison, J. & Heslop-Harrison, Y. (1982). The pollen–stigma interaction in grasses. 4. An interpretation of the self-incompatibility response. *Acta Bot. Neerl.* 31:429–439.

Huang, S., Lee, H.S., Karunanandaa, B. & Kao, T.-h. (1994). Ribonuclease activity of *Petunia inflata* S proteins is essential for rejection of self-pollen. *Plant Cell* 6:1021–1028.

Ioerger, T.R., Gohlke, J.R., Xu, B. & Kao, T.-h. (1991). Primary structural features of the self-incompatibility proteins in the *Solanaceae*. *Sex. Plant Reprod.* 4:81–87.

Kamboj, R.K. & Jackson, J.K. (1986). Self-incompatibility alleles control a low molecular weight basic protein in pistils of *Petunia hybrida*. *Theor. Appl. Genet.* 71:815–819.

Kanno, T. & Hinata, K. (1969). An electron microscopic study of the barrier against pollen tube growth in self-incompatible *Cruciferae*. *Plant Cell Physiol.* 10:213–216.

Karunanandaa, B., Huang, S. & Kao, T.-h. (1994). Carbohydrate moiety of the *Petunia inflata* S_3 protein is not required for self-incompatibility interactions between pollen and pistil. *Plant Cell* 6:1933–1940.

Kaufmann, H., Salamini, F. & Thompson, R.D. (1991). Sequence variability and gene structure at the self-incompatibility locus in *Solanum tuberosum*. *Mol. Gen. Genet.* 226:457–466.

Kawata, Y., Sakiyama, F. & Tamaoki, H. (1990). Identification of two essential histidine residues of ribonuclease T2 from *Aspergillus oryzae*. *Eur. J. Biochem.* 176:683–697.

Kheyr-Pour, A., Bintrim, S.B., Ioerger, T.R., Remy, R., Hammond, S.A. & Kao, T.-h. (1990). Sequence diversity of pistil S-proteins associated with gametophytic self-incompatibility in *Nicotiana alata*. *Sex Plant Reprod.* 3:88–97.

Kirch, H.H., Uhrig, H., Lottspeich, F., Salamini, F. & Thompson, R.D. (1989). Characterization of proteins associated with self-incompatibility in *Solanum tuberosum*. *Theor. Appl. Genet.* 78:581–588.

Lalonde, B.A., Nasrallah, M.E., Dwyer, K.G., Chen, C.H., Barlow, B. & Nasrallah, J.B. (1989). A highly conserved *Brassica* gene with homology to the S-locus specific glycoprotein gene structure. *Plant Cell* 1:249–258.

Lee, H.S., Huang, S., & Kao, T.-h. (1994). S-proteins control rejection of incompatible pollen in *Petunia inflata*. *Nature* 367:560–563.

Lewis, D. (1949). Structure of the incompatibility gene. II. Induced mutation rate. *Heredity* 3:339–355.

Lundqvist, A. (1956). Self-incompatibility in rye. I. Genetic control of the diploid. *Hereditas* 42:239–348.

Lundqvist, A. (1964). The nature of the two-locus incompatibility system in grasses. IV. Interaction between loci in relation to pseudo-compatibility in *Festuca pratensis*. *Hereditas* 52:221–234.

Mascarenhas, J.P. (1990). Gene action during pollen development. *Ann. Rev. Plant Physiol. Plant Mol. Biol.* 41:317–338.

Mather, K. & De Winton, D. (1941). Adaptation and counter-adaptation of the breeding system in *Primula*. *Ann. Bot.* II 5:297–311.

Mau, S.L., Williams, E.G., Atkinson, M.A., Cornish, E.C., Grego, B., Simpson, R.J., Kyeur-Pour, A. & Clarke, A.E. (1986). Style proteins of a wild tomato (*Lycopersicum peruvianum*) associated with the expression of self-incompatibility. *Planta* 169:184–191.

McClure, B.A., Gray, J.E., Anderson, M.A. & Clarke, A.E. (1990). Self-incompatibility in *Nicotiana alata* involves degradation of pollen rRNA. *Nature* 347:757–760.

McClure, B.A., Haring, V., Ebert, P.R., Anderson, M.A., Simpson, R.J., Sakiyama, F. & Clarke, A.E. (1989). Style self-incompatibility gene products in *Nicotiana alata* are ribonucleases. *Nature* 342:955–957.

Murfett, J., Atherton, T.L., Mou, B., Gasser, C.S., McClure, B.A. & Clarke, A.E. (1994). S-RNase expressed in transgenic *Nicotiana* causes S-allele specific pollen rejection. *Nature* 367:563–566.

Nasrallah, J.B., Doney, R.C. & Nasrallah, M.E. (1985a). Biosynthesis of glycoproteins involved with the pollen/stigma interaction in developing flowers of *Brassica oleracea*. *Planta* 165:100–107.

Nasrallah, J.B., Kao, T.-h., Goldberg, M.L. & Nasrallah, M.E. (1985b). A cDNA clone encoding an S-locus specific glycoprotein from *Brassica oleracea*. *Nature* 318:263–267.

Nasrallah, J.B., Kao, T.-h., Chen, C.H., Goldberg, M.L. & Nasrallah, M.E. (1987). Amino-acid sequence of glycoproteins encoded by three S-alleles of the S-locus of *Brassica oleracea*. *Nature* 326:617–619.

Nasrallah, J.B. & Nasrallah, M.E. (1993). The molecular genetics of self-incompatibility in *Brassica*. *Plant Cell* 5:1325–1335.

Nasrallah, J.B., Nishio, T. & Nasrallah, M.E. (1991). The self-incompatibility genes of *Brassica*: expression and use in genetic ablation of floral tissues. *Ann. Rev. Plant Physiol. Plant Mol. Biol.* 42:393–422.

Nasrallah, J.B., Yu, S.M. & Nasrallah, M.E. (1988). Self-incompatibility genes of *Brassica*: expression, isolation and structure. *Proc. Natl. Acad. Sci. USA* 85:5551–5555.

Nasrallah, M.E., Barber, J.T. & Wallace, D.H. (1970). Self-incompatibility proteins in plants: genetics and possible mode of action. *Heredity* 25:23–27.

Nasrallah, M.E. & Wallace, D.H. (1967). Immunogenetics of self-compatibility in *Brassica oleracea* L. *Heredity* 22:519–527.

Nettancourt, D. de (1977). *Incompatibility in Angiosperms*. Berlin: Springer-Verlag.

Nettancourt, D. de, Devreux, M., Bozzini, A., Cresti, M., Pacini, E. & Sarfatti, G. (1973). Ultrastructural aspects of the self-incompatibility mechanism in *Lycopersicum peruvianum* Mill. *J. Cell Sci.* 12:403–419.

Obrig, T.G., Moran, T.P. & Colinas, R.J. (1985). Ribonuclease activity associated with the 60 S ribosome inactivating proteins ricin A, phytoalaccin and Shinga toxin. *Biochem. Biophys. Res. Commun.* 130:879–884.

Ockendon, D.J. (1974). Distribution of self-incompatibility alleles and breeding structure of open pollinated cultivars of Brussels sprouts. *Heredity* 33:159–171.

Ornduff, R. (1979). Pollen flow in *Primula vulgaris*. *Bot. J. Linn. Soc.* 78:1–10.

Pandey, K.K. (1956). Mutation of the self-incompatibility alleles in *Trifolium pratense* and *T. repens*. *Genetics* 41:353–366.

Pandey, K.K. (1965). Centric chromosome fragments and pollen part mutation of the incompatibility gene in *Nicotiana alata*. *Nature* 206:792–795.

Pandey, K.K. & Troughton, J.H. (1974). Scanning E.M. observations of pollen grains and stigma in the heteromorphic species *Primula malacoides* Franch. and *Forsythia* x *intermedia* and genetics of sporopollenin deposition. *Euphytica* 23:337–344.

Richards, A.J. (1986). *Plant Breeding Systems*. London: George Allen and Unwin (Pub.) Ltd.

Richards, A.J. & Ibrahim, H. (1982). The breeding system in *Primula veris* L. II. Pollen tube growth and seed set. *New Phytol.* 90:305–314.

Sampson, D.R. (1957). The genetics of self-incompatibility in the radish. *J. Hered.* 48:26–29.

Scutt, C. & Croy, R.R.D. (1992). An S_5 self-incompatibility allele specific cDNA sequence from *Brassica oleracea* shows high homology to the SLR2 gene. *Mol. Gen. Genet.* 232:240–246.

Scutt, C.P., Gates, P.J., Gatehouse, J.A., Boulter, D. & Croy. R.R.D. (1990). A cDNA encoding an S-locus specific glycoprotein from *Brassica oleracea* plants containing the S_5 self-incompatibility allele. *Mol. Gen. Genet.* 220:409–413.

Shih, C.Y. & Rappaport, L. (1971). Regulation of bud rest in tubers of potato *Solanum tuberosum* L. VIII. Early effects of gibberellin A3 and abscisic acid on ultrastructure. *Plant. Physiol.* 48:31–55.

Shivanna, K.R., Heslop-Harrison, J. & Heslop-Harrison, Y. (1983). Heterostyly in *Primula*. 3. Pollen water economy: a factor in the intermorph incompatibility response. *Protoplasma* 117:175–184.

Singh, A., & Kao, T.-h. (1992). Gametophytic self-incompatibility: biochemistry; molecular genetics and evolutionary aspects. In *Sexual Reproduction in Flowering Plants*, eds. S.D. Russel and C. Dumas, pp. 449–483. New York: Academic Press.

Stebbins, G.L. (1950). *Variation and Evolution in Plants*. New York: Columbia University Press.

Stein, J.C., Howlett, B., Boyes, D.C., Nasrallah, M.E. & Nasrallah, J.B. (1991). Molecular cloning of a putative receptor protein kinase gene encoded at the self-incompatibility locus of *Brassica oleracea*. *Proc. Natl. Acad. Sci. USA* 88:8816–8820.

Stevens, V.A.M. & Murray, B.G. (1982). Studies on heteromorphic self-incompatibility: physiological aspects of the incompatibility system of *Primula obonica*. *Theor. Appl. Genet.* 61:245–256.

Stout, A.B. (1920). Further experimental studies on self-incompatibility in hermaphrodite plants. *J. Genet.* 9:85–129.

Takayama, S., Isogai, A., Tsukamoto, C., Ueda, Y., Hinata, K., Okazaki, K., Koseki, K. & Suzuki, A. (1986). Structure of the carbohydrate side chains of S-glycoproteins in *Brassica campestris* associated with self-incompatibility. *Agr. Biol. Chem.* 50:1673–1676.

Tantikanjana, T., Nasrallah, M.E., Stein, J.C., Chen, C.H. & Nasrallah, J.B. (1993). An alternative transcript of the S-locus glycoprotein gene in a class II pollen recessive self-incompatibility haplotype of *Brassica oleracea* encodes a membrane anchored protein. *Plant Cell* 5:657–666.

Thompson, K.F. & Taylor, J.P. (1966). Non-linear dominance relationships between S-alleles. *Heredity* 21:345–362.

Thompson, R.D., Uhrig, H., Hermsen, J.G.Th., Salamini, F. & Kaufman, H. (1991). Investigation of a self-compatible mutation in *Solanum tuberosum* clones inhibiting S-allele activity in pollen differentially. *Mol. Gen. Genet.* 226:283–288.

Trick, M. & Flavell, R.B. (1989). A homozygous S-genotype of *Brassica oleracea* expresses two S-like genes. *Mol. Gen. Genet.* 218:112–117.

Umbach, A.L., Lalonde, B.A., Kandaswamy, M.K., Nasrallah, J.B. & Nasrallah, M.E. (1990). Immunodetection and post-translational modification of two products encoded by two independent genes of the self-incompatibility multigene family of *Brassica*. *Plant Physiol.* 93:739–747.

Walker, J.C. & Zhang, R. (1990). Relationship of a putative receptor kinase from maize to S-locus glycoproteins of *Brassica*. *Nature* 345:743–746.

Williams, M.E. & Levings, C.S. III (1992). Molecular biology of cytoplasmic male sterility. In *Plant Breeding Reviews*, *Vol. 10*, ed. J. Janick, pp. 23–51. New York: Wiley.

Xu, B., Grun, P., Kheyr-Pour, A. & Kao, T.-h. (1990a). Identification of pistil specific proteins associated with three S-alleles in *Solanum chacoense*. *Sex. Plant. Reprod.* 3:54–60.

Xu, B., Mu, J., Nevins, D.L., Grun, P. & Kao, T-h. (1990b). Cloning and sequencing of cDNA encoding two self-incompatibility associated proteins in *Solanum chacoense*. *Mol. Gen. Genet.* 224:341–346.

10

Chemical induction of male sterility

JOHN W. CROSS and PATRICIA J. SCHULZ

Summary   218
Introduction: Tools and insights into pollen development   219
The logic of chemical hybridization   220
CHAs and pollen development   221
    Plant-growth regulators and substances that disrupt floral development   222
    Metabolic inhibitors   224
    Inhibitors of microspore development   224
    Inhibitors of pollen fertility   231
Phloem mobility and CHA activity   232
CHAs and anther culture   232
Prospects   233
References   234

Summary

Chemicals capable of selectively inhibiting pollen development and thus blocking male fertility have been known for some time. Most were identified in screening programs designed to discover chemical hybridizing agents (CHAs) for the large-scale commercial production of hybrid seed, particularly in small grains. Driven by practical requirements, this research has identified newer generations of CHAs with increasingly better selectivity and effectiveness. However, research into the mode of action of these substances has lagged their practical applications in breeding and hybrid production. Compounds are known that cause a range of effects, including feminization of male florets and inhibition of early anther development, interference with tapetal functions and microspore development, and defective germination of the mature pollen. Several CHAs also have demonstrated a stimulatory effect on the induction of androgenic plants from anther cultures derived from treated plants. The strength of this response varies among genotypes. The molecular and physiological mechanisms of these effects are unknown. The purpose of this review is to encourage further research with these substances.

Introduction: Tools and insights into pollen development

The study of complex natural events requires development of tools suitable for their analysis and manipulation. Effective use of such tools then becomes the basis of practical new technologies. So it is with efforts to probe the development and function of the male gametophyte. Descriptive methods have led just so far into understanding the development of pollen and the process of pollination. Beyond description, one must functionally separate its essential from its accidental elements and identify the points of regulation. The classic analytical tools for dissection and disruption of developmental processes are surgical, chemical, and genetic. This discussion considers chemical methods.

Chemicals that selectively inhibit pollen development or reduce pollen fertility have been known for the past 20 or more years. Most have been identified by directly screening for substances with this activity. The objective was to discover chemical hybridizing agents (CHAs) that could be used in the large-scale commercial production of hybrid seed. Many of these agents are reported only in the patent literature or are mentioned briefly in progress reports and newsletters concerned with the development of commercial hybrids, particularly the identification of heterotic combinations. This chapter focuses on the select few for which there is more detailed information or which are currently undergoing development for commercial hybrid production.

The heaviest research investment has favored discovery of chemical hybridizing agents for field crops that naturally reproduce by self-pollination but lack the profit margin per seed to justify hand emasculation. For these reasons – and because alternative technology based on the existing genetic male sterility has not been completely satisfactory for commercial hybrid seed production – wheat has been a particularly attractive candidate (McRae 1985). These economic biases should be kept in mind when reading the literature. Although the bulk of CHA research has concerned wheat, one should not assume that the reported chemicals are effective only on wheat and related small grains. Other species of lesser economic interest may also respond to these substances, if applied at the proper developmental stage.

The utility of certain chemicals as hybridizing agents should not be taken as an implication that they have true molecular specificity as inhibitors of pollen development. Most CHAs are applied to plants only at certain critical stages of male gametophyte development. Their action could result from a range of mechanisms, including (1) inherently selective action as male gametocides or inhibitors of anther development, (2) selective transport of generally toxic or growth-inhibitory substances to the anthers during these periods, or (3) metabolic detoxification of generally toxic or growth-inhibitory substances after they have suppressed male fertility. A successful CHA might also combine these attributes.

No known CHA has demonstrated unique specificity towards a distinct process required for male fertility. Most of the CHAs are associated to varying degrees with other forms of growth-regulating activity or sites of phytotoxicity, at least when overdosed or applied at other times during the life cycle (McRae 1985). Stunting, chlorosis of the flag leaf, and reduced seed set are encountered with some regularity. Because the flag leaf is developing rapidly at the premeiotic stage when most of these substances are applied (McRae 1985), this may be an indication that these compounds are active against a process that is common to both systems, but it is hard to determine if male sterility and these other forms of phytotoxicity have the same cause. The roles of differential transport and metabolism in providing selective pollen suppression are largely unexplored. Conversely, a highly selective inhibitor of pollen development would fail as a chemical hybridizing agent if it is excessively expensive, if it cannot be applied effectively in the field, or if it is environmentally unsound or unsafe for workers or consumers.

The logic of chemical hybridization

Any essential step in the development or function of the male gametophyte that can be prevented is fair game for inventors wanting to develop new hybrid varieties. The means can likewise be selected broadly from mechanical, genetic, or biochemical principles (Table 10.1).

Chemical induction of male sterility has been considered desirable because it has the unique potential to provide for the development of hybrids directly out of elite germplasm, without the time and effort required to regressively backcross male-sterility genes and fertility restoration systems. In several species, the available male-sterile genes or the known restoration systems have not been sufficiently effective or reliable for commercial production of hybrid seed. There has been hope that chemical induction of male sterility would solve these problems (McRae 1985).

Genetic engineering of male-sterility and restoration genes has attracted great attention recently as a clean application of genetic engineering (Goldberg et al. 1993; also see Chapter 11). However, the same problems of effectiveness, restoration efficiency, and time lost in regressing these genes from easily transformed lines into elite germplasm also apply to genetically engineered genes.

Commercially viable chemical hybridization, however, places a number of stringent requirements on the inventor in the form of practical and regulatory hurdles. These include high degrees of efficacy and developmental selectivity, persistence during the development of later flowers or spikes, low cost, and acceptable levels of toxicity to people and the environment. The commercial requirement of high female fertility in the treated plants going into production fields had the force of a strong spur, ensuring that compounds with commercial potential would have a high degree of selectivity and relatively low gen-

Table 10.1. *Means of producing male-sterile plants*

Means of intervention	Examples
Mechanical	Hand or machine emasculation
Genetic	Temporal separation of male and female fertility
	Cytoplasmic or nuclear male sterility
	Self-incompatibility
Biochemical	Feminizing hormones
	Inhibitors of anther or pollen development:
	– acting on sporophytic tissue
	– acting on gametophytic tissues (gametocides)
	Inhibitors of pollen fertility

eral phytotoxicity (McRae 1985). Although certain early CHAs had adverse effects on seed yield or seed quality from the treated plants, numerous studies have confirmed that the agronomic performance of hybrid seed produced with leading chemical pollen suppressants is not inferior to equivalent crosses produced by genetic methods (McRae 1985; Oury et al. 1990). Although very few compounds have been discovered that can satisfy all of these requirements for commercial success, a greater number could be employed as breeder's aids and in research on pollen development and function.

CHAs and pollen development

Chemical inhibitors of pollen development are not a familiar topic to the majority of academic scientists. For example, a recent major publication devoted to plant reproduction did not mention the existence of growth regulators capable of selective inhibition of pollen development (Goldberg and Chasan 1993). The most likely explanation for the unfamiliarity is that these substances have been identified and developed almost entirely within industrial research laboratories for a specific practical application: the production of hybrid seed. The availability of inhibitory substances capable of selectively interfering with specific events in the developmental program of the male gametophyte could have great utility in dissecting out the essential features and their biochemical regulation.

Pollen comes into being through a sequential and determinate program within the central cavity or locule of the anther (see Chapter 2). The internal cells of the anther divide by mitosis to produce the pollen mother cells and a surrounding layer of differentiated nurse cells, the tapetal cells. There are also several outer cell layers, and the anther is attached to the base of the flower by the filament, which contains the vascular tissue. Each of these cell types has an important role in pollen development or in release of the mature pollen, and each is affected by one or more chemical hybridizing agents.

There are at least four classes of chemical agents in the current literature: (1) plant-growth regulators and substances that disrupt floral development, (2) metabolic inhibitors, (3) inhibitors of microspore development, and (4) inhibitors of pollen fertility. These categories have considerable conceptual overlap and do not address the molecular actions of chemical male sterilants. However, they do allow a somewhat orderly discussion of the subject.

The overlapping nature of these categories is apparent in the discussion of individual compounds. Certain compounds may grade from feminization and disrupting floral development at higher dosages to inhibiting aspects of anther function at threshold treatment levels. Because many of these compounds have been identified as plant-growth regulators only in CHA screening programs, their activity has been reported only in this context. Other growth-regulatory effects may also occur but are minimal when the compound is applied during early anther development. Phytotoxicity observed when a CHA is applied too early or too late or at an excessive dosage may be rather an indication of effects on other processes sharing common requirements (somatic cell division, chloroplast development, and so forth), than evidence that the chemical is biochemically nonspecific. Until we know the actual molecular mechanism of action of one of these compounds, assertions of toxicological cause and effect cannot be drawn from such observations.

Plant-growth regulators and substances that disrupt floral development

Plant hormones/hormone antagonists. Many of the earliest known chemical hybridizing agents were plant hormones and their antagonists. These include several auxin-like herbicides (indole butyric acid, 2,4-D, NAA, TIBA) and maleic hydrazide, which may be either an anti-auxin or a growth inhibitor (McRae 1985). Most of these substances have not been commercially successful as CHAs because of their many inhibitory and phytotoxic effects unrelated to male fertility. Because these substances have not enjoyed commercial attention, recent research has been minimal. More promising results have been obtained with ethrel, gibberellins, and abscisic acid. Except for ethrel (discussed in detail as an inhibitor of microspore development), little is known about their mechanism of action.

Auxins and auxin antagonists. Two publications (Awasthi and Dubey 1985a, b) are instructive. Lentil plants were treated with naphthalene acetic acid (NAA), indole butyric acid (IBA), maleic hydrazide (MH), and coumarin. Vegetative growth, flowering, pollen sterility, fruit set, and seed yield were studied. A variety of side effects were observed, including general reduction of plant growth and seed yield. They also noted enhancement of branching and floral abnormalities. Male sterility induced by the treatments was expressed in several ways, including in situ pollen germination, in situ exudation of pollen cytoplasm, and modification of certain stamens into staminodes (Awasthi and Dubey 1985b). With MH and IBA, the tapetum failed to enlarge and develop into the normal peripheral plasmodium (Awasthi and Dubey 1985a), whereas with coumarin, the tapetal cells enlarged atypically and were persistent.

There is no common thread to these findings. These substances may have multiple independent actions, or each may differently affect some far-reaching process, such as blockade of nutrient transport to the developing anthers. When several effects are reported, it is often unclear if these effects represent multiple specific effects, or if they are the result of applications at inexactly controlled rates and developmental stages. Stage-dependent differences have been claimed with a synthetic growth regulator, RH-531 (Colhoun and Steer 1982, 1983), but these effects were later found to be associated with excessive or multiple applications (McRae 1985).

Gibberellins and antagonists. Gibberellins (GAs) are well known for their effects on sexual determination and floral development, which is the basis for chemical hybridization in sunflowers (McRae 1985). The response of plants to gibberellins and GA antagonists varies by species. In some species, GA interferes with development of male floral organs or promotes feminization, whereas in others these effects are absent. Effects on pollen production in several crops have been demonstrated experimentally (see McRae 1985 for a review). For example, van der Meer and van Bennekom (1982) reported reduced pollen shed in onion treated with a GA_{4+7} mixture. This growth-regulator mixture did not affect the viability of the pollen that was shed. Although used experimentally to produce hybrid onion for breeding purposes, the GA_{4+7} mixture was not commercially practical because of cost. For maize, another negative factor was the excessively short window of efficacy (McRae 1985).

In Solanaceae, the effects of GA applications vary by species. In tomato, normal plants treated with GA_3 produce fertile flowers with separate stamens and split pistils (Chandra Sekhar and Sawhney 1990). In pepper (*Capsicum annuum*), however, GA_3 application produces feminization of flowers and inhibition of pollen development (Sawhney 1981). In tomato, the gibberellin-synthesis inhibitor CCC (2-chloroethyl-trimethyl ammonium chloride), at certain concentrations, selectively inhibits the development of the stamen or otherwise suppresses pollen development (see Rastogi and Sawhney 1988 for references). These effects are not sufficiently selective to serve as the basis for chemical hybridization in Solanaceae.

Abscisic acid. In tomato, abscisic acid (ABA) caused effects on developing floral buds similar to those produced by CCC (Rastogi and Sawhney 1988; Chandra Sekhar and Sawhney 1991). In wheat, ABA caused male sterility (Saini and Aspinall 1982) if applied to plants just prior to or during meiosis of the pollen mother cells. Thus, ABA may cause male sterility through more than one mechanism. It does not appear that ABA is currently under development as a CHA.

Other substances that disrupt floral development. Two synthetic growth regulators, LY195259 and TD1123, are believed to interfere primarily with overall male-flower development. The molecular action of these substances has not been reported.

LY195259. 5-(Aminocarbonyl)-1-(3-methylphenyl)-1H-pyrazole-4-carboxylic acid – was reported as an effective chemical hybridizing agent for wheat (Tschabold et al. 1988). This compound was considered commercially promising at the time, but more recent publications are lacking. LY195259 was applied when the flowering spike was quite short, averaging 0.5 cm. High application rates at this time prevented anther development, whereas lower dosages resulted in progressively reduced inhibition. Sterility at lower dosages was associated with smaller, abnormally twisted, and intensively pigmented locules. Rates as low as 1.12 kg/ha produced greater than 95% male sterility. Seed set as high as 80% of the male parent was reported with treated cultivars. The hybrid seed appeared normal, and no other phytotoxic effects were visually evident from rates as high as three times the mean effective dose. ^{14}C-uptake and translocation experiments indicated that much of the compound remains at the site of application. Only a small percentage is translocated to plant organs above a treated leaf. It was found that uptake of LY195259 from soil was particularly effective in producing male sterility.

TD1123. Underdeveloped anthers that fail to dehisce are also produced by the cotton CHA, TD1123 – potassium 3,4-dichloro-5-isothiocarboxylate (Olvey et al. 1981). A variety of morphological effects were observed at higher treatment levels.

Metabolic inhibitors

Screening evaluations of off-the-shelf chemicals have identified several compounds that selectively repress male fertility. These earlier findings are adequately reviewed by McRae (1985), and include the effects of halogenated aliphatic acids (for example, alpha,beta-dichloroisobutyrate and 2,2-dichloropropionate salts) and arsenicals (methanearsonate salts). Although studies of these substances as hybridizing agents have been discontinued (with a few exceptions: for example, Sun et al. 1988; Berzy et al. 1990), investigation of their selectivity may help explain the physiological basis of cytoplasmic male-sterile mutants. Many of these mutants are associated with defects in mitochondrial proteins (Levings 1993). Arsenates, in particular, would be expected to reduce the efficiency of normal metabolic processes.

Inhibitors of microspore development

Copper chelators. Studies of copper deficiency syndrome in small grains indicate that pollen development is irregular or absent (Jewell et al. 1988).

Abortive microspore development is associated with irregular or reduced formation of the outer (exine) wall, which contains sporopollenin. The tapetal cells, normally suppliers of nutrients and sporopollenin precursors to the developing microspores, become disoriented and expanded, and they contain irregular, endopolyploid nuclei. It appears that their secretory functions are severely disrupted. It has been asserted that copper deficiency exerts these effects by inhibiting copper-requiring oxidases that function in auxin metabolism (Jewell et al. 1988, reviewed by Cross and Ladyman 1991). However, this assertion has not been investigated further, even though, as seen already, auxin analogues are known to depress male fertility,

Graham (1976) proposed that copper deficiency could be manipulated for the production of hybrid wheat. Using copper chelators, particularly benzotriazole, Graham (1986) was able to demonstrate selective induction of male sterility. However, practical application of these findings to hybridization has not been achieved.

Ethylene. Ethylene has been implicated as a natural regulator of the development and maturation of several floral organs. Filament and corolla growth in flowers of *Ipomoea nil* are inhibited by ethylene production (Kiss and Koning 1989). Anthers were identified as a source of an early filament-growth inhibitor in vitro. Anthers contained increased levels of the ethylene precursor, aminocyclopropane carboxylate (ACC), from 15 to 6 hours before anthesis. Ethylene peaked about 6 hours nearer anthesis than did ACC levels in this organ. ACC and ethylene levels in filaments fluctuated along with those in the anthers. This paper also reported basipetal transport of endogenous ACC from the anther through the filaments. It was proposed that stamens act as source tissues and transport vectors for ACC to regulate corolla growth, particularly unfolding and senescence.

Fenridazon. Fenridazon (RH-0007, Hybrex) is 1-(4-chlorophenyl)-1,4-dihydro-6-methyl-4-oxopyridazine-3-carboxylic acid and was the latest compound identified in a long-term project at Rohm and Haas (McRae 1985). Mizelle et al. (1989) and El-Ghazaly and Jensen (1990) described pollen and tapetum development in wheat treated with fenridazon. Mizelle et al. found no effect of fenridazon on development through the meiosis stage. In young microspores, the exine wall was thin and irregular compared with the control. In many cases the treated microspores had wavy surfaces. This progressed to plasmolysis and abortion with the onset of the microspore vacuolation stage. Development of tapetal cells was also normal through meiosis. Less sporopollenin was deposited in the Ubisch bodies and their pattern was irregular. The tapetal cells often expanded into the locule, as has been observed with several other agents. El-Ghazaly and Jensen (1990) further found that the pollen wall was 80% thinner in treated plants. They speculated that an accumulation of unpolymerized sporopollenin precursors increased the osmotic potential of the locular fluid and drew water out of the microspores, causing them to collapse. However, there are several problems with this hypothesis. There is no evidence that unpolymerized sporopollenin precursors accumulate in the locular fluid rather than in the tapetum. Since the tapetum is often expanded and vacuolate (Mizelle et al. 1989), these unpolymerized precursors may have accumulated within these cells. The bulk of sporopollenin appears to represent hydrocarbon residues (Guilford et al. 1988; Wiermann and Gubatz 1992), whose unpolymerized precursors would not be expected to have a significant effect on water potential.

Goasdoue et al. (1993) studied the lipid composition of plants treated with fenridazon and their seeds. There were small but significant variations in the fatty acid composition of seeds from treated plants, and the total lipid content of these seeds decreased markedly compared with the control. Fatty acid composition and polar lipids of anthers were not drastically altered by the treatment. Fenridazon did induce drastic decreases in chlorophyll and carotenoid contents of treated leaves. Changes in fatty acid composition were not so marked as changes in pigment levels. These changes in anthers and leaves were more likely related to the phytotoxicity of fenridazon rather to its activity as a pollen suppressant.

Adler et al. (1984) provided an HPLC method for the analysis of fenridazon in wheat grain and straw. An improved residue method has been developed for detection of fenridazon in grain, leaves, straw, and glumes of treated wheat at low levels (Dardoize et al. 1993). This method provides 60%–99+% recovery in grain fortified to the 0.5–1.0×10^{-7} M level, allowing demonstration that fenridazon residues are present in the mature gain. Thus, this CHA is still mobile at the time of grain fill, potentially accounting for its ability to produce male sterility in late tillers.

Phenylcinnoline carboxylates (SC-1058, SC-1271 and SC-2053). The phenylcinnoline carboxylates (Guilford et al. 1992) were synthesized and developed by Sogetal, Inc., a research laboratory belonging to Orsan, a French biotechnology group. Hybrinova, a related company, has undertaken commercialization of hybrid wheat using the Sogetal lead compound SC-2053.

Hybrinova has taken advantage of the earlier advanced research at Sogetal and Western Plant Breeders (USA) with this compound and its predecessors, SC-1058 and SC-1271. Hybrinova is producing new hybrid combinations for testing in France. Published data have centered on three lead compounds:

(a) SC-1058, 1-(4'-trifluoromethylphenyl)-4-oxo-5-fluorocinnoline-3-carboxylic acid
(b) SC-1271, 1-(4'-chlorophenyl)-4-oxo-5-propoxycinnoline-3-carboxylic acid
(c) SC-2053, 1-(4'-chlorophenyl)-4-oxo-5-(methoxyethoxy)cinnoline-3-carboxylic acid

The initial lead compound (SC-1058) was identified from screening a series of 1,4-dihydro-4-oxo-1-phenylcinnoline-3-carboxylic acids. The unsubstituted compound (1-phenyl-4-oxocinnoline-3-carboxylic acid) was without pollen suppressant activity. Although hybrid wheat could be made with SC-1058 compound, screening for agents with improved characteristics continued. A range of substitutions on the cinnoline and phenyl rings were prepared and evaluated (Guilford et al. 1992). Significant pollen-suppressant activity was identified for compounds substituted at C-5 on the cinnoline ring with fluorine or alkoxy groups. Compounds bearing a 5-amino substituent showed little or no activity, and compounds with methyl substitution at C-5 were inactive. Phenyl ring substitutions were also explored systematically and found to modulate the activity of the compounds. Symptoms associated with the more phytotoxic compounds of this series, or with excessively high dosage levels, typically included chlorosis, plant height reduction, or necrosis. However, no effect on seed quality was seen. The lead compounds developed from this program (SC-1271 and, later, SC-2053) for use as chemical hybridization agents are members of the alkoxy series containing less than four carbon atoms, which have the best balance of high male sterility and low phytotoxicity (Guilford et al. 1992).

SC-2053 is most effective when applied to wheat having spike lengths from 9 to 21 mm (Batreau et al. 1991). There is essentially no CHA activity if the compound is applied to plants with spikes shorter than 9 mm or greater than 30 mm in length. Wong et al. (1992), working in growth chambers under controlled conditions, reported the best results with treatments of plants having spikes ranging from 11 to 20 mm at 700 and 1,000 g/ha. This window of sensitivity corresponds to the treatments used for the earlier Sogetal compounds (for example, Cross et al. 1992) and corresponds to anthers at the premeiotic to early vacuolate microspore stages of development.

Batreau et al. (1991) also present dose–response data for SC-2053 obtained in the field indicating that treatments of 0.9 kg/ha or more are required for high levels of male sterility (>90%). At intermediate levels (0.3 and 0.6 kg/ha), there is sufficient residual male fertility, and the majority of seeds were the result of selfing. Lower dosages are sufficient for treatments in the greenhouse. The higher dosages of SC-2053 that produce male sterility approaching 100% also caused some reduction in hybrid seed yield (Batreau et al. 1991). This reduction in seed yield was variable among female parental lines (Batreau et al. 1991). Therefore, identification of female parents with enhanced seed production in treated fields would be important for successful commercialization of hybrid wheat with this technology. However, further optimization of the dosage by treatment stage interaction eliminated the problem of reduction in seed yield from a practical standpoint, without compromising the level of hybrid purity (Alain Gervais, personal communication, September 15, 1994).

The relationship between male sterility, seed yield, and hybrid purity has been reported to be complex (Patterson et al. 1991). Male pollinator lines differed significantly in their effect on quantitative hybrid purity, even at constant high levels of male sterility. Furthermore, Patterson et al. could not predict hybrid purity from male sterility or seed yield data. It was necessary to develop direct analytical methods for determination of hybrid seed purity, such as analysis of seed storage proteins.

A complete study of the ultrastructural effects of two phenylcinnoline carboxylates has been reported (Schulz et al. 1993). The two compounds studied (SC-1058 and SC-1271) are no longer lead compounds in the Hybrinova project, having been replaced by SC-2053. No study of the ultrastructural effects of SC-2053 has been reported, but the similarities of the molecular structure of SC-2053 to that of SC-1058 and SC-1271, and their parallel macroscopic effects on wheat anther development, suggest that the mode of action for all three compounds is the same. SC-1271 and SC-2053 differ only by the insertion of an oxygen into the side chain at the cinnoline ring 5-position in SC-2053. These compounds are all capable of producing complete male sterility with minimal phytotoxicity and loss of seed yield when applied just prior to meiosis. Earlier application causes increased phytotoxicity, and later application risks incomplete male sterility, diminished female fertility, or both.

Applied just prior to meiosis, SC-1058 and SC-1271 cause a general retardation of anther development (Schulz et al. 1993). This retardation is expressed gradually and makes it difficult to make stage comparisons between treated and control anthers. Anthers in the different florets of the main spike, progressing from the base to the tip, vary slightly in their developmental stage. Because of this retardation, comparisons were made between structures that had reached the same stage of development, based on overall anther appearance and such criteria as completion of meiosis, callose dissolution, microspore vacuolation, and microspore mitosis.

Pollen development was generally arrested in the late prevacuolate or early vacuolate microspore stage (Figure 10.1). At this point, the shape of these microspores often becomes wavy or wrinkled (Schulz et al. 1993). Occasionally, the first pollen mitosis was observed. Microspore cytoplasm degenerates and the cells become collapsed ghosts. The tapetal cells, in contrast, do not immediately degenerate, but swell, expanding into the locule (Figure 10.2). These cells show increased early vacuolation not seen in the control. These vacuoles are likely derived from the fusion of secretory vesicles. Some of these vacuoles are filled with electron-dense material. Attempts to isolate this substance were not successful (J. W. Cross, unpublished results), but we believe that these deposits may represent lipidlike sporopollenin precursors (Guilford et al. 1988; Wiermann and Gubatz 1992). Eventually, the swollen tapetal cells degenerate (Figure 10.2).

The secretion or formation of sporopollenin was reduced both in the exine walls of the microspores and in the sporopollenin-containing orbicular wall and Ubisch bodies of the locular wall (Cross et al. 1989; Schulz et al. 1993). These structures were thinner and less electron-dense than in the controls. Further, the sporopollenin ghosts remaining after acetolysis were thinner and more flexible in the anthers from treated plants (Schulz et al. 1993). The treated anthers were also visibly less reactive to diazotized dye reagents (Schulz et al. 1993). The authors concluded that less sporopollenin was

Pollen Development

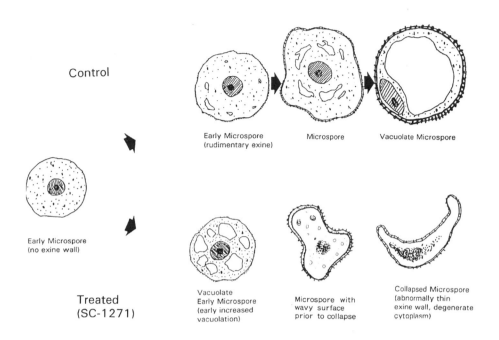

Control

Early Microspore (rudimentary exine)   Microspore   Vacuolate Microspore

Early Microspore (no exine wall)

Treated (SC-1271)

Vacuolate Early Microspore (early increased vacuolation)

Microspore with wavy surface prior to collapse

Collapsed Microspore (abnormally thin exine wall, degenerate cytoplasm)

Figure 10.1 Effects of the pollen suppressant SC-1271 on pollen development. The effects of the other phenylcinnoline carboxylates and fenridazon appear to be similar.

formed and that it was less rigid. This is in accord with the observations for fenridazon (Mizelle et al. 1989; El-Ghazaly and Jensen 1990).

Genesis® (MON 21200). HybriTech, a Monsanto-affiliated company, is developing a chemical hybridizing agent coded MON 21200 for commercial production of hybrid wheat. HybriTech has also developed elite germplasm for wheat hybridization using *Triticum timopheevi* cytoplasm and specially bred restorer lines. This group has also acquired the former wheat breeding and chemical hybridization program of Rohm and Haas. Despite the prior availability of the CMS system originated from *Triticum timopheevi*, the chemical has accelerated Hybritech's efforts to identify parental lines with superior combining ability for hybrid production.

In addition to identification of strongly heterotic combinations and other standard agronomic characteristics, the HybriTech program also emphasizes selection of parents with characteristics that promote hybridization. These include pollen donor characteristics such as anther extrusion and anther length, and maternal qualities such as degree of flower opening and female seed yield. These characteristics are selected in breeding populations by selec-

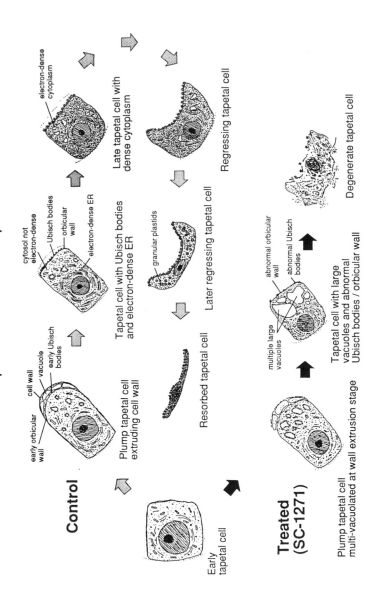

Figure 10.2 Effects of SC-1271 on tapetal development. Swelling of tapetal cells is elicited by a variety of chemical pollen suppressants.

tion based on test-cross performance. The CHA has allowed the elimination of hand emasculation from HybriTech's greenhouse crossing program and production of thousands of new hybrid combinations in the field each year, using crossing block designs.

HybriTech reports that MON 21200 provides good CHA activity over a very diverse range of genotypes, geographic regions, and growing conditions. Seed production with Genesis has provided a high and reliable level of outcrossing. Outcrossing percentages for soft red winter wheat have averaged 79% over seven years of testing; for hard red western wheat, the percentages averaged 87% over six years. Seed set ranges from 75% to 99% in material selected for outcrossing ability, depending on local conditions. Yield trials indicate that hybrids produced with the aid of Genesis are equivalent to, or better than, conventional hybrids based on CMS technology.

HybriTech intends to market the Genesis hybridizing agent as soon as regulatory approval is received. According to company representatives, registration of the product is progressing satisfactorily. HybriTech has not encouraged investigational use of the Genesis active ingredient, and its chemical structure has not been reported. In keeping with these business policies, HybriTech and Monsanto have not disclosed its molecular structure, mode of action, and environmental fate.

Monsanto earlier disclosed research on a series of pyridine monocarboxylates and benzoic acid analogues (Ciha and Ruminski 1991). These substances were tested in growth chambers as potential chemical hybridizing agents for wheat. The 3-pyridinecarboxylic acid, 4-hydroxy-2,6-bis(trifluoromethyl) methyl ester, and 2,4-bis(trifluoromethyl) benzoic acid were the only substances in each series that provided complete spike sterility. A total loss of activity accompanied minor changes in either molecule. With the pyridine monocarboxylate, only substitutions at the 4-position that contain an acidic proton functionality, or that are subject to hydrolysis to the 4-hydroxyl derivative, were active. The relationship of these chemical series to MON 21200, if any, has not been reported.

Inhibitors of pollen fertility

Azetidine-3-carboxylate (A3C, CHA™) effectively induces male sterility in small grains, particularly wheat (reviewed in Cross and Ladyman 1991; Kofoid 1991). Anther development and pollen structure appear normal under the light microscope (Mogensen and Ladyman 1989) and, if applied after the flag leaf has emerged from the boot, there is little general phytotoxicity. In treated ryegrass, A3C also inhibits stem elongation substantially in parallel with pollen fertility (McGuinness et al. 1991). It also increased tillering but had no effect on the accumulation of neutral-detergent–soluble substances.

Ultrastructural studies by Mogensen and Ladyman (1989) indicated that the major effect on mature pollen is a structural alteration of cell wall precursor vesicles (wp-vesicles). These vesicles are a major proportion of the contents of mature pollen in wheat. From pollination studies, Mogensen and Ladyman reported that only 10% of the pollen grains showed normal pollen tube growth in the first hour after pollination and none of these penetrated the secondary stigmatic branch. In the controls, 50% germinated normally in the first 15 minutes and 80% of these had entered the secondary branch. Although effects on steroid biosynthesis were proposed to account for these effects (Mogensen and Ladyman 1989), no further evidence has been presented.

Kelly (1991) reported that *Chlorella zofingiensis* cell division was selectively blocked by azetidine-3-carboxylate. Biomass in the treated cells continued to accumulate. Some of the cells become very large, up to 26 times the cell volume of the largest cells in the control cultures. That the target is cell division rather than DNA synthesis is indicated by the continued accumulation of DNA in treated cells and a rapid catch-up cell division in treated cells after transfer to control medium. The results are preliminary, but appear to warrant detailed investigation in the case of A3C. How these results relate to the action of A3C in treated plants is unclear, since pollen formation in these plants appears to be normal except for abnormal wall precursor vesicles (Mogensen and Ladyman 1989). Kelly (1991) suggested that the *Chlorella* test system might be suitable for screening other potential gametocides and assessing their modes of action. However, it is clear that other effective pollen suppressants do not inhibit cell division, as evidenced by tapetal cell division (Mizelle et al. 1989), and by high female fertility and grain yield.

Phloem mobility and CHA activity

Discovery of new chemical hybridizing agents has been through a process of screening, although good theoretical hunches have played a role in the identification of starting points (McRae 1985), particularly in the discovery of proline-like CHAs (Kerr 1984). More recently, Hsu and Kleier (1990) presented evidence that good mobility in the phloem is a prerequisite for CHA activity. They postulate that phloem mobility is necessary to overcome the low transpiration rates and the discontinuity of xylem in many reproductive tissues.

CHAs and anther culture

In 1972, Bennett and Hughes reported additional mitoses in microspores of plants that had been sprayed with ethrel. Work at that time was not successful in taking advantage of this effect to stimulate the frequency of androgenic callus formation. In the years that followed, the practice of conventional anther culture made strides, and few phytotoxic chemical hybridizing agents were discovered. Picard et al. (1987) took up the clue and reinvestigated the use of a chemical hybridizing agent in wheat anther culture. Using fenridazon potassium (Hybrex), this group was able to obtain a substantial improvement in the total yield of androgenetic green plants.

Picard et al. (1987) sprayed intact plants with the CHA, then removed the spikes for anther culture at intervals. The fenridazon treatments that were used (0.75–1.0 kg/ha) induced 95% to 100% male sterility. Following the treatment, the rate of embryo induction per anther was typically increased 5x, the percentage of spikes yielding at least one embryo was increased 2x, the regeneration rate increased 2x, and the overall plantlet yield was increased 10x. Embryos were obtained more quickly after initiation of anther cultures (as early as 9 days vs. 16 days) and continued to be formed over a longer period at high frequency. Similar results were obtained in a variety of F_1 genotypes. In many of these tests, respectable numbers of green plantlets were obtained from the treated anthers, but none were obtained from the untreated controls.

Other groups have also found improved rates of anther culture response using plants treated with chemical hybridizing agents. Beaumont and Courtois (1990), working with rice (*Oryza sativa*, japonica x indica hybrids), noted a three-fold increase in anther callus induction frequency and more rapid callus initiation. These results were obtained despite a low percentage of overall pollen sterility (16%) in the treated plants. The compound used (a Sogetal compound coded G_5) was not specified, but was almost certainly one of the phenylcinnoline carboxylates. Qualitatively, the results were analogous to those obtained in wheat with fenridazon by Picard et al. (1987), but quantitatively they were less dramatic.

Genotypic differences in the androgenic response probably exist. This has been observed in wheat. Hybrex significantly augmented the rate of callus induction in several breeding lines of wheat (O. Hewstone et al. 1992), but this rate was dramatically improved in just one treated line. The rate of plantlet formation (plantlets per anther explant) was also dramatically improved only in this same line. The numbers of calli and plantlets produced in the other cases were too small to allow identification of significant increases in the rate of plantlet formation.

Prospects

Chemical induction of male sterility has been developed primarily for the production of hybrid crops. The highly competitive hybrid seed market has narrowed the field to a few active agents currently under commercial development. Because prior knowledge of their biochemical actions was not required for commercialization, their mechanisms are largely unknown. The recent discovery of their utility in anther culture may, however, stimulate renewed academic interest in these compounds as tools for research.

Acknowledgment

We thank E. Almeida for his insightful drawings summarizing the effects of SC-1271.

References

Adler, I.L., Hofmann, C.K. & Stavinski, S.S. (1984). High-performance liquid chromatographic analysis of fenridazon-potassium in wheat grain and straw. *J. Agric. Food Chem.* 32:1358–1361.

Awasthi, N.N.C. & Dubey, D.K. (1985a). Pollen abortion in chemically-induced male sterile lentil [*Lens culinaris*]. *LENS Newsletter (ICARDA)* 12:12–16.

Awasthi, N.N.C. & Dubey, D.K. (1985b). Effects of some phytogametocides on growth, fertility and yield of lentil [*Lens culinaris*]. *LENS Newsletter (ICARDA)* 12:16–22.

Batreau, L., Gervais, A., Quandalle, C. & Sunderworth, S. (1991). Le SC-2053, nouvel agent chimique d'hybridation pour le blé. *ANPP Annales,3ème Colloque sur les Substances de Croissance et leurs Utilisations dans les Productions Végétales.* Ministère de la Recherche et de la Technologie, Paris, pp. 75–82.

Beaumont, V. & Courtois, B. (1990). Comportement en androgenèse d'anthères de riz provenant de plantes traitées avec un agent chimique d'hybridation. *Agronomie Tropicale (France)* 45:95–100.

Bennett, M.D. & Hughes, W G. (1972). Additional mitosis in wheat pollen induced by ethrel. *Nature* 240:566–568.

Berzy, T., Szundy, T., Barnabas,B. Bauer, K. & Matolcsy, G. (1990). [The biological effect of gametocides on sexual processes and individual plant development in maize.] (in Hungarian). *Novenytermeles* 39:97–110.

Ciha, A.J. & Ruminski, P.G. (1991). Specificity of pyridinemonocarboxylates and benzoic acid analogues as chemical hybridizing agents in wheat. *J. Agric. Food Chem.* 39:2072–2076.

Chandra Sekhar, K.N. & Sawhney, V.K. (1990). Regulation of the fusion of floral organs by temperature and gibberellic acid in the normal and solanifolia mutant of tomato (*Lycopersicon esculentum*). *Can. J. Bot.* 68:713–718.

Chandra Sekhar, K.N. & Sawhney, V.K. (1991). Role of ABA in stamen and pistil development in the normal and solanifolia mutant of tomato (*Lycopersicon esculentum*). *Sex. Plant Reprod.* 4:279–283.

Colhoun, C.W. & Steer, M.W. (1982). Gametocide induction of male sterility: a review and observations on the site of action in the anther. *Rev. Cytol. Veget. Bot.* 5:283–302.

Colhoun, C.W. & Steer, M.W. (1983). The cytological effect of the gametocides ethrel and RH-531 on microsporogenesis in barley (*Hordeum vulgare* L.). *Plant Cell Environ.* 6:21–29.

Cross, J.W., Herzmark, P., Guilford, W.L., Patterson, T.G. & Labovitz, J.N. (1992). Transport and metabolism of pollen suppressant SC-1271 in wheat. *Pest. Biochem. & Physiol.* 44:28–39.

Cross, J.W. & Ladyman, J.A.R. (1991). Chemical agents that inhibit pollen development: tools for research. *Sex. Plant Reprod.* 4:235–243.

Cross, J.W., Patterson, T., Labovitz, J., Almeida, E. & Schulz, P.J. (1989). Effects of chemical hybridizing agents SC-1058 and SC-1271 on the ultrastructure of developing wheat anthers (*Triticum aestivum*, L. var Yecora rojo). In *Current Topics in Plant Physiology, Vol. 1.*, eds. E. Lord and G. Bernier, pp. 187–188. Rockville: Amer. Soc. Plant Physiol.

Dardoize, F., Goasdoue, N., Goasdoue, C. & Couffignal, R. (1993). High-performance liquid chromatography of chemical hybridizing agent in wheat. *J. Liquid Chromatog.* 16:1517–1528.

El-Ghazaly, G.A. & Jensen, W.A. (1990). Development of wheat (*Triticum aestivum*) pollen wall before and after effect of a gametocide. *Can. J. Bot.* 68:2509–2516.

Goasdoue, N., Goasdoue, C., Dardoize, F. & Couffignal, R. (1993). Changes in fatty acid and lipid composition of wheat when treated with a chemical hybridizing agent. *Phytochemistry* 34:375–380.

Goldberg, R.B. & Chasan, R., eds. (1993). Special review issue on plant reproduction. *Plant Cell* Vol. 5(10).

Goldberg, R.B., Beals, T.P. & Sanders, P.M. (1993). Anther development: basic principles and practical applications. *Plant Cell* 5:1217–1229.

Graham, R.D. (1976). Physiological aspects of time of application of copper to wheat plants. *J. Exp. Biol.* 27:717–724.

Graham, R.D. (1986). Induction of male sterility in wheat (*Triticum aestivum* L.) using organic ligands with high specificity for binding copper. *Euphytica* 35:621–630.

Guilford, W.J., Schneider, D.M., Labovitz, J. & Opella, S.J. (1988). High resolution solid state ^{13}C NMR spectroscopy of sporopollenins from different plant taxa. *Plant Physiol.* 86:134–136.

Guilford, W.J., Patterson, T.G., Vega, R.O., Fang, L., Liang, Y., Lewis, H.A. & Labovitz, J.N. (1992). Synthesis and pollen suppressant activity of phenylcinnoline-3-carboxylic acids. *J. Agric. Food Chem.* 40:2026–2032.

Hewstone O. N., Nitsch, C., Hewstone, M. & Munoz, S.C. (1992). Uso del gametocida Hybrex para aumentar la androgenesis en trigo. (The use of the gametocide Hybrex to augment androgenesis in wheat.) *Agricultura Tecnica (Chile)* 52:101–104.

Hsu, F.C. & Kleier, D.A. (1990). Phloem mobility of xenobiotics. III. Sensitivity of unified model to plant parameters and application to patented chemical hybridizing agents. *Weed Sci.* 38:315–323.

Jewell, A.W., Murray, B.G. & Alloway, B.J. (1988). Light and electron microscopic studies on pollen development in barley (*Hordeum vulgare* L.) grown under copper-sufficient and deficient conditions. *Plant Cell Environ.* 11:273–281.

Kelly, H.A. (1991). Use of *Chlorella* as a test system for understanding the mode of action of a chemical hybridizing agent. *J. Appl. Phycol.* 3:147–152.

Kerr, M.W. (1984). Rational design of a pollen suppressant for small-grained cereals. *Brit. Plant Growth Regul. Group Vol. 11, Biochemical Aspects of Synthetic and Naturally Occurring Plant Growth Regulators*, pp. 59–72. Wantage Oxfordshire: The Group.

Kiss, H.G. & Koning, R.E. (1989). Endogenous levels and transport of 1-aminocyclopropane-1-carboxylic acid in stamens of *Ipomoea nil* (Convolvulaceae). *Plant Physiol.* 90:157–161.

Kofoid, K.D. (1991). Selection for seed set in a wheat population treated with a chemical hybridizing agent. *Crop Sci.* 31:277–281.

Levings, C.S. (1993). Thoughts on cytoplasmic male sterility in cms-T maize. *Plant Cell* 5:1285–1290.

McGuinness, S.F., Pearce, G.R., Dalling, M.J. & Simpson, R.J. (1991). The effect of a chemical hybridizing agent on the morphology and chemical composition of annual ryegrass. *Grass Forage Sci.* 46:107–111.

McRae, D.H. (1985). Advances in chemical hybridization. *Plant Breed. Rev.* 3:169–191.

Meer, Q.P. van der & Bennekom, J.L. van (1982). Gibberellins as gametocides for the common onion (*Allium cepa* L.), 3: GA_{4+7} and pollen viability. *Euphytica* 31:503–506.

Mizelle, M.B., Sethi, R., Ashton, M.E. & Jensen, W.A. (1989). Development of the pollen grain and tapetum of wheat (*Triticum aestivum*) in untreated plants and plants treated with chemical hybridizing agent RH0007. *Sex. Plant Reprod.* 2:231–253.

Mogensen, H.L. & Ladyman, J.A.R. (1989). A structural study on the mode of action of CHA™ Chemical Hybridizing Agent in wheat. *Sex Plant Reprod.* 2:173–183.

Olvey, J.M., Fisher, W.D. & Patterson, L.L. (1981). TD1123: a selective male gametocide. In *Proc. Beltwide Cotton Prod. Res. Conf.*, ed. J.M. Brown, p. 84. Memphis, TN: National Cotton Council of America.

Oury, F.X., Koenig, J., Berard, P. & Rousset, M. (1990). A comparison between wheat hybrids produced using a chemical hybridizing agent and their parents' levels of heterosis and yield elaboration. *Agronomie (Paris)* 10:291–304.

Patterson, T.G., Batreau, L., Lewis, H., Vega, R. & Sunderworth, S. (1991). Factors influencing the purity of hybrid wheat seed produced with a chemical hybridizing agent, SC-2053. *Agronomy Abstracts* 169.

Picard, E., Hours, C., Gregoire, S., Phan, T.H. & Meunier, J.P. (1987). Signficant improvement of androgenetic haploid and diploid induction from wheat plants treated with a chemical hybridization agent. *Thero. Appl. Genet.* 74:289–297.

Rastogi, R. & Sawhney, V.K. (1988). Suppression of stamen development by CCC and ABA in tomato floral buds cultured in vitro. *J. Plant Physiol.* 133:620–624.

Saini, H.S. & Aspinall, D. (1982). Sterility in wheat (*Triticum aestivum* L.) induced by water deficit or high temperature: possible mediation by abscisic acid. *Aust. J. Plant Physiol.* 9:529–537

Sawhney, V.K. (1981). Abnormalities in pepper (*Capsicum annuum*) flowers induced by gibberellic acid. *Can. J. Bot.* 59:8–16.

Schulz, P.J., Cross, J.W. & Almeida, E. (1993). Chemical agents that inhibit pollen development: effects of the phenylcinnoline carboxylates SC-1058 and SC-1271 on the ultrastructure of developing wheat anthers (*Triticum aestivum* L. var *Yecora rojo*). *Sex. Plant Reprod.* 6:108–121.

Sun Y., Wang C. & Li Y. (1988). Using male-gametocide in producing primary Triticale. *Acta Agronomica Sinica (China)* 14:60–65.

Tschabold, E.E., Heim, D.R., Beck, J.R., Wright, F.L., Rainey, D.P., Terando, N.H. & Schwer, J.F. (1988). LY195259, new chemical hybridizing agent for wheat. *Crop Sci.* 28:583–588.

Wiermann, R. & Gubatz, S. (1992). Pollen wall and sporopollenin. *Int. Rev. Cytol.* 140:35–69.

Wong, M., Blouet, A. & Guckert, A. (1992). Effect of a chemical hybridizing agent on winter wheat. In *Proc. Second Congress, Euro. Soc. Agron.*, ed. A. Scaife, pp. 152–153.

11

Male sterility through recombinant DNA technology

MARK E. WILLIAMS, JAN LEEMANS,
and FRANK MICHIELS

Summary   237

Introduction: The importance and production of F_1 hybrid varieties   238

Anther- and pollen-specific gene expression   239

Pollination control systems: I. Dominant male-sterility genes   240

    Dominant genes for male sterility   240

    Fertility restoration   243

    Production of 100% male-sterile populations   244

Pollination control systems: II. Recessive male sterility   248

Pollination control systems: III. Targeted gametocide   250

Pollination control systems: IV. Dual method   251

Genetically engineered pollination control system versus CMS   251

Conclusions: Future prospects for genetically engineered pollination
    control systems   252

References   253

Summary

Hybrid varieties of crop plants are grown when the increased productivity gained from heterosis offsets the extra cost of their development and seed production. One important factor in economically viable hybrid seed production is the availability of a practical and effective pollination control system, which is employed to prevent sib- or self-pollination of the female parent. Such a system is not available in many important crops and, in others, difficult and costly manual emasculation of the male flower or flower parts from the female parent plants is the only possibility. Alternative methods of pollination control have long been desired in these crops. The development of recombinant DNA technologies has opened new possibilities for creating and manipulating male sterility for pollination control. These possibilities are the subject of this chapter. Recombinant DNA concepts devised only in the last five years have already developed into practical tools for hybrid seed production and will be one of the first and most important contributions of biotechnology to plant agriculture.

237

Introduction: The importance and production of
F₁ hybrid varieties

The exploitation of heterosis through the use of hybrid varieties is arguably
the single most important contribution of genetic research to agriculture
(Peacock 1992). The term "heterosis," or hybrid vigor, was defined by Shull
(1952) as the increase in vigor and productivity resulting from the differences
in parental gametes. Thus, in contrast to the open-pollinated or inbred varieties they replace, hybrid (F_1) varieties are derived from controlled crossings
between two genetically distinct groups of parents, usually inbred lines. An
added premium in the case of hybrids of inbred lines is uniformity, which is
often a major parameter of quality and can also facilitate mechanical harvesting. Hybrid varieties are used when the increased yield from heterosis will
more than pay for their development and the higher cost of seed production.
When this is the case, both the seed producer and farmer benefit.

Economically viable hybrid seed production requires a practical and
effective means of pollination control. Another important factor in seed production economics for species that are primarily self-pollinating is the transfer of pollen from the male to female parent; this is discussed in Chapters 2
and 4. Pollination control refers to the practices employed to prevent sib- or
self-pollination of the female parent, thereby ensuring hybridization between
the male and female parents. The two most widely used methods of pollination control are physical emasculation of male flower or flower parts and
incorporation of genetic mutations that prevent normal pollen development
(male sterility). Because maize has separate male and female flowers, and the
male flowers are located at the top of the plant in the tassel, pollination control by physical emasculation (detasseling) is feasible and widely practiced,
although it is costly and logistically complicated. Pollination control to produce several high-value vegetable crop seeds, such as tomato, is also done by
hand emasculation of stamens from the small bisexual flowers. Such seed is
largely produced in countries with low labor cost. Most major crops, however, have small bisexual flowers, making physical emasculation impractical,
and hybrid varieties are not commercially feasible without some form of
male sterility.

Seed producers have exploited several biological, primarily genetic, systems for producing male or self-sterility (Wricke 1989), many of which are
described in Chapters 7–9. Of these, cytoplasmic male sterility (CMS) is by
far the most used mechanism of pollination control in hybrid seed production
of major field and vegetable crops (Riggs 1987; Poehlman and Sleper 1995).
CMS is the subject of Chapter 7; however, it is relevant to mention here that
although CMS has been found in, or introduced into, nearly all crops (Kaul
1988), many CMS types have unfavorable characteristics that restrict their
utility in such important crops as maize (Gracen 1982), oilseed rape
(Downey and Rakow 1987; Röbbelen 1987), and wheat (Allan 1987;

Edwards 1987). Consequently, much effort is being, and has been, made to improve existing genetic systems for producing hybrid seeds and to find effective systems in species that up to now have proved difficult to handle (Wricke 1989). Over the last five years, new genetic approaches toward pollination control have been proposed based on the tools and techniques available from biotechnology. These approaches, with emphasis on a system reaching commercial deployment, are the subject of this chapter.

Anther- and pollen-specific gene expression

The development of pollen, the male gametophyte, takes place within the anther compartment of the stamen (see Chapter 2). The anther represents an outstanding and convenient system to study the differentiation of plant cell types, and considerable work has been done in anther and pollen biology (reviewed in Goldberg et al. 1993 and McCormick 1993). The whole process of male gametogenesis, from stamen primordia initiation to dehiscence and pollen shed, is a developmentally complex process. Within the differentiated anther itself, there are several specialized cells and tissues that are responsible for carrying out both nonreproductive and reproductive functions (Koltunow et al. 1990; Scott et al. 1991a). "Differentiation" is a term used to describe developmental processes that cause structural or functional distinction between parts of an organism; the primary mechanism of differentiation is qualitative or quantitative changes in gene expression (Strickberger 1985). Comparison of floral and vegetative organ RNA populations showed that ~10,000 diverse anther mRNAs are undetectable in the nuclear RNA and mRNA populations of other organs and are anther specific (Kamalay and Goldberg 1980, 1984). Subsequent experiments by several groups identified specific cDNA clones that represent RNAs that are expressed exclusively, or at elevated levels, in the anther (Koltunow et al. 1990; Smith et al. 1990; Evrard et al. 1991; McCormick 1991; Nacken et al. 1991; Scott et al. 1991b; Shen and Hsu 1992). Further study with these anther-specific cDNAs revealed their expression could be further partitioned with respect to both cell type and developmental time (Koltunow et al. 1990). For example, several different anther-specific mRNAs were found to be localized within the tapetum (Koltunow et al. 1990), a sporophytic (2n) tissue comprising the innermost layer of the anther wall that surrounds the locule containing the sporogenous cells. Experiments by several groups have shown that the temporal and spatial gene expression programs occurring in the anther are controlled primarily at the transcriptional level (Kamalay and Goldberg 1984; Koltunow et al. 1990; McCormick 1991, 1993; Scott et al. 1991a). In addition to genes that are expressed in other sporophytic anther tissue(s) (review in Goldberg et al. 1993), genes have also been identified whose expression is gametophyte (n)-specific (see Chapter 3 and McCormick 1993).

Pollination control systems: I. Dominant male-sterility genes

Dominant genes for male sterility

An important practical spin-off of anther gene-expression studies was the conception and creation of a "genetic laser" approach to engineer male sterility (Mariani et al. 1990; Goldberg et al. 1993). The basic concept is to target the expression of a gene encoding a cytotoxin by placing it under the control of an anther-specific promoter. Specifically, the 5' regulatory (promoter) region of the *TA29* gene, characterized by its tapetal-cell specificity (Mariani et al. 1990; Koltunow et al. 1990), was used to target the expression of two genes encoding ribonucleases. One gene was a chemically synthesized *RNase-T1* gene from the fungus *Aspergillus oryzae* (Quaas et al. 1988), and the other was a natural gene, called *barnase*, from the bacterium *Bacillus amyloliquefaciens* (Hartley 1988, 1989). Uncontrolled breakdown of RNA in a plant cell would be expected to be lethal, as is the case with barnase production in *E. coli* (Paddon and Hartley 1987). Both chimeric *TA29-RNase* genes were introduced individually into tobacco. In many of these transformants, RNase production led to precocious degeneration of the tapetum cells, the arrest of microspore development, and male sterility (Mariani et al. 1990). Except for the male-sterile phenotype, these transformants were identical to the untransformed control plants with respect to growth rate, height, morphology of vegetative and floral organ systems, time of flowering, flower coloration pattern, and female fertility (Mariani et al. 1990). *TA29-RNase*–induced male sterility was also produced in oilseed rape (Mariani et al. 1990), in maize (Mariani et al. 1993) (see Figure 11.1), and in several vegetable species (Reynaerts et al. 1993). In practice, a genetically engineered male-sterility gene had been created. In addition, the importance of the tapetum in microsporogenesis, long inferred by cytological and genetic observations (Rick 1948; Kaul 1988), was confirmed beyond dispute (Bedinger 1992).

Subsequently, numerous chimeric genes have been introduced into various plants, resulting in male sterility (Table 11.1). Many investigators follow the same concept of controlling the expression of a gene encoding a cytotoxic protein by means of a promoter whose expression is confined to cells and tissues involved in male gametophyte development (Table 11.1; Koltunow et al. 1990; Spena et al. 1992; Paul et al. 1992; An et al. 1994; Day et al. 1994). In one case, where creation of male sterility was not the intention, a constitutive promoter was used and it reduced female fertility (Table 11.1; Schmülling et al. 1988). Other investigators have used antisense or cosuppression of endogenous genes that are essential for pollen formation or function (Table 11.1; Taylor and Jorgensen 1992; van der Meer et al. 1992; Ylstra et al. 1994; Xu et al. 1995). Male sterility was also created by reproducing a specific phenotype observed in a petunia CMS – premature callose wall dissolution around the microsporogenous cells (Table 11.1; Worrall et al. 1992) – as well as by reproducing mitochondrial dysfunction, a general phenotype observed in many CMSs (Table 11.1; Hernould et al. 1993).

Figure 11.1 Genetically engineered male-sterility and fertility-restorer genes in the maize inbred line H99. These genes are based on the tapetal-specific transcriptional activity of the tobacco *TA29* gene (Koltunow et al. 1990) and an RNase (barnase)/RNase-inhibitor (barstar) pair of the bacterium *Bacillus amyloliquefaciens* (Hartley 1989). The male-sterility gene is composed of the *TA29* gene promoter and the *barnase* coding region. Targeted expression of *barnase* by the *TA29* promoter selectively ablates tapetum, inhibiting pollen formation and causing male sterility with no effect on any other characteristic of the plant (center figure; Mariani et al. 1990, 1993). Barstar is a specific inhibitor of barnase. Inhibition involves the formation of a stable, noncovalent, one-to-one complex of the two proteins that has no residual RNase activity (Hartley 1989). The presence of barstar alone has no effect on tapetal development and the plants are male fertile (not shown). When male-sterile plants containing *TA29-barnase* are crossed with male-fertile plants containing *TA29-barstar*, progeny are produced that contain and express both genes. The two proteins form the inactive complex, tapetum development is normal, and the plants are male fertile (left), indistinguishable from wild type (right).

Building on these concepts, the list of genetically engineered male-sterility genes could in the future be extended considerably. However, the genes all share some common characteristics that must be addressed before a workable pollination control system can be produced with them. *TA29-RNase*–induced male sterility is an active process, with the cytotoxic protein actively involved in inhibiting pollen development. Consequently, *TA29-RNase* is, as are all of the other transgenes listed in Table 11.1, a dominant nuclear-encoded, or genetic male-sterile (GMS), gene. Although in the majority of cases of endogenous GMS the male-sterile phenotype is recessive, in some cases it is

Table 11.1. *Dominant male-sterile transgenes*

Transgene	Recipient plant	Effect produced on female fertility	Restorer gene?	References
c p35S/*rolC*	Tobacco, potato, *Arabidopsis*	Reduced	Yes	Schmülling et al. (1988) Schmülling et al. (1993)
t-s pTA29/*DTA*	Tobacco	N.A.	No	Koltunow et al. (1990)
t-s pTA29/*barnase* RNase	Tobacco, oilseed rape, maize	None	Yes	Mariani et al. (1990) Mariani et al. (1992) Mariani et al. (1993)
t-s pTA29/*T1* RNase	Tobacco, oilseed rape	None	No	Mariani et al. (1990)
t-s pA3 or t-s pA9/*β-1,3-glucanase*	Tobacco	None	No	Worrall et al. (1992)
t-s pA9/*barnase* RNase	Tobacco	None	No	Paul et al. (1992)
c p35S:anther expression box or pCHS/antisense CHS cDNA	Petunia	None	No	van der Meer et al. (1992) Ylstra et al. (1994)
c p35S/CHS cDNA (cosuppression)	Petunia	None	No	Taylor and Jorgensen (1992)
t-s pTAP1/*rolB(indole β-glucosidase)*	Tobacco	N.A.	No	Spena et al. (1992)
c p35S/yeast *coxIV* mitochondrial transit peptide:unedited wheat *atp9*	Tobacco	None	No	Hernould et al. (1993)
psp-p pAP3/*DTA*	*Arabidopsis*	None	No	Day et al. (1994)
p-s p/*rolB*	Tobacco	None	No	An et al. (1994)
t&p-s pBcp1/antisense *Bcp1* cDNA	*Arabidopsis*	None	No	Xu et al. (1995)

Note: Abbreviations: p, promoter; c, constitutive; t-s, tapetum-specific; psp-s, petal/stamen primordia-specific; p-s, pollen-specific; DTA, diphtheria toxin A-chain; CHS, chalcone synthase. N.A., not analyzed.

242

dominant (Kaul 1988). In either case, there is the difficulty that a population of 100% male-steriles cannot be produced by normal crossing procedures; the greatest percentage of male-sterile plants that can be expected is 50%. With a dominant GMS gene (the recessive case will be discussed in another section), male-fertile segregation results from the nonmutant allele (in the case of trans-genics, no allele) carried by the dominant male-sterile plant, since a dominant male-sterile plant cannot be made homozygous. Subsequently, when a dominant (sporophytically acting) male-sterile plant is crossed with a fertile plant, the resulting progeny will segregate 1:1 for male-fertiles and male-steriles (Figure 11.2). In hybrid seed production, the 50% male-fertile plants would need to be rogued out before they shed pollen, a laborious and expensive procedure, which is why endogenous GMS genes are not widely used for pollination control. Consequently, in and of itself, a transgenic GMS gene is no better than an endogenous one. However, with transgenic GMSs, several ways to circumvent this difficulty have been demonstrated or proposed. To better understand some of these concepts, however, it is first necessary to discuss fertility restoration.

Fertility restoration

When hybrids are produced with dominant male-sterility genes, 50% of the F_1 hybrid plants will inherit the gene and be male sterile (Figure 11.2). In crop plants that are primarily self-pollinated and in which the seed or fruit is the harvested product, it is necessary to restore full male fertility in the F_1 hybrid. Consequently, restorer (RF) genes must be devised that can suppress the action of the male-sterility gene.

Mariani et al. (1992) reported the first genetically engineered RF gene. It produces a protein called barstar, which is a specific inhibitor of barnase (one of the RNases used to engineer male sterility) under the control of the *TA29* gene promoter (Mariani et al. 1992). Barstar is also derived from *B. amy-loliquefaciens* and serves to protect the bacterium from its own RNase by forming a diffusion-dependent, extremely stable one-to-one complex with it, which is devoid of residual RNase activity (Hartley 1988, 1989). Expressing the *barstar* gene under the same promoter as the *barnase* gene was done to ensure that it would be activated in tapetal cells at the same time and to max-imize the chance that barstar molecules would accumulate in amounts at least equal to barnase. In oilseed rape, the majority of crosses between *TA29-barstar* and *TA29-barnase* plants produced progeny with both chimeric genes that were male-fertile (Mariani et al. 1992). Anthers restored to male fertility are indistinguishable from those of wild-type plants: They develop and dehisce normally, have well-differentiated tapetal cell layers, and produce large amounts of functional pollen grains (Mariani et al. 1992). Thus, barstar is able to complex efficiently with barnase in anther tapetal cells. Restoration has also been demonstrated in tobacco and maize (Figure 11.1), indicating that barstar is functional in different plants.

In other cases, inhibiting the male-sterility gene by antisense could be a feasible approach, and it has been demonstrated in one case (Schmülling et al. 1993). In cases where the male-sterility gene is itself an antisense gene, designing a restorer counterpart is more problematic. One proposal is to use a gene, from another species for example, that is sufficiently divergent in nucleotide sequence so as to be unaffected by the antisense male-sterility gene, yet it encodes a protein capable of substituting for the endogenous function (van Tunen et al. 1994a). The feasibility of this approach has not been demonstrated.

Production of 100% male-sterile populations

As stated previously, when using a dominant GMS gene, a means to produce 100% male-sterile populations is required in order to produce a practical pollination control system. Several possibilities are discussed here.

Linkage to a selectable marker

The first possibility for producing 100% male-sterile populations was demonstrated by Mariani et al. (1990), who linked the dominant *TA29-barnase* male-sterility gene to a dominant selectable marker gene, in this case the *bar* gene, whose expression confers tolerance to the herbicide glufosinate ammonium (De Block et al. 1987). Treatment at an early stage with glufosinate ammonium during the female parent increase and hybrid seed production phases eliminates the 50% male-fertile and herbicide-sensitive plants, resulting in a population of 100% male-sterile plants at flowering (Figure 11.2).

To compensate for the elimination of the $A^{-/-}$ plants (Figure 11.2), female rows are planted at twice the normal density. Given the multiplication rates of plants per generation, ranging from ~35:1 for wheat (Stroike 1987) to ~1,000:1 for oilseed rape (Downey and Rakow 1987), the area required for female parent seed increase is much smaller than the area needed for hybrid seed production, so a doubling of the female parent seed requirement has little impact on the total cost of hybrid seed production.

Since the 50% male-fertiles and 50% male-steriles would be distributed at random (Figure 11.2), double-density planting followed by elimination of the male-fertiles by herbicide treatment produces a more variable intrarow plant distribution than does a conventional planting. The effect of this, if any, on seed yield or quality will probably depend on the growth characteristics of each crop. In oilseed rape, no effect is observed (Plant Genetic Systems [PGS] NV, unpublished). In maize, preliminary tests also show no effect (van Mellaert 1993), although further testing is necessary.

Pollen lethality

Another possible method to produce 100% male-sterile populations is to add a second transgenic locus to the female parent line, consisting of an RF gene linked to a pollen-lethality transgene (Williams and Leemans 1993). One example is expressing *barnase* with a pollen-specific promoter, such as that of the *ZM13* gene of maize (Figure 11.3; also Hanson et al. 1989). A key dif-

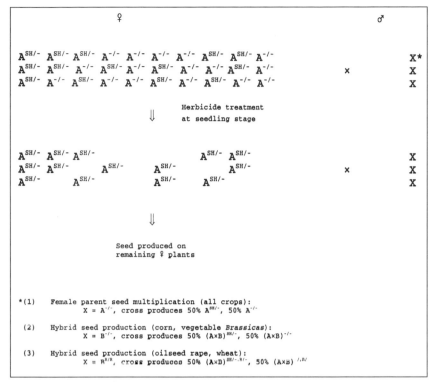

Figure 11.2 An example of hybrid seed production using a dominant male-sterility gene linked to a selectable marker gene. S is a dominant male-sterility gene, *TA29-barnase* (Mariani et al. 1990), linked to the selectable marker gene (H) *35S-bar*, which confers tolerance to the herbicide glufosinate ammonium (De Block et al. 1987). Most male-sterile female lines will be produced by backcrossing from another SH-containing line. (1) To multiply the A$^{SH/-}$ version of the female parent line A, it must be crossed with its nontransgenic counterpart (A$^{-/-}$); this results in a 1:1 segregation of A$^{SH/-}$ and A$^{-/-}$ plants in the progeny. In both female parent increase (1) and hybrid seed production (2, 3), glufosinate ammonium treatment of the female rows at an early stage eliminates the A$^{-/-}$ plants, resulting in 100% A$^{SH/-}$ plants at flowering. To compensate for the elimination of the A$^{-/-}$ plants, female rows are planted at twice the normal density.

Half of the F$_1$ plants will inherit the S gene. For crops that do not require 100% male-fertile plants in the F$_1$ hybrid, the male parent does not require modification (2). For crops that do require 100% male-fertile plants, the fertility-restorer gene, *TA29-barstar*, is incorporated into the male parent (3).

ference between lethality induced in sporophytic (2n) tissues, such as the tapetum, and lethality induced in gametophytic (n) cells is the phenotype of heterozygotes. With *TA29-barnase*, a heterozygous plant is completely male sterile (Figure 11.1), whereas a heterozygous *ZM13-barnase* plant retains 50% pollen viability (Figure 11.3). A "maintainer" plant can be produced that is homozygous for the male-sterility gene (which is possible because a restored plant can be self-pollinated) and heterozygous for the pollen-lethality/RF gene locus. Such a plant would be male fertile with 50% viable pollen;

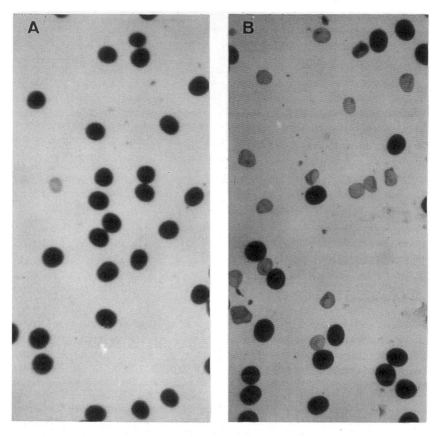

Figure 11.3 A pollen-lethal transgene. The transgene consists of the promoter of the maize pollen-specific gene *Zm13* (Hanson et al. 1989) and the *barnase* RNase coding region (Hartley 1989). It was transformed into the maize inbred line H99 linked to a selectable marker gene, *35S-bar*, which confers tolerance to glufosinate ammonium (De Block et al. 1987). Wild-type H99 shows nearly 100% pollen viability (A), based on Alexander's staining (Alexander 1969). Primary transformants are heterozygous for the linked transgenes and show ~50% pollen lethality (B). Pollen lethality was confirmed in crosses to wild-type H99 using the primary transformant as a male. These crosses resulted in complete seed set but, as expected, none of the progeny were glufosinate ammonium tolerant, the character linked to the pollen-lethality transgene.

however, none of the viable pollen would contain the RF gene. When such a plant is used in crosses as a male, it produces 100% male-sterile progeny. Barnase is produced only in the male gametophyte, so this locus does not interfere with female gametophyte development and thus can be transmitted and maintained through the egg. One requirement for this system to work is high expression of the RF gene, since a heterozygous restorer must effectively neutralize a homozygous male-sterility gene. Effective pollen-lethality genes have also been made by expressing antisense RNA of essential gametophytically expressed genes (Muschietti et al. 1994; Xu et al. 1995).

Chemically induced restoration

Another proposed alternative to the use of a linked selectable marker gene is chemically induced fertility restoration, of which two different types have been proposed.

One type applies compound(s) that are necessary for pollen function, but whose synthesis is blocked by the male-sterility gene (van der Meer et al. 1992; Dirks et al. 1993). The best example is flavonoids. Flavonoids, which can be extracted from the outer wall of pollen (the exine) of all higher plant species examined, comprise up to 2%–4% of the pollen dry weight (Wiermann and Vieth 1983) and confer the yellow color typical of the pollen of most plant species. Tapetal cells are the main producers of flavonoids (van Tunen et al. 1994b). In 1981, Coe et al. described the white pollen phenotype of maize, which was shown to be determined by the double recessive condition at two loci, *c2* and *whp*; self-pollination of white pollen plants was unsuccessful. Importantly, white pollen is nonfunctional, but not aborted (Coe et al. 1981). It was subsequently found that both loci that are involved (*c2* and *whp*) encode chalcone synthase (CHS) (Franken et al. 1991), an enzyme that catalyzes the first step in flavonoid biosynthesis. The association of the absence of flavonoids and self-sterility suggested that the synthesis or deposition of flavonoids is vital to pollen function (Coe et al. 1981). To further address the question of whether flavonoids have a function during fertilization or in the development of the male gametophyte, van der Meer et al. (1992) performed an experiment in petunia in which flavonoids were depleted by an antisense approach. Antisense CHS cDNA was expressed behind a *35SCaMV* promoter modified to direct gene expression in tapetum tissues. Transgenic petunia plants displayed the white pollen phenotype and numerous attempts at self-pollination resulted in no seed set. Based on crosses with wild-type pollen, female fertility was unaffected. Cosuppression of CHS also resulted in white pollen plants (Taylor and Jorgensen 1992).

Subsequently, it was shown that application of the missing flavonoids, such as kaempferol, to the stigmas (transgenic petunia) or mixed with the pollen (maize mutant) restored pollen tube growth and self-fertilization became possible (Taylor and Jorgensen 1992; Mo et al. 1992; Ylstra et al. 1994). Thus, inhibition of CHS expression acts as a gametostat rather than a gametocide, and can be reversed (Mo et al. 1992). Consequently, a reversible male-sterile system for the production of hybrid seed based on the pollen-rescue ability of some flavonoids was proposed (Mo et al. 1992; van Tunen et al. 1994b). In the presence of the flavonoid(s), normally male-sterile lines can be reproduced by sib- or self-pollination and thus 100% male-sterile populations can be produced. Practical and effective recovery of male fertility by flavonoid application on a scale needed to produce female parent seed of a major crop has not been demonstrated. Also, in the case of petunia, mixing white pollen with wild-type ("mentor") pollen of other species, such as tobacco and maize, restored its functionality, apparently by diffusion of flavonols (Ylstra et al. 1994; Pollack et al. 1995). However, the effect of wild-type petunia pollen on white petunia pollen has not been reported. Consequently, the feasibility of this proposal for pollination control in a given crop may depend on the response of its white pollen produced by the female parent to the wild-type pollen of the male parent.

Another proposal (Bridges et al. 1990) utilizes chemically induced promoters, several of which have been identified or introduced into plants (Mett et al. 1993; Ward et al. 1993). In this approach, an operator sequence is added to the promoter controlling expression of the male-sterility gene and a gene encoding a transcriptional repressor protein specific for that operator sequence is placed under the control of the chemically induced promoter. During female parent increase, the inducing chemical is applied, the repressor protein is produced, and the protein binds to the operator sequence. Transcription of the male-sterility gene is inhibited, and the plants are male fertile. During hybrid seed production, no chemical is applied, the repressor gene is silent, and the plants are male sterile. The relative activity of any chemically induced promoters in the cells and tissues involved in pollen development has not been reported.

Pollination control systems: II. Recessive male sterility

GMS mutants have been found in nearly all crop species examined (Kaul 1988; also see Chapter 8). In maize, for example, more than one hundred GMS loci have been found (Kaul 1988; Albertsen et al. 1993). The male-sterile phenotype is predominantly, although not exclusively, recessive; therefore, most male-sterility genes are simply mutations in genes essential for male fertility (Albertsen et al. 1993). As mentioned previously, recessive GMS suffers from the same problem as dominant GMS, in that the greatest percentage of male-sterile plants that can be produced is 50%. This is because at least one nonmutant (dominant) male-fertility allele is required for pollen production. When heterozygous plants are used to pollinate homozygous recessive male-sterile plants, 50% of the progeny inherit the dominant allele and are male fertile. The other 50% inherit the recessive allele and are male-sterile (Albertsen et al. 1993). In the past, various schemes designed to circumvent this problem and to allow recessive GMS to be utilized in hybrid seed production have been proposed in barley, maize, sorghum, and wheat, but none has proved practical (Poehlman and Sleper 1995). Again, biotechnology can provide new solutions to these problems.

There have been a number of proposals for using naturally occurring linkages between a GMS and a marker gene as a means of detecting or isolating male-sterile plants or seeds from a mixture of male-steriles and male-fertiles.

For example, Wiebe (1960) proposed finding and utilizing a linkage between a recessive GMS locus and recessive DDT resistance as a means of removing male-fertiles in a segregating population. However, this concept and others like it have not been exploited because of the lack of appropriate markers for many crops and, for those crops for which markers are known, because of the lack of sufficient linkage (Jorgensen 1986). Jorgensen (1986) proposed the use of transformation to create linkages of marker genes to endogenous GMS loci. Assuming random integration of marker genes into the genome, an acceptably tight linkage should be found in most crops if several hundred transformants can be produced (Jorgensen 1986). However, this assumes that the marker gene is always expressed at an acceptable level, which is generally not the case for transgenes, and the number of transformants required may be much higher.

Two other potential solutions to the problems associated with the use of GMS are similar to concepts discussed in relation to dominant male-sterile transgenes. Both involve manipulating the expression or transmission of the dominant (fertile) allele; this requires the cloning of the GMS locus. Two GMS loci have been cloned to date, both by transposon tagging: the first in *Arabidopsis* (Aarts et al. 1993), and the second, soon after, in maize (Albertsen et al. 1993). Continuing work in this area should result in the cloning of many more GMS loci, not only by transposon tagging, but by other means such as T-DNA mutagenesis and chromosome walking (Bedinger 1992; Albertsen et al. 1993; Chaudhury 1993). Albertsen et al. (1993) proposed placing the functional coding region of a cloned GMS gene under the control of a chemically induced promoter. As in the case of dominant GMS genes, female parent increase is done with chemical application, and hybrid seed production is not. Gene activation does not have to be perfect; it is thought that even a 20% (viable pollen) success rate relative to wild type would be more than sufficient to ensure complete pollination and seed set (Albertsen et al. 1993).

Another possibility for effective use of GMS is to link the dominant (fertile) allele to a pollen-lethality transgene (Williams and Leemans 1993). A plant heterozygous for the pollen-lethality/dominant (fertile) allele and homozygous for the recessive allele (at the endogenous locus) would produce 100% male-sterile progeny when crossed to a male-sterile plant. This biotechnological approach is similar in effect to proposals put forward in the 1970s (Kaul 1988), which were based on the observation that some chromosomal defects are egg viable but not transmitted through pollen. These proposals required that an endogenous GMS locus be tightly linked to the defect. These methods have not been used for pollination control because of the need for the tight linkage between the two characters as well as the difficulty of incorporating and maintaining chromosomal defects in many female lines. The transgenic approach would avoid the problem of linkage and should be easier to introgress and maintain than a chromosomal abnormality.

The most attractive feature of recessive GMS-based pollination control systems is that all normal lines are automatically fertility restorers, because they carry the intact dominant male-fertility gene. It has also been suggested that recessive sterility would be more reliable because it is achieved passively, through the absence or nonfunctionality of a gene product (recessive allele) from an endogenous fertility gene (dominant allele), rather than through the active disruptions of the dominant genes described previously (Albertsen et al. 1993).

Pollination control systems: III. Targeted gametocide

Considerable research has been conducted on the use of chemicals to suppress pollen development. Ideally, these chemicals, commonly known as gametocides or chemical hybridizing agents (CHAs), completely inhibit the production of viable pollen but do not injure the pistillate reproductive organs or reduce seed set in all genetic backgrounds and environments. In reality, this has not been the case and CHAs are not in widespread use for pollination control in major crops (Kaul 1988). Although chemical induction of sterility is the subject of Chapter 10, a pollination control system that combines chemical application and recombinant DNA technology deserves mention here. The difference between this work and other research is that the location and target of the CHA is specified with a transgene (O'Keefe et al. 1994). When expressed in the chloroplast, a cytochrome $P450_{SU1}$ from *Streptomyces griseolus* converts the sulfonylurea compound R7402 from its normally nonherbicidal form to one that is highly phytotoxic. Tobacco plants were transformed with a $P450_{SU1}$ gene driven by the *TA29* promoter to confine the phytotoxic conversion to tapetum cells. In the absence of R7402, pollen developed normally, whereas treatment with R7402 resulted in nonviable pollen. In the most responsive transformant, pollen sampled from newly opened flowers showed germination rates of 0% at 7 and 11 days posttreatment and <1% at 18 days, but the rates had recovered to 80% at 21 days (O'Keefe et al. 1994). Recovery of pollen viability probably occurred because these later flowers were at a pre-*TA29* expression stage at the time of treatment (J. T. Odell, personal communication, 1995). Prior to treatment, all flowers thought to be at a post-*TA29* stage on the basis of external morphology had been removed from the plant. Consequently, crops in which flowering is not determinate may require multiple applications to maintain sterility, and the duration of *TA29* expression would fix the window of effectiveness, an important consideration for practical application. Recently, it has been reported that the use of hybrid promoters can extend the duration of anther-specific gene expression (Huffman et al. 1995).

In contrast to chemically induced restoration, where a 20% success rate is thought to be acceptable, there is a much higher burden of performance for chemically induced sterility. In maize, for example, a chemical treatment to prevent pollen shed must have a >99% success rate. System failure for chemically induced sterility occurs at the hybrid seed production level, whereas with chemically induced restoration it would occur at female parent increase, where the costs and consequences of failure are much less.

Because male sterility occurs only in the presence of the appropriate chemical, female lines can be increased by sib- or self-pollination; also, RF genes are not necessary for 100% male-fertile hybrids.

In comparison to other CHAs, chemically induced restoration has both advantages and disadvantages. Common past problems with CHAs, such as genotype–chemical interaction, the risk of affecting the male parent with the CHA, and effects on female fertility or general phytotoxicity, can be avoided. However, one of the big advantages of CHAs is that the time and effort of converting female lines (to CMS, for example) is avoided. With the targeted CHA approach, female lines must be converted to contain the transgenic locus.

Pollination control systems: IV. Dual method

A dual method for producing male-sterile plants has been proposed (Crossland and Tuttle 1993). Two different versions of the female parent are created, each containing a different transgenic component. One component consists of an anther-specific promoter driving expression of a transactivating protein such as the T7 RNA polymerase. The second component consists of the target sequence of the transactivating protein linked to an antisense RNA or a polypeptide that will disrupt the formation of viable pollen. Separately, the individual components have no effect on male fertility. However, crossing the two different versions of the female parent puts together the male-sterility gene and its activator, resulting in 100% male-sterile progeny. As a result of segregation and crossing to the wild-type male parent, 75% of the F_1 hybrid plants would be male fertile. Because the final female parent seed multiplication step must be done using manual emasculation, use of this method is probably limited to maize and a few vegetable crops, such as tomato. Successful deployment of this method has not yet been reported.

Genetically engineered pollination control systems versus CMS

The value of genetically engineered systems of pollination control in a given crop depends on the quality and availability of alternatives, usually CMS. In maize and oilseed rape, where the existing CMSs are in limited use, genetically engineered pollination control systems will be very valuable. What about crops in which CMS is in widespread use? In general, the use of the same sterilizing cytoplasm in a crop increases genetic vulnerability to diseases and other pests. The case of the near-universal use of Texas (T) cytoplasm for pollination control in maize, its specific susceptibility to the fungal pathogen *Bipolaris maydis* Race T, and the resulting Southern corn leaf blight epidemic of 1970 are the best-known example; another case has also been described in pearl millet (Kaul 1988). Pollination control systems in which the transgenes are expressed specifically in the tapetum, a very small and unexposed tissue, or recessive-based systems that use endogenous genes already present in most lines, are very unlikely to increase vulnerability to diseases and pests. Fertility restoration allows the male-sterility transgene to "escape" the cytoplasm of the original transformed line so that all converted female lines can possess their original cytoplasm.

The use of CMS requires that if the necessary restorer gene(s), and in many cases modifier gene(s), are not present in a prospective male parent, they must be introgressed. Conversely, if they are present in a prospective female parent, they must be removed. For example, only ~8% of the elite lines developed by conventional rice breeding programs were found to maintain the sterility of CMS (Virmani 1987). Unlike CMS, the expression of the male-sterility gene will not be affected by the presence of endogenous restorer or modifier gene(s) in the female parent line. Only ~20% of the rice varieties and breeding lines developed at IRRI (International Rice Research Institute, Manila, Philippines) have the capacity to restore CMS lines (Virmani 1987). In most cases, such as sunflower, rice, and sugarbeet, there are two restorer genes (Fehr 1987). Even in cases such as sorghum, which is considered to have one major fertility-restoring gene, modifying gene(s) are required to obtain complete fertility in a wide array of environments and genotypic backgrounds (Poehlman and Sleper 1995). In the genetically engineered pollination control systems described here, only one or no restorer gene is required. These time-consuming and often difficult conversions delay the introduction of new hybrids or limit the germplasm base from which acceptable hybrids can be produced. Consequently, even in crops where CMS is widely used for pollen control, transgenic systems could be a safer and more efficient alternative.

Conclusions: Future prospects for genetically engineered pollination control systems

The value of genetically engineered pollination control systems will be affected by breakthroughs in other areas, such as the development of new or improved CMS types (Delourme et al. 1995; Prakash et al. 1995) and CHAs (Mabbett 1992; Kidd and Dvorak 1995). Given the current difficulties and costs involved in developing, testing, and commercializing transgenic plants, recombinant DNA–based approaches must demonstrate a clear advantage over nontransgenic alternatives. Several genetically engineered systems of pollination control have been produced or proposed; more are likely to be forthcoming.

A genetically engineered pollination control system nearing commercial use is the SeedLink™ system of PGS NV. Its components include the *TA29-barnase* male-sterility gene, the *TA29-barstar* RF gene, and the *35S-bar* selectable marker gene (Figure 11.2; Mariani et al. 1990, 1992). Promoters other than *TA29* and *35S* have also been used successfully for the male-sterility/RF and selectable marker genes, respectively (PGS NV, unpublished results).

Results to date in oilseed rape indicate that many of the drawbacks associated with currently used CMSs, such as yield penalty and the linkage of the RF gene to high glucosinolate levels, are avoided (Downey and Rimmer 1993; De Both 1995; Delourme 1995). The first two oilseed rape hybrids produced with the SeedLink™ system entered official registration trials in Canada in 1994. In maize, results to date indicate that male sterility is effective in all inbred backgrounds tested (PGS NV, unpublished results). Due to the presence of endogenous RF or modifier gene(s), this is not the case with CMS. Currently, CMS-produced hybrid seed is used in blends with hybrid seed produced by detasseling. This is because seed producers want to avoid the time, expense, and difficulty of converting male lines to restorers. The F_1 hybrid plants produced by detasseling are male fertile, and they supply pollen to the male-sterile F_1 hybrid plants produced with CMS. With SeedLink™, segregation of the male-sterility gene in the F_1 (Figure 11.2) results in 50% male-fertile plants, so detasseling can be avoided completely. Hybrid maize seed produced with SeedLink™ technology may become available in 1997.

Thus, recombinant DNA concepts devised only in the last five years have already developed into practical tools for F_1 hybrid seed production. Male sterility through recombinant DNA technology, by providing new, effective, and efficient means of producing hybrid seed, is an example of the tremendous potential impact of biotechnology on agriculture.

Acknowledgments

We thank Stephan Jansens for photography (Figure 11.1) and Anne-Marie Bouckaert for review and discussion.

References

Aarts, M.G.M., Dirkse, W.G., Stiekema, W.J. & Pereira, A. (1993). Transposon tagging of a male sterility gene in *Arabidopsis*. *Nature* 363:715–717.

Albertsen, M.C., Fox, T.W., & Trimell, M.R. (1993). Cloning and utilizing a maize nuclear male sterility gene. *Proc. Annu. Corn Sorghum Res. Conf.* 48:224–233.

Alexander, M.P. (1969). Differential staining of aborted pollen and non-aborted pollen. *Stain Technol.* 44:117-122.

Allan, R. E. (1987). Wheat. In *Principles of Cultivar Development, Vol. 2*, ed. W.R. Fehr, pp. 699–748.

An, G., Costa, M.A., Tepfer, D. & Gupta, H. S. (1994). Induction of male sterility by pollen-specific expression of *rolB* gene. *Abstr. 4th Intl. Congr. Plant Mol. Biol.*, #1850. Athens, GA: The International Society for Plant Molecular Biology.

Bedinger, P. (1992). The remarkable biology of pollen. *Plant Cell* 4:879–887.

Bridges, I.G., Bright, S.W.J., Greenland, A. J. & Schuch, W.W. (1990). Hybrid seed production. *PCT Pat. Appl.*, WO 90/08830.

Chaudhury, A.M. (1993). Nuclear genes controlling male fertility. *Plant Cell* 5:1277–1283.

Coe, E.H., McCormick, S.M. & Modena, S.A. (1981). White pollen in maize. *J. Hered.* 72:318–320.

Crossland, L.D. & Tuttle, A. (1993). Methods for the production of hybrid seed. *Eur. Pat. Appl.*, Pub. No. 0589 841 A2.

Day, C., Miller, R. & Irish, V. (1994). Genetic cell ablation to analyse cell interactions during *Arabidopsis* floral development. *Abstr. 4th Intl. Cong. Plant Mol. Biol.*, #738. Athens, GA: The International Society for Plant Molecular Biology.

De Block, M., Botterman, J., Vandewiele, M., Dockx, J., Thoen, C., Gosselé, V., Rao Mouva, N., Thompson, C., Van Montagu, M. & Leemans, J. (1987). Engineering herbicide resistance in plants by expression of a detoxifying enzyme. *EMBO J.* 6:2513–2518.

De Both, G. (1995). SeedLink™ technology. *Proc. 9th Intl. Rapeseed Congr.* 1:64–69.

Delourme, R., Eber, F. & Renard, M. (1995). Breeding double low restorer lines in radish cytoplasmic male sterility of rapeseed (*Brassica napus* L.). *Proc. 9th Intl. Rapeseed Congr.* 1:6–8.

Dirks, R., Trinks, K., Uijtewall, B., Bartsch, K., Peeters, R., Höfgen, R. & Pohlenz, H.D. (1993). Process for generating male sterile plants. *Eur. Pat. Appl.*, Pub. No. 0 628 635 A1.

Downey, R.K. & Rakow, G.F. (1987). Rapeseed and mustard. In *Principles of Cultivar Development, Vol. 2*, ed. W.R. Fehr, pp. 437–486.

Downey, R.K. & Rimmer, S.R. (1993). Agronomic improvement in oilseed brassicas. In *Advances in Agronomy* 50:1–66.

Edwards, I.B. (1987). Breeding hybrid varieties of wheat. In *Hybrid Seed Production of Selected Cereal and Vegetable Crops*, eds. W.P. Fiestritzer and A.F. Kelly, pp. 3–33. Rome: Food and Agriculture Organization of the United Nations.

Evrard, J.L., Jako, C., Saint-Guily, A., Weil, J.H. & Kuntz, M. (1991). Anther-specific, developmentally regulated expression of genes encoding a new class of proline-rich proteins. *Plant Mol. Biol.* 16:271–281.

Fehr, W.R. (1987). *Principles of Cultivar Development. Vol. 2, Crop species*. New York: MacMillan Pub. Co.

Franken, P., Niesbach-Klosgen, U., Weydemann, U., Marechal-Drouard, L., Saedler, H. & Wienand, U. (1991). The duplicated chalcone synthase genes *C2* and *Whp* (*white pollen*) of *Zea mays* are independently regulated; evidence for translational control of *Whp* expression by the anthocyanin intensifying gene *In*. *EMBO J.* 10:2605–2612.

Goldberg, R.B., Beals, T.P. & Sanders, P.M. (1993). Anther development: basic principles and practical applications. *Plant Cell* 5:1217–1229.

Gracen, V.E. (1982). Types and availability of male sterile cytoplasms. In *Maize for Biological Research*, ed. W.F. Sheridan, pp. 221–224. Grand Forks, ND: University Press, North Dakota State University.

Hanson, D.D., Hamilton, D.A., Travis, J.L., Bashe, D.M. & Mascarenhas, J.P. (1989). Characterization of a pollen-specific cDNA clone from *Zea mays* and its expression. *Plant Cell* 1:173–179.

Hartley, R.W. (1988). Barnase and barstar, expression of its cloned inhibitor permits expression of a cloned ribonuclease. *J. Mol. Biol.* 202:913–915.

Hartley, R.W. (1989). Barnase and barstar: two small proteins to fold and fit together. *Trends Biochem. Sci.* 14:450–454.

Hernould, M., Suharsono, S., Litvak, S., Araya, A. & Mouras, A. (1993). Male-sterility induction in transgenic tobacco plants with an unedited *atp9* mitochondrial gene from wheat. *Proc. Natl. Acad. Sci. USA* 90:2370–2374.

Huffman, G., Garnatt, C., Phillips, V., Kendall, T., Betz, S., Sandahl, J., Ross, M., Malone-Schoneberg, J-B. & Sonstad, D. (1995). Use of hybrid promoters to extend anther-specific gene expression during tassel development. *Abstr. 37th*

Annu. Maize Genetics Conf., ed. W.F. Sheridan, p. 78. Grand Forks, ND: University Press, North Dakota State University.

Jorgensen, R.A. (1986). Transformation of plants to introduce closely linked markers. *Eur. Pat. Appl.*, Pub. No. 0 198 288 A3.

Kamalay, J.C. & Goldberg, R.B. (1980). Regulation of structural gene expression in tobacco. *Cell* 19:935–946.

Kamalay, J.C. & Goldberg, R.B. (1984). Organ-specific nuclear RNAs in tobacco. *Proc. Natl. Acad. Sci. USA* 81:2801–2805.

Kaul, M.L.H. (1988). *Male Sterility in Higher Plants.* Berlin: Springer-Verlag.

Kidd, G. & Dvorak, J. (1995). A persistent Monsanto tames hybrid wheat. *Bio/Technology* 13:15–18.

Koltunow, A.M., Truettner, J., Cox, K.H., Wallroth, M. & Goldberg, R.B. (1990). Different temporal and spacial gene expression patterns occur during anther development. *Plant Cell* 2:1201–1224.

Mabbett, T.H. (1992). Chemical hybridizing agent cuts costs. *Prophyta* 2:30–31.

Mariani, C., De Beuckeleer, M., Truettner, J., Leemans, J. & Goldberg, R.B. (1990). Induction of male sterility in plants by a chimaeric ribonuclease gene. *Nature* 347:737–741.

Mariani, C., Gossele, V., De Beuckeleer, M., De Block, M., Goldberg, R.B., De Greef, W. & Leemans, J. (1992). A chimaeric ribonuclease-inhibitor gene restores fertility to male sterile plants. *Nature* 357:384–387.

Mariani, C., D'Halluin, K., Dickburt, C., De Beuckeleer, C. & Williams, M.E. (1993). The production and analysis of genetically-engineered male-sterile plants of maize. In *Abstr. 35th Annu. Maize Genetics Conf.*, ed. W.F. Sheridan, p. 45. Grand Forks, ND: University Press, North Dakota State University.

McCormick, S. (1991). Molecular analysis of male gametogenesis in plants. *Trends Genet.* 7:298–303.

McCormick, S. (1993). Male gametophyte development. *Plant Cell* 5:1265–1275.

Mett, V.L., Lochhead, L.P. & Reynolds, P.H.S. (1993). Copper-controllable gene expression system for whole plants. *Proc. Natl. Acad. Sci. USA* 90:4567–4571.

Mo, Y., Nagel, C., & Taylor, L.P. (1992). Biochemical complementation of chalcone synthase mutants defines a role for flavonols in functional pollen. *Proc. Natl. Acad. Sci. USA* 89:7213–7217.

Muschietti, J., Dircks, L., Vancanneyt, G. & McCormick, S. (1994). LAT52 protein is essential for tomato pollen development: pollen expressing antisense *LAT52* RNA hydrates and germinates abnormally and cannot achieve fertilization. *Plant J.* 6:321–338.

Nacken, W.K.F., Huijser, P., Beltràn, J.P., Saedler, H. & Sommer, H. (1991). Molecular characterization of two stamen-specific genes, *tap1* and *fil1*, that are expressed in wild type, but not in the *deficiens* mutant of *Antirrhinum majus*. *Mol. Gen. Genet.* 229:129–136.

O'Keefe, D.P., Tepperman, J.M., Dean, C., Leto, K.J., Erbes, D.L. & Odell, J. T. (1994). Plant expression of a bacterial cytochrome P450 that catalyzes activation of a sulfonylurea pro-herbicide. *Plant Physiol.* 105:473–482.

Paddon, C.J. & Hartley, R.W. (1987). Expression of *Bacillus amyloliquefaciens* extracellular ribonuclease (barnase) in *Escherichia coli* following an inactivating mutation. *Gene* 53:11–19.

Paul, W., Hodge, R., Smartt, S., Draper, J. & Scott, R. (1992). The isolation and characterization of the tapetum-specific *Arabidopsis thaliana* A9 gene. *Plant Mol. Biol.* 19:611–622.

Peacock, J. (1992). Twenty-first century crops. *Nature* 357:358.

Poehlman, J.M. & Sleper, D.A. (1995). *Breeding Field Crops*, 4th ed. Ames: Iowa State University Press.

Pollack, P.E., Hansen, K., Astwood, J.D. & Taylor, L.P. (1995). Conditional male fertility in maize. *Sex. Plant Reprod.* 8:231–241.

Prakash, S., Kirti, P.B. & Chopra, V.L. (1995). Cytoplasmic male sterility (CMS) systems other than *ogu* and *Polima* in *Brassicaceae*: current status. *Proc. 9th Intl. Rapeseed Congr.* 1:44–48.

Quaas, R., McKeown, Y., Stanssens, P., Frank, R., Blocker, H. & Hahn, U. (1988). Expression of the chemically synthesized gene for ribonuclease T1 in *Escherichia coli* using a secreting cloning vector. *Eur. J. Biochem.* 173:617–622.

Reynaerts, A., Van de Wiele, H., De Sutter, G. & Jansens, J. (1993). Engineered genes for fertility control and their application in hybrid seed production. *Scientia Horticulturae* 55:125–139.

Rick, C.M. (1948). Genetics and development of nine male-sterile mutants. *Hilgardia* 18:599–633.

Riggs, T.J. (1987). Breeding F₁ hybrid varieties in vegetables. In *Hybrid Seed Production of Selected Cereal and Vegetable Crops*, eds. W.P. Fiestritzer and A.F. Kelly, pp. 149–173. Rome: Food and Agriculture Organization of the United Nations.

Röbbelen, G. (1987). Breeding of hybrid varieties of rape. In *Hybrid Seed Production of Selected Cereal and Vegetable Crops*, eds. W.P. Fiestritzer and A.F. Kelly, pp. 133–148. Rome: Food and Agriculture Organization of the United Nations.

Schmülling, T., Schell, J. & Spena, A. (1988). Single genes from *Agrobacterium rhizogenes* influence plant development. *EMBO J.* 7:2621–2629.

Schmülling, T., Röhrig, H., Pilz, S., Walden, R. & Schell, J. (1993). Restoration of fertility by antisense RNA in genetically engineered male sterile tobacco plants. *Mol. Gen. Genet.* 237:385–394.

Scott, R., Hodge, R., Paul, W. & Draper, J. (1991a). The molecular biology of anther differentiation. *Plant Sci.* 80:167–191.

Scott, R., Dagless, E., Hodge, R., Paul, W., Soufleri, I & Draper, J. (1991b). Patterns of gene expression in developing anthers of *Brassica napus*. *Plant Mol. Biol.* 17:195–207.

Shen, J.B. & Hsu, F.C. (1992). *Brassica* anther-specific genes: characterization and in situ location of expression. *Mol. Gen. Genet.* 234:379–389.

Shull, G.H. (1952). Beginnings of the heterosis concept. In *Heterosis*, ed. J.D. Gowen, pp. 14–48. Ames: Iowa State College Press.

Smith, A.G., Gasser, C.S., Budelier, K. & Fraley, R. T. (1990). Identification and characterization of stamen- and tapetum-specific genes from tomato. *Mol. Gen. Genet.* 222:9–16.

Spena, A., Estruch, J.J., Prensen, E., Nacken, W., Van Onckelen, H. & Sommer, H. (1992). Anther-specific expression of the *rolB* gene of *Agrobacterium rhizogenes* increases IAA content in anthers and alters anther development in whole flower growth. *Theor. Appl. Genet.* 84:520–527.

Strickberger, M.W. (1985). *Genetics*, 3rd ed. New York: Macmillan.

Stroike, J.E. (1987). Technical and economic aspects of hybrid wheat seed production. In *Hybrid Seed Production of Selected Cereal and Vegetable Crops*, eds. W.P. Fiestritzer and A.F. Kelly, pp. 177–185. Rome: Food and Agriculture Organization of the United Nations.

Taylor, L.P., & Jorgensen, R. (1992). Conditional male fertility in chalcone synthase–deficient petunia. *J. Hered.* 83:11–17.

van der Meer, I.M., Stam, M.E., van Tunen, A.J., Mol, J.N.M. & Stuitje, A.R.

(1992). Antisense inhibition of flavonoid biosynthesis in petunia anthers results in male sterility. *Plant Cell* 4:253–262.

van Mellaert, H. (1993). Engineered pollen control in corn. *Proc. Annu. Corn Sorghum Res. Conf.* 48:234–240.

van Tunen, A.J., Mol, J.N.M. & van den Elzen, P.J.M. (1994a). Genetic moderation or restoration of plant phenotypes. *PCT Pat. Appl.*, WO 94/09143.

van Tunen, A.J., van der Meer, I.M. & Mol, J.N.M. (1994b). Flavonoids and genetic modification of male fertility. In *Genetic Control of Self-Incompatibility and Reproductive Development in Flowering Plants*, eds. E.G. Williams, A.E. Clarke, and R. B. Knox, pp. 423–442. Dordrecht, Netherlands: Kluwer.

Virmani, S.S. (1987). Hybrid rice breeding. In *Hybrid Seed Production of Selected Cereal and Vegetable Crops*, eds. W.P. Fiestritzer and A.F. Kelly, pp. 35–53. Rome: Food and Agriculture Organization of the United Nations.

Ward, E.R., Ryals, J. A. & Miflin, B.J. (1993). Chemical regulation of transgene expression in plants. *Plant Mol. Biol.* 22:361–366.

Wiebe, G.A. (1960). A proposal for hybrid barley. *Agron. J.* 52:181–182.

Wiermann, R. & Vieth, K. (1983). Outer pollen wall, an important accumulation site for flavonoids. *Protoplasma* 118:230–233.

Williams, M.E. & Leemans, J. (1993). Maintenance of male-sterile plants. *PCT Pat. Appl.*, WO 93/25695.

Worrall, D., Hird, D.L., Hodge, R., Paul, W., Draper, J.& Scott, R. (1992). Premature dissolution of the microsporocyte callose wall causes male sterility in transgenic tobacco. *Plant Cell* 4:759–771.

Wricke, G. (1989). Genetic mechanisms for hybrid seed production. *Vortr. Pflanzenzüchtg.* 16:369–378.

Xu, H., Knox, R.B., Taylor, P.E. & Singh, M.B. (1995). *Bcp1*, a gene required for male fertility in *Arabidopsis*. *Proc. Natl. Acad. Sci. USA* 92:2106–2110.

Ylstra, B., Busscher, J., Franken, J., Hollman, P.C.H., Mol, J.N.M. & van Tunen, A.J. (1994). Flavonols and fertilization in *Petunia hybrida*: localization and mode of action during pollen tube growth. *Plant J.* 6:201–212.

Part IV
Pollen biotechnology and plant breeding

12

Barriers to hybridization

K. R. SHIVANNA

Summary 261
Introduction 262
Classification of barriers 262
I. Temporal and spatial isolation of parental species 262
II. Prefertilization barriers 263
    Unilateral incompatibility 263
    Active versus passive inhibition 264
    Inhibition on the stigma surface 265
    Inhibition in the stigma and style 266
    Inhibition in the ovary 268
III. Postfertilization barriers 269
Concluding remarks 269
References 269

Summary

Incompatibility barriers are major impediments in crop improvement programs. Nonsynchronous flowering and/or geographical isolation of parental
species is common, particularly in wide crosses, and is critical in tree species.
Postpollination barriers may operate before and/or after fertilization.
Prefertilization barriers act either on the surface of the stigma (by inhibiting
pollen germination or pollen tube entry into the stigma) or in the transmitting
tissue of the stigma and style. Occasionally, pollen tubes may be inhibited in
the ovary or in the ovule. More often, prefertilization barriers are not
restricted to a particular level but may be active at all levels. The proportion
of pollen grains that complete sequential postpollination events is reduced at
each level, with the result that very few or no pollen tubes reach the ovule. In
most of the interspecific crosses, pollen inhibition is passive (not as a result
of active recognition of the pollen) because of the lack of co-adaptation
between the pollen and the pistil. The most common postfertilization barrier
is the abortion of the hybrid embryo at different developmental stages. In
many of the crosses, this is a result of the lack of endosperm development or

of its early breakdown. An understanding of the details of barriers at different levels is important for the application of effective techniques to overcome such barriers.

Introduction

Hybridization is one of the most effective methods of crop improvement programs. Most of the hybridization work carried out so far has used genetic variability within the species, and thus crossability barriers were not the main constraints in breeding programs. However, extensive breeding for yield and genetic uniformity over the years has led to reduced genetic variability and consequent loss of many useful genes in crop species. In recent years, the breeding programs have been increasingly dependent on interspecific and intergeneric crosses, often covering wild and weedy species. Wide hybridization refers to the crosses made between distantly related species or genera. Crossability barriers are very frequent in interspecific and intergeneric hybridization programs. This chapter describes such barriers operating at different levels, with particular reference to prefertilization barriers. An understanding of the nature of these barriers is essential for the application of effective techniques to overcome such barriers.

Classification of barriers

Crossability barriers prevent the fusion of male and female gametes originating from individuals of different species/genera and/or the development of a fertilized ovule into viable seed. The barriers also include those that limit effective utilization of the hybrids for gene introgression. The first comprehensive discussion on crossability barriers was presented by Stebbins (1958). Since then, the subject has been reviewed by different workers (Maheshwari and Rangaswamy 1965; de Nettancourt 1977; Hadley and Openshaw 1980; Shivanna and Johri 1985; Raghavan 1986; Khush and Brar 1992).

The crossability barriers may be grouped into three major categories (Table 12.1). The first is physical and the other two are physiological barriers.

I. Temporal and spatial isolation of parental species

Nonsynchronous flowering of the parental species is quite common, particularly in wide hybridization programs. Unless an effective method, that is, early or staggered sowing or suitable photoperiodic treatment, is available to achieve synchronous flowering of the parents, nonsynchronous flowering is a major barrier. Similarly, hybridization between parental species growing in different geographical areas is sometimes necessary (see Hadley and Openshaw 1980). When species that normally are geographically isolated from one another are grown together, their flowering response may be

Table 12.1. *Major interspecific crossability barriers*

I. Temporal and spatial isolation of species — *physical*
II. Prefertilization barriers
1. On the surface of the stigma before pollen tube entry
2. Inside the tissues of the stigma and style
3. Inside the ovary and embryo sac   *physiological.*
III. Postfertilization barriers
1. Nonviability of hybrid embryos
2. Failure of hybrid to flower
3. Hybrid sterility
4. Lack of recombination
5. Hybrid breakdown in F_2 or later generations

affected, resulting in nonsynchronous flowering, particularly when the parental species grow in different ecological conditions. In tree species, this approach is not convenient as the plants raised from seeds could take many years to flower.

II. Prefertilization barriers

Fertilization is the result of the successful completion of a series of sequential events following pollination (see Chapter 2). Prefertilization barriers prevent fertilization by arresting postpollination events at one or many levels. These barriers can be conveniently described under three categories, depending on the level of the barrier in the pistil (Table 12.1). Cytological details of prefertilization barriers have been investigated in only a limited number of crosses. Until the 1960s, this was probably due to the lack of suitable techniques to study pollen germination and pollen tube growth in the pistil. The methods of clearing and/or staining used by early investigators to study postpollination events in the pistil (see Shivanna and Johri 1985) were also not very effective.

The aniline blue fluorescence method developed to study pollen tube growth during the 1950s (Linskens and Esser 1957; Martin 1959; Shivanna and Rangaswamy 1992) is the most effective technique. The method is based on the affinity of the fluorochrome, water-soluble aniline blue, for callose. Because pollen tubes invariably contain callose deposition along the wall, they can be readily observed under the fluorescence microscope following staining with aniline blue.

Unilateral incompatibility

In many interspecific and intergeneric crosses, incompatibility operates in one direction, whereas the reciprocal cross is successful. Such incompatibility is generally referred to as unilateral incompatibility (UI). Most of the authors restrict the term "UI" to the barriers operating before fertilization (Dhaliwal 1992). UI has been reported in a large number of crosses (Lewis and Crowe 1958; Pandey 1968; Abdella 1974; de Nettancourt 1977; Dhaliwal 1992; Harder et al. 1993; Sorensson and Brewbaker 1994). UI is more common when the cross includes a self-compatible (SC) and a self-incompatible (SI) species. The crosses, in general, show incompatibility when an SI species is used as the female parent (SI x SC). However, in many crosses UI also operates in other combinations (SC x SI, SC x SC, and SI x SI). Many recent studies have reported UI in crosses between wild and cultivated species. Irrespective of the presence or absence of SI in one of the species, the pistil of the cultivated species generally inhibits pollen of the wild species, whereas the pistil of the latter permits good pollen germination and satisfactory tube growth of the cultivated species (Batra et al. 1990; Gundimeda et al. 1992; Nanda Kumar and Shivanna 1993).

Two hypotheses have been put forward to explain UI. According to Lewis and Crowe (1958), the SI allele has a dual function; in addition to inhibiting the pollen carrying an identical SI allele, it also inhibits pollen carrying the self-compatible (SC) allele. Lewis and Crowe (1958) consider self-compatible species to have evolved from self-incompatible species through mutation of the SI allele in stages: SI → Sc → Sc' → SC. They explained the growth of pollen tubes of self-compatible species through the style of self-incompatible species in some crosses on the basis of the fact the pollen has an Sc allele and has not yet stabilized to an SC allele. Studies of Pandey (1969, 1973) and de Nettancourt (1977) further supported the dual function of the S-gene. Often, S-gene–mediated response of UI may be modified by other genes (Pandey 1968; de Nettancourt 1977).

According to the second hypothesis, SI genes do not play a role in UI (Grun and Aubertin 1966; Abdella and Hermsen 1972). The inhibition of pollen carrying the SC allele is brought about by specific genes not belonging to the S-locus. There are hardly any physiological studies on UI. A few cytological studies available do not shed much light on the genetics of UI. In many crosses, UI has been reported to be stronger than self-incompatibility. The extent of pollen tube growth following UI is much less than that following SI (Pandey 1964; Grun and Aubertin 1966; Hogenboom 1973).

Active versus passive inhibition

Self-incompatibility inhibition is the result of active recognition of the pollen. Self-pollen is positively recognized as a result of the interaction of S-allele products in the pollen and the pistil; positive recognition results in the activation of metabolic processes in the pollen and/or the pistil to bring about pollen inhibition (see Chapter 9).

Many investigators have also implicated active recognition mediated through the S-allele in interspecific incompatibility. As mentioned in the previous section, the SI gene has been implicated in UI. Pandey (1968,

1969) suggested that the *S*-gene imparts two types of specificity: (a) primary specificity, which controls interspecific incompatibility, and (b) secondary specificity, which controls self-incompatibility. According to de Nettancourt et al. (1974), the loci inhibiting pollen tube growth in the cross *Lycopersicon peruvianum* × *L. esculentum* are either closely linked to, or allelic to, the *S*-locus.

According to Sampson (1962), pollen–pistil interaction depends upon the complementation of pollen and stigmatic molecules. There are two areas in which complementation can occur: the *S*-allele area and the species area. Complementation in the *S*-allele area results in self-incompatibility due to *S*-allele identity. Interspecific compatibility results when there is complementation at the species area but not in the *S*-allele area. Thus, complementation at both the areas or none of the areas results in incompatibility (see also Heslop-Harrison 1975).

Most of the available evidence indicates that in interspecific crosses between closely related species involving one or both SI parents, pollen inhibition seems to be the result of active recognition of the pollen mediated through *S*-gene action (Shivanna and Johri 1985). This is well illustrated in heteromorphic species of *Linum* (Ghosh and Shivanna 1984), which exhibit dimorphic self-incompatibility (see Chapter 9). Intramorph pollinations (pin × pin and thrum × thrum) both within, as well as between, species are incompatible, whereas intermorph pollinations (pin × thrum, thrum × pin) are compatible even between species. Thus, intramorph incompatibility and intermorph compatibility transgress species limits. Obviously, pollen inhibition in such systems is mediated by the *S*-gene, and the mechanism of inhibition is similar to that of self-incompatibility.

In a majority of interspecific crosses, however, the arrest of postpollination events seems to be passive (not as a result of active recognition of pollen) and a result of lack of co-adaptation between the pollen and the pistil. It is like a "lock and key" mechanism; absence of suitable key(s) with the pollen for the lock(s) present in the pistil results in incompatibility. Such passive inhibition resulting from lack of genetic information in one partner for some relevant character of the other partner has been termed "incongruity" (Hogenboom 1973, 1984). The extent of completion of postpollination events depends on the degree of reproductive isolation of the parental species; the greater the isolation, the earlier would be the barrier. Figure 12.1 identifies the stages of passive inhibition and probable mechanisms.

Inhibition on the stigma surface

Prefertilization barriers on the stigma surface result in the arrest of pollen germination or pollen tube entry into the stigma. This is one of the frequent barriers, particularly in distantly related species (Röbbelen 1960; Martin 1970; Knox et al. 1976; Stettler et al. 1980; Batra et al. 1990, Barone et al. 1992; Gundimeda et al. 1992). The causative factors for the failure of pollen germination are (a) lack of effective adhesion, (b) lack of full hydration, and (3) absence of pollen germination factors on the stigma.

Pollen adhesion and hydration are prerequisites for germination (see Chapter 2). Pollen adhesion largely depends on the nature and extent of surface components of the pollen and the stigma. Pollen adhesion is not a constraint in species having wet stigma. In species with dry stigma, pollen adhesion may be a problem, particularly when pollen grains do not possess sufficient pollen coat materials (Ghosh and Shivanna 1984).

Failure of pollen germination may be the result of insufficient or uncontrolled hydration. Pollen hydration is the result of the transfer of water from the stigma to the pollen through an osmotic gradient. Insufficient hydration may result in crosses in which the osmotic potential of the pollen does not match that of the stigma. Because pollen grains of many species require gradual hydration for germination, the rapid hydration that occurs on a wet stigma covered with aqueous exudate may lead to the failure of pollen germination.

Even after effective adhesion and hydration, pollen grains require suitable conditions on the stigma for germination. Pollen of many species requires calcium and boron for germination, and it has been shown that the stigma provides these elements to the pollen (Bednarska 1991). In *Vitis vinifera,* boron-deficient stigmas fail to support pollen germination (Gartel 1974). Only those stigmas containing 2–5 ppm boron in the stigmatic secretion permit pollen germination. Further, pollen grains of many species require a specific pH for germination (Ganeshaiah and Shaanker 1988). The stigmatic exudates of different species often show marked variation in their pH. For example, different species of *Rosa hybrida* show a variation in pH from 5.0 to 9.0 (Gudin and Arene 1991). Such differences in the stigmatic exudate may act as an important barrier. Absence of any of these components or conditions on the stigma of the female parent may be a factor in the failure of pollen germination.

The next critical step in postpollination events is pollen tube entry into the stigma. The surface cuticle of the stigma is disrupted during the secretion of the exudate in a wet stigma, and thus pollen tube entry into such a stigma does not seem to be critical. In a dry stigma, however, the entry of the pollen tube into the stigma is an important barrier because pollen tubes have to digest the cuticle by activation of cutinases (Batra et al. 1990; Gundimeda et al. 1992). Cutinases present in the pollen require suitable activators present on the surface of the stigma. In many systems, removal of the stigma components prevents pollen tube entry even in compatible pollinations (Heslop-Harrison and Heslop-Harrison 1975, 1981; Knox et al. 1976). Thus, pollen tube entry into the stigma may be arrested due to the lack of cutinase activators on the stigma of the female parent.

Inhibition in the stigma and style

Failure of the pollen tube to reach the ovary is perhaps the most common interspecific prefertilization barrier. This is due to the arrest of pollen tubes in the

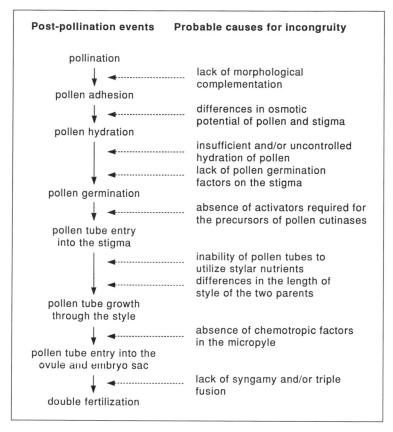

Post-pollination events	Probable causes for incongruity

pollination
↓ ◄······· lack of morphological complementation
pollen adhesion
↓ ◄······· differences in osmotic potential of pollen and stigma
pollen hydration
↓ ◄······· insufficient and/or uncontrolled hydration of pollen
↓ ◄······· lack of pollen germination factors on the stigma
pollen germination
↓ ◄······· absence of activators required for the precursors of pollen cutinases
pollen tube entry into the stigma
↓ ◄······· inability of pollen tubes to utilize stylar nutrients
↓ ◄······· differences in the length of style of the two parents
pollen tube growth through the style
↓ ◄······· absence of chemotropic factors in the micropyle
pollen tube entry into the ovule and embryo sac
↓ ◄······· lack of syngamy and/or triple fusion
double fertilization

Figure 12.1 Stages of pollen inhibition and probable mechanisms.

stigma, or just below the stigma, or further down the style. Such arrested pollen tubes often show abnormalities in the form of thicker tubes, excessive deposition of callose, swollen tips, and branching of tubes (Lewis and Crowe 1958; Röbbelen 1960; Hogenboom 1973; de Nettancourt 1977; Pundir and Singh 1985; Fritz and Hanneman 1989; Barone et al. 1992; Gundimeda et al. 1992).

It is now well established that growing pollen tubes utilize stylar nutrients (Labarka and Loewus 1973). One of the possible reasons for the arrest of pollen tube growth is an inability of the pollen tubes to utilize stylar nutrients. This may be due to lack of suitable nutrients in the transmitting tissue or lack of suitable enzymes in the pollen tubes.

There are many examples (such as *Nicotiana*, *Datura*, and *Lilium*) in which the length of the style in the two parental species is significantly different. The crosses in which short-styled parents are used as female parents are often successful, but the reciprocal crosses are unsuccessful (Avery et al. 1959; Maheshwari and Rangaswamy 1965; Gopinathan et al. 1986; Potts et al. 1987). In such unsuccessful crosses, the failure of pollen tubes of the short-styled parent to reach the ovary of the long-styled parent appears to be due to the intrinsic inability of pollen tubes to grow beyond the length of their own pistils.

Inhibition in the ovary

Unlike the inhibition of pollen tubes in the style, inhibition of pollen tubes in the ovary is not well investigated. However, there are many examples of the phenomenon, particularly in crosses involving Graminaceous species in which pollen tubes grow normally and reach the ovary. In the cross maize × sorghum, for example, no inherent barriers exist for the growth of pollen tubes through the style, but pollen tube entry into the ovule seems to be a problem (Heslop-Harrison et al. 1985). Because pollen tube entry into the ovule seems to be dependent on the presence of a chemotropic substance in the micropyle, failure of pollen tubes to enter the micropyle may be due to the absence of such a substance. Another abnormality reported in maize × sorghum cross is the hypertrophy and lysis of nucellar tissue after the entry of pollen tubes into the upper part of the ovary (Heslop-Harrison et al. 1985).

Even after the entry of pollen tubes into the embryo sac, there may be disturbances in double fertilization. In wheat × maize cross, 80% of the ovules showed development of the embryo but not of the endosperm; 8% of the ovules showed development of the endosperm but not of the embryo; and only 12% showed development of both the embryo and the endosperm (Laurie and Bennett 1988).

In multiovulate systems such as tomato, tobacco, and petunia, fertilization of a minimum number of ovules is needed to initiate fruit development. In many interspecific crosses, fertilization may not occur in the required number of ovules and hence fruit development is not initiated.

Often, prefertilization barriers are not restricted to a particular level but may be active at all levels. The proportion of pollen grains that complete sequential postpollination events is reduced at each level, with the result that very few or no pollen tubes reach the ovary/embryo sac (Fritz and Hanneman 1989). In the cross *Vigna unguiculata* × *V. vexillata* (Barone et al. 1992), for example, there was no pollen germination in 6.1% of pistils. Pollen tubes were inhibited on the stigma surface in 53% of the pistils, in the stigma tissue in 32.6% of the pistils, and pollen tubes reached the base of the style in only 8.2% of the pistils. Unlike the intraspecific compatible pollinations in which fertilization was observed in 76.6% of the ovules, in interspecific pollinations only 18.3% of the ovules were fertilized. Similarly, the cross *Brassica fruticulosa* × *B. campestris* showed germination of about 50% of the pollen grains. However, pollen tubes entered only 30% of the ovules, in contrast to 80% of the ovules' receiving pollen tubes in intraspecific pollinations (Nanda Kumar et al. 1988). Different accessions of the same species show considerable variation in the intensity of prefertilization barriers. In the cross *Vigna vexillata* × *V. unguiculata*, acc. TVnu 72 as the male parent resulted in fertilization of 18.6% of the ovules, whereas the acc. TVnu 73 resulted in 31.7% fertilized ovules (Barone et al. 1992). Similarly, out of over a dozen clones of *Tripsacum dactyloides* in crosses with maize, only one was found to be effective (Harlan and de Wet 1977; see also Snape et al. 1979; Zeven and Keijzer 1980).

III. Postfertilization barriers

Postfertilization barriers result in the failure of fertilized ovules to develop into mature seeds. Postfertilization barriers are more prevalent than prefertilization barriers in interspecific crosses. These barriers may operate at different stages of embryo development or during germination and subsequent growth of the F_1 hybrid. Many causative factors have been identified for the breakdown of embryo development at different stages. These include the presence of lethal genes, genic disharmony in the embryo, and failure or early breakdown of the endosperm (Khush and Brar 1992). Arrest of normal development of the embryo and endosperm may also result from unfavorable interactions between the embryo sac and the surrounding ovular tissues. Often, embryo breakdown is associated with the proliferation of the nucellar or integumentary cells (Cooper and Brink 1940). Detailed discussion on postfertilization barriers is beyond the scope of this chapter; for further information the reader may refer to Stebbins (1958), Maheshwari and Rangaswamy (1965), de Nettancourt (1977), Hadley and Openshaw (1980), Raghavan (1986), and Khush and Brar (1992).

Concluding remarks

Interspecific hybridization continues to be one of the important approaches for crop improvement. A range of techniques are now available to overcome the barriers operating at different levels. Rational application of these techniques for crop improvement, however, requires basic knowledge of the aspects of these barriers. Nevertheless, only limited information is available on the structural and functional aspects of the barriers. Most studies conducted so far are on the cytology of pollen inhibition. Structural details of postfertilization barriers are much less investigated. Further, there are hardly any genetic, physiological, or biochemical studies on crossability barriers. In the absence of such basic information, most of the techniques used to overcome the barriers so far have remained arbitrary and the results unpredictable. Comprehensive studies on the structural and functional details of the barriers would greatly facilitate the breeder's ability to apply the most effective techniques to overcome hybridization barriers.

References

Abdella, M.M.F. (1974). Unilateral incompatibility in plant species. *Egypt. J. Genet. Cytol.* 3:133–154.

Abdella, M.M.F. & Hermsen, J.G.Th. (1972). Unilateral incompatibility: hypothesis, debate and its implications for plant breeding. *Euphytica* 21:32–47.

Avery, A.G., Satina, S. & Reitsma, J. (1959). *Blakeslee: The Genus Datura*. New York: The Ronald Press Co.

Barone, A., del Giudice, A. & Ng, N.Q. (1992). Barriers to interspecific hybridization between *Vigna unguiculata* and *Vigna vexillata*. *Sex. Plant Repro.* 5:195–200.

Batra, V., Prakash, S. & Shivanna, K.R. (1990). Intergeneric hybridization between *Diplotaxis siifolia*, a wild species, and crop brassicas. *Theor. Appl. Genet.* 80:537–541.

Bednarska, E. (1991). Calcium uptake from the stigma by the germinating pollen in *Primula officinalis* L. and *Ruscus aculeatus* L. *Sex. Plant Reprod.* 4:36–38.

Cooper, D.V. & Brink, R.A. (1940). Somatoplastic sterility as a cause of seed failure after interspecific hybridization. *Genetics* 25:593–617.

Dhaliwal, H.S. (1992). Unilateral incompatibility. In *Distant Hybridization of Crop Plants*, eds. G. Kalloo and J.B. Chowdhury, pp. 32–46. Berlin, Heidelberg, New York: Springer-Verlag.

Fritz, N.K. & Hannaman, Jr., R.E. (1989). Interspecific incompatibility due to stylar barriers in tuber-bearing and closely related non-tuber-bearing Solanums. *Sex. Plant Reprod.* 2:184–192.

Ganeshaiah, K.N. & Shaanker, R.U. (1988). Regulation of seed number and female incitation of male competition by a pH dependent proteinacious inhibitor of pollen grain germination in *Leucaena leucocephala*. *Oecologia* 75:110–113.

Gartel, W. (1974). Micronutrients: their significance in vein nutrition with special regard to boron deficiency and toxicity. *Weinberg and Keller* 21: 435–508.

Ghosh, S. & Shivanna, K.R. (1984). Interspecific incompatibility in *Linum*. *Phytomorphology* 34:128–135.

Gopinathan, M.C., Babu, C.R. & Shivanna, K.R. (1986). Interspecific hybridization between rice bean (*Vigna umbellata*) and its wild relative (*Vigna minima*): fertility–sterility relationships. *Euphytica* 35:1017–1022.

Grun, P. & Aubertin, M. (1966). The inheritance and expression of unilateral incompatibility in Solanum. *Heredity* 21:131–138.

Gudin, S. & Arene, L. (1991). Influence of the pH of the stigmatic exudate on male female interaction in *Rosa hybrida* L. *Sex. Plant Reprod.* 4:110–112.

Gundimeda, H.R., Prakash, S. & Shivanna, K.R. (1992). Intergeneric hybrids between *Enarthocarpus lyratus*, a wild species, and crop brassicas. *Theor. Appl. Genet.* 83:655–662.

Hadley, H.H. & Openshaw, S.S. (1980). Interspecific and intergeneric hybridization. In *Hybridization in Crop Plants*, eds. R.W.Fehr and H.H.Hadley, pp 133–159.Madison, WI: American Soc. Agron-Crop Sci. Soc. America.

Harder, L.D., Cruzan, M.B. & Thomson, J.D. (1993). Unilateral incompatibility and the effects of interspecific pollination for *Erythronium americanum* and *Erythronium albidum* (Liliaceae). *Can. J. Bot.* 71:353–358.

Harlan, J.R. & de Wet, J.M.J. (1977). Pathways of genetic transfer from *Tripsacum* to *Zea mays*. *Proc. Natl. Acad. Sci. USA* 74:3494–3497.

Heslop-Harrison, J. (1975). Incompatibility and the pollen stigma interaction. *Ann. Rev. Plant Physiol.* 26:403–425.

Heslop-Harrison, J. & Heslop-Harrison, Y. (1975). Enzymic removal of proteinaceous pellicle of the stigma papillae prevents pollen tube entry in the Caryophyllaceae. *Ann. Bot.* 39:163–165.

Heslop-Harrison, J. & Heslop-Harrison, Y. (1981). Pollen–stigma interaction in the grasses. 2. Pollen tube penetration and the stigma response in *Secale*. *Acta Bot. Neerl.* 30:289–307.

Heslop-Harrison, Y., Reger, B.J. & Heslop-Harrison, J. (1985). Wide hybridization: pollination of *Zea mays* L. by *Sorghum bicolor* (L.) Moench. *Theor. Appl. Genet.* 70:252–258.

Hogenboom, N.G. (1973). A model for incongruity in intimate partner relationships. *Euphytica* 22:219–233.

Hogenboom, N.G. (1984). Incongruity: non-functioning of intercellular and intracellular partner relationship through non-matching information. In *Cellular Interactions,* eds. H.F. Linskens and J. Heslop-Harrison, pp. 640–654. Berlin, Heidelberg, New York: Springer-Verlag.

Khush, G.S. & Brar, D.S. (1992). Overcoming the barriers in hybridization. In *Distant Hybridization of Crop Plants*, eds. G. Kalloo and J.B. Chowdhury, pp. 47–61. Berlin, Heidelberg, New York: Springer-Verlag.

Knox, R.B., Clarke, A.E., Harrison, S., Smith, P. & Marchalonis, J.J. (1976). Cell recognition in plants: determinants of the stigma surface and their pollen interactions. *Proc. Natl. Acad. Sci. USA* 73:2788–2792.

Labarka, C. & Loewus, F. (1973). The nutritional role of pistil exudate in pollen tube wall formation in *Lilium longiflorum*. II. Production and utilization of exudate from stigma and stylar canal. *Plant Physiol.* 52:87–92.

Laurie, D.A. & Bennett, M.D. (1988). Cytological evidence for fertilization in hexaploid wheat x Sorghum crosses. *Plant Breed.* 100:73–82.

Lewis, D. & Crowe, L.K. (1958). Unilateral interspecific incompatibility in flowering plants. *Heredity* 12:233–256.

Linskens, H.F. & Esser, K. (1957). Uber eine specifische Anfarbung der Pollenschlauche in Griffel und die Zahl der Kallosepfropfen nach Selbstung und Fremdung. *Naturwissenschaften* 44:16.

Maheshwari, P. & Rangaswamy, N.S. (1965). Embryology in relation to physiology and genetics. In *Advances in Botanical Research Vol. 2,* ed. R.D. Preston, pp. 219–312. London, New York: Academic Press.

Martin, F.W. (1959). Staining and observing pollen tubes in the style by means of fluorescence. *Stain. Technol.* 34:125–128.

Martin, F.W. (1970). Pollen germination on foreign stigmas. *Bull. Torrey Bot. Club.* 97:1–6.

Nanda Kumar, P.B.A. & Shivanna, K.R. (1993). Intergeneric hybridization between *Diplotaxis siettiana* and crop brassicas for the production of alloplasmic lines. *Theor. Appl. Genet.* 85:770–776.

Nanda Kumar, P.B.A., Prakash, S. & Shivanna, K.R. (1988). Wide hybridization in Brassica: crossability barriers and studies on the hybrids and synthetic amphidiploids of *B. fruticulosa* x *B. campestris. Sex. Plant Reprod.* 1:234–239.

Nettancourt, D. de (1977). *Incompatibility in Angiosperms.* Berlin, Heidelberg, New York: Springer-Verlag.

Nettencourt, D. de, Devreux, M., Laneri, U., Cresti, M., Pacini, E. & Sarfatti, G. (1974). Genetical and ultrastructural aspects of self- and cross-incompatibility in interspecific hybrids between self-compatible *Lycopersicon esculentum* and self-incompatible *L. peruvianum. Theor. Appl. Genet.* 44:278–288.

Pandey, K.K. (1964). Elements of S-gene complex: the SFL alleles in *Nicotiana. Genet. Res.* 5: 397–409.

Pandey, K.K. (1968). Compatibility relationships in flowering plants: role of S-gene complex. *Amer. Nat.* 102:475–489.

Pandey, K.K. (1969). Elements of S-gene complex. V. Interspecific cross compatibility relationships and theory of the evolution of S-complex. *Genetica* 40:447–474.

Pandey, K.K. (1973). Phases in the S-gene expression and S-allele interaction in the control of interspecific incompatibility. *Heredity* 31:381–400.

Potts, B.M., Potts, W.C. & Cauvin, B. (1987). Inbreeding and interspecific hybridization in *Eucalyptus gunnii. Silvae Genetica* 30:194–199.

Pundir, R.P.S. & Singh, R.B. (1985). Crossability relationships among *Cajanus*, *Atylosia* and *Rhynchosia* species and detection of crossability barriers. *Euphytica* 34:303–308.

Raghavan, V. (1986). Variability through wide crosses and embryo rescue. In *Cell Culture and Somatic Cell Genetics of Plants*, ed. I.K. Vasil, pp. 613–63. New York, London: Academic Press.

Röbbelen, G. (1960). Uber die Kreuzungsunvertraglichkeit verschiedener *Brassica*: Arten als Folge eines gehemmten Pollenschlauchwachstums. *Zuchter*. 30:300–312.

Sampson, D.R. (1962). Intergeneric pollen–stigma incompatibility in Crucifereae. *Can. J. Genet. Cytol.* 4: 38–49.

Shivanna, K.R. & Johri, B.M. (1985). *The Angiosperm Pollen: Structure and Function*. New Delhi: Wiley Eastern.

Shivanna, K.R. & Rangaswamy, N.S. (1992). *Pollen Biology: A Laboratory Manual*. Berlin, Heidelberg, New York: Springer-Verlag.

Snape, J.W., Chapman, V., Moss, J., Blanchard, C.E. & Miller, T.E. (1979). The crossability of wheat varieties with *Hordeum*. *Heredity* 42:291–298.

Sorensson, C.T. & Brewbaker, J.L. (1994). Interspecific compatibility among 15 *Leucaena* species (Leguminoseae: Mimosoideae) via artificial hybridization. *Amer. J. Bot.* 81:240–247.

Stebbins, G.L. (1958). The inviability, weakness, and sterility of interspecific hybrids. *Adv. Genet.* 9:147–215.

Stettler, R.F., Koster, R. & Steenackens, V. (1980). Interspecific crossability studies in poplars. *Theor. Appl. Genet.* 58:273–282.

Zeven, A.C. & Keijzer, C.J. (1980). The effect of the number of chromosomes in the rye on its crossability with wheat. *Cereal Res. Commun.* 8:491–494.

13

Methods for overcoming interspecific crossing barriers

J. M. VAN TUYL and M. J. DE JEU

Summary 273
Introduction 274
Techniques for overcoming stigmatal and stylar barriers 275
    Genetic variation in interspecific crossability 275
    Use of mixed and mentor pollen 275
    Influence of environmental conditions 275
    Style and ovary manipulations 275
    Chemical treatments 277
Techniques for overcoming postfertilization barriers 279
    Ovary culture and ovary-slice culture 279
    Ovule culture 279
    Embryo culture 281
Integrated techniques for overcoming pre- and postfertilization barriers 282
Techniques for overcoming F_1 sterility 285
    Chromosome doubling 285
    Application of 2n-gametes 285
Concluding remarks 286
References 287

Summary

Crossing barriers occur frequently when intra- or interspecific crosses are attempted. These barriers are the result of incompatibility and incongruity. Sexual barriers preventing interspecific hybridization have been distinguished into pre- and postfertilization barriers. The nature of the barrier governs the method to be used to overcome the specific barrier. A range of techniques, such as bud pollination, stump pollination, use of mentor pollen, and grafting of the style, have been applied successfully to overcome prefertilization barriers. In vitro methods in the form of ovary, ovary-slice, ovule, and embryo culture are being used to overcome postfertilization barriers that cause endosperm failure and embryo abortion. An integrated method of in vitro pollination and fertilization followed by embryo rescue has been applied in many crosses. Vital hybrid plants may display lack of flowering or male and female sterility, resulting in failure of sexual reproduction. If sterility is caused by a lack of chromosome pairing during meiosis, fertility may be restored by polyploidization, enabling pairing of homologous chromosomes in the allopolyploid hybrid. Integration of these techniques into the breeding programs would enable the breeder to introgress genes across the species barriers.

Introduction

The phenomena underlying crossing barriers are incompatibility and incongruity. Incompatibility operates in intraspecific crosses and is the result of the activity of S-alleles. Incongruity occurs in interspecific crosses as a result of lack of genetic information in one partner necessary to complete pre- and postpollination processes in the other (Hogenboom 1973). This chapter focuses on different methods used to overcome incongruity.

Interspecific and intergeneric crosses are made to introduce new genetic variation into cultivated plants. In breeding ornamental crops, interspecific hybridization is the most important source of genetic variation. Many of the cultivars have originated from complex species crosses that have given rise to a broad range of shapes and colors to plants and flowers (Ohri and Khoshoo 1983; Van Eijk et al. 1991; Ramanna 1992; Van Creij et al. 1993). Many examples in this chapter are from research on ornamental plants, especially *Lilium*, because lily is a model plant for style manipulations and in vitro culture methods. For an overview of the entire field of reproductive barriers, the reader may refer to Chapter 12, and to reviews by Frankel and Galun (1977), Raghavan and Srivastava (1982), Shivanna (1982), Sastri (1985), Williams et al. (1987), Khush and Brar (1992), Liedl and Anderson (1993), and Pickersgill (1993).

The sexual barriers hampering interspecific hybridization have been distinguished into pre- and postfertilization barriers (Stebbins 1958). Many studies deal with methods for overcoming prefertilization barriers. Once fertilization has occurred, hybrid embryo growth may be affected by postfertilization barriers. Both embryo and endosperm have to develop in equilibrium for sharing nutrients in an undisturbed developmental process. In general, the first division of the zygote is delayed to favor the initial development of the endosperm. When the equilibrium in the development of the zygote and endosperm is disturbed, an abortion of the young embryo or disintegration of the endosperm follows. This abortion can take place at various stages of development of the young seed. Depending on the stage of embryo abortion, various in vitro techniques can be applied to rescue the abortive embryo.

Crossability barriers imposed by temporal and spatial isolation of the parents (see Chapter 12) can be effectively overcome through pollen storage. Because pollen storage is discussed in detail in Chapter 14, it is not covered here.

Techniques for overcoming stigmatal and stylar barriers

Genetic variation in interspecific crossability

The statement that "two species are not crossable, is controversial unless a broad genetic variation of the parental species has been used and the cross combinations have been carried out on a large scale under a wide range of environmental conditions." This quotation of Hermsen (1984a, p. 468) implies that crossability is determined by both genetic and environmental factors. It is therefore necessary to test different accessions of both the parents for hybridization programs.

Unilateral incongruity is the phenomenon in which a cross is successful only in one direction, whereas the reciprocal cross fails. In lily, crossing barriers can be overcome using cut-style pollination, but mostly in one direction (Van Creij et al. 1993). In interspecific crosses between *Sorghum* and maize, Laurie and Bennett (1989) found the variation of pollen tube inhibition to be due to a single gene.

Use of mixed and mentor pollen

The use of mixed pollen (a mixture of compatible and incompatible pollen) (Brown and Adiwilaga 1991) and of mentor pollen (compatible pollen genetically inactivated by irradiation but still capable of germination) mixed with incompatible pollen is reported to overcome inhibition on the stigma and in the style in many plant species (see Chapter 15). For lily, mentor pollen was effective in overcoming self-incompatibility but not in interspecific crosses (Van Tuyl et al. 1982).

Influence of environmental conditions

The presence of an optimal level of receptivity of the stigma can vary from several hours (in mango) to more than one week (in lily). It determines the optimal time of pollination.

A positive effect of high temperature in overcoming self-incompatibility and incongruity has been detected and has been applied in breeding programs of lily by heating the style (Ascher and Peloquin 1968) or by pollinating at high temperatures (Van Tuyl et al. 1982; Okazaki and Murakami 1992). Heat-sensitive inhibitors of pollen tube growth are probably inactivated at high temperature. Comparable effects of floral aging on pollen tube growth are reported by Ascher and Peloquin (1966).

Style and ovary manipulations

As it was first demonstrated more than 50 years ago in *Datura* (Blakeslee 1945), pollen tube growth inhibition in the style can be overcome using different pollination techniques in which style and ovary are manipulated

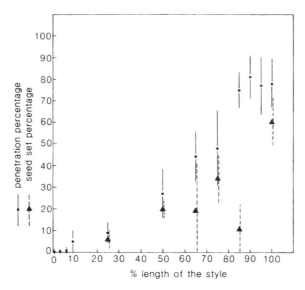

Figure 13.1 The percentage of ovules with a pollen tube in the micropyle (● = experiment with 69 flowers, solid lines represent S.D.), and the percentage of seeds with an embryo (▲ = experiment with 30 flowers, dotted lines represent S.D.) achieved after cut-style pollination in *Lilium longiflorum* with compatible pollen and at different style lengths: stigma present only in 100% (uncut).

(*Fritillaria*: Wietsma et al. 1994; *Lathyrus*: Davies 1957; Herrick et al. 1993; *Lilium*: Myodo 1963; Van Tuyl et al. 1988, 1991; Janson et al. 1993; *Nicotiana*: Swaminathan and Murty 1957). One of these manipulations involves removal of the stigma and a part or whole of the style and pollinating the cut end. This is referred to as "stump pollination," or "cut-style," "intrastylar," or "amputated-style" pollination.

In lily, two types of incongruity are distinguished, based on the arrest of pollen tube growth at different levels of the style. In the first type, the site of arrest of pollen tube growth is just below the stigma. This "upper inhibition" occurs 12–24 hours after pollination and results in short pollen tubes. In the other type, pollen tube growth ceases halfway in the style 3–4 days after pollination. This "lower inhibition" results in medium-sized pollen tubes (Asano 1980b, 1985). In a comparison of several pollination methods, it was shown that prefertilization barriers in lily can be circumvented by using the cut-style technique (Van Tuyl et al. 1991; Janson et al. 1993). Following stump pollination, many pollen tubes of lily, *Lathyrus,* and *Fritillaria,* for example, grow normally into the ovary. In this way, pollen circumvents stylar and stigmatal barriers that inhibit pollen tube growth. However, a complication associated with this method in lily is the low seed set, probably caused by the premature arrival of pollen tubes in the ovary (Janson et al. 1993). A majority of pollen tubes either grow past the inner integument or grow along, but not into, the

Table 13.1. *Effect of different pollination techniques on the number of seeds.*
A comparison between three types of pollinations: GSM (grafted-style
method), CSM (cut-style method), and N (normal pollination) in a compatible
cross between Lilium longiflorum *"Gelria" and* L. longiflorum *"Albivetta."*

Pollination technique	Number of ovaries pollinated	Number of fruits developed	Total number of seeds	Seeds per fruit
GSM exudate	26	1	15	15.0
GSM straw exudate	26	7	96	13.7
GSM straw medium	26	5	36	7.2
CSM	25	19	85	4.5
N	25	19	451	23.7

micropyle after cut-style pollination. Activation of the ovary by stigmatic pollination preceding intrastylar pollination did not result in an increase in the percentage of ovule penetration. The percentage of ovule penetration after cut-style pollination did increase when a longer part of the style was retained on the ovary (Figure 13.1). Despite the low seed set, a large number of unique interspecific lily hybrids were obtained using this method (Asano and Myodo 1977; Asano 1980a; Okazaki et al. 1992; Van Creij et al. 1993). Also, in crosses between *Fritillaria imperialis* and *F. ruddeana*, Wietsma et al. (1994) were able to obtain interspecific hybrids using the cut-style technique.

The stylar graft technique was applied successfully to improve the results of the cut-style technique (Van Tuyl et al. 1991). In this method, pollen grains are deposited on a compatible stigma. After one day, the style of the pollen donor is cut 1–2 mm above the ovary and grafted onto the ovary of another plant. Style and stigma are joined in vivo using a piece of a straw filled with *L. longiflorum* stigmatic exudate or are stuck together with only the exudate. In vitro, a piece of "water agar" was placed on the style (Figure 13.2a, b). Table 13.1 shows the effect of different pollination methods on average seed set. It can be observed that cut-style pollinations result in a low seed set, whereas the grafted-style method gives a better seed set but is not successful in a high percentage of the pollinated ovaries.

Chemical treatments

Application of growth regulators, such as auxins, cytokinins, and gibberellins, to the pedicel or the ovary at the time of, or soon after, pollination may improve fruit and seed set after interspecific pollination (Emsweller and Stuart 1948; Dionne 1958; Al Yasiri and Coyne 1964; Pittarelli and Stavely 1975). Application of growth regulators to delay abscission of the style shows positive effects on the development of young fruits (Larter and Chaubey 1965; Kruse 1974; Islam et al. 1975; Fedak 1978; Alonso and Kimber 1980; Mujeeb-Kazi

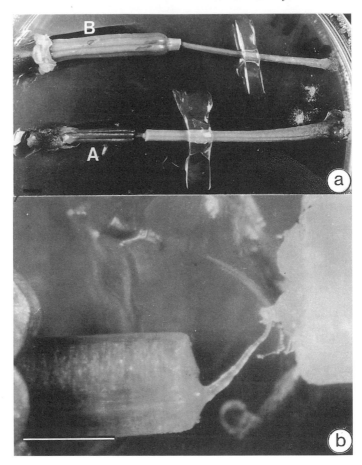

Figure 13.2 **(a)** GSM (grafted-style method of pollination). (A) Style of Oriental hybrid "Star Gazer" compatibly pollinated with pollen of another Oriental hybrid and attached to an ovary of the Asiatic hybrid "Esther." (B) Style of "Esther," compatibly pollinated with pollen of "Connecticut King" and attached to an ovary of the Oriental hybrid "Star Gazer" Bar = 1 cm. **(b)** Details of the graft in Figure 13.2a. Both style ends are slightly separated to show the bundle of pollen tubes growing through the graft. Bar = 0.1 cm.

1981). In many crosses, application of growth substances promotes postpollination development up to a stage when hybrid embryos can be excised and cultured (Islam 1964; Subrahmanyam 1979; Sastri et al. 1981, 1983). In interspecific crosses of *Populus*, treatment of the stigma with organic solvents such as hexane and ethyl acetate before pollination has been reported to be effective in overcoming prefertilization barriers (Willing and Pryor 1976).

Immunosuppressors such as amino-n-caproic acid, salicylic acid, and acriflavin have been used to produce wide hybrids in many cereals (Baker et al. 1975; Tiara and Larter 1977; Bates et al. 1979; Mujeeb-Kazi and Rodriguez

1980; Mujeeb-Kazi 1981) and legumes (Baker et al. 1975; Chen et al. 1978). The female parent is treated with immunosuppressors before and/or after pollination for many days, and the resulting embryo is cultured on a suitable medium. It has been suggested (Bates and Deyoe 1973) that crossability barriers between distantly related taxa are mediated through a specific inhibition reaction analogous to immunochemical mechanism of animals; treatment with immunosuppressors inactivates these immunochemical reactions. However, there is so far no evidence to indicate the involvement of immunochemical reactions in interspecific crosses.

Techniques for overcoming postfertilization barriers

A range of in vitro methods have been developed to overcome postfertilization barriers in a number of plant species. When abortion occurs at a very early stage and maternal tissue has no negative influence on the development of seeds, ovary culture can be applied: Young fruits can be grown in vitro to a stage at which dissection of embryos is possible. In some crops, the ovary is large, and slicing the ovary into small parts and then culturing them is a better option for rescuing hybrids in vitro. This technique is referred to as "ovary-slice culture." When the mismatch between embryo and endosperm development starts very early and ovary culture and/or ovary-slice culture fails, ovules can be dissected out of the ovaries and cultured in vitro. Since Hännig (1904) employed embryo culture for the first time 90 years ago, these techniques have been applied in numerous crops (Williams et al. 1987).

Ovary culture and ovary-slice culture

Ovary culture has been applied in many species: *Brassica* (Inomata 1980; Kerlan et al. 1992; Gundimeda et al. 1992; Sarla and Raut 1988; Takeshita et al. 1980), *Eruca-Brassica* hybrids (Agnihotri et al. 1990), *Lilium, Nerine,* and *Tulipa* (Van Tuyl et al. 1993), and *Phaseolus* (Sabja et al. 1990). Ovary-slice culture was applied by Kanoh et al. (1988), Straathof et al. (1987), and Van Tuyl et al. (1991) for the production of interspecific *Lilium* hybrids. Ovaries were harvested 7–40 days after intrastylar pollination and, after surface sterilizing, sliced into 2-mm-thick disks. Seed germination occurred 30–150 days after pollination (DAP). By this method, plantlets were obtained from very small embryos.

Ovule culture

In those crops in which the fruit is aborted before embryo culture can be applied, ovule culture is an easy and fast method. This technique is applied in *Alstroemeria* (Bridgen et al. 1989), *Cyclamen* (Ishizaka and Uematsu 1992), *Lycopersicon* (Neal and Topoleski 1983), *Nicotiana* (Iwai et al. 1986), and *Vitis* (Gray et al. 1990). In *Alstroemeria,* for example, fertilization occurs 24

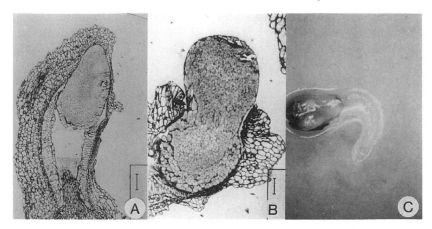

Figure 13.3 (**A**) Ovule culture in *Alstroemeria* 35 DAP: Transverse section showing embryo development without the development of cellular endosperm in an incongruous cross between a tetraploid cultivar and *A. aurea*. Bar = 400 µm. (**B**) Ovule culture in *Alstroemeria* 42 DAP: The embryo emerges out of the ovule; histological investigation in an incongruous cross between a tetraploid cultivar and *A. aurea*. Bar = 200 µm. (**C**) Ovule culture in *Alstroemeria* 42 DAP: Germination in vitro of an ovule in an incongruous cross between *A. pelegrina* × *A. aurea*.

hours after pollination (De Jeu et al. 1992). Ovaries are harvested 2 DAP, the ovules are dissected and placed on a Murashige and Skoog (MS) medium containing 9% sucrose. Six weeks later the ovules are transferred to an MS medium with 4%–5% sucrose. Germination of the ovules starts 1–2 weeks later (De Jeu et al. 1992). Histological studies of the in vitro cultured ovules revealed that the ovules within the first two weeks of culture enlarged to twice their size due to proliferation of the inner and outer integuments. No development of cellular endosperm in vitro was found, whereas the embryo cells divided almost normally (Figure 13.3A). By 42 DAP, the first plantlets arose from the cultured ovules (Figure 13.3B, C), whereas normal seed development in *Alstroemeria* takes 2½–3 months (De Jeu et al. 1992). Depending on the genotypic combination of the interspecific crossing, the percentage of seedlings obtained from ovule culture varied from 0.5% to 22.5%, whereas in the in vivo situation on the plant, no seeds could be harvested.

Monnier and Lagriffol (1985) compared the growth of *Capsella* embryos in cultured ovules with the growth of cultured embryos and with embryos grown in vivo. They concluded that embryos grown in vivo were more differentiated than those raised from in vitro culture. Embryos from cultured ovules showed better growth and survival than cultures of isolated embryos.

Interspecific hybrids between *Nicotiana tabacum* × *Nicotiana acuminata* were obtained through ovule culture in a liquid Nitsch H medium (Nitsch and Nitsch 1969) after culturing the ovules 2–10 DAP (Iwai et al. 1986). Ovules from 4–6 DAP onward gave a good germination rate; younger ovules died. In *Zea mays*, Campenot et al. (1992) isolated embryo sacs (from ovules 1 DAP) that were partially surrounded by nucellar tissue and contained the zygote and a few endosperm nuclei, and cultured them in vitro. They achieved germination of the embryos from 7 DAP onward using basic media. Modification of the basic medium by the addition of the growth regulator BAP (6-benzylaminopurine) resulted in germination of 1 DAP zygotes. Through this method, a zygote resulting from an in vitro fertilization as described by Kranz (Chapter 19) could be developed into a mature plant (Campenot et al. 1992).

Embryo culture

Embryo culture can be applied successfully in crosses in which pollinated flowers stay on the plant for a considerable time before natural abscission occurs. This method has been applied in a large number of crops (see Raghavan 1986 for details). Some of the recent examples are *Allium* (Nomura and Oosawa 1990), *Alstroemeria* (Buitendijk et al. 1992), *Freesia* (Reiser and Ziessler 1989), *Howea* (Moura and Carneiro 1992), *Lilium* (Van Tuyl et al. 1991), *Lycopersicon* (Imanishi 1988), and *Solanum* (Singsit and Hanneman 1991).

To overcome the problems involved in isolating the young embryos from ovules and providing suitable conditions for their growth, embryo culture has been modified in some systems (Harberd 1969; Buitendijk et al. 1992). The ovule is cut in half and the cut halves, or only the halves containing the embryo, are cultured in a liquid medium. Out of these half ovules, germinating embryos emerged, which could be raised to plantlets on a solid medium.

A large number of interspecific and intergeneric hybrids have been produced in *Brassica* through sequential culture of ovary, ovules, and often embryos (Nanda Kumar et al. 1988; Agnihotri et al. 1990; Gundimeda et al. 1992; Nanda Kumar and Shivanna 1993; Vyas et al. 1995). In sequential culture, ovaries are initially cultured for 6–10 days. Enlarged ovules are excised from cultured ovaries and recultured on a fresh medium. Hybrids are realized either directly from cultured ovules or after excising and culturing the embryos. Przywara et al. (1989) followed sequential culturing of ovules and embryos to raise hybrids between *Trifolium repens* and *T. hybridum*.

In interspecific hybridization of *Phaseolus vulgaris* x *P. acutifolius* (Sabja et al. 1990), pods were collected from 10 DAP and cultured upright in a modified liquid (MS) medium (Murashige and Skoog 1962) supported by glasswool. In the in vivo situation, the hybrid embryos ceased growth by 21 DAP and soon thereafter the pods abscised. The ovary culture resulted in the development of small and weak embryos; 90% of them survived when they were excised and cultured. Only the ovaries 14 DAP and older gave viable hybrid plants (Sabja et al. 1990). In most cases, ovary culture is applied during the growth of a small, undifferentiated embryo inside the ovule, and the differentiated embryo is taken out and cultured.

In many of the distant crosses, the number of hybrids realized is rather limited. However, hybrids can be multiplied through in vitro culture techniques (Agnihotri et al. 1990; Nanda Kumar and Shivanna 1991; Chen and

Adachi 1992). In some hybrids, embryonal callus and subsequent differentiation of hybrid plantlets have been achieved. In others, culture of single node segments or shoot tips have been used to propagate the hybrid. The use of embryonal callus for multiplication of hybrids is also of importance for induction of variations in the hybrid and for experiments on genetic transformation.

Integrated techniques for overcoming pre- and postfertilization barriers

In many interspecific and intergeneric crosses, integrated techniques to manipulate both pre- and postfertilization barriers have been applied. In vitro pollination and fertilization is one such technique. Unlike the other techniques, which retain the zone of inhibition (stigma and style) and manipulate pollen germination and pollen tube growth to overcome prefertilization barriers, in vitro pollination brings pollen grains in direct contact with the ovules and is, therefore, considered more effective (see Rangaswamy 1977).

Kameya and Hinata (1970) obtained hybrids between *Brassica chinensis* and *B. pikenensis* through pollination of excised, cultured ovules (see also Kanta et al. 1962). This approach has not been successful in other systems, but a modified technique of in vitro pollination, termed placental pollination, has been successful. Placental pollination involves removal of the stigma, style, and ovary wall, pollination of the intact ovules on the placenta, and culturing of the ovule mass on a suitable medium (Rangaswamy and Shivanna 1967). In this method, ovules do not come in contact with the medium because only the pedicel is inserted into the medium. Keeping the pollen and ovules free from moisture is critical for achieving success in many systems. Using this technique, hybrids have been produced between *Melandrium album* and *M. rubrum*, *M. album* and *Silene schafta*, and *M. album* and *Viscaria vulgaris* (Zenkteler 1990), *Nicotiana tabacum* × *N. oesoebila* (Reed and Collins 1978), *Petunia parodii* × *P. inflata*, and *Zea mays* × *Z. mexicana* (Dhaliwal and King 1978). In some of the crosses between species of *Melandrium* and *Datura*, although fertilization was achieved through placental pollination, proembryos aborted.

In *Brassica* and related taxa, a part of the ovary wall was removed to expose the ovules, and pollen grains were deposited on the surface of the ovules; the pollinated ovary was implanted through the pedicel in the culture medium. Developing embryos were subsequently dissected from the enlarged ovules (15–21 DAP) and cultured. Through this integrated technique of in vitro pollination and embryo rescue many hybrids were produced: *B. napus* × *B. campestris* (Zenkteler et al. 1987), *B. napus* × *Diplotaxis tenuifolia*, *B. napus* × *Moricandia arvensis*, *B. oleracea* × *D. tenuifolia,* and *D. tenuifolia* × *B. napus* (Zenkteler 1990). In vitro pollination has also been successful in overcoming incongruity barriers between *Nicotiana* species (DeVerna et al. 1987).

In *Lilium,* various combinations of in vitro pollination (cut-style and grafted-style methods) and embryo rescue (ovary, ovule, and embryo culture, and placental pollination, Figures 13.4A, B) were applied in order to control

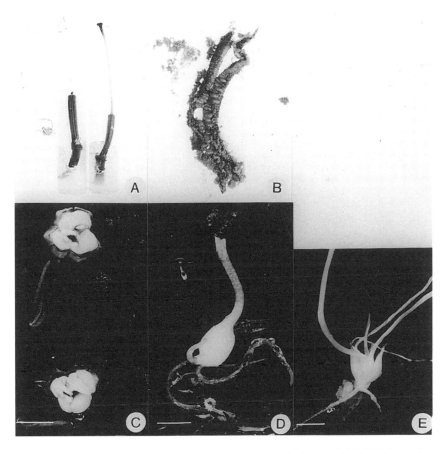

Figure 13.4 (A) In vitro pollinated ovaries from *L. longiflorum* "Gelria" 10 days after cut-style (left) and stigmatal pollination (right) with an Asiatic hybrid lily. **(B)** Placental pollination of *L. longiflorum* "Gelria" with the oriental hybrid "Star Gazer": pollen directly pollinated on placentas, some swollen ovules can be observed. **(C)** Asiatic hybrid lily "Enchantment," ovary slices, 4 weeks after compatible pollination, on MS medium supplemented with 9% sucrose and 1 mg/l NAA, in the presence of anthers. Note the swollen ovules inside the ovary slices. Bar = 1 cm. **(D)** Seedling of a directly germinated ovule of the Asiatic hybrid lily "Enchantment," 12 weeks after compatible pollination, on MS medium supplemented with 5% sucrose and 0.1 mg/l NAA. Bar = 0.5 cm. **(E)** Hybrid plantlet with bulb formation, from an ovule obtained after the interspecific cross *L. longiflorum* × *L. dauricum*, 40 weeks after pollination, ready to be transferred to soil. Bar = 1 cm.

the whole fertilization process (Van Tuyl et al. 1991; Janson 1993). This resulted in the realization of a range of new interspecific hybrids (Van Creij et al. 1993). In crosses between *L. longiflorum* cultivars, ovaries with complete pistil were cultured three days before anthesis and were pollinated with aseptic pollen when the stigma was receptive (Figure 13.4C). Mature seeds were harvested out of these in vitro pollinated cultured ovaries 60–90 DAP. In some

Table 13.2. *Comparison of the number of plantlets obtained per ovary from the incongruous combinations* Lilium longiflorum × L. henryi *and* L. longi-florum × L. dauricum *using three pollination methods (GSM = grafted-style method, CSM = cut-style method, and N = normal pollination). As a control, the intraspecific cross* L. longiflorum *cv "Gelria" × cv "White American" was studied both in the climate room (cl) and in the greenhouse (gr). Interspecific pollinations were carried out in the climate room; 8 DAP, ovaries were cultured; 42 DAP, ovules were excised and the swelling score was determined (on a scale from 1 to 9 in which 1= no swelling, 9 = very strong swelling). Subsequently, swollen ovules were cultured; plantlet development was scored from 10 to more than 40 weeks after the start of ovary culture (nd = not determined).*

L. longiflorum hybrids	Pollination method	Number of ovaries cultured	Swelling score of ovules	Number of ovules cultured	Plantlets per ovary
Intraspecific	N (gr)	41	nd	nd	143
Intraspecific	N (cl)	24	nd	nd	19
× L. henryi	N (cl)	10	4.8	2,206	0
× L. henryi	CSM (cl)	10	5.6	2,622	2.1
× L. henryi	GSM (cl)	10	3.2	1,195	0.1
× L. dauricum	N (cl)	13	3.9	3,475	0
× L. dauricum	CSM (cl)	14	5.1	4,568	3.5
× L. dauricum	GSM (cl)	14	2.7	2,446	4.8

crosses, ovaries were sliced 30–45 DAP and the embryos were dissected from the ovules for embryo culture. In other combinations between *Lilium* species, an ovary-slice culture in combination with ovule culture was applied. Ovary-slice culture was initiated 5–8 DAP, and the swollen ovules were excised 42 DAP and cultured on a medium until germination started (Figure 13.4D, E).

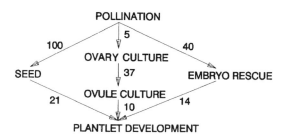

Figure 13.5 A diagram of alternative routes for application of in vitro culture methods in *Lilium*. Numbers represent minimal number of days in a compatible intraspecific situation (crosses between different *L. longiflorum* cultivars). Note that in the normal *in* vivo situation, seedlings are produced after 121 DAP, whereas in the embryo rescue and ovary–ovule route, plantlets emerge from 52 to 54 DAP.

The emerged plantlets were transferred to an embryo culture medium. Using this method, plantlets were obtained 52–54 DAP, whereas in the normal in vivo situation seedlings were not produced before 121 DAP (Table 13.2). Figure 13.5 shows a diagram of alternative routes for application of in vitro culture methods for *Lilium* (Van Tuyl et al. 1991). Similar results were obtained from interspecific crosses in *Tulipa* and intergeneric crosses between *Nerine* and *Amaryllis* (Van Tuyl et al. 1990, 1993). Recently, in vitro fertilization has been achieved using isolated sperms and eggs; this aspect is discussed in Chapter 19.

Techniques for overcoming F_1 sterility

Barriers occurring after a successful embryo rescue are hybrid breakdown and F_1 sterility. Hybrid breakdown results in the loss of the hybrid before flowering and is a result of the unbalanced new gene combinations. F_1 sterility of interspecific hybrids is very common, and may be the consequence of reduced chromosome pairing during meiosis. Breeding at polyploid levels is widely used in interspecific hybridization programs of many ornamental crops, such as *Alstroemeria*, *Chrysanthemum*, *Freesia*, *Gladiolus*, and *Lilium*.

Chromosome doubling

Interspecific F_1 hybrids may display sterility owing to lack of chromosome pairing during meiosis. This sterility hampers further breeding. Somatic (mitotic) chromosome doubling may induce homologous pairing of chromosomes and restores fertility (Hermsen 1984a, b). Colchicine has been used successfully to produce fertile allotetraploids in many crops, such as *Anigozanthos* (Griesbach 1990), *Arachis* (Singh 1985), *Brassica* (Nanda Kumar and Shivanna 1991), *Lilium* (Asano 1982; Van Tuyl 1993), *Phaseolus* (Weilenmann et al. 1986), and *Tagetes* (Bolz 1961). As an alternative for colchicine, oryzalin – a herbicide with antimitotic activity – was recently used successfully (Van Tuyl et al. 1992). Crossability between parents may be improved in the process of plant breeding by equalizing their functional ploidy level. Allopolyploids may function as fertile bridges for gene introgression into the cultivar assortment (Hermsen 1984a, b; Nanda Kumar and Shivanna 1993; Vyas et al. 1995).

Application of 2n-gametes

Application of meiotic polyploidization in interspecific breeding programs may be of great importance for the introgression of characters from diploids to tetraploids (Hermsen 1984a, b; Veilleux 1985; Ramanna 1992). In many species, polyploidization results in functional 2n-gametes in one or both parents. Such gametes may result from mechanisms of meiotic restitution. Normally, the frequency of 2n-gametes is low. Wide interspecific lily hybrids are usually completely male and female sterile. In rare cases, however, some fertile pollen is detected. In a group of more than 50 hybrids from the cross *L. longiflorum* × *L. candidum*, raised through embryo culture, only one hybrid showed pollen fertility of 25%. Meiosis in this hybrid was highly irregular and all pollen contained 2n-gametes (Van Tuyl 1989, 1993). Comparable cases were found in lily by backcrossing oriental hybrids with "Shikayama" × *L. henryi* and *L. auratum* × *L. henryi* (Asano 1982), resulting in triploid progenies. Backcrossing these triploids with *L. auratum* × *L. henryi* resulted in aneuploids with chromosome numbers ranging between 36 and 48 (Van Tuyl 1989, 1993). In contrast to the wide interspecific crosses, the cross between the Asiatic hybrid "Enchantment" and the related *Lilium pumilum* produced fertile pollen. Meiotic studies of several of these hybrids showed the formation of not only haploid pollen but also relatively high percentages of 2n-pollen (Van Tuyl et al. 1989).

In *Alstroemeria*, interspecific hybrids produce 2n-pollen of up to 60%–70% in some of the genotypes of crosses between Chilean and Brazilian species (Ramanna, personal communication). In this crop, sexual polyploidization appears to be a convenient system for breeders to produce triploid and tetraploid interspecific hybrids. For example, if a trispecific triploid generates 2n-gametes, then in one step of unilateral sexual polyploidization, a tetraploid with four different genomes can be produced (Ramanna 1992).

Concluding remarks

For many crops, especially for ornamentals, interspecific hybridization is a very effective method for the introduction of desired characters into breeding material and for achieving crop improvement (Uhlinger 1982). The possibilities for interspecific crosses are nevertheless limited, due to various crossing barriers. Many diverse techniques have been developed to overcome these barriers. A relatively underexplored aspect in this field of research is the use of genetic variation to overcome these barriers.

The application of in vitro pollination and other pollination methods combined with embryo rescue is a very powerful breeding approach. More controlled environmental conditions during the processes of pollination, fertilization, and embryo development would result in repeatability of experiments almost independently of the season. Conditions for each process can be optimized, and crossing barriers can be studied more systematically. In vitro methods make it possible to develop an integrated procedure for overcoming pre- and postfertilization barriers.

In the future, tissue-specific promoters along with RNase genes may enable breeders to eliminate the recognition barriers in the stigma and style by preventing the formation of these organs. In this way, Goldman et al. (1994) have developed a female-sterile tobacco plant by eliminating the stigmatic tissue. This stigma does not permit pollen germination, but when pollen, together with the stigmatic exudate, is used for pollination, pollen germination occurs and normal fertilization and seed set take place. By this method, the whole pollen tube pathway and thus the recognition and barrier functions could be eliminated, which opens a way for new applications. These techniques provide a very powerful tool to the plant breeders to transfer genes across species barriers.

Acknowledgments

We thank Drs. R. J. Bino, J. G. T. Hermsen, M. C. G. van Creij, H. M. C. van Holsteijn, and L. W. D. van Raamsdonk for critical reading of the manuscript.

References

Al Yasiri, A. & Coyne, D.F. (1964). Effects of growth regulators in delaying pod abscission and embryo abortion in the interspecific cross *Phaseolus vulgaris* × *P. acutifolius*. *Crop Sci.* 4:433–435.

Alonso, L.C. and Kimber, G. (1980). A haploid between *Agropyron junceum* and *Triticum aestivum*. *Cereal Res. Commun.* 8:355–358.

Agnihotri, A., Gupta, V., Lakshmikumaran, M.S., Shivanna, K.R., Prakash, S. & Jagannathan, V. (1990). Production of *Eruca-Brassica* hybrids by embryo rescue. *Plant Breeding* 104:281–289.

Asano, Y. (1980a). Studies on crosses between distantly related species of lilies. *J. Japan. Soc. Hort. Sci.* 49:114–118.

Asano, Y. (1980b). Studies on crosses between distantly related species of lilies. VI. Pollen-tube growth in interspecific crosses on *Lilium longiflorum*. *J. Japan. Soc. Hort. Sci.* 49(3):392-396.

Asano, Y. (1982). Overcoming interspecific hybrid sterility in *Lilium*. *J. Japan. Soc. Hort. Sci.* 51:75–81.

Asano, Y. (1985). Interspecific pollen-tube growth behavior and a model for the explanation in *Lilium*. *Plant Cell Incompatibility Newslett.* 17:4–7.

Asano, Y. & Myodo, H. (1977). Studies on crosses between distantly related species of lilies. I. For the intrastylar pollination technique. *J. Japan. Soc. Hort. Sci.* 46(1):59–65.

Ascher, P.D. & Peloquin, S.J. (1966). Effects of floral aging on the growth of compatible and incompatible pollen tubes in *Lilium longiflorum*. *Amer. J. Bot.* 53:99–102.

Ascher, P.D. & Peloquin, S.J. (1968). Pollen tube growth and incompatibility following intra- and interspecific pollinations in *Lilium longiflorum*. *Amer. J. Bot.* 55:1230–1234.

Baker, L.R., Chen, N.C. & Park, H.G. (1975). Effect of an immunosuppressant on an interspecific cross of the genus *Vigna*. *Hort. Sci.* 10:313.

Bates, L.S. & Deyoe, C.W. (1973). Wide hybridization and cereal improvement. *Econ. Bot.* 27:401–412.

Bates, L.S., Campos, V.A., Rodriguez, R.R. & Anderson, R.G. (1979). Progress towards novel cereal grains. *Cereal Sci. Today* 19:283–284, 286.

Blakeslee, A.F. (1945). Removing some of the barriers to crossability in plants. *Proc. Amer. Philosoph. Soc.* 89:561–574.

Bolz, G. (1961). Genetisch-züchterische Untersuchungen bei *Tagetes*. III. Artkreuzungen. *Z. Planzenzüchtg.* 46:169–211.

Bridgen, M.P., Langhans, R. & Graig, R. (1989). Biotechnological breeding techniques for *Alstroemeria*. *Herbertia* 45(1&2):93–96.

Brown, C.R. & Adiwilaga, K.D. (1991). Use of rescue pollination to make a complex interspecific cross in potato. *Amer. Potato J.* 68:813–820.

Buitendijk, J.H., Ramanna, M.S. & Jacobsen, E. (1992). Micropropagation ability: towards a selection criterion in *Alstroemeria* breeding. *Acta Hort.* 325:493–498.

Campenot, M.K., Zhang, G., Cutler, A.J. & Cass, D.D. (1992). *Zea mays* embryo sacs in culture. I. Plant regeneration from 1 day after pollination embryos. *Amer. J. Bot.* 79(12):1368–1373.

Chen, L. & Adachi, T. (1992). Embryo abortion and efficient rescue in interspecific hybrids, *Lycopersicon esculentum* and the "*peruvianum*-complex." *Japan. J. Breed*ing 42:65–77.

Chen, N.C., Parrot, J.F., Jacobs, T., Baker, L.R. & Carlson, P.S. (1978). Interspecific hybridization of food grain legumes by unconventional methods of breeding. In *International Mungbean Symposium,* pp. 247–252. Taiwan: AVRDC.

Davies, A.J.S. (1957). Successful crossing in the genus *Lathyrus* through stylar amputation. *Nature* 180:61.

De Jeu, M.J., Sasbrink, H., Garriga Calderé, F. & Piket, J. (1992). Sexual reproduction biology of *Alstroemeria*. *Acta Hort.* 325:571–575.

DeVerna, J.W., Myers, J.R. & Collins, G.B. (1987). Bypassing prefertilization barriers to hybridization in *Nicotiana* using *in vitro* pollination and fertilization. *Theor. Appl. Genet.* 73:665–671.

Dhaliwal, A.S. & King, P.J. (1978). Direct pollination of *Zea mays* ovules in vitro with *Z. mays, Z. mexicana* and *Sorghum bicolor* pollen. *Theor. Appl. Genet.* 53:43–46.

Dionne, L.A. (1958). A survey of methods for overcoming cross-incompatibility between certain species of the genus *Solanum*. *Amer. Potato J.* 35:422–423.

Emsweller, S.L. & Stuart, N.W. (1948). Use of growth regulating substances to overcome incompatibilities in *Lilium*. *Proc. Amer. Soc. Hort. Sci.* 51:581–589.

Fedak, G. (1978). Barley-wheat hybrids. In *Interspecific Hybridization in Plant Breeding,* eds. E. Sanchez-Monge and F. Garcia-Omedo.pp. 261–267. Proceedings of the VIIth Congress Eucarpia, Escuela Technica Superior de Inginieros Agronomos, Madrid.

Frankel, R. & Galun, E. (1977). *Pollination Mechanisms, Reproduction and Plant Breeding*. Berlin: Springer-Verlag.

Goldman, M.H., Goldberg, R.B., Mariani, C. (1994). Female sterile tobacco plants are produced by stigma specific cell ablations. *EMBO J.* 13:2976–2984.

Gray, D.J., Mortensen, J.A., Benton, C.M., Durham, R.E. & Moore, G.A. (1990). Ovule culture to obtain progeny from hybrid seedless bunch grapes. *J. Amer. Soc. Hort. Sci.* 115(6):1019–1024.

Griesbach, R.J. (1990). A fertile tetraploid *Anigozanthos* hybrid produced by *in vitro* colchicine treatment. *Hort Sci.* 25(7):802–803.

Gundimeda, H.R., Prakash, S. & Shivanna, K.R. (1992). Intergeneric hybrids between *Enathrocarpus lyratus*, a wild species, and crop brassicas. *Theor. Appl. Genet.* 83:655–662.

Hännig, E. (1904). Zur Physiologie pflanzenlicher Embryonen. I. Über die Kultur von *Cruciferen*-Embryonen außerhalb des Embryosacks. *Z. Bot.* 62:45–80.

Harberd, A. (1969). A simple effective embryo culture technique for *Brassica*. *Euphytica* 18:425–429.

Hermsen, J.G.T. (1984a). Nature, evolution, and breeding of polyploids. *IOWA State J. Res.* 58(4):411–420.

Hermsen, J.G.T. (1984b). Some fundamental considerations on interspecific hybridization. *IOWA State J. Res.* 58(4):461–474.

Herrick, J.F., Murray, B.G. & Hammett, K.R.W. (1993). Barriers preventing hybridization of *Lathyrus odoratus* with *L. chloranthus* and *L. chrysanthus*. *N. Zeal. J. Crop Hort. Sci.* 21(2):115–121.

Hogenboom, N.G. (1973). A model for incongruity in intimate partner relationships. *Euphytica* 22:219–233.

Imanishi, S. (1988). Efficient ovule culture for the hybridization of *Lycopersicon esculentum* and *L. peruvianum, L. glandulosum. Japan. J. Breeding* 38:1–9.

Inomata, N. (1980). Production of interspecific hybrids in *Brassica campestris* × *B. oleracea* by culture *in vitro* of excised ovaries. I. Development of excised ovaries in the crosses of various cultivars. *Japan. J. Genet.* 53:161–173.

Ishizaka, H. & Uematsu, J. (1992). Production of interspecific hybrids of *Cyclamen persicum* Mill. and *C. hederifolium* Aiton. by ovule culture. *Japan. J. Breeding* 42:353–366.

Islam, A.K.M.R., Shepherd, K.W., Sparrow, D.H.B. (1975). Addition of individual barley chromosomes to wheat. In *Barley Genetics III*, pp. 260–270.

Islam, A.S. (1964). A rare hybrid combination through application of hormone and embryo culture. *Nature* 201:320.

Iwai, S., Kishi, C., Nakata, K. & Kawashima, N. (1986). Production of *Nicotiana tabacum* × *Nicotiana acuminata* hybrid by ovule culture. *Plant Cell Reports* 5:403–404.

Janson, J. (1993). Placental pollination in *Lilium longiflorum* Thunb. *Plant Sci.* 99:105–115.

Janson, J., Reinders, M.C., Van Tuyl, J.M. & Keijzer, C.J. (1993). Pollen tube growth in *Lilium longiflorum* following different pollination techniques and flower manipulations. *Acta Botanica Neerlandica* 42(4):461–472.

Kameya, T. & Hinata, K. (1970). Test tube fertilization of excised ovules in *Brassica. Japan. J. Breeding* 20:253–260.

Kanoh, K., Hayashi, M., Serizawa, Y. & Konishi, T. (1988). Production of interspecific hybrids between *Lilium longiflorum* and *L.* × *elegance* by ovary slice culture. *Japan. J. Breeding* 38:278–282.

Kanta, K., Rangaswamy, N.S. & Maheshwari, P. (1962). Test-tube fertilization in a flowering plant. *Nature* 194:1214–1217.

Kerlan, M.C., Chèvre, A.M., Eber, F., Baranger, A. & Renard, M. (1992). Risk assessment of outcrossing of transgenic rapeseed to related species. I. Interspecific hybrid production under optimal conditions with emphasis on pollination and fertilization. *Euphytica* 62:145–153.

Khush, G.S. & Brar, D.S. (1992). Overcoming barriers in hybridization. In *Distant Hybridization of Crop Plants*, eds. G. Kalloo and J.B. Chowdhury, pp. 47–62. Berlin, Heidelberg, New York: Springer-Verlag.

Kruse, A. (1974). A 2,4-D treatment prior to pollination eliminates the haplontic (gametic) sterility in wide intergeneric crosses with 2-rowed barley, *Hordeum vulgare* subsp. *distichum* as maternal parent. *Hereditas* 78:319.

Larter, E. & Chaubey, C. (1965). Use of exogenous growth substances in promoting pollen tube growth and fertilization in barley–rye crosses. *Can. J. Genet. Cytol.* 7:511–518.

Laurie, D.A. & Bennett, M.D. (1989). Genetic variation in *Sorghum* for the inhibition of maize pollen tube growth. *Ann. Bot.* 64:675–681.

Liedl, B.E. & Anderson, N.O. (1993). Reproductive barriers: identification, uses, and circumvention. *Plant Breeding Rev.* 11:11–154.

Monnier, M. & Lagriffol, J. (1985). Development of embryos in *Capsella* ovules cultured *in vitro*. In *Experimental Manipulation of Ovule Tissues*, eds. G.P Chapman, S.H. Mantell, and R.W. Daniels, pp. 117–134. London, New York: Longman.

Moura, I. & Carneiro, M.F.N. (1992). *In vitro* culture of immature embryos of *Howea forsteriana* Becc. *Plant Cell, Tissue, Organ Cult.* 31:207–209.

Mujeeb-Kazi, A. (1981). *Triticum timopheevii* × *Secale cereale* crossability. *J. Hered.* 72:227–228.

Mujeeb-Kazi, A. & Rodriguez, R. (1980). Some intergeneric hybrids in the Triticeae. *Cereal Res. Commun.* 8:469–475.

Murashige, T. & Skoog, F. (1962). A revised medium for rapid growth and bioassays with tobacco tissue cultures. *Physiol. Plant.* 15:473–497.

Myodo, H. (1963). Experimental studies on the sterility of some *Lilium* species. *J. Fac. Agric. Univ. Sapporo* 52:70–122.

Nanda Kumar, P.B.A. & Shivanna, K.R. (1991). In vitro multiplication of sterile interspecific hybrid, *Brassica fruticulosa* × *B. campestris*. *Plant Cell, Tissue, Organ Cult.* 26:17–22.

Nanda Kumar, P.B.A. & Shivanna, K.R. (1993). Intergeneric hybridization between *Diplotaxis siettiana* and crop brassicas for production of alloplasmic lines. *Theor. Appl. Genet.* 85:770–776.

Nanda Kumar, P.B.A., Prakash, S. & Shivanna, K.R. (1988). Wide hybridization in crop Brassicas. In *Sexual Reproduction in Higher Plants*, eds. M. Cresti, P. Gori and E. Pacini, pp. 95–100. Berlin, Heidelberg, New York: Springer-Verlag.

Neal, C.A. & Topoleski, L.D. (1983). Effects of the basal medium on growth of immature tomato hybrids *in vitro*. *J. Amer. Soc. Hort. Sci.* 110:869–873.

Nitsch, J.P. & Nitsch C. (1969). Haploid plants from pollen grains. *Science* 163:85–86.

Nomura, Y. & Oosawa, K. (1990). Production of interspecific hybrids between *Allium chinense* and *A. thunbergii* by in ovulo embryo culture. *Japan. J. Breeding* 40:531–535.

Ohri, D. & Khoshoo, T.N. (1983). Cytogenetics of garden gladiolus. III. Hybridization. *Z. Pflanzenzüchtg.* 91:46–60.

Okazaki, K. & Murakami, K. (1992). Effects of flowering time (in forcing culture), stigma excision, and high temperature on overcoming of self incompatibility in tulip. *J. Japan. Soc. Hort. Sci.* 61:405–411.

Okazaki, K., Umada, Y., Urashima, O., Kawada, J., Kunishige, M. & Murakami, K. (1992). Interspecific hybrids of *Lilium longiflorum* and *L.* x *formolongi* with *L. rubellum* and *L. japonicum* through embryo culture. *J. Japan. Soc. Hort. Sci.* 60:997–1002.

Pickersgill, B. (1993). Interspecific hybridization by sexual means. In *Plant Breeding*, eds. M.D. Hayward, N.O. Bosemark, and I. Romagosa, pp. 63–78. London: Chapman and Hall.

Pittarelli, G.W. & Stavely, J.R. (1975). Direct hybridization of *Nicotiana repanda* × *N. tabacum*. *J. Hered.* 66:281–284.

Przywara, L., White, D.W.R., Sanders, P.M. & Maher, D. (1989). Interspecific hybridization of *Trifolium repens* with *T. hybridum* using *in ovulo* embryo and embryo culture. *Ann. Bot.* 64:613–624.

Raghavan, V. (1986). *Experimental Embryology of Vascular Plants*. London: Academic Press.

Raghavan, V. & Srivastava, P.S. (1982). Embryo culture. In *Experimental Embryology of Vascular Plants*, ed. B.M. Johri, pp. 195–230. Berlin: Springer-Verlag.

Ramanna, M.S. (1992). The role of sexual polyploidization in the origins of horticultural crops: *Alstroemeria* as an example. In *Gametes with Somatic Chromosome Number in the Evolution and Breeding of Polyploid Polysomic Species: Achievements and Perspectives, 2nd International Symposium Chromosome Engineering in Plants*, eds. A. Mariani and S. Tavoletti, pp. 83–89.

Rangaswamy, N.S. (1977). Applications of in vitro pollination and in vitro fertilization. In *Applied and Fundamental Aspects of Plant Cell, Tissue and Organ Culture*, eds. J. Reinert and Y.P.S. Bajaj, pp. 412–425. Berlin, Heidelberg, New York: Springer-Verlag.

Rangaswamy, N.S. & Shivanna, K.R. (1967). Induction of gametic compatibility and seed formation in axenic cultures of a diploid self-incompatible species *Petunia*. *Nature* 216:937–939.

Reed, S.M. & Collins, G.B. (1978). Interspecific hybrids in *Nicotiana* through in vitro culture of fertilized ovules. *J. Hered.* 69:311–315.

Reiser, W., Ziessler, C.M. (1989). Die Überwindung postgamer Inkompatibilität bei *Freesia*-Hybriden. *Tag.-Ber., Akad. Landwirtsch. -Wiss.* DDR, Berlin 281:135–138.

Sabja, A.M., Mok, D.W.S., Mok, M.C. (1990). Seed and embryo growth in pod cultures of *Phaseolus vulgaris* and *P. vulgaris* x *P. acutifolius*. *Hort. Sci.* 25(10):1288–1291.

Sarla, N. & Raut, R.N. (1988). Synthesis of *Brassica carinata* from *Brassica* x *Brassica oleracea* hybrids obtained by ovary culture. *Theor. Appl. Genet.* 76:846–849.

Sastri, D.C. (1985). Incompatibility in Angiosperms: significance in crop improvement. *Adv. Appl. Biol.* 10:71–111.

Sastri, D.C., Moss, J.P. & Nalini, M.S. (1983). The use of in vitro methods in groundnut improvement. In *Proceedings of the International Symposium on Plant Cell Culture in Crop Improvement*, eds. S.K. and K.L. Giles, pp. 365–370. New York: Plenum Press.

Sastri, D.C., Nalini, M.S. & Moss, J.P. (1981). Tissue culture and prospects of crop improvement in *Arachis hypogaea* and other oilseeds crops. In *Proceedings of the Symposium on Tissue Culture of Economically Important Plants in Developing Countries, National Univ. Singapore, Singapore*, cd. A.N. Rao, pp. 42-57.

Shivanna, K.R. (1982). Pollen–pistil interaction and control of fertilization. In *Experimental Embryology of Vascular Plants*, ed. B.M. Johri, pp. 131–174. Berlin: Springer-Verlag.

Singh, A.K. (1985). Utilization of wild relatives in the genetic improvement of *Arachis hypogaea* L. *Theor. Appl. Genet.* 72:1654–169.

Singsit, C. & Hanneman Jr., R.E. (1991). Rescuing abortive inter-EBN potato hybrids through double pollination and embryo culture. *Plant Cell Rep.* 9:475–478.

Stebbins, G.L. (1958). The inviability, weakness, and sterility of interspecific hybrids. *Adv. Genet.* 9:147–215.

Straathof, T.P., Van Tuyl, J.M., Keijzer, C.J., Wilms, H.J., Kwakkenbos, A.A.M. & Van Diën, M.P. (1987). Overcoming post-fertilization barriers in *Lilium* by ovary- and ovule culture. *Plant Cell Incompatibility Newslett.* 19:69–74.

Subrahmanyam, N.C. (1979). Haploidy from *Hordeum* interspecific crosses. Part 2. Dihaploids of *H. brachyantherum* and *H. depressum*. *Theor. Appl. Genet.* 55:139–144.

Swaminathan, M.S. & Murty, B.R. (1957). One-way incompatibility in some species crosses in the genus *Nicotiana*. *Indian J. Genet. Plant Breeding* 17:23–26.

Takeshita, M., Kato, M. & Tokumasu, S. (1980). Ovule culture in the production of intergeneric or interspecific hybrids in *Brassica* and *Raphanus*. *Japan. J. Genet.* 55:373–387.

Tiara, T. & Larter, E.N. (1977). Effects of E-amino-n-caproic acid and L-lysine on the development of hybrid embryo of *Triticale* (x-*Triticale-Secale*). *Can. J. Bot.* 55:2330–2334.

Uhlinger, R.D. (1982). Wide crosses in herbaceous perennials. *Hort. Sci.* 17(4):570–574.

Van Creij, M.G.M., Van Raamsdonk, L.W.D. & Van Tuyl, J.M. (1993). Wide interspecific hybridization of *Lilium*: preliminary results of the application of pollination and embryo-rescue methods. *Lily Yearbook N. Amer. Lily Soc.* 43:28–37.

Van Eijk, J.P., Van Raamsdonk, L.W.D., Eikelboom, W. & Bino, R.J. (1991). Interspecific crosses between *Tulipa gesneriana* cultivars and wild *Tulipa* species: a survey. *Sex. Plant Reprod.* 4(1):1–5.

Van Tuyl, J.M. (1989). Research on mitotic and meiotic polyploidization in lily breeding. *Herbertia* 45(1, 2):97–103.

Van Tuyl, J.M. (1993). Survey of research on mitotic and meiotic polyploidization at CPRO-DLO. *Lily Yearbook N. Amer. Lily Soc.* 43:10–18.

Van Tuyl, J.M., Marcucci, M.C. & Visser, T. (1982). Pollen and pollination experiments. VII. The effect of pollen treatment and application method on incompatibility and incongruity in *Lilium. Euphytica* 31:613–619.

Van Tuyl, J.M., Straathof, T.P., Bino, R.J. & Kwakkenbos, A.A.M. (1988). Effect of three pollination methods on embryo development and seed set in intra- and interspecific crosses between seven *Lilium* species. *Sex. Plant Reprod.* 1:119–123.

Van Tuyl, J.M., De Vries, J.N., Bino, R.J. & Kwakkenbos, A.A.M. (1989). Identification of 2n-pollen producing interspecific hybrids of *Lilium* using flow cytometry. *Cytologia* 54:737–745.

Van Tuyl, J.M., Bino, R.J. & Custers, J.B.M. (1990). Application of *in vitro* pollination, ovary culture, ovule culture and embryo rescue techniques in breeding of *Lilium, Tulipa* and *Nerine*. In *Integration of in Vitro Techniques in Ornamental Plant Breeding*, ed. J. de Jong, pp. 86–97. Wageningen: CPO.

Van Tuyl, J.M., Van Diën, M.P., Van Creij, M.G.M., Van Kleinwee, T.C.M., Franken, J. & Bino, R.J. (1991). Application of *in vitro* pollination, ovary culture, ovule culture and embryo rescue for overcoming incongruity barriers in interspecific *Lilium* crosses. *Plant Sci.* 74:115–126.

Van Tuyl, J.M., Meijer, H. & Van Diën, M.P. (1992). The use of oryzalin as an alternative for colchicine in vitro chromosome doubling of *Lilium* and *Nerine*. *Acta Hort.* 325:625–630.

Van Tuyl, J.M., Van Creij, M.C.M., Eikelboom, W., Kerckhoffs, D.M.F.J. & Meijer, B. (1993). New genetic variation in the *Lilium* and *Tulipa* assortment by wide hybridization. In *Proceedings of the XVIIth Eucarpia Symposium*, eds. T. Schiva and A. Mercuri, pp. 141–149. Sanremo, Italy:.Istituto sperimentale per la Floricoltura.

Veilleux, R. (1985). Diploid and polyploid gametes in crop plants: mechanisms of formation and utilization in plant breeding. *Plant Breeding Rev.* 3:253–288.

Vyas, P., Prakash, S. & Shivanna, K.R. (1995). Production of wide hybrids and backcross progenies between *Diplotaxis erucoides* and crop brassicas. *Theor. Appl. Genet.* 90:549–553.

Weilenmann, E., Baudoin, J.P. & Maréchal, R. (1986). Obtention d'allopolyploides fertiles chez le croisement entre *Phaseolus vulgaris* et *Phaseolus filiformis*. *Bull. Rech. Agron. Gembloux* 21(1):35–46.

Wietsma, W.A., De Jong, K.Y. & Van Tuyl, J.M. (1994). Overcoming prefertilization barriers in interspecific crosses of *Fritillaria imperialis* and *F. raddeana*. *Plant Cell Incompatibility Newslett.* 26:89–92.

Williams, E.G., Maheswaran, G. & Hutschinson, J.F. (1987). Embryo and ovule culture in crop improvement. *Plant Breeding Rev.* 5:181–236.

Willing, R.R. & Pryor, L.D. (1976). Interspecific hybridization in poplar. *Theor. Appl. Genet.* 47:141–151.

Zenkteler, M. (1990). *In vitro* fertilization and wide hybridization in higher plants. *Critical Rev. Plant Sci.* 9:267–279.

Zenkteler, M., Maheshwaran, G. & Willims, E.G. (1987). In vitro placental pollination in *B. campestris* and *B. napus*. *J. Plant Physiol.* 128:245–250.

14

Storage of pollen

B. BARNABÁS and G. KOVÁCS

Summary 293

Introduction 294

Taxonomic differences in the longevity of pollen 295

Factors affecting pollen viability and storability 295

Causes of the loss of pollen viability 297

Biochemical changes in pollen 297

Desiccation and cellular changes of pollen 297

Short-term pollen storage 300

Effects of temperature and humidity 300

Pollen storage in organic solvents 300

Long-term storage of pollen 301

Storage at sub-zero temperatures 301

Freeze- or vacuum-drying (lyophilization) 301

Cryopreservation by deep-freezing 302

Storage of graminaceous pollen in deep-frozen condition 304

Estimation of pollen viability after storage 305

Quick methods based on enzyme assay 305

Pollen germination in vitro 307

Pollen germination on the stigma 307

Fruit and seed set 307

References 309

Summary

The pollen longevity of different species varies between minutes and years depending primarily on the taxonomic status of the plant and on abiotic environmental conditions. For a number of agronomically important taxa, including the short-lived graminaceous pollen, special storage conditions are needed to preserve the viability and fertilizing ability of pollen for a long period. Recent sophisticated methods such as nuclear magnetic resonance (NMR) spectrometry, Fourier transform infrared spectroscopy (FTIR), and different ultramicrotechniques for electron microscopy have helped to carry out precise studies on the water regime in pollen, parallel to the molecular changes occurring in membranes during pollen dehydration and rehydration. Cryopreservation seems to be the most efficient method for the long-term preservation (now up to 10 years) of partly dehydrated pollen grains. Beyond the classical role of "pollen banks," the promising application of the modern in vitro techniques of plant biotechnology (isolation and fusion of reproductive cells, and DNA transformation of artificially produced zygotes and embryos) has opened new, challenging opportunities for germplasm cryopreservation in the near future.

Introduction

The artificial maintenance of the viability and fertilizing ability of pollen over a long period is an important problem from both the theoretical and practical points of view. Historically, the need to retain functional pollen for practical breeding and experimental research has launched detailed studies into pollen morphology, genetics, and physiology. As early as 1885, William King referred to the necessity for learning how to store pollen and stated, "Nothing could tend more to the speedy termination of an experiment than that we had control over the supply of pollen, so that we might use it when and where convenient to ourselves."

There are a large number of crop species, including vegetables, fiber and fruit crops, forages and cereals, for which pollen storage strategies are desirable. Within a species, there exist numerous genotypes, some of cultivar status, others as lines or genetic stocks, for which preservation is necessary. It should be possible to preserve the genetic resources of the plant kingdom in an unaltered state in pollen banks, as has already been done for human and animal sperms and embryos. This means that scientists and breeders could use preserved pollen of various plant species at any time and any place.

Up to now, pollen storage has been used to satisfy the following practical needs:

1. To hybridize plants that flower at different times and locations or show nonsynchronous flowering
2. To provide a constant supply of short-lived (recalcitrant) pollen
3. To facilitate supplementary pollination for improving yields
4. To eliminate the need to grow male lines continuously in breeding programs
5. To obviate the variability incidental to the daily collection of pollen samples
6. To study pollen allergens and the mechanism of self-incompatibility
7. To provide materials for international germplasm exchange
8. To ensure the availability of pollen throughout the year without using nurseries or artificial climate for plant growth

Because the importance of pollen storage has been recognized for more than a century, a large number of scientists have attempted to store pollen grains of numerous taxa, particularly of agronomically important crop species. Earlier literature on the storage of pollen is reviewed by Knowlton

(1922), Holman and Brubaker (1926), Visser (1955), Johri and Vasil (1961), Linskens (1964), J. R. King (1965), Stanley and Linskens (1974), and Shivanna and Johri (1985).

Because the storability of pollen is highly dependent on different biotic and abiotic environmental factors, there have been a number of scientific and technical problems to be solved for the successful pollen preservation of various plant species and genotypes. A detailed knowledge of pollen viability is needed to be able to elaborate efficient pollen storage strategies.

Taxonomic differences in the longevity of pollen

Systematic research into the physiology of pollen began at the turn of this century. Rittinghaus (1886), Mangin (1886), Molisch (1893), Goff (1901), and Pfundt (1910) measured the life-span of pollen from various plant species under natural conditions. These authors and, later, Knowlton (1922), Vasil (1961), and Kozaki (1975) concluded that the life-span of pollen is primarily determined by the plant genome but is also influenced by external environmental conditions. The plant taxa examined could be classified into three main groups on the basis of their pollen longevity (Harrington 1970):

1. Long-lived pollen (6 months to a year):
   Gingkoaceae, Pinaceae, Palmae, Saxifragaceae, Rosaceae,
   Leguminosae, Anacardiaceae, Vitaceae, and Primulaceae
2. Pollen with a medium life-span (approximately 1–3 months):
   Liliaceae, Amaryllidaceae, Salicaceae, Ranunculaceae, Cruciferae,
   Rutaceae, Solanaceae, and Scrophulariaceae
3. Short-lived pollen (ranging from a few minutes to a couple of days):
   Alismataceae, Gramineae, Cyperaceae, Commelinaceae, and Juncaceae.

Factors affecting pollen viability and storability

Comparative studies on pollen morphology and physiology (Brewbaker 1959; Hoekstra and Bruinsma 1975; Johri and Shivanna 1977; Hoekstra 1979; Singh and Sawhney 1992) showed that the binucleate and trinucleate pollen grains show differences in their physiological and structural characters at the time of pollen dispersal. Under natural conditions, the two-celled pollen grains have a much longer life-span because of their protective structure, low plasma water content, and reduced metabolic activity, whereas the trinucleate pollens are short-lived due to their less resistant wall construction and high moisture content, which can easily be lost by desiccation. This type of pollen has a high rate of metabolism.

The retention of pollen viability after shedding varies significantly from species to species. Environmental factors, particularly humidity and temperature, greatly affect viability. The relationship between air humidity and pollen

Table 14.1. *Prolonged pollen viability of different plant species under non-cryogenic temperatures and various relative air humidities*

Taxa	Storage temperature (°C)	RH (%)	Duration of storage	References
Aesculus hippocastanum	17–20	30	72 days	Pfundt 1910
Carica papaya	1	10	153 days	Traub 1936
Citrus sp.	5	20–30	3 years	Kobayashi et al. 1978
Cocos nucifera	5	40	1.5 years	Whitehead 1963
Fragaria spec.	2–4	10–30	3 years	MacFarlene et al. 1989
Gladiolus spec.	10	50	100 days	Pfeiffer 1939
Lycopersicon esculentum	2–4	10	252 days	McGuire 1952
Olea europea	20	28–33	1 year	Pinney & Polito 1990
Papaver hybridum	17–20	30	49 days	Pfundt 1910
Papaver rhoeas	4	Desiccator	200 days	Dhingra & Varghese 1990
Pinus silvestris	2	25–75	1 year	Johnson 1943
Pistachia atlantica	0	25	550 days	J. R. King & Hesse 1938
Pyrus malus	2–8	10	673 days	Nebel & Ruttle 1936
Rhododendron sp.	2	25	124 days	Visser 1955
Vitis vinifera	10	25	2 years	Olmo 1942

longevity was investigated in detail for the first time by Pfundt (1910). Pollen of more than 140 species was kept at 17°–22°C with relative humidities of 0%, 30%, 60%, and 90%. These investigations showed that the pollen of the majority of species maintained its vitality best at low relative air humidities (0%–30% RH). These results were then confirmed by Holman and Brubaker (1926), who examined the pollen longevity of 52 other species at 18°C with a wide range of relative humidities. They gave a review of the literature published until that time on the pollen longevity of 231 species belonging to 175 genera and 23 families under air dry conditions. With regard to storage temperatures above zero (between 0° and 10°C), several experts discovered that the pollen of numerous species maintained its viability for approximately one year or longer if kept at low humidities.

Table 14.1 gives a survey of pollen storage for different taxa under various controlled temperature and humidity conditions. In many cases it is feasible to standardize conditions (low temperature and/or low humidity) for prolonging pollen viability of two-celled taxa. However, little progress has been made with the preservation of the three-celled pollen taxa, including the Gramineae, Umbelliferae, Cruciferae, Araceae, Caryophyllaceae, and Chenopodiaceae families. Although prolonging the viability of pollen is a routine technique for a number of plant species, the tri-cellular pollen of important crops such as *Asteraceae* (Compositae) and *Poaceae* (Gramineae) is still recalcitrant.

In contrast to the relatively long life-span of the pollen of species listed in Table 14.1, the longevity of Gramineae pollen appears to be short under all conditions. Low relative humidities are harmful, and pollen stored at 0°–10°C remains viable only for a couple of days. Even under high relative humidity (80%–100%), pollen viability is prolonged 1–3 weeks at the most (Table 14.2).

Causes of the loss of pollen viability

Biochemical changes in pollen

The major cause of the loss of viability during storage appears to be the deficiency of metabolites due to the continued metabolic activity of the pollen, even though this activity is going on at a very reduced rate (Wilson et al. 1979). Nielsen (1956), Johri and Vasil (1961), Stanley and Poostachi (1962), and Linskens and Pfahler (1973) reported considerable changes in the carbohydrate and organic acid content of the pollen of various species during long-term storage. Deficiency of respiratory substrates can strongly contribute to the loss of pollen viability, especially in the case of three-celled pollen, where the respiration rate is much higher than in the two-celled type (Hoekstra and Bruinsma 1975). Pollen grains of grasses, however, contain a satisfactory amount of starch grains even after losing viability (Hoekstra and Bruinsma 1980). Consequently, viability decrease might be associated with the inactivation of enzymes, such as the amylases and phosphatases (Hoeckel 1951) involved in the degradation of reserves stored in pollen grains.

Desiccation and cellular changes of pollen

The water content of living pollen grains varies considerably between different families, with most recorded values between 15% and 35% of fresh weight at the time of shedding (Linskens 1974; Heslop-Harrison and Heslop-Harrison 1992). The water content of Gramineae pollen, however, is generally high (35%–60%) and the metabolism is not retarded greatly. To some extent,

Table 14.2. *Maximum longevity of some important tricellular pollen species*

Species	Storage temperature (°C)	RH (%)	Duration of storage	References
Brassica campestris	4	Air dry state	25 days	Chiang 1974
Buffalo grass	4	90	6 days	Jones & Newell 1948
Dieffenbachia maculata	5	90	5 days	Henny 1980
Elephant grass	4	100	7–14 days	Aken'ova & Chhedda 1970
Helianthus annuus	4–5	<40	20–25 days	Frank et al. 1982
Pennisetum typhoides	4	Sealed jars	4 days	Cooper & Burton 1965
	14	100	14 days	Pokhriyal & Mangth 1979
Saccharum spontaneum	4	90–100	10 days	Sartoris 1942
Secale cereale	17–21	<30	12 hours	Pfundt 1910
	4	80	7 days	Barnabás 1982
Sorghum sp.	4	75	4 days	Patil & Goud 1980
Spathiphyllum floribundum	7	65	24 weeks	Henny 1978
Triticum aestivum	20	90	1–3 hours	Barnabás 1982
	4	80	1 day	Barnabás 1982
Triticosecale	20	90	7 hours	Barnabás 1982
Zea mays	4	90	9–11 days	Jones & Newell 1948
	2	100	6 days	Pfahler & Linskens 1973
Xanthosoma sagittifolium	5	30	28 days	Aguegnia & Fatakum 1988

the original pollen moisture content depends upon the environmental conditions, particularly temperature, air humidity, and the water supply to the pollen donor plant (Barnabás and Rajki 1976). The initial pollen moisture content is generally expressed as a percentage of the fresh weight. Recently, nuclear magnetic resonance (NMR) spectrometry, which is more accurate and nondestructive than the conventional method, has been used to measure the water content of pollen (Dumas and Gaude 1981; Dumas et al. 1985; Kerhoas et al. 1990).

The regulation of pollen water content might be an important adaptive mechanism for survival after pollen dispersal. Among cereals, the pollen of cross-pollinating species can survive longer in nature than the pollen of self-pollinating species (Fritz and Lukaszewski 1989). Pollen grains that remain viable after dehydration are termed desiccation tolerant; those that lose viability parallel to dehydration are termed desiccation sensitive (Roberts 1973). In contrast to the research on the molecular mechanism of membrane deterioration during the dehydration and storage of seeds (Roberts 1973; Simon 1974, 1978; Thomson 1979; Pristley and de Kruijff 1982), little was published in this field for pollen. The correlation between the condition of the vegetative cell membrane and pollen viability was first discussed by Heslop-Harrison (1979) and by Shivanna and Heslop-Harrison (1981). Studies on the membrane state using the fluorescein diacetate test in the pollen of several taxa indicated that most of the pollen samples exposed to dry conditions had lost their membrane integrity, and invariably they failed to germinate. Investigations by Barnabás (1985) showed that pollen grain viability (germinability) in maize was closely and positively correlated with desiccation tolerance. In fact, maize pollen could sustain up to a limit of 80% water loss without detrimental changes in normal pollen functions. If the molecular reorganization taking place in the pollen membranes as a result of dehydration does not lead to an irreversible alteration in the structure, the membranes of the desiccated grains can be restored by the uptake of water while on the stigma.

Different ultramicrotechniques, including transmission electron microscopy (anhydrous fixation, freeze fracturing) and ^{32}P-NMR measurements, have been used to follow the behavior of plasma membranes in dehydrating pollen grains of various species (Dumas et al 1984; Elleman and Dickinson 1986; Platt-Aloia et al. 1986; Kerhoas et al. 1987; Barnabás et al. 1988). The plasma membrane may undergo a gel-phase transition during water loss, after increasing van der Waals interaction or free-radical damage. Water plays a key role in maintaining the structural integrity and stability of the pollen membranes, acting through hydrophobic and hydrophilic interactions. A positive significant correlation between the loss of viability and a reduction in the amount of membrane phospholipids irrespective of the storage conditions (Jain and Shivanna 1989) indicates that depletion of membrane components constitutes the primary cause for the loss of viability.

The period and extent of desiccation after which pollen membranes can recover their structure vary from species to species. The way in which desiccated pollen is rehydrated prior to viability assessment is important for the restoration of membrane integrity. For most of the desiccated pollen systems, gradual hydration in humid air is favorable for restoration of membrane integrity (Shivanna and Heslop-Harrison 1981). Hoekstra (1984), Hoekstra and van der Val (1988), and Crowe et al. (1989a, b) suggested that the gradual rehydration of dry pollen causes a shift in membrane lipids from the gel phase to the liquid crystalline phase and that imbibitional damage and leakage could be reduced or eliminated by adequate hydration.

The pollen grains of grasses represent an exception, because they are highly sensitive to greater degrees of desiccation and the restoration of normal pollen function seems to be difficult even in the course of controlled rehydration. The most recent studies on cytoskeletal elements in wheat pollen (Heslop-Harrison and Heslop-Harrison 1992) indicate that dehydration after dispersal rapidly disrupts the actin cytoskeleton and leads to a loss of germination capacity in the pollen. Therefore, for the successful preservation of pollen that loses its viability in a short time, special conditions should be provided before and during storage.

Short-term storage of pollen

Effect of temperature and humidity

Studies on the water regime in pollen indicate that the loss of pollen viability in the course of short-term storage is related to irreversible structural changes in the plasma membranes and cytoskeletal elements of the pollen rather than to a deficiency of essential metabolites. As was described previously, the pollen of a large number of taxa can be successfully stored for a limited period of time through the manipulation of storage temperature and humidity. Pollen can conveniently be stored in glass or polyethylene vials or in other suitable small containers. The unsealed containers are frequently put into desiccators containing suitable dehydrating agents (dried silica gel, saturated solutions of different salts, or various concentrations of sulfuric acid) to maintain the desired relative air humidity. To prevent sticking, powdered diluents, such as *Lycopodium* powder, talc, or corn or wheat flour, are mixed with the pollen. In general, low temperatures and relative humidities are favorable for most taxa (see Table 14.1). Tricellular pollen, especially that of Gramineae, requires sophisticated environmental conditions to preserve viability and fertility even for a short period of storage.

Pollen storage in organic solvents

It was first demonstrated by Iwanami (1972a, b) and Iwanami and Nakamura (1972) that pollen grains of various taxa could be successfully preserved in different organic solvents. The pollen of *Camellia japonica*,

Impatiens balsamina, and a *Lilium* sp. was able to survive storage in benzene, petroleum, diethyl ether, acetone, and chloroform. Pollen of *Citrus* sp. maintained viability in various organic solvents for three months (Kobayashi et al. 1978). Liu et al. (1985) found that of the 11 species examined (*Camellia japonica, Armenica vulgaris, Prunus salicina, Pyrus ussuriensis, Prunus triloba, Malus pumila, Prunus persica, Zea mays, Juglans regia, Salix babylonica,* and *Gingko biloba*), the pollen of all the insect-pollinated species showed viability after storage at 4°C in a suitable organic solvent for 35–40 days.

A correlation between the polarity of the organic solvent and its efficacy for pollen storage has been established (Jain and Shivanna 1989, 1990). Pollen grains stored in nonpolar organic solvents such as hexane, cyclohexane, and diethyl ether retained viability and showed very little leaching of phospholipids, sugars, and amino acids into the solvents. Pollen grains stored in polar solvents, on the contrary, lost viability and showed extensive leaching of these substances (Jain and Shivanna 1988a, b).

The efficiency of individual organic solvents varies greatly for different plant species; however, solvent has proved to be better for the storage of any pollen than have low temperature and humidity (Mishra and Shivanna 1982). Also, whereas the technique of storing pollen in organic solvents seems to be fairly simple, this method can be used with advantage only for short-term storage. Serious attempts should be made to extend this pollen storage method to a large number of taxa.

Long-term storage of pollen

During storage at temperatures above 0°C, the metabolic activity in pollen, although slow, will nevertheless continue, resulting in a gradual decrease in, and finally total loss of, pollen viability. Thus, cryogenic conditions seem to show promise for the really long-term preservation of pollen viability.

Storage at sub-zero temperatures

The longevity of bicellular pollen that is considered to be desiccation tolerant or that has a low original water content has successfully been prolonged using storage temperatures between -10°C and -34°C. All the data listed in the review by Towill (1985) come from the direct exposure of pollen to the storage temperatures. Pollen survival varied with the species examined, but ranged between 1 and 3 years.

Freeze- or vacuum-drying (lyophilization)

The technique of freeze-drying involves the rapid freezing of pollen to a sub-zero temperature (frequently to -60°C or -80°C) and then the gradual removal of water under vacuum sublimation. Vacuum-drying does not include the initial freezing step; the pollen is directly exposed to a vacuum and simultaneous cooling, and the moisture is withdrawn by evaporative cooling. To develop useful procedures, the optimum pollen water content, cooling method, and duration and extent of drying and rehydration must be determined. Pollen storage at sub-zero temperatures, especially if combined with lyophilization, can be very effective for a number of taxa (see J. R. King 1965). Pollen samples may respond even better if they are air dried before freeze-drying (Walden 1967). The storage atmosphere may influence the longevity of freeze-dried or vacuum-dried pollen. Both vacuum and an inert gas, such as helium or nitrogen, have been used and compared so far (Snope and Ellison 1963; Nath and Anderson 1975).

Thus, pollen of different taxa, especially desiccation-tolerant pollen, can be successfully preserved for fairly long periods of time by freeze- or vacuum-drying. The viability of stored pollen often remains high if the crucial factors are evaluated and optimized for the given species. Unless pollen is stored at low temperatures, there is considerable reduction in pollen survival over periods of one or a few years (see Towill 1985 for details).

Cryopreservation by deep-freezing

Ultra low temperatures (from -70°C to -196°C) can be used for preserving pollen in an unaltered condition with great potential, especially in cases where other storage methods are not successfully applicable, or where long-term preservation is desired. Knowlton (1922) was the first to observe that pollen could survive extremely low temperatures. *Anthirrinum majus* pollen that had a low water content (approximately 10%) was exposed to -180°C in liquid air without loss of germination ability. Since the early 1950s, numerous studies have been conducted on the cryopreservation of pollen (Griggs et al. 1950, 1953; Visser 1955; Ichikawa and Shidei 1972). Not long ago, detailed reviews were published on pollen storage that included cryopreservation (Towill 1985, 1991; Shivanna and Johri 1985; Bajaj 1987). A selected list of important crop species whose pollen survived exposure and storage at cryogenic temperatures lower than -70°C is presented in Table 14.3.

Pollen of most species does not require any specific rate of cooling or warming when the moisture content is satisfactory for freeze storage. A reduction in the pollen water content to below a threshold level prior to low temperature exposure seems to be crucial for achieving survival (Ching and Slabaugh 1966; Ichikawa et al. 1970; Ichikawa and Shidei 1972; Barnabás and Rajki 1976; Kerhoas et al. 1987). These observations suggest that partially dehydrated pollen possesses less freezable water and can survive deep-freezing. Further experiments are necessary to determine the critical moisture level (pollen water content below which the fertilizing ability of pollen is retained after low temperature exposure) for successful, long-term cryostorage of the pollen of a wide range of species (Connor and Towill 1993). Recent

Table 14.3. *Successful cryostorage of pollen from various crop species*

Taxa	Storage temperature (°C)	Duration of storage	Quality of stored pollen[a]	References
Beta vulgaris	-196	1 year	fertile	Hecker et al. 1986
Brassica oleracea	-196	16 months	fertile	Crips & Grout 1984
Capsicum annuum	-196	42 months	fertile	Alexander et al. 1991
Carya illinoensis	-196	1 year	fertile	Yates & Sparks 1990
Carica papaya	-196	485 days	viable	Ganeshan 1986
Diospyros kaki	-80	1 year	fertile	Wakisaka 1964
Glycine max	-192	21 days	fertile	Collins et al. 1973
Gossypium hirsutum	-192	10 days	viable	Collins et al. 1973
Helianthus annuus	-76	4 years	fertile	Frank et al. 1982
	-196	4 years	fertile	Frank et al. 1982
Humulus lupulus	-196	2 years	fertile	Haunold & Stanwood 1985
Juglans nigra	-196	2 years	viable	Farmer & Barnett 1974
Lycopersicon esculentum	-190	1,062 days	fertile	Visser 1955
Narcissus cv.	-196	351 days	fertile	Bowes 1990
Persea sp.	-196	1 year	fertile	Sedgley 1981
Pistacia sp	-196	1 year	fertile	Vithanage & Alexander 1985
Prunus persica	-196	1 year	fertile	Jiang & Gao 1989
Pyrus communis	-190	1,062 days	viable	Visser 1955
Pyrus communis	-70	7 months	viable	Griggs et al. 1950
Pyrus malus	-190	673 days	viable	Visser 1955
Rosa spp.	-196	8 weeks	fertile	Marchant et al. 1993
Solanum tuberosum	-196	9 months	viable	Weatherhead et al. 1978
		24 months	fertile	Towill 1984
Trifolium pratense	-196	24 weeks	fertile	Collins et al. 1973
Vicia faba	-196	1 month	viable	Telaye et al. 1990
Vitis vinifera	-196	64 weeks	viable	Ganeshan 1985
	-196	5 years	fertile	Ganeshan & Alexander 1988

[a] viable = stored pollen showed in vitro germination
fertile = stored pollen induced seed set

investigations on the molecular structures of membranes using Fourier transform infrared spectroscopy (FTIR) indicate that lipid transitions in membranes may cause major damage during freezing or warming after freeze–thaw has been completed (Crowe et al. 1989a, b). Membrane phase transitions seem to be responsible for imbibitional damage in dry pollen; damage can therefore be avoided by gradual hydration or by raising the tempera-

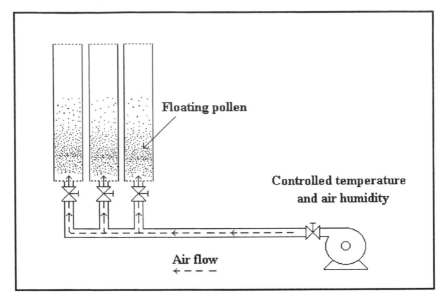

Figure 14.1   A diagrammatic sketch of the "pollen dryer."

ture to effect transition to the liquid crystalline phase. FTIR spectroscopy might be a useful tool in examining the biochemical properties of differently treated pollen, and it has a potential for evaluating optimum cryopreservation protocols. Such special pollen handling methods are needed for the cryostorage of the desiccation-sensitive type of pollen.

Storage of graminaceous pollen in deep-frozen conditions

Since Gramineae pollen contains a high amount of water (35%–60%) when shed, immediate freezing, even when it is rapid enough, would cause irreversible structural changes in the pollen as a consequence of ice formation (Barnabás et al. 1988). This is why earlier attempts to expose freshly collected grass pollen to ultra low temperature led to negative results (Knowlton 1922; Collins et al. 1973). These results indicated that it would be desirable to reduce the originally high pollen water content before cryostorage. However, the dehydration of Gramineae pollen seemed to be rather problematic due to the rapid loss of viability. This phenomenon has been fully described in the previous sections of this chapter. For maize and rye, different methods of pollen drying (desiccator, laboratory conditions, and air flotation) were compared with respect to pollen viability (Barnabás and Rajki 1976, 1981; Barnabás 1982, 1985). It was observed in all cases that in the course of drying, the water content of the pollen decreased exponentially with different intensities, depending on the temperature and air humidity. High temperature

(25°–30°C) and long duration of drying resulted in a serious reduction in pollen germination capacity. Further experiments were thus required to determine the optimum conditions of controlled dehydration for bigger quantities of pollen.

A "pollen-dryer" was constructed (Figure 14.1), which consisted of a vertical glass tube with a filter at each end (45 μm pore size). In a humidity incubator, air with controlled temperature and humidity can be blown into the tube from below, thus floating the pollen grains. This method (Barnabás and Rajki 1981; Barnabás 1982, 1985, 1994), using air flow at a temperature of 20°C and 20%–40% air humidity, ensured sufficiently rapid, but gentle and uniform, drying for maize pollen grains. The same method could be utilized for the dehydration of pollen from other Gramineae species and was of assistance in multidisciplinary studies on the pollen membranes during controlled dehydration (Kerhoas et al. 1987). Pollen dried in this way to a suitable degree could regain its normal functions after cryopreservation and warming up for a couple of minutes in a water bath at 40°C. After a prolonged lag period of adhesion and rehydration on the stigma, pollen tube initiation and growth occurred normally (Barnabás and Fridvalszky 1984).

The amount of water that can be lost without detrimental changes in pollen viability seems to range between 50% and 80%, depending on the Gramineae species and the genotype. Maize and rye pollen can tolerate higher degrees of desiccation better than can triticale (Barnabás 1982; Kovács and Barnabás 1993). Wheat pollen showed intolerance to any degree of dehydration. At an optimum stage of dehydration with 15%–20% actual moisture content, which eliminates mechanical damage during the freezing procedure and still ensures the reversible molecular reorganization of the membranes, the pollen of several agronomically important Gramineae species can be successfully stored at cryogenic temperatures in liquid nitrogen or in a deep freezer for long periods of time (see examples in Table 14.4).

Estimation of pollen viability after storage

A number of cytological and cytochemical methods have been developed to determine the viability of preserved pollen. Detailed surveys on a wide range of useful protocols can be found in recent books (Shivanna and Johri 1985; Shivanna and Rangaswamy 1992).

Quick methods based on enzyme assay

Several tests have been standardized to demonstrate cytochemically the activity of certain enzymes in living pollen, such as dehydrogenases by the triphenyl tetrazolium chloride (TTC) test (Stanley and Linskens 1974), catalases by the benzidine test (J. R. King 1960), or esterases by the fluorescein diacetate (FDA) test (Heslop-Harrison and Heslop-Harrison 1970). Of these, the FDA

Table 14.4. *Survival of deep-frozen Gramineae pollen*

Species	Storage temperature (°C)	Duration of storage	Comments	References
Avena sativa	-192	1 day	TTB[+]	Collins et al. 1973
Pennisetum americanum	-73	185 days	Seed set	Hanna et al. 1983
	-80	200 days	Seed set	Chaudhury & Shivanna 1986
Pennisetum glaucum	-73	3 years	Germination	Hanna 1990
Saccharum spontaneum	-80	30–140 days	Seed set	Tai 1988
Secale cereale	-192	7 days	Seed set	Collins et al. 1973
	-196	2 years	Seed set	Barnabás 1982
	-196	1 year	Seed set	Shi and Tian 1989
	-196	10 years	Seed set	Barnabás et al. 1992
	-76	10 years	Seed set	Barnabás et al. 1992
Triticum aestivum	-192	1 day	TTB[+]	Collins et al. 1973
	-196	Few weeks	Germination	Andreica et al. 1988
Triticosecale	-196	10 years	Germination, low seed set	Barnabás et al. 1992
Zea mays	-196	180 days	Germination	Nath & Anderson 1975
	-76	363 days	Seed set	Barnabás & Rajki 1981
	-196	92 days	Seed set	Filipova 1985
	-196	10 years	Seed set	Barnabás et al. 1992, Barnabás 1994

test has shown positive correlation with in vitro germination as well as seed set (Heslop-Harrison et al. 1984; Shivanna et al. 1991; Rao et al. 1992).

Viability tests based on the determination of enzyme activity may often lead to false estimations, because the presence of a functional enzyme in the cell itself does not guarantee that the pollen, though recovered, is still viable and functionable. Enzyme assays are better used only for a rough estimation of pollen survival.

Pollen germination in vitro

In vitro germination of pollen is the most frequently used method for checking pollen viability. It is relatively simple and quantitative, especially when applied to bicellular pollen. In numerous taxa, the percentage of in vitro germination of stored pollen is correlated with its fertilizing ability (Visser 1955). The BK medium (Brewbaker and Kwack 1963) is suitable for in vitro germination in more than 86 species. Gradual, controlled rehydration before in vitro germination is recommended to avoid imbibitional damage to the pollen (Hoekstra 1984; Crowe et al. 1989a, b). However, application of this test is limited because of difficulties in achieving in vitro germination in three-celled taxa. Efficient germination media were composed by Pfahler (1965, 1967) for the pollen of rye and maize. Wheat is still considered recalcitrant with regard to germination in vitro (Heslop-Harrison and Heslop-Harrison 1992).

Pollen germination on the stigma

Intact or excised stigmas pollinated with stored pollen are able to provide a true picture of pollen function. A "semi in vitro germination test" was devised for the efficient examination of viability in rehydrated and deep-frozen stored Gramineae pollen (Barnabás 1982, 1985). The percentage of germinating pollen and the length of the pollen tubes developed can be measured after staining with cotton blue (D'Souza 1972) or aniline blue (Linskens and Esser 1957). Figure 14.2 represents the semi in vitro germination of graminaceous pollen (stored for a long period) after 3 hours' incubation at room temperature and stained with cotton blue.

Fruit and seed set

Fruit set and seed set give the most exact estimation of pollen viability and fertilizing capacity. Seed set is determined after the artificial pollination of well-protected stigmas in previously emasculated flowers. Because the environmental conditions existing at the time of pollination may have a strong influence on the success of fertilization, a complex comparative analysis of the data from in vitro and semi in vitro pollen germination and from seed set is recommended. Especially in the case of stress-sensitive pollen, only a full exami-

Figure 14.2 Germination of maize (a), rye (b), and triticale (c) pollen on the stigma after 10-year storage in liquid nitrogen. sp: stigma papilla; pt: pollen tube; pg: pollen grain.

nation can provide a reliable idea of the changes in pollen functions occurring during dehydration and storage.

The ultimate aim of pollen storage has been the establishment of "pollen banks" through which the pollen of any species can be obtained at any time of the year and at any place in the world. Until now, assessment of the quality of stored pollen was based exclusively on viability (in vitro germination/seed set). Extensive studies carried out on stored seeds have clearly shown that the assessment of vigor (speed of germination) is a better indication of seed quality than germinability (Perry 1978). Recent studies by Shivanna and associates (Nanda Kumar et al. 1988; Shivanna and Cresti 1989; Shivanna et al. 1991) have shown that in pollen, as in seeds, storage and other stresses affect vigor before affecting viability. Loss of vigor may affect the pollen's competitive ability and the quality of the subsequent progeny. Thus, it is important to study the vigor of the stored pollen in addition to its viability.

References

Aguegnia, A. & Fatakum, C.A. (1988). Pollen storage in cocoyam (*Xanthosoma sagittifolium* (L.) Schott. *Euphytica* 39:195–198.

Aken'ova, M.E. & Chhedda, H.R. (1970). Effects of pollen storage on *Pennisetum purpureum* (Schum.) pollen. *Niger. J. Agric.* 7:111–114.

Alexander, M.P., Ganeshan, S. & Rajasekharan, P.E. (1991). Freeze preservation of capsicum pollen (*Capsicum annuum* L.) in liquid nitrogen (-196°C) for 42 months: effect on viability and fertility. *Plant Cell Incompatibility Newslett.* 23:1–4.

Andreica, A., Sparchez, C. & Soran, V. (1988). Wheat pollen germination in normal and cryopreservation condition. *Studii si Cercetari de Biologie Vegetala* 40:55–57.

Bajaj, Y.P.S. (1987). Cryopreservation of pollen and pollen embryos, and the establishment of pollen banks. *Int. Rev. Cytol.* 107:397–420.

Barnabás, B. (1982). Possibilities for storage of graminaceous pollen. Ph.D. thesis Hungarian Academy of Sciences Budapest (in Hungarian), pp. 1–156.

Barnabás, B. (1985). Effect of water loss on germination ability of maize (*Zea mays* L.) pollen. *Ann. Bot.* 48:861–864.

Barnabás, B. (1994). Preservation of maize pollen. In *Biotechnology in Agriculture and Forestry, Vol. 25, "Maize,"* ed. Y.P.S. Bajaj, pp. 607–618. Berlin, Heidelberg: Springer-Verlag.

Barnabás, B. & Fridvalszky, L. (1984). Adhesion and germination of differently treated maize pollen grains on the stigma. *Acta Botanica Hungarica* 30:329–332.

Barnabás, B. & Rajki, E. (1976). Storage of maize (*Zea mays* L.) pollen in liquid nitrogen at -196°C. *Euphytica* 25:747–752.

Barnabás, B. & Rajki, E. (1981). Fertilization ability of deep-frozen maize (*Zea mays* L.) pollen. *Ann. Bot.* 48:861–864.

Barnabás, B., Kieft, H., Schel, J.H.N. & Willemse, M.T.M. (1988). Ultrastructure of freeze-substituted maize pollen after cold storage. *Annales Scientifiques, Reims* 23:100–103.

Barnabás, B., Kovács, M. & Kovács, G. (1992). Long term cryopreservation of graminaceous pollen. *XIIIth EUCARPIA Congress on Reproductive Biology and Plant Breeding. Book of Poster Abstracts,* pp. 409–410.

Bowes, S.A. (1990). Long term storage of *Narcissus* anthers and pollen in liquid nitrogen. *Euphytica* 48:275–278.

Brewbaker, J.L. (1959). Biology of the angiosprm pollen grains. *Indian J. Genet. Plant Breeding* 19:121–135.

Brewbaker, J.L. & Kwack, B.H. (1963). The essential role of calcium ion in pollen germination and pollen tube growth. *Amer. J. Bot.* 50:859–865.

Chaudhury, R. & Shivanna, K.R. (1986). Studies on pollen storage of *Pennisetum typhoides*. *Phytomorphology* 36:211–218.

Chiang, M.S. (1974). Cabbage pollen germination and longevity. *Euphytica* 23:579.

Ching, T.M. & Slabaugh, W.H. (1966). X-ray diffraction analysis of ice crystals in coniferous pollen. *Cryobiology* 2:321.

Collins, F.C., Lertmongkol, V. & Jones, J.P. (1973). Pollen storage of certain agronomic species in liquid nitrogen. *Crop Sci.* 13:493–494.

Connor, K.F. & Towill, E.L. (1993). Pollen-handling protocol and hydration/dehydration characteristics of pollen for application to long-term storage. *Euphytica* 68:77–84.

Cooper, R.B. & Burton, G.W. (1965). Effect of pollen storage and hour of pollination on seed set in pearl millet (*Pennisetum typhoides*). *Crop Sci.* 5:18–20.

Crips, P. & Grout, B.W.W. (1984). Storage of broccoli pollen in liquid nitrogen. *Euphytica* 33:819–823.

Crowe, J.H., Hoekstra, F.A. & Crowe, L.M. (1989a). Membrane phase transitions are responsible for imbibitional damage in dry pollen. *Proc. Natl. Acad. Sci.* 86:520–523.

Crowe, J.H., Hoekstra, F.A., Crowe, L.M., Achordoguy, T.J. & Drobnis, E. (1989b). Lipid phase transitions measured in intact cells with Fourier transform infrared spectroscopy. *Cryobiology* 26:76–83.

D'Souza, L. (1972). A comparative study of size and receptivity of the stigma in wheat, rye, *Triticale* and *Secaloctricum*. *Z. Pflanzenzucht.* 68:73–82.

Dhingra, H.R. & Varghese, T.M. (1990). Pollen storage in poppy (*Papaver rhoeas* L.). *Acta Bot. Indica* 18:309–311.

Dumas, C. & Gaude, T. (1981). Stigma–pollen recognition and pollen hydration. *Phytomorphology* 31:191–201.

Dumas, C., Knox, R.B. & Gaude, T. (1984). Pollen–pistil recognition: new concepts from electron microscopy and cytochemistry. *Int. Rev. Cytol.* 90:239–272.

Dumas, C., Kerhoas, C., Gay, G. & Gaude, T. (1985). Water content, membrane state and pollen physiology. In *Biotechnology and Ecology of Pollen,* eds. D.L. Mulcahy, G. Bergamini Mulcahy, and F. Ottaviano, pp. 333–337. Berlin-Heidelberg: Springer-Verlag.

Elleman, C.J. & Dickinson, H.G. (1986). Pollen stigma interactions in *Brassica.* IV. Structural reorganisation in the pollen grains during hydration. *J. Cell. Sci.* 80:141–157.

Farmer, R.E. & Barnett, P.E. (1974). Low-temperature storage of black walnut pollen. *Cryobiology* 11:366–367.

Filipova, T.V. (1985). [Possibility for long-term storage of maize pollen at -196°C temperature.] (in Russian). *Nauka Novosibirsk*, pp. 64–68.

Frank, J., Barnabás, B., Gál, E. & Farkas, J. (1982). Storage of sunflower pollen. *Z. Pflanzenzüchtg.* 89:341–343.

Fritz, S.E. & Lukaszewski, A.J. (1989). Pollen longevity in wheat, rye and triticale. *Plant Breeding* 102:31–34.

Ganeshan, S. (1985). Cryogenic preservation of grape (*Vitis vinifera* L.) pollen. *Vitis.* 24:169–173.

Ganeshan, S. (1986). Cryogenic preservation of papaya pollen. *Sci. Hort.* 28:65–70.

Ganeshan, S. & Alexander, M.P. (1988). Fertilizing ability of cryopreserved grape (*Vitis vinifera* L.) pollen. *Genome* 30, suppl. 1, p. 464.

Goff, E.S. (1901). A study of certain conditions affecting the setting of fruits. *Wiscon. Agr. Exp. Sta. Rep.* 18:289–303.

Griggs, W.H., Vansell, G.H. & Reinhardt, J.F. (1950). The germinating ability of quick-frozen bee-collected apple pollen stored in a dry ice container. *J. Econ. Ent.* 43:549.

Griggs, W.H., Vansell, G.H. & Iwakiri, B.T. (1953). Pollen storage. *Cal. Agr. Exp. St.* 7:12.

Hanna, W.W. (1990). Long-term storage of *Pennisetum glaucum* (L) R. Br. pollen. *Theor. Appl. Genet.* 79:605–608.

Hanna, W.W., Wells, H.D., Burton, G.W. & Monson, W.G. (1983). Long-term pollen storage of pearl millet. *Crop Sci.* 1:174–175.

Harrington, J.F. (1970). Seed and pollen storage for conservation of plant gene resources. In *Genetic Resources in Plants: Their Exploration and Conservation,* eds. O.H. Frankel and E. Bennett, pp. 501–521. Oxford and Edinburgh: Blackwell.

Haunold, A. & Stanwood, P.C. (1985). Long-term preservation of hop pollen in liquid nitrogen. *Crop Sci.* 25:194–196.

Hecker, R.J., Stanwood, P.C. & Soulis, C.A. (1986). Storage of sugarbeet pollen. *Euphytica* 35:777–783.

Henny, R.J. (1978). Germination of *Spathiphyllum* and *Vriesea* pollen after storage at different temperatures and relative humidities. *Hort. Sci.* 13:596.

Henny, R.J. (1980). Germination of *Dieffenbachia maculata* "Perfection" pollen after storage at different temperatures and relative humidity regimes. *Hort. Sci.* 15:191.

Heslop-Harrison, J. (1979). An interpretation of the hydrodynamics of pollen. *Amer. J. Bot.* 66:737–743.

Heslop-Harrison, J. & Heslop-Harrison, Y. (1970). Evaluation of pollen viability by enzymatically induced fluorescence; intracellular hydrolysis of fluorescein diacetate. *Stain Tech.* 45:115–120.

Heslop-Harrison, J. & Heslop-Harrison, Y. (1992). Intracellular motility, the actin cytoskeleton and germinability in the pollen of wheat (*Triticum aestivum* L.). *Sex Plant Reprod.* 5:247–255.

Heslop-Harrison, J., Heslop-Harrison, Y. & Shivanna, K.R. (1984). The evaluation of pollen quality and further appraisal of the fluorochromatic (FCR) test procedure. *Theor. Appl. Genet.* 67:367–379.

Hoeckel, A. (1951). Beitrag zen Kenntnis Pollen-fermente. *Planta* 39:431–459.

Hoekstra, F.A. (1979). Mitochondrial development and activity of binucleate and trinucleate pollen during germination in vitro. *Planta* 145:25–36.

Hoekstra, F.A. (1984). Inhibitional chilling injury in pollen. *Plant Physiol.* 74:815–821.

Hoekstra, F.A. & Bruinsma, J. (1975). Respiration and vitality of binucleate and trinucleate pollen. *Physiol. Plant.* 34:221–225.

Hoekstra, F.A. & Bruinsma, J. (1980). Control of respiration of binucleate and trinucleate pollen under humid conditions. *Physiol. Plant.* 48:71–77.

Hoekstra, F.A. & van der Wal, E.W. (1988). Initial moisture content and temperature of imbibition determine extent of imbibitional injury in pollen. *Plant Physiol.* 133:257–262.

Holman, R.M. & Brubaker, F. (1926). On the longevity of pollen. *Univ. Calif. Publ. Bot.* 13:179–204.

Ichikawa, S. & Shidei, T. (1972). Fundamental studies on deep-freezing storage of tree pollen. III. *Bull. Kyoto Univ. For.* 44:47.

Ichikawa, S., Kaji, K. & Kubota, Y., (1970). Studies on the storage of larch (*Larix leptolepis*) pollen at superlow temperatures. *Bull. Hokkaido For. Exp. Stn.* 8:11.

Iwanami, Y. (1972a). Viability of pollen grains in organic solvents. *Botanique* 3:61–68.

Iwanami, Y. (1972b). Retaining the viability of *Camellia japonica* pollen in various organic solvents. *Plant Cell Physiol.* 13:1139–1141.

Iwanami, Y. & Nakamura, N. (1972). Storage in an organic solvent as a means for preserving viability of pollen grains. *Stain Technol.* 47:137–139.

Jain, A. & Shivanna, K.R. (1988a). Storage of pollen grains in organic solvents: effects of organic solvents on leaching of phospholipids and its relationship to pollen viability. *Ann. Bot.* 61:325–330.

Jain, A. & Shivanna, K.R. (1988b). Storage of pollen grains in organic solvents: effects of solvents on pollen viability and membrane integrity. *J. Plant Physiol.* 132:499–502.

Jain, A. & Shivanna, K.R. (1989). Loss of viability during storage is associated with changes in membrane phospholipid. *Phytochemistry* 28:999–1002.

Jain, A. & Shivanna, K.R. (1990). Membrane state and pollen viability during storage in organic solvents. In *Proceedings of the International Congress Plant Physiology, New Delhi Soc. Plant Physiol. & Biochem.*, eds. S.K. Sinha, P.V. Sane, S.C. Bhargava, and P.K. Agrawal, pp. 1341–1349.

Jiang, Y.S. & Gao, Z.J. (1989). Ultra-low temperature (-196°C) storage of peach and pear pollen. *Acta Agriculturae Shanghai* 5:1–8.

Johnson, L.P.V. (1943). The storage and artificial germination of forest tree pollens. *Can. J. Res.* 21:332–342.

Johri, B.M. & Shivanna, K.R. (1977). Differential behaviour of 2- and 3-celled pollen. *Phytomorphology* 27:98–106.

Johri, B.M. & Vasil, I.K. (1961). Physiology of pollen. *Bot. Rev.* 27:325–381.

Jones, M.D. & Newell, L.C. (1948). Longevity of pollen and stigma of grasses. *J. Amer. Soc. Agron.* 40:195–204.

Kerhoas, C., Gay, G. & Dumas, C. (1987). Multidisciplinary approach to the study of the plasma membrane of *Zea mays* pollen during controlled dehydration. *Planta* 171:1–10.

Kerhoas, D., Gay, G., Heslop-Harrison, J. & Heslop-Harrison, Y. (1990). Pollen quality: definition and estimation. *Bulletin de la Société Botanique de France, Actualités Botaniques* 137:97–100.

King, J.R. (1960). Peroxidase reaction as an indicator of pollen viability. *Stain Technol.* 35:225–227.

King, J.R. (1965). The storage of pollen particularly by the freeze-drying method. *Bull. T. Bot. Club* 92:270–278.

King, J.R. & Hesse, C.O. (1938). Pollen longevity studies with deciduous fruits. *Proc. Amer. Soc. Hort. Sci.* 36:310–313.

King, W.M. (1885). Report of chief on seed divisions. In *Report of the Commissioner of Agriculture (Yearbook)*, pp. 47–61. Washington, DC: GPO.

Knowlton, E.H. (1922). Studies in pollen with special reference to longevity. *Cornell Univ. Agr. Exp. Sta. Memoir* 52:747–794.

Kobayashi, S., Ikeda, I. & Nakatani, M. (1978). Long term storage of *Citrus* pollen. In *Long Term Preservation of Favourable Germ Plasm in Arboreal Crops*, eds. T. Akihama and K. Nakajima, pp. 8–12. Japan: The Fruit Tree Research Station.

Kovács, G. & Barnabás, B. (1993).[Long term storage of rye and triticale pollen in liquid nitrogen.] (in Hungarian). *Növénytermelés* 42:301–305.

Kozaki, I. (1975). Storage methods of pollen. *HIBP Synthesis, Tokyo* 5:14.

Linskens, H.F. (1964). Pollen physiology. *Ann. Rev. Physiol.* 15:255.

Linskens, H.F. (1974). *Fertilization in Higher Plants*. Amsterdam: North-Holland Publ.

Linskens, H.F. & Esser, K. (1957). Uber eine spezifische Anfarbung der Pollenschläuche im Griffel und die Zahl der Kallosepfropfen nach Selbstung und Fremdung. *Naturwissenschaften* 44:16.

Linskens, H.F. & Pfahler, P.L. (1973). Biochemical composition of maize (*Zea mays* L.) pollen. III. Effects of allele x storage interactions at the waxy (wx), sugary (su$_1$) and shrunken (sh$_2$) loci, on the amino acid content. *Theor. Appl. Genet.* 43:49–53.

Liu, W., Ma, L. & Cao, Z. (1985). Studies on the viability of pollen of 11 species stored in some organic solvents. *Acta Hort. Sinica* 12:69–71.

MacFarlene Smith, W.H., Jones, J.K. & Sebastiampillai, A.R. (1989). Pollen storage of *Fragaria* and *Potentilla*. *Euphytica* 41:65–69.

Mangin, L. (1886). Recherches sur le pollen. *Soc. Bot. France* 33:512–517.

Marchant, R., Power, J.B., Davey, M.R., Chartier-Hollis, J.M. & Lynch, P.T. (1993). Cryopreservation of pollen from two rose cultivars. *Euphytica* 66:235–241.

McGuire, D.C. (1952). Storage of tomato pollen. *Proc. Amer. Soc. Hort. Sci.* 60:419–425.

Mishra, R. & Shivanna, K.R. (1982). Efficacy of organic solvents for storing pollen grains of some leguminous taxa. *Euphytica* 31:991–995.

Molisch, H. (1893). Zur Physiologie des Pollens, mit besonderer Rücksicht auf die Chemotropichen Bowegungen der Pollenschläuche. *Sitzber. Ak. Wiss. Wien, Nath. Naturv. Cl.* 102:423–449.

Nanda Kumar, P.B.A., Chaudhury, R. and Shivanna, K.R. (1988). Effect of storage on pollen germination and pollen tube growth. *Curr. Sci.* 57:557–559.

Nath, J. & Anderson, J.O. (1975). Effect of freezing and freeze drying on viability and storage of *Lilium longiflorum* and *Zea mays* L. pollen. *Cryobiology* 12:81–88.

Nebel, B.R. & Ruttle, M.L. (1936). Storage experiments with pollen of cultivated fruit trees. *J. Pomol.* 14:347–359.

Nielsen, N. (1956). Vitamin content of pollen after storing. *Acta Chem. Scand.* 10:332–333.

Olmo, H.P. (1942). Storage of grape pollen. *Proc. Amer. Soc. Hort. Sci.* 41:219–224.

Patil, R.C. & Goud, J.V. (1980). Viability of pollen and receptivity of stigma in sorghum. *Indian J. Agric. Sci.* 50:522–526.

Perry, D.A. (1978). Report of the vigor test committee, 1974–1977. *Seed Sci. Technol.* 6:159–181.

Pfahler, P.L. (1965). In vitro germination of rye pollen. *Crop Sci.* 5:597–598.

Pfahler, P.L. (1967). In vitro germination and pollen tube growth of maize (*Zea mays* L.) pollen. 1. Calcium and boron effects. *Can. J. Bot.* 45:839–845.

Pfahler, P.L. & Linskens, H.F. (1973). In vitro germination and pollen tube growth of maize (*Zea mays* L.) pollen. VIII. Storage temperature and pollen source effects. *Planta* 111:253–259.

Pfeiffer, N.E. (1939). Life of *Gladiolus* pollen prolonged by controlled conditions of storage. *Contr. B. Thom. Inst.* 10:429–440.

Pfundt, M. (1910). Der Einfluss der Luftfeuchtigkeit auf die Lebensdauer des Blütenstaubes. *Jahrb. Wiss. Bot.* 47:1–40.

Pinney, K. & Polito, U.S. (1990). Olive pollen storage and in vitro germination. *Acta Hort.* 286:207–210.

Platt-Aloia, K.A., Lord, E.M., De Mason, D.A. & Thomson, W.W. (1986). Freeze fracture observations on membranes of dry and hydrated pollen from *Colomia*, *Phoenix* and *Zea*. *Planta* 168:291–198.

Pokhriyal, S.C. & Mangth, K.S. (1979). Effect of pollen storage on seed set in pearl millet. *Seed Res.* 7:131–135.

Pristley, D.A. & de Kruijff, B. (1982). Phospholipid motional characteristics in a dry biological system. *Plant Physiol.* 70:1075–1078.

Rao, G.U., Jain, A. and Shivanna, K.R. (1992). Effects of high temperature stress on *Brassica* pollen: viability, germination and ability to set fruits and seeds. *Ann. Bot.* 69:193–198.

Rittinghaus, P. (1886). Der Einfluss der Luftfeuchtigkeit auf die Lebendsdauer des Blütenstaubes. *Verh. Naturwiss. Ver. Rheinl.* 43:123–166.

Roberts, E.H. (1973). Predicting the storage life of seeds. *Seed Sci. Technol.* 1:499–514.

Sartoris, G.B. (1942). Longevity of sugar cane and corn pollen: a method for long distance shipment of sugar cane pollen by airplane. *Amer. J. Bot.* 29:395–400.

Sedgley, M. (1981). Storage of avocado pollen. *Euphytica* 30: 595-599.

Shi, S.X. & Tian, Y. (1989). Fertility of maize pollen stored in liquid nitrogen for a year. *Acta Agron, Sinica* 15:283–286.

Shivanna, K.R. & Cresti, M. (1989). Effects of high humidity and temperature stress on pollen membrane integrity and pollen vigor in *Nicotiana tabacum*. *Sex. Plant. Reprod.* 2:137–141.

Shivanna, K.R. & Heslop-Harrison, J. (1981). Membrane state and pollen viability. *Ann. Bot.* 47:759–770.

Shivanna, K.R. & Johri, B.M. (1985). *The Angiosperm Pollen Structure and Function*. New Delhi: Wiley Eastern Ltd.

Shivanna, K.R. & Rangaswamy, N.S. (1992). *Pollen Biology: A Laboratory Manual.* Berlin, Heidelberg: Springer-Verlag.

Shivanna, K.R., Linskens, H.F. & Cresti, M. (1991). Pollen viability and pollen vigor. *Theor. Appl. Genet.* 81:38–42.

Simon, E.W. (1974). Phospholipids and plant membrane permeability. *New Phytol.* 73:337–420.

Simon, E.W. (1978). Membranes in dry and imbibing seeds. In *Dry Biological Systems,* eds. J.H. Crowe and J.S. Clegg, pp. 205–224. New York: Academic Press.

Singh, S. & Sawhney, V.K. (1992). Plant hormones in *Brassica napus* and *Lycopersicum esculentum* pollen. *Phytochemistry* 12:4051–4053.

Snope, A.J. & Ellison, J.H. (1963). Storage of asparagus pollen under various conditions of temperature, humidity and pressure. *Proc. Am. Soc. Hort. Sci.* 83:447–452.

Stanley, R.G. & Poostachi, L. (1962). Endogenous carbohydrates, organic acids and pine pollen viability. *Silvae Genet.* 11:1–3.

Stanley, R.G. & Linskens, H.F. (1974). *Pollen: Biology, Biochemistry and Management.* Berlin: Springer-Verlag.

Tai, P.Y.P. (1988). Long-term storage of *Saccharum spontaneum* L. pollen at low temperature. *Sugar Cane, Spring supplement,* pp. 12–16.

Telaye, A., Bemiwal, S.P.S. & Gates, P. (1990). Effect of desiccation and storage on the viability of *Vicia faba* pollen. *FABIS Newslett.* 26:6–10.

Thomson, W.W. (1979). Ultrastructure of dry seed tissue after a non-aqueous primary fixation. *New Phytol.* 82:207–212.

Towill, L.E. (1984). Seed set with potato pollen stored at low temperatures. *American Potato J.* 61:569–575.

Towill, L.E. (1985). Low temperature and freeze- (vacuum-) drying preservation of pollen. In *Cryopreservation of Plant Cells and Organs,* ed. K.K. Kartha, pp. 171–198. FL: CRC Press.

Towill, L.E. (1991). Cryopreservation. In *In Vitro Methods for Conservation of Plant Genetic Resources,* ed. J.H. Dodds, pp. 41–70. London: Chapman and Hall.

Traub, H.P. (1936). Papaya pollen germination and storage. *Proc. Amer. Soc. Hort. Sci.* 34:18.

Vasil, I.K. (1961). Physiology of pollen. *Bot. Rev.* 27:326–381.

Visser, T. (1955). Germination and storage of pollen. *Meded. Landb. Wageningen* 55:1–68.

Vithanage, H.I.M.V. & Alexander, D.M. (1985). Synchronous flowering and pollen storage techniques as aids to artificial hybridization in pistachio (*Pistacia* ssp.). *J. Hort. Sci.* 60:107–113.

Wakisaka, I. (1964). Ultra low temperature of pollens of Japanese persimons (*Diospyros kaki* L. f). *Engei Gakkai Zasshi* 33:291.

Walden, D.B. (1967). Male gametophyte of *Zea mays* L. 1. Some factors influencing fertilization. *Crop Sci.* 7:441–444.

Weatherhead, M.A., Grout, B.W.W. & Henshaw, G.G. (1978). Advantage of storage of potato pollen in liquid nitrogen. *Potato Res.* 21:331.

Whitehead, R.A. (1963). The processing of coconut pollen. *Euphytica* 12:167–177.

Wilson, A.T., Vickers, M. & Mann, L.R.B. (1979). Metabolism in dry pollen: a novel technique for studying anhydrobiosis. *Naturwissenschaften* 66:53–54.

Yates, I.E. & Sparks, D. (1990). Three-year-old pecan pollen retains fertility. *J. Amer. Soc. Hort. Sci.* 115:359–363.

15

Mentor effects in pistil-mediated pollen–pollen interactions

M. VILLAR and M. GAGET-FAUROBERT

Summary   315
Introduction   315
Review of cases reported   320
    Overview of cases by crops   320
    Mentor treatments and their efficiency   322
Mentor pollen functions   323
    Pistil-mediated pollen–pollen interactions   323
    Hypotheses on the mechanism of mentor pollen action   326
Conclusions   329
References   329

Summary

Recent results on the use of mentor pollen to overcome self- or interspecific incompatibility are reviewed. The different treatments used to prepare the mentor pollen and their efficiency are analyzed, especially on tree, fruit, and vegetable crops. Four hypotheses have been put forward to explain the action of mentor pollen: (1) It provides recognition substances; (2) it provides pollen growth promoting substances; (3) it activates the style and/or the ovary; and (4) it promotes fruit retention.

Introduction

Hybridization among closely related species and selfing of self-incompatible species have always been a challenge in plant improvement. Therefore, breeders are constantly searching for methods of modifying or bypassing incompatibility barriers. Many techniques have been developed to overcome such barriers (see Chapter 13, and Knox 1984).

More than a century ago, the Russian plant breeder Michurin (1855–1935) developed a method for improving wide hybridization. He added to the foreign pollen a small amount of pollen of the maternal species to stimulate the pistil for fertilization with the alien pollen (see history in Stettler and Ager 1984). This phenomenon is now referred to as "mentor effect" and has been reviewed by Stettler and Ager (1984) and Knox et al. (1987). Incompatibility barriers can occur during different stages of the progamic phase: on the stig-

315

Table 15.1. *Synopsis of studies reporting mentor effects since March 1982 (for previous studies, see Stettler and Ager 1984)*

Attempted mating	Pollen treatment	Outcome	Reference
INTERGENERIC			
Malus sp. × *Pyrus* sp.	PMX with UT pollen (mix / double pollinations)	No hybrids	Visser & Marcucci 1986
Salix sp. × *Vetrix* sp.	PMX with IRR pollen, PMX with MeOH pollen	Verified hybrids	Hathaway 1987
INTERSPECIFIC			
(*Populus alba* × *P. glandulosa*) × *P. maximowiczii*	PMX with MeOH pollen	Verified hybrids	Koo et al. 1987
Brassica campestris × *B. oleracea*	PMX with IRR pollen	Stigmatic barrier overcome, one hybrid	Sarla 1988
Populus nigra × *P. alba*	PMX with IRR or FT pollen	Stylar barriers overcome, no hybrids	Gaget et al. 1989
INTRASPECIFIC Gametophytic self-incompatibility			
Malus sp. (Starkrimson)	SPMX with IRR pollen of Golden Delicious	No seed	Marcucci et al. 1984
Malus sp. (several varieties)	SPMX with UT marker pollen (DP)	Self–seed set	Visser et al. 1983
Malus sp. (several varieties)	SPMX with UT marker pollen (different ratios, SP or DP)	4% to 19.4% self-seed	Visser & Marcucci, 1984

Attempted mating	Pollen treatment	Outcome	Reference
Malus pumila (Macspur)	PMX with foreign pollen of another genus (DP)	Fruit set stimulated	Lane 1983
Malus sp. (several varieties)	SPMX with UT marker pollen (comparison SP/DP)	4% to 10% self-seed	Visser & Marcucci 1986
Prunus avium	PMX with foreign pollen of another genus (DP)	Fruit set stimulated	Lane 1983
Pyrus sp. (Doyenné du Comice)	SPMX with UT marker pollen (DP)	Self–seed set	Visser et al. 1983
Pyrus sp. (Bonne Louise, Doyenné du Comice)	SPMX with UT marker pollen (different ratios, SP or DP)	Self–seed set from 0% to 16.8%	Visser & Marcucci 1984
Pyrus sp.	SPMX with UT marker pollen (comparison SP/DP)	No seed	Visser & Marcucci 1986
Theobroma cacao	SPMX with IRR pollen	Self–seed	Adu-Ampomah et al. 1991

Abbreviations: PMX = pollen mix containing pollen of the desired cross and compatible mentor pollen from the maternal species. Instances where mentor is from another species are specifically noted.
SPMX = self-pollen mix containing self-pollen and compatible mentor pollen from the maternal species
IRR = pollen subjected to gamma irradiation
FT = pollen subjected to repeated freezing and thawing
MeOH = pollen washed in anhydrous methanol
UT = untreated pollen
SP = single pollination
DP = double pollinations with intervals of hours or days

matic surface; or during the growth of the pollen tube through the style, in the ovary, in the ovules, or in the embryo sac (Knox 1984). Mentor pollen techniques have been developed mainly in order to circumvent barriers on the stigmatic surface and barriers occurring during the growth of pollen tubes in the style. The concept of mentor effect depends on the combination of two sets of pollen. In the original experiments, incompatible pollen was mixed with viable compatible pollen (Stettler and Ager 1984). To eliminate the fertilization capacity of the compatible pollen, mentor pollen is now rendered genetically ineffective by treatments that block its fertilizing capacity. Therefore, this treatment represents a critical factor in the mentor effect. Ionizing radiation, washing in solvents, and repeated cycles of freezing and thawing of the compatible mentor pollen have been used. The original method is to apply the pollen mix (mentor pollen + incompatible pollen) on the stigmatic surface once (mainly in the proportion 1:1). Breeders dealing with overcoming self-incompatibility (SI) of apple and pear have introduced a delay in the application of the two pollen sets (Visser and Verhaegh 1980). Extracts of lipid or protein materials of the compatible pollen have also been used in combination with the incompatible pollen. The efficiency of these treatments will be analyzed in the first part of this chapter.

Table 15.1 lists the last 12 years of experiments dealing with mentor pollen (Stettler and Ager's 1984 review was concluded in March 1982). The table is a follow-up of Stettler and Ager's review and contains the same abbreviations. From the data in Table 15.1 and Stettler and Ager's review, two comments can be made.

First, in terms of the number of taxa tested, there are two distinct phases. In the first phase, from 1968 to 1981, experiments dealt with a wide array of genera and species. These studies involved a few species representing sporophytic self-incompatibility (SSI), but the vast majority of experiments dealt with gametophytic self-incompatibility (GSI) or interspecific incompatibility (Stettler and Ager 1984). In the second phase, from 1982 up to now, very few experiments were reported in the literature; most of them dealt with tree species of Salicaceae and Rosaceae (Table 15.1). It is interesting to note that overcoming SSI has never been reported during the recent period. In the case of SSI, the pollen is arrested on the stigma surface, whereas in the case of GSI, the pollen tubes are arrested in the style. From the experiments reported, it is clear that mentor pollen is rather ineffective in circumventing SSI; that is, the mentor pollen action is ineffective on the stigmatic surface.

Second, in terms of efficiency, the results of 66 experiments reported in the literature are presented in Table 15.2; about one-half were successful in terms of self-seed and interspecific hybrids produced.

What are the reasons for the low popularity of the mentor pollen technique? Although the technique is simple, it is not always successful; the results have been generally unpredictable due to the lack of precise experimental conditions for obtaining the maximum seed set. Moreover, in some experiments, overcoming stigmatic/stylar barriers has been successful, but breeders have

Table 15.2. *Efficiency of different treatments used to prepare mentor pollen (from Stettler and Ager's overview and Table 15.1). The success is expressed in terms of self-seed or interspecific hybrids obtained. The cases involving untreated compatible pollen are not reported.*

Pollen treatment	Number of experiments	Success
Irradiation	41	20
Solvent	15	5
Extracts of proteins or lipid materials	6	4
Thermic treatment	4	3
Total	66	32

pointed out other barriers, such as abortion of the selfed embryos (as for apple and pear: Visser and Marcucci 1986). Another argument in favor of the existence of additional incompatibility barriers or of hybrid inviability is reported in many cases, where hybrids or selfs are never obtained in numbers approaching those of compatible matings, even after a barrier to hybridization has supposedly been overcome (see Knox et al. 1987). Furthermore, in instances of intraspecific incompatibility, an effective screening is required to distinguish between the seeds/seedlings obtained from incompatible pollen and those obtained from compatible pollen.

During recent years, plant geneticists have also been much more enthusiastic about other methods of transforming and combining genomes, such as genetic engineering (Jouanin et al. 1993) and in vitro fertilization (see Chapter 19; also see reviews in Dumas and Gaude 1993).

After a period of extensive experimentation with the use of mentor pollen techniques, there remain only two groups of plants, fruit species, and tree species in which these techniques are effective. The first group includes apple and pear to obtain high yields of self-seed. The second is poplar, in which the objectives are to cross different species that are isolated by interspecific incompatibility barriers. These two models, plus the situation in vegetable crops, will be discussed in the first part of this chapter.

Although there has been a decline in the use of the mentor pollen technique, researchers have emphasized the need to explain the biological functions of the mentor pollen. Because the concept of the mentor effect depends on the combination of two pollen sets, pollen–pollen interactions have been investigated. These interactions were difficult to study, largely due to the difficulty in differentiating one pollen from the other in the pollen mix. The use of vital dyes or of pollen carrying marker genes has been very useful in this context (Visser et al. 1983; Mulcahy and Mulcahy 1986a). In the second part of this chapter, hypotheses on the action of the mentor pollen are presented with the new concepts on pollen–pollen (pollen tube–pollen tube) interactions.

Review of cases reported

This review integrates Table 15.1 and Stettler and Ager's (1984) survey.

Overview of cases by crops

Poplar species

Historically, the mentor method was applied for the first time in a wide hybridization trial in *Populus* (Stettler 1968) and used extensively by this research group (Guries and Stettler 1976; Stettler and Guries 1976; Stettler et al. 1980). These experiments were successfully repeated with other species of poplars by Knox et al. (1972b), Willing and Pryor (1976), Hamilton (1976), and Koo et al. (1987). The efficiency of these experiments in poplars is reported in terms of a Mentor index: the ratio of mentor seed set divided by compatible seed set (Knox et al. 1987). Values range from 0 to 0.6, demonstrating the variable success depending on the treatment and the species involved. Such data are also dependent on the authenticity of the hybrids produced; the hybrids have largely been characterized by phenotypic markers. However, mentor pollen was ineffective in the cross *P. nigra* × *P. alba* (Gaget et al. 1989).

A variety of treatments were used for these experiments, irradiation being the most frequent. Alternating cycles of freezing and thawing of mentor pollen have been used in different interspecific crosses, giving variable results (successful for Knox et al. 1972b, but not for Gaget et al. 1989). Although this technique did not result in interspecific hybrids, Gaget et al. have demonstrated the promise of this treatment – in terms of incompatible pollen tube enhancement in the style – as compared to irradiation or compatible pollen extracts. Methylated pollen was also used, with the same inconsistency of results (successful for Knox et al. 1972b and Koo et al. 1987, but not for Stettler and Guries 1976). Knox and his colleagues also extracted diffusible proteins from the pollen walls of compatible pollen, mixed this lyophilized dialysate with incompatible pollen, and were able to obtain interspecific hybrids.

These contradictory results cannot be explained by the difference in the species involved in these experiments. Methylated pollen and pollen extracts are mainly employed to bypass stigmatic barriers, but interspecific barriers in *Populus* sp. are known to be located primarily in the stylar tissues (Guries and Stettler 1976; Villar et al. 1993).

Apple and pear

The mentor effect has been extensively used in apple and pear in trying to enhance the production of self-seeds. These species are considered highly self-incompatible; in most varieties the SI pollen tubes stop growing in the upper half of the style. Nevertheless, self-seeds can sometimes be obtained on a few cultivars, but with a very low yield (Visser 1981; Visser and Marcucci 1984).

The first successful experiment by Dayton (1974) using a mix of methanol-treated pollen and untreated self-pollen was not confirmed by others on another cultivar (Williams and Church 1975).

From 1981 onwards, extensive studies have been carried out with different varieties of apple and pear by Visser's group. The first experiments with methylated or irradiated mentor pollen did not significantly improve the seed set by self-pollen or that after intercrossing (Visser 1981). One explanation of these variable results could be the differences in the genotypes used and the variation in field conditions, especially temperature at the time of pollination (demonstrated by Visser and Marcucci 1986).

Numerous experiments have been subsequently reported on the use of the pioneer pollen method, consisting of double pollination in which the first pollen (compatible pollen) is applied 1–2 days ahead of the second, self-incompatible pollen (Visser and Verhaegh 1980). Most of the time, the compatible pollen was not treated. A comparison of pioneer and mentor pollen in incompatible matings showed no differences in effectiveness between the two techniques in apple (Visser 1981). Thus, the use of the pioneer method has been widely used to increase self–seed set in apple and pear (Visser et al. 1983; Visser and Marcucci 1984, 1986). Variable results have, however, been found between trials in different years (Table 15.1).

Vegetable crops

The situation in vegetable crops is very different from that in tree and fruit crops, as one has to deal with a wide array of genera and species with different incompatibility systems.

COMPOSITEAE. First experiments by Howlett et al. (1975) in *Cosmos bipinnatus* reported the successful use of recognition pollen in overcoming a strong SSI barrier. Pollen mixes of irradiated compatible and live self-pollen, or a protein extract from exine diffusate from the compatible pollen mixed with self-pollen, were equally successful in setting seed at levels as high as one-quarter of those obtained from compatible matings.

BRASSICACEAE. Attempts to overcome SI with methanol treatment in *Brassica campestris* were unsuccessful (Sastri and Shivanna 1980). On the other hand, the interspecific cross *B. campestris* × *B.oleraceae* produced at least one true hybrid using irradiated mentor pollen (Sarla 1988).

CUCURBITACEAE. In order to introduce resistance genes to green mottle mosaic virus in *Cucumis* sp., interspecific hybridization was attempted with mentor pollen (den Nijs and Oost 1980). Irradiated pollen in mixture with foreign pollen increased fruit set and, in a few cases, embryogenesis. Due to early embryo degeneration, the hybrid nature of the progeny could not be checked.

SOLANACEAE. Experiments on *Nicotiana* sp. were reported by Pandey (1977) and Sree Ramulu et al. (1979). Irradiated mentor pollen was used in attempts to overcome both intra- and interspecific incompatibility. Interspecific and self-incompatibility were partially overcome. According to Pandey (1977), the role of mentor pollen may be to provide extra free pollen growth promoting substances. Furthermore, in other experiments, Pandey reported an apparent instance of specific gene transfer in this species via irradiated pollen: These experiments of "egg transformation" have been a subject of debate (Pandey 1986).

Mentor treatments and their efficiency

Two points must be considered for an efficient treatment of mentor pollen. First, the mentor pollen gametes must be inactivated or eliminated. This can be achieved by three treatments: accurate doses of gamma radiation, solvent treatment (during which the pollen is "killed"), or circumvention of intact pollen by the use of pollen wall extracts. For freeze-thawed pollen, no data are available on their viability. Second, the treated pollen must take part in the male–female dialogue within the pistil, exchanging signals and/or growth substances so as to allow the incompatible pollen to take part in the fertilization process.

In many cases, seeds have been produced, but not the self-seed or hybrid expected. Therefore, although being "treated," the mentor pollen has maintained its ability of fertilization, demonstrating that the treatment was not efficient. In that sense, the dose of irradiation must be well defined, because irradiation effects are variable between species (Marcucci et al. 1984; Falque et al. 1992). In other cases, no seeds have been produced due to too drastic a treatment or due to competition between the compatible mentor pollen and the incompatible pollen, the latter blocking the ovules (Knox et al. 1972b; Stettler and Guries 1976). The use of methanol-treated pollen or the use of protein and lipoidal extracts of the compatible pollen in the pollen mix can eliminate this problem: In these studies, the pollen was killed or not present in its intact form, and thus was not competing with incompatible pollen for available ovules.

Of the different treatments listed in Table 15.2, gamma radiation has been the most common treatment used. This is the oldest technique used to sterilize pollen (Stettler 1968) by strongly inhibiting the second pollen mitosis (review in Falque et al. 1992). About one-half of the attempts using this method have been successful. Even when it is not always easy to have access to radiation sources, this technique has been very popular. One of the consistent features of this treatment is that it permits pollen germination and pollen tube growth in vitro and in vivo (examples in Marcucci et al. 1984; Gaget et al. 1989; Falque et al. 1992).

Use of methanol to prepare the mentor pollen was introduced by Knox et al. (1972a, b), with their objective being pollen lethality with minimal damage to pollen wall macromolecules. The denaturing action of this alcohol has been observed in vitro and in vivo (Knox et al. 1972a, b; Dayton 1974; Visser 1981) and on apple pollen membranes by Calzoni and Speranza (1982). Overall, the solvent treatment has given the worst results, with only one-third of the attempted studies being successful (reviewed in Stettler and Ager 1984).

The use of protein or lipoidal extracts of the pollen has given good results, but this treatment is limited to barriers located on the stigmatic surface. These compounds may act as informational molecules during the pollen–pistil interactions (Nasrallah and Nasrallah 1993).

The use of cycles of freezing and thawing seems promising (Knox et al. 1972b; Gaget et al. 1989). The results are variable, because the damage inflicted by this treatment on the pollen (for example, ice crystal formation) remains unknown at the cellular and molecular levels. However, poplar pollen remains viable after 14 cycles of $-196°C/+37°C$ (Gaget et al. 1989).

Mentor pollen functions

Pistil-mediated pollen–pollen interactions

As a practical tool to the plant geneticist, mentor pollen technology provides a system that in some cases circumvents self- or interspecific incompatibility. To the plant physiologist, mentor pollen, and especially its relation with the incompatible pollen, provides an interesting system where cell–cell interactions can be studied. These pollen–pollen or pollen tube–pollen tube interactions take place within the female tissues and can be studied by experiments in which one pollen has a marker, such as resistance to a disease or a marker for red leaf color (Visser et al. 1983), or by the use of dyes (Mulcahy and Mulcahy 1986a). These interactions parallel the more classical pollen–pistil interactions (see Chapters 2 and 12, and Mascarenhas 1993). It is important to keep in mind that the pollen tube interacts intimately with the tissues of the pistil and that the stylar tissue serves as an ideal structure to facilitate the separation of fast and slow pollen; therefore, pollen tubes must possess elaborate signaling processes that allow the tubes to establish cell–cell communications with female tissues (see Chasan 1993). Pollen–ovary dialogue has also been demonstrated ("cross-talk": Jensen et al. 1983; "ovary activation": Mulcahy and Mulcahy 1986b).

Why not consider that dialogue could also be initiated with other pollen tubes? Evidence is presented here of male–male interactions (pollen tubes interacting with other pollen tubes, mediated by the pistil) that have been demonstrated to occur in natural and experimental conditions.

Gametophytic competition in natural conditions

Gametophytic competition has been proposed to have played an important role in the rapid evolution of angiosperms (Mulcahy 1979). Since 1979, information has been accumulated in support of that view, giving evidence for selection at the male gametophytic level in natural populations (reviewed in Hormaza and Herrero 1992). This idea is further supported by Snow and Spira (1991), who demonstrated that in *Hibiscus moscheutos* one of the two pollen donors was consistently delivering the faster pollen on all maternal genotypes tested. According to Hormaza and Herrero (1992), pollen competition could occur at two levels: either through direct pollen competition (physical competition based on the rate of pollen tube growth) or through interactions between the haploid and diploid genes (male–female interactions). Both phenomena, mutual stimulation among pollen tubes and mediation by the style, may be occurring in *Erythronium grandiflorum* (Cruzan 1990).

Density-dependent pollen tube growth in vivo

When many pollen tubes are present within the style, competition for limited stylar resources could reduce pollen tube growth rate. Conversely, Cruzan (1986) has reported increased growth rates with increasing numbers of pollen tubes. This density-dependent response (facilitation) could be due to greater availability of substrates among pollen tubes (Cruzan 1986). Such a population effect had been earlier recorded at the germination level, where large accumulations of pollen grains have a synergistic effect on germination (Brewbaker and Majumder 1961, cited in Herrero 1992). According to Herrero (1992), those pollen grains destined not to fertilize ovules could play an important part in assisting other pollen grains to do so.

Mentor pollen experiments

COMPARISON BETWEEN KILLED AND LIVING MENTOR POLLEN. According to the various mentor pollen experiments, it makes a marked difference whether dead or living pollen is used. Comparison of pollen treated with methanol (dead) and irradiated pollen (viable) has been made in pear (Visser and Oost 1982). Irradiated pollen appeared to break down GSI barriers. Dead pollen did not bring about any continued growth beyond this point. Apparently, the viable pollen (after irradiation or after freezing and thawing treatment) did not function primarily as a provider of "recognition" but rather "paved the way" in the style, mechanically and/or biochemically, as suggested with respect to the role played by the pioneer pollen (Visser and Verhaegh 1980). On the other hand, the stimulus resulting from the growth of the pollen tubes may be negated by the adverse effects of competition or by the presence of a lethal factor causing abortion of the selfed embryo.

COMPETITION BETWEEN MENTOR AND INCOMPATIBLE POLLEN FOR AVAILABLE OVULES. Untreated and irradiated pollen compete for available ovules on equal terms. In other words, irradiated pollen tubes reach their destination but their radiation-killed nuclei cannot effect normal fertilization (Knox et al. 1972b; Stettler and Guries 1976; Pandey 1977). In this respect, irradiation and freezing and thawing differ in their effect. Whereas irradiation does reduce pollen tube growth, freezing and thawing seems to retard it as well as affect the state of the nucleus (Pandey 1977). This may allow the slow-growing incompatible tubes to compete more successfully and thus result in more seed set (hypothesis from Pandey 1977). This is consistent with the results of Knox et al. (1972a), where the average number of capsules was 11.0 after freeze–thaw treatment, instead of 5.0 and 4.5 after the use of gamma irradiated or methanol-killed mentor pollen, respectively. Thus, the freezing and thawing treatment may be more effective for the preparation of the mentor pollen to bypass stylar barriers (Gaget et al. 1989).

PROPORTION OF MENTOR POLLEN IN THE MIX. The proportion of mentor pollen to incompatible pollen may be an important factor. In poplar experiments, Willing and Pryor (1976) have shown that until the mentor pollen is about 40% of the mixture, there is little or no overcoming of the incompatibility barrier. Beyond this level and up to 90%, the viable incompatible pollen produces increasing amounts of seeds (unfortunately there is no information of the hybrid status of the seeds). Dayton (1974) found that in apples the most favorable proportion of mentor to incompatible pollen was 2:1. Furthermore, in apple and pear, Visser and Marcucci (1984) tested mixtures of compatible and self-incompatible pollen in ratios of 1:1, 1:2, and 1:9 applied once or twice or followed by self-pollination. Fruit set tended to decrease and the percentage of self-seed to increase (a maximum of 16.8% in pear and 19.4% in apple) with increasing amounts of self-pollen in the pollinations.

Pioneer pollen experiments

Pollinating apple cultivars twice with compatible pollen at an interval of one or two days produced twice as many seeds per pollinated flower when compared to single pollination (Visser and Verhaegh 1980). Furthermore, these authors found that the second pollen (carrying marker genes) was twice as effective as the pollen applied first. The first pollen appeared to promote the efficiency of the second pollen, but this promotion was dependent on the temperature and the interval between pollinations (Visser and Marcucci 1983). It may "pave the way" through the stylar tissue, allowing the tubes of the second pollen to grow faster: That is why the first pollen is called the pioneer pollen. "The tubes have a greater reserve for subsequent growth than the relatively exhausted tubes of the first" (Visser and Verhaegh 1980, p. 390). In overcoming SI, the advantage of the pioneer pollen may be that the interval between pollinations allows time to condition the style for the passage of the self-tubes (Visser and Oost 1982).

In studying the mixture of incompatible and untreated compatible pioneer pollen, Visser and Marcucci (1986) have reported similar trends between double pollination (at a one-day interval) of self-incompatible/compatible pollen and incongruous/compatible pollen, in terms of the failure to produce viable seeds. In both cases, in the presence of the compatible pollen, the self-pollen or incongruous pollen tube was able to reach and fertilize the egg, although abortion occurred afterwards. This applies only when self- or

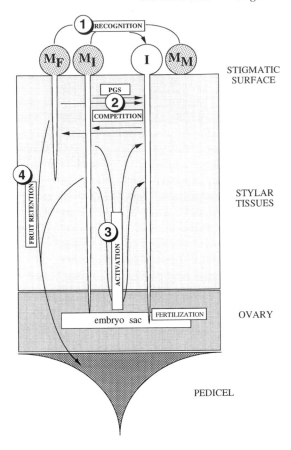

Figure 15.1 Diagrammatic presentation of the different functions of the mentor pollen on the stigmatic surface and in the stylar tissues, the ovary, and the pedicel of the flower: (1) providing recognition substances; (2) providing pollen growth promoting substances (the stimulus resulting from the presence of these substances may be negated by the adverse effect of competition); (3) activating the style and/or the ovary; and (4) promoting fruit retention. *Abbreviations:* I: incompatible pollen; MF: mentor pollen subjected to freezing and thawing; MI: mentor pollen subjected to gamma irradiation; MM: mentor pollen washed in anhydrous methanol.

incompatible pollen was applied first. In the reverse case, self- or incongruous pollen could not compete with the viable compatible pollen (Visser and Marcucci 1986).

Hypotheses on the mechanism of mentor pollen action

At least four hypotheses on the mechanism of mentor pollen action can be presented (Figure 15.1).

1. Mentor pollen provides recognition substances or stimulus for pollen tube growth

Proteins located in the pollen wall could serve as "recognition substances." According to Knox et al. (1972a), in *Populus* sp., mobile surface proteins are the factors involved in the regulation of the mentor effect. Foreign ("incompatible") pollen could borrow them to gain access to the ovules. Since these experiments were conducted, the role of proteins in pollen–pistil interactions in poplars has been confirmed by the characterization of glycoproteins in pollen extracts and diffusates in vitro and in vivo (reviewed in Villar et al. 1993).

Willing and Pryor (1976) and Hamilton (1976) have pointed out another group of compounds serving the mentor function, namely lipid materials. The former authors suggested that pollen–pistil recognition in *Populus* sp. involves lipid factors on the pollen (P-factor) and on the stigmatic surface (S-factor). Treatment of incompatible pollen with organic solvents or coating of incompatible pollen with solvent-extracted tryphine from mentor pollen wall resulted in successful hybridization. Moreover, Hamilton demonstrated the presence of lipid materials within waxy mounds on the stigmatic papillae. Hamilton's hypothesis is that these stigmatic lipids are degraded by pollen enzymes, thus providing stimulus for pollen tube growth. Thus, mentor pollen could provide a P-factor that interacts with an S-factor on the stigma to render the latter accessible to incompatible pollen.

2. Mentor pollen provides pollen growth promoting substances

In some species, pollination stimulates the transmitting tissue to release carbohydrates that are later used to support pollen tube growth (reviewed in Herrero 1992). In that sense, compatible pollen tubes may provide pollen growth promoting substances to incompatible tubes (Pandey 1977). In poplars, our recent work on the physiology of pollen tube growth in vivo has clearly demonstrated the involvement of a pollen enzyme (β-galactosidase) in heterotrophic nutrition of the pollen tube (Villar et al. 1993). Such a system, active in the mentor pollen tube after treatment by freezing and thawing or by irradiation (Gaget 1988, in Villar et al. 1993), could provide growth substances to the tubes of the surrounding incompatible pollen. Furthermore, in mentor experiments on fruit crops involving foreign pollen, Lane (1983) suggested that pollen growth could be promoted by a diffusible growth regulator, such as gibberellin.

3. Mentor pollen activates the style and/or the ovary

Activation of the style by compatible and incompatible grains has been hypothesized in mixed pollinations in *Petunia hybrida*, because the number of pollen tubes reaching the ovary was considerably higher than would be anticipated (Herrero and Dickinson 1980). This total population could induce growth of a much larger number of compatible tubes than would occur if solely the compatible grains had been placed on the stigma; the incompatible pollen in some ways "primes" the style for growth of compatible pollen tubes. Similar observations have been made by Mulcahy and Mulcahy (1986a). Using a vital dye (pollen was presoaked in medium with Hoechst 33258), they have shown that mentor pollen has a positive effect on the pollen tube length of the incompatible pollen in *Nicotiana alata*.

A possible explanation of mentor pollen may be the interaction between the pollen and the ovary (Ottaviano and Mulcahy 1989). In cotton, the first contact between stigma and pollen leads to the transmission of a signal to the ovary (Jensen et al. 1983). Then, the ovary undergoes multiple changes, including synergid breakdown and polar nuclei fusion. Furthermore, Mulcahy and Mulcahy (1986b, 1987) have demonstrated that in *Nicotiana alata*, a countersignal is emitted from the ovary to the approaching pollen tubes, thus strengthening the idea that compatibility is the ability to activate the ovary. In the cases of mentor and pioneer effects, incompatible pollen could be stimulated by a signal emitted by the ovary in response to compatible pollen that had been applied either simultaneously (as in mentor effect) or previously (as in pioneer effects). From these three mechanisms of action, we can quote Mulcahy and Mulcahy (1986a, p. 177): "Once the compatible pollen established the conversation, will the incompatible, always around, join in?"

4. Mentor pollen promotes fruit retention

Poplar breeders have observed that one direct phenomenon of interspecific incompatibility is the early abscission of flowers (and sometimes of the whole catkin) soon after pollination (Guries and Stettler 1976). In this light, the primary function of the irradiated mentor pollen could be to promote catkin retention, thus allowing the hybrid embryo to mature (Stettler and Guries 1976). The presence of pollen tubes is necessary in the female tissues to prevent abscission and to trigger fruit development (irradiation and freezing and thawing treatments do not inhibit pollen germination). This effect was also reported in *Cucumis* sp. by den Nijs and Oost (1980) and in apple and pear by Visser (1981). Recent studies have demonstrated that abscission of flowers in the interspecific cross *Populus nigra* (female) x *P. deltoides* (male) is correlated with a threefold increase of abscisic acid (ABA) in the pedicel of the flower five days after pollination (Label et al. 1994). Thus, the role of the mentor pollen (via the pollen tube in the pistil) could be to provide stimuli for fruit development, preventing the accumulation of ABA in the pedicel and thus allowing the maturation of the hybrid embryos. Experiments are currently under way in the authors' lab to test this hypothesis.

The survey of various experiments points out that the mentor effects could occur at different levels of the progamic phase (Figure 15.1). Mentor pollen functions are mediated via pollen–pistil and pollen–pollen interactions, as well as through effects on stylar tissues and ovaries before the arrival of pollen tubes.

Conclusions

This chapter has clearly pointed out that there are no general rules in terms of overcoming SI or interspecific incompatibility with mentor pollen. It is a case-to-case study and, even if the mentor concept is less popular, it is a simple technique to bypass sexual incompatibility barriers. For such attempts, the choice of treatment is a critical factor and information on the reproductive biology of the species is a prerequisite.

This chapter has also emphasized the gametophytic competition that occurs in natural populations. This competition is based on male–male interactions, adding a new level of biocommunication to the classical pollen–pistil dialogue. Pollen–stigma and/or pollen tube–style interactions have been widely studied and have been characterized by expressions such as "cross-talk" (Jensen et al. 1983, p. 72), "pollen–stigma signaling" (Nasrallah and Nasrallah 1993, p. 1325), or "signal-receptor like interactions" (Dumas and Gaude 1993, p. 203). Similarly, but under more empirical terms, pollen–pollen interactions mediated by the pistil have been defined, such as "mentor pollen could pave the way in the style" (Visser and Verhaegh 1980, p. 390), "the incompatible pollen thus effectively borrows recognition material from the killed pollen" (Knox et al. 1972b, p 68), and incompatible pollen could "prime the style for growth of compatible pollen tubes" (Herrero and Dickinson 1980, p. 217).

In conclusion, evidence is emerging on how the mentor and the incompatible pollen (or pollen tube) interact and affect each other. Further research on these interactions is needed to fully understand the role of gametophytic competition in natural and experimental conditions.

References

Adu-Ampomah, Y., Novak, F.J., Klu, G.Y.P. & Lamptey, T.V.O. (1991). Use of irradiated pollen as mentor pollen to induce self-fertilization of two self-incompatible Upper Amazon cacao clones. *Euphytica* 51:219–225.

Calzoni, G.L. & Speranza, A. (1982). Effect of methanol and gamma irradiation on enzymic activities of apple pollen. *Scientia Hort.* 17:231–239.

Chasan, R. (1993). Papillar cells and pollen tubes: a tale of two plants. *Plant Cell* 5:237–239.

Cruzan, M.B. (1986). Pollen tube distributions in *Nicotiana glauca*: evidence for density dependent growth. *Amer. J. Bot.* 73:902–907.

Cruzan, M.B. (1990). Pollen–pollen and pollen–style interactions during pollen tube growth in *Erythronium grandiflorum* (*Liliaceae*). *Amer. J. Bot.* 77:116–122.

Dayton, D.F. (1974). Overcoming self-incompatibility in apple with killed compatible pollen. *J. Amer. Soc. Hort. Sci.* 99:190–192.

Dumas, C. & Gaude, T. (1993). Progress towards understanding fertilization in Angiosperms. In *The Molecular Biology of Flowering*, ed. B.R. Jordan, pp. 185–217. Lijttlchampton: C.A.B. International.

Falque, M., Kodia, A.A., Sounigo, O., Eskes, A.B. & Charrier, A. (1992). Gamma-irradiation of cacao (*Theobroma cacao* L.) pollen: effect on pollen grain viability, germination and mitosis and on fruit set. *Euphytica* 64:167–172.

Gaget, M., Villar, M. & Dumas, C. (1989). The mentor pollen phenomenon in poplars: a new concept. *Theor. Appl. Genet.* 78:129–135.

Guries, R.P. & Stettler, R.F. (1976) Pre-fertilization barriers to hybridization in the poplars. *Silvae Genet.* 25:37–44.

Hamilton, D. (1976). Intersectional incompatibility in *Populus*. Ph.D. Thesis, Australian National Univ. Canberra.

Hathaway, R.L. (1987). Inter-subgeneric hybridisation in Willows (*Salix*). In *IEA Task II - Biomass Growth and Production Technology*. Technical Advisory Committee Meeting and Activity Workshops, 08/24–09/04/1987, Oulu, Finland and Uppsala, Sweden.

Herrero, M. (1992). From pollination to fertilization in fruit trees. *Plant Growth Regul.* 11:27–32.

Herrero, M. & Dickinson, H.G. (1980). Pollen tube growth following compatible and incompatible intraspecific pollination in *Petunia hybrida*. *Planta* 148:217–221.

Hormaza, J.I. & Herrero, M. (1992). Pollen selection. *Theor. Appl. Genet.* 83:663–672.

Howlett, B.J, Knox, R.B., Paxton, J.B. & Heslop-Harrison, J. (1975). Pollen-wall proteins: physiochemical characterization and its role in self-incompatibility in *Cosmos bipinnatus*. *J. Proc. R. Soc. B.* 188:167–182.

Jensen, W.A., Ashton, M. & Beasley, C.A. (1983). Pollen tube–embryo sac interaction in cotton. In *Pollen: Biology and Implications in Plant Breeding*, eds. D.L. Mulcahy and E. Ottaviano, pp. 67–72. New York: Elsevier.

Jouanin, L., Brasileiro, A.C.M., Leplé, J.C., Pilate, G. & Cornu, D. (1993). Genetic transformation: a short review of methods and their applications, results and perspectives for forest trees. *Ann. Sci. For.* 50:325–338.

Knox, R.B. (1984). Pollen–pistil interactions. In *Encyclopedia of Plant Physiology, Vol. 17, Cellular Interactions*, eds. H.F. Linskens and J. Heslop-Harrison, pp. 508–608. Berlin: Springer-Verlag.

Knox, R.B., Gaget, M. & Dumas, C. (1987). Mentor pollen techniques. *Int. Rev. Cytol.* 107:315–332.

Knox, R.B., Willing, R.R. & Ashford, A.E. (1972a). Role of pollen wall proteins as recognition substances in interspecific incompatibility in poplars. *Nature* 237:381–383.

Knox, R.B., Willing, R.R. & Pryor, L.D. (1972b). Interspecific hybridization in poplars using recognition pollen. *Silvae Genet.* 21:65–69.

Koo, Y.B., Noh, E.R. & Lee, S.K. (1987). Hybridization between incompatible poplar species using mentor pollen. *Res. Rep. Inst. For. Gen. Korea* 23:23–29.

Label, P., Imbault, N. & Villar M. (1994). ELISA quantitation and GC-MS identification of abscisic acid in stigma, ovary and pedicel of pollinated poplar flowers (*Populus nigra* L.). *Tree Physiol.* 14:521–530

Lane, W.D. (1983). Fruit-set after pre-treatment with foreign compared with killed compatible pollen. *Can. J. Bot.* 62:1678–1681.

Marcucci, M.C., Ragazzini, D. & Sansavini, S. (1984). The effects of gamma and laser rays on the functioning of apple pollen in pollination and mentor pollen experiments. *J. Hort. Sci.* 59:57–61.

Mascarenhas, J.P. (1993). Molecular mechanisms of pollen tube growth and differentiation. *Plant Cell* 5:1303–1314.

Mulcahy, D.L. (1979). The rise of Angiosperms: a genecological factor. *Science* 206:20–23.

Mulcahy, G.B. & Mulcahy, D.L. (1986a). Pollen–pistil interactions. In *Biotechnology and Ecology of Pollen.*, eds. D.L. Mulcahy, G.B. Mulcahy, and E. Ottaviano, pp. 173–178. Berlin: Springer-Verlag.

Mulcahy, G.B. & Mulcahy, D.L. (1986b). More evidence on the preponderant influence of the pistil on pollen tube growth. In *Biology of Reproduction and Cell Motility in Plants and Animals,* eds. M. Cresti and R. Dallai, pp. 139–144. Siena, Italy: University of Siena.

Mulcahy, G.B. & Mulcahy, D.L. (1987). Induced pollen tube directionality. *Amer. J. Bot.* 74:1458–1459.

Nasrallah, J.B. & Nasrallah, M.E. (1993). Pollen–stigma signaling in the sporophytic self-incompatibility response. *Plant Cell* 5:1325–1335.

Nijs, A.P.M. den & Oost, E.H. (1980). Effect of the mentor pollen on pistil–pollen incongruities among species of *Cucumis. Euphytica* 29:267–271.

Ottaviano, E. & Mulcahy, D.L. (1989). Genetics of angiosperm pollen. *Advances in Genetics* 26:1–64.

Pandey, K.K. (1977). Mentor pollen: possible role of wall-held pollen growth promoting substances in overcoming intra- and interspecific incompatibility. *Genetica* 47:219–229.

Pandey, K.K. (1986). "Egg transformation" induced by irradiated pollen in *Nicotiana*: critical appraisal of Chyi and Sanford's observations. *Theor. Appl. Genet.* 72:739–742.

Sarla, N. (1988). Overcoming interspecific incompatibility in the cross *Brassica campestris* ssp. japonica × *Brassica oleracea* var. *botrytis* using irradiated mentor pollen. *Biologia Plantarum* 30:384–386.

Sastri, D.C. & Shivanna, K.R. (1980). Efficacy of mentor pollen in overcoming intraspecific incompatibility in *Petunia, Raphanus* and *Brassica. J. Cytol. Genet.* 15:107–112.

Snow, A.A. & Spira, T.P. (1991). Pollen vigour and the selection for sexual selection in plants. *Nature* 352:796–797.

Sree Ramulu, K., Bredemeijer, G.M.M., Dijkhuis, P., Nettancourt, D. de & Schibilla, H. (1979). Mentor pollen effects on gametophytic incompatibility in *Nicotiana, Oenothera* and *Lycopersicon. Theor. Appl. Genet.* 54:215–218.

Stettler, R.F. (1968). Irradiated mentor pollen: its use in remote hybridization of black cottonwood. *Nature* 219:746–747.

Stettler, R.F. & Ager, A.A. (1984). Mentor effects in pollen interactions. In *Cellular Interactions*, eds. H.F. Linskens and J. Heslop-Harrison, pp. 609–623. Berlin: Springer-Verlag

Stettler, R.F. & Guries, R.P. (1976). The mentor pollen phenomenon in black cottonwood. *Can. J. Bot.* 54:820–830.

Stettler, R.F., Koster, R. & Steenackers, V. (1980). Interspecific crossability studies in poplars. *Theor. Appl. Genet.* 58:273–282.

Villar, M., Gaget, M., Rougier, M. & Dumas, C. (1993). Pollen–pistil interactions in *Populus:* ß-galactosidase activity associated with pollen tube growth during the crosses *Populus nigra* × *P. nigra* and *P. nigra* × *P. alba. Sex Plant Reprod.* 6:249–256.

Visser, T. (1981). Pollen and pollination experiments. IV. "Mentor pollen" and "Pioneer pollen" techniques regarding incompatibility and incongruity in apple and pear. *Euphytica* 30:363–369.

Visser, T. & Marcucci, M.C. (1983). Pollen and pollination experiments. IX. The pioneer pollen effect in apple and pear related to the interval between pollinations and the temperature. *Euphytica* 32: 703-709.

Visser, T. & Marcucci, M.C. (1984). The interaction between compatible and self-incompatible pollen of apple and pear as influenced by their ratio in the pollen cloud. *Euphytica* 33:699–704.

Visser, T. & Marcucci, M.C. (1986). The performance of double and mixed pollinations with compatible and self-incompatible or incongruous pollen of pear and apple. *Euphytica* 35:1011–1015.

Visser, T. & Oost, E.H. (1982). Pollen and pollination experiments. V. An empirical basis for mentor pollen effect observed on the growth of incompatible pollen tubes in pear. *Euphytica* 31:305–312.

Visser, T. & Verhaegh, J.J. (1980). Pollen and pollination experiments. II. The influence of the first pollination on the effectiveness of the second one in apple. *Euphytica* 29:385–390.

Visser, T., Verhaegh, J.J., Marcucci, M.C. & Uijtewaal, B.A. (1983). Pollen and pollination experiments. VIII. The effect of successive pollinations with compatible and self-incompatible pollen in apple and pear. *Euphytica* 32:57–64.

Williams, R.R. & Church, R.M. (1975). The effect of killed compatible pollen on self-compatibility in apple. *J. Hort. Sci.* 50:457–461.

Willing, R.R. & Pryor, L.D. (1976). Interspecific hybridization in poplar. *Theor. Appl. Genet.* 47:141–151.

16

Pollen tube growth and pollen selection

M. SARI-GORLA and C. FROVA

Summary 333
Introduction 333
  In vitro pollen germination and tube growth 334
  Pollen germination and tube growth in vivo 336
Genetic control of pollen competitive ability 336
Pollen–pollen and pollen–pistil interaction 338
  Role of the pistil in pollen function 338
  Genetic control of pollen–pistil interaction 340
Applications in plant breeding 342
  Pollen assay for the identification of plants with desirable traits 342
  Pollen selection 345
Concluding remarks 347
References 348

Summary

Pollen tube growth is the result of rapid polar extension of a single cell and is fundamentally different from the standard model of cellular growth in plants. The pollen tube emerges from the germpore, penetrates the stigma, and grows within the intercellular matrix of the style. As the pollen tube elongates, the living cytoplasm is confined within the tip of the pollen tube, isolated by callose plugs. The process concludes when the tube tip reaches the micropyle and discharges the sperm nuclei into the embryo sac. This chapter reviews the information available concerning the genetic control of these processes, the role of the pistil in supporting and modulating pollen function, and the genetic basis of the interaction between pollen and pistil. As well, since a positive correlation between pollen and sporophytic responses to certain traits has been demonstrated in many systems, the use of pollen assays and pollen selection for the identification of plants with desirable traits is discussed. These technologies, when integrated into the conventional breeding program, form powerful breeding tools.

Introduction

The male gametophyte has a very short life-span in the biological cycle of higher plants. Nevertheless, it is able to exist as a free organism and to express a large portion of the plant genome, including many specific genes. These genes control the development and function of a highly specialized structure, the pollen tube, which can grow at a very fast rate and respond to a sophisticated cell–cell communication system. Pollen tube growth is a unique model of plant cell growth, which more resembles fungal growth than the cell growth in a higher organism. Pollen tube growth is the result of a rapid polar extension of a single cell, fundamentally different from the standard model of cellular growth in plants.

In vitro pollen germination and tube growth

In vitro pollen germination is an important technique not only for understanding the fundamental problems concerning pollen function, but also in many areas of pollen biotechnology, particularly pollen selection.

A number of culture media and methods have been used for the culture of pollen in vitro. These have been described in detail by Pfahler (1992), Shivanna and Rangaswamy (1992), and Kearns and Inouye (1993). In general, the media contain calcium and boron, combined with 10%–20% sucrose as an osmoticum and nutritional source (Brewbaker and Kwack 1963; Dickinson 1967; Schrauwen and Linskens 1967; Pfahler 1967a). The use of more complex media by addition of nutrients, such as dicarboxylic acid (Iwanami 1980), stigma extracts (Hodgkin 1987), or the substitution of sucrose with other sugars or polyethylene glycol (Zhang and Croes 1982) does not significantly improve the rate of germination and tube elongation in many systems.

A widely used method for culturing pollen grains is the hanging drop technique (Shivanna and Rangaswamy 1992). A small droplet of liquid germination medium is suspended from a coverslip that is supported over the well of a cavity slide; the edge is sealed with lanolin or mineral oil, providing a germination chamber that minimizes evaporation. Pollen grains can also be cultured on the surface of solid media in a closed container such as a petri dish; the tubes grow just beneath the medium surface and are easily observed and measured with a microscope. For physiological and biochemical studies that require a large amount of pollen, a convenient method is liquid shake cultures, normally set up in small conical flasks.

The extent of pollen tube elongation in vitro often does not equal that obtained in vivo (Vasil 1987). Nonetheless, pollen of many plant species, typically binucleate pollen, germinates and grows well in culture. Trinucleate pollen and, in particular, the pollen of cereals (with rare exceptions such as pearl millet) are very difficult to grow in vitro and the extent of tube elongation is not comparable to the in vivo tube length. In maize, for instance, maximum in vitro elongation is 2 mm, whereas in vivo it can reach 20 cm or more. The different behavior of the two pollen types probably reflects metabolic differences between them. Whereas binucleate pollen starts its growth autotrophically, and switches later to heterotrophic nutrition, trinucleate pollen grows heterotrophically from the start (D. L. Mulcahy and Mulcahy 1988).

In vitro germination of maize pollen is often limited by the frequent bursting of the grains or of the emerged tube tips. Although a high sugar concentration prevents bursting, it reduces viability and germinability. A method that gives good results for this species has been developed by Cheng and Freeling (1976); the culture medium is allowed to solidify in a glass cylinder for 3–4 days at 4°C, after which thin slices of medium (1–2 mm thick) are cut and placed in a petri dish immediately before pollen culture. Kranz and Lorz (1990) developed a new procedure for in vitro pollen germination by adding small amounts of distilled water locally to grains overlaid by mineral or silicone oil.

A major problem when pollen cannot be grown in liquid medium is the difficulty of recovering germinated pollen from the medium for use in biochemical studies or fertilization. A simple way to overcome this problem is to cover the solid medium with a membrane, on the surface of which pollen is dusted; after germination, the membrane can be removed and the pollen recovered from it. A more sophisticated system was developed by Kranz and Lorz (1990) for in vitro fertilization of maize; silks were inserted into glass capillary tubes and a suspension of pollen grains, pregerminated in mineral oil, was drawn into the capillary. The oil was immediately sucked out, leaving most of the pollen grains directly in contact with the silk hairs.

For some purposes, such as gametophytic selection, the ability to separate germinated and ungerminated pollen after selective treatment would provide a tool for identifying stress-tolerant and stress-sensitive alleles carried by the pollen grains. A relatively simple, nontoxic and effective method based on a screen and column system has been developed by Hi Zhang et al. (1993) for maize pollen. Macrofiltration screens are laid on top of a column filled with culture solution; the basal part of the column is packed with sea sand of such a size that the separation of germinated and ungerminated pollen is allowed. Recovery was 18% of the total germinated grains; recovered pollen retained viability and continued tube growth when placed in a culture medium.

A very elegant procedure to study many aspects of pollen tube growth is the semi vivo technique (G. B. Mulcahy and Mulcahy 1985; Shivanna et al. 1988, 1991). Pollinated styles are incubated in a saturated atmosphere for about 24 hours, the styles are cut just ahead of the growing pollen tubes and the cut end is dipped in a suitable medium; pollen tubes emerge from the cut end of the styles and can be easily studied or excised and collected (see Shivanna and Rangaswamy 1992). Unfortunately, this technique cannot be applied to all plant species; in maize, for instance, the wound at the cut end of the silk heals soon after the cut, so that the growing tube tips cannot emerge.

Often, application of pollen selection is more effective if applied to the developing pollen instead of to mature pollen. For this, culturing of microspores and their maturation in vitro would be very effective (Heberle-Bors 1989; Stauffer et al 1991; Barnabás and Kovacs 1992; Pareddy and Petolino 1992).

Pollen germination and tube growth in vivo

Germination in vivo begins with rehydration, mainly by means of water transfer from the stigma. Pollen tubes emerge from the germpore; in grasses this
occurs within a few minutes after pollination, whereas in other species the
time of tube emergence varies. In some gymnosperms, germination and tube
development may take months. Pollen tubes penetrate the stigma and grow by
tip extension within the intercellular matrix at an astounding rate, in comparison with all the other plant cell types; in maize, for instance, the rate of tube
growth is about 1 cm/hr (Barnabás and Fridvalszky 1984).

As the pollen tube elongates, the pollen cytoplasm, containing the vegetative nucleus and sperm cells, moves into the tip of the pollen tube; behind the
tip, callose plugs are formed, which isolate the vital tip from the old tube wall.
Thus, after the initial period of growth, the cytoplasm occupies only a small
part of the tube. The process has been compared to a peculiar case of cell
migration in plant development (Sanders and Lord 1989), which terminates
with the entrance of the tube tip into the micropyle and the release of sperm
cells into the embryo sac.

Genetic control of pollen competitive ability

Pollen competitive ability (PCA) describes the reproductive success of a
pollen grain and can therefore be considered as equivalent to pollen fitness; it
is the result of various pre- and postshedding components, among which
pollen germinability and tube growth rate are the most important. In fact, the
most intense competition occurs because of the great difference in the number of male and female gametes; many pollen grains germinate on the same
stigma and many pollen tubes grow in the same style in order to fertilize a few
ovules present in the ovary.

In several species it has been observed that PCA shows a large amount of
genetic variability. Its biological significance is based on the assumptions that
(a) pollen performance, in terms of germinability and tube growth rate, is determined, at least in part, by the genotype of the pollen itself, so that genetically
determined differences in pollen performance lead to nonrandom fertilization;
and (b) many of the genes expressed in pollen are also expressed during the
sporophytic phase of the life cycle, so that nonrandom fertilization modifies the
genetic composition of the sporophytic generation (see Chapter 3).

The strategies adopted to study PCA are based on the experimental variation of either pollination intensity (amount of pollen deposited on the stigma)
or the distance between the stigma and the ovule, using the pollen population
produced by a single heterozygous plant. The reduction, in the resulting
sporophytic population, of the variance and the increase in the means of characters expressing plant vigor and fertility indicate that (a) only the gametophytes having greater competitive ability were able to reach the ovules and
achieve fertilization, and (b) major components of PCA are controlled to a

significant extent by gametophytic genes that are expressed also in the sporophyte, where they contribute to the determination of plant vigor (Ottaviano and Sari-Gorla 1993 for review).

A more detailed analysis of the phenomenon has been carried out in maize, in which the length of the styles varies according to the position of the ovules on the ear, increasing from the apex to the base. When a mixture of two genetically marked pollen types is used for pollination, the relative pollen tube growth rates can be estimated on the basis of the proportion of ovules fertilized by one of the two competing pollen types from the apex to the base of the ear. The results so obtained have revealed the existence of a large amount of variability for male gametophytic fitness, at least a portion of which is due to postmeiotic gene expression. The data have also indicated that the pollen tube growth rate can be significantly affected by gametophytic selection and results in a correlated response for some sporophytic traits (Ottaviano et al. 1988a).

Pollen tube competition and its effects on the resultant sporophytic generation have also been observed in natural populations (Ottaviano and Mulcahy 1989). The experimental demonstrations of gametophytically controlled pollen competition have been carried out using pollen from a single donor plant. However, in natural populations, nonrandom fertilization and selection of "superior" pollen genotypes are expected to be the result of competition between different donor plants (Snow and Spira 1991). The phenomenon is expected to be particularly marked in crop species (Ottaviano and Mulcahy 1989); because of the crop species' high population density and compressed flowering time, pollination with a large amount of pollen is expected to create strong competition between pollen grains.

The extent of the observed variability in pollen competition, apart from physiological considerations, suggests that PCA is under the control of a large set of genes, and that these genes can be divided in two categories: genes controlling pollen functions, and genes controlling metabolic processes common to pollen and sporophyte.

As to the first category, very few mutants of pollen function specifically affecting pollen germination and growth have been described. Chalcone synthase was shown to be required for normal pollen tube growth (Coe et al. 1981). Due to the difficulties of detecting gametophytic mutants, most of the available data concern sporophytic mutants that are also expressed in the gametophyte. In the analysis of a large collection of defective endosperm (*de*) mutants of maize, Ottaviano et al. (1988b) found that 65% of these mutants were also expressed in the gametophyte. Precise analysis of the distorted segregation from the Mendelian ratios permitted the demonstration of gametophytic gene expression and the identification of the phase in which the genes are expressed. One class of *de-ga* mutants affects pollen development processes, whereas a second class affects the pollen tube growth rate. Of the same type are lethal embryo mutations in *Arabidopsis*, expressed during tube growth (Meinke 1982, 1985). However, the biological function of these genes remains unknown.

To our knowledge, the only data on the genetic system involved in the control of pollen function have been produced by RFLP analysis carried out on a recombinant inbred line population of maize, which allows a precise quantitative estimate (see previous discussion) of both early and late pollen tube growth (Sari-Gorla et al. 1992, 1994b). In these studies, different components of PCA – namely, germination and tube growth in vitro, and early and late tube growth in vivo – were analyzed. The results allowed the identification and the chromosome localization of the genes involved in the control of these traits. On the whole, the results suggested that, in maize, different sets of genes are involved, acting according to a precise timing system. A first group of genes affecting the germination rate, as evaluated in vitro, is active during microsporogenesis. The second set controls the early phases of tube growth, and a third set of genes regulates later pollen tube growth until the tube enters the embryo sac.

Pollen–pollen and pollen–pistil interaction

Pollen germination and tube growth, both in vitro and in vivo, are very sensitive to, and dependent on, the surrounding environment. In vitro studies first revealed the existence of a "population effect": in many plant species, pollen germination is strictly dependent on pollen density (Brewbaker and Kwack 1963; Iwanami 1970), and below a critical threshold no germination occurs. The dependence of germination on pollen density was interpreted as due to calcium leakage from pollen grains, the optimum level of which is reached in the presence of a minimum number of germinating grains.

Population effect has also been observed in vivo in many species, such as *Passiflora* (Snow 1982), *Nicotiana* (Cruzan 1986), *Pyrus, Malus,* and *Rosa* (Visser et al. 1988). In some cases the biological basis of these effects has been established. In *Leucaena leucocephala,* stigmatic inhibition of pollen germination, when the pollen grains are fewer than a critical number, is due to a pH-dependent proteinaceous inhibitor, which is inactivated by the change of the pH when the pollen load exceeds a minimum number (Ganeshaiah and Uma Shaanker 1988).

Role of the pistil in pollen function

The relationship between the pollen and the pistil is not simply that between a growing entity and a passive nutrient substratum. This is indicated by the fact that the tubes generally do not grow as fast or as long in culture as they do in vivo. Furthermore, in many plant species – typically those such as cereals having long tubes – the stylar environment is not artificially reproducible. A dynamic role of the stylar environment is clearly indicated by the results reported by Sanders and Lord (1989). These authors demonstrated that in three species with either open or closed stylar types, latex beads moved along the same secretory matrix pathway and at the same rate as normal pollen tubes in mature styles. Thus the extracellular matrix actively facilitated bead movements, just as it normally facilitates pollen tube extension. However, the movement of the beads into the stylar tissues was not autonomous; it called for the progressive cutting off of the style after the beads' passage, suggesting that in natural conditions the pollen itself plays an active role in this process.

Recent studies on specific proteins of the stylar matrix suggest some interesting hypotheses concerning the functional role of the matrix. The most thoroughly studied floral extracellular matrix proteins known to be important in pollen–pistil interaction are the self-incompatibility–related glycoproteins (see Chapter 9). Identification and characterization of other extracellular matrix components from both pollen and pistil, not directly involved in self-incompatibility processes, is in progress.

The extracellular matrix in animal systems has been shown to play an active role in developmental processes, in particular in cell migration. Molecules that are thought to confer dynamic capabilities to the extracellular matrix are "substrate adhesion molecules" (SAMs), a family of secretory adhesive glycoproteins (among them, vitronectin) that interact with cell surface receptors called integrins (Adair and Mecham 1990). Detection of SAM-like genes and gene products in flowering plants (*Lilium longiflorum, Vicia faba, Lycopersicon esculentum, Glycine max*), specifically a vitronectin or a vitronectin-like compound, suggests that this protein is involved in pollen tube guidance, possibly by means of an active and specific recognition mechanism, relying on an integrin-like protein located in the plasma membrane of the pollen tube tip (Sanders and Lord 1992).

A stylar-specific glycosylated protein, sp41, has been characterized in tobacco (Ori et al. 1990); it constitutes the major soluble protein of the transmitting tissue in this species and is localized mainly in the extracellular matrix. The sequence of the corresponding cDNA revealed high homology with pathogen-induced (1-3)-ß-glucanases of the leaf; thus a subfamily of pathogenesis-related proteins constitutes one of the major stylar matrix proteins. The developmental regulation of sp41 expression suggests that it plays a part in the reproductive physiology. sp41 is expressed almost exclusively in the style; its product is first detected 8 days before anthesis and reaches the maximum by anthesis.

In tobacco, two transmitting tissue specific (TTS) cDNAs have been characterized (Wang et al. 1993); the TTS proteins are extracellular matrix glycoproteins that are developmentally regulated and quantitatively affected by pollination; the amount of mRNA and proteins increases with pistil development and reaches its highest level in the style at anthesis, maintaining very high levels for 3–4 days after pollination. Moreover, when young pistils, which have levels of these gene products insufficient to sustain pollen tube growth, are hand pollinated, TTS mRNA levels are stimulated. Thus, although the function of TTS proteins remains unclear, they appear to be involved in the process of pollen tube growth.

Tomato *LAT56* and *LAT59* (pollen-expressed genes) gene products show sequence similarities to a region of a pollen allergen of Japanese cedar (Wing et al. 1989) and ragweed, and to the predicted amino acid sequence encoded by a tomato cDNA of a gene predominantly expressed in the pistil (Boudelier et al. 1990). This suggests that the products of the pollen-expressed genes somehow interact with the pistil-expressed gene products during compatible pollination (McCormick 1991).

S-related sequences were detected in *Arabidopsis thaliana*, a crucifer that does not have a self-incompatibility system, by cross-hybridization with *Brassica* DNA probes of a multigene family involved in the self-incompatibility reaction (Dwyer et al. 1992). Dwyer and colleagues suggested that the gene is representative of S-locus–related genes specifically expressed in the male and female structures of the flower, the product of which may be necessary for germination and/or subsequent development of pollen tubes through the pistil tissues. Thus, it is possible that the system regulating pollen–pistil interactions has a common, general basis for both incompatible and compatible reproductive systems.

Genetic control of pollen–pistil interaction

A direct approach for the identification and the isolation of genes specifically involved in the control of pollen–pistil interaction would be the analysis of mutations affecting the trait. However, the availability of such genes is rather scarce, since in general the variability observed for pollen adaptability to different stylar environments is of a quantitative type. The tube growth rate of a pollen genotype is affected by the genotype of the stylar tissues. In many plant species, such as maize (Pfahler 1967b; Sari-Gorla et al. 1976; Ottaviano et al. 1980), pearl millet (Sarr et al. 1988; Robert et al. 1991), and barley (Pedersen 1988), it has been found that pollen performance varies according to the male–female genetic combination.

An attempt to resolve this quantitative variability into discrete genetic factors was made in maize by means of molecular markers (Sari-Gorla et al. 1995). RFLP analysis was performed on a population of recombinant inbred lines, characterized for about two hundred mapped loci, the pollen of which was used to pollinate female plants of different genotypes. Significant association between specific pollen–pistil interaction effects and allelic composition of each molecular locus allowed the identification and localization of some genetic factors involved in the control of these processes.

Recent evidence indicates that some flavonols are essential for pollen function. A lack of chalcone synthase (CHS) activity, an enzyme catalyzing the initial step of flavonoid biosynthesis, has a pleiotropic effect in maize and petunia – it disrupts not only flavonoid synthesis, but also pollen function. In both maize and petunia, CHS-deficient mutants are self-sterile; pollen is apparently normal but is unable to germinate and grow, since mutant pistils are unable to sustain pollen growth. However, CHS-deficient pistils set seed when pollinated with wild-type pollen, and the mutant pollen is partially functional on wild-type stigmas (Coe et al. 1981; van der Meer et al. 1992). The same phenomenon was observed in transgenic petunia plants with suppressed CHS gene expression; pollen with greatly reduced function both in vitro and in vivo is partially restored on wild-type stigmas (Taylor and Jorgensen 1992). Kaempferol was identified as a pollen germination–inducing constituent in wild-type petunia stigma extracts (Mo et al. 1992). Adding micromolar quantities of this compound to the germination medium or to the stigma is sufficient to restore mutant pollen germination and tube growth in vitro and seed set in vivo. In tobacco, flavonols (quercetin, kaempferol, myricetin), when added to the germination medium, strongly promote germination and tube growth of in vitro matured pollen (Ylstra et al. 1992). These results indicate that some type of flavonoid is essential for normal pollen function and can be provided either by the pollen or by the stigma. The fact that the compound is effective at very low dosages (micromolar range) and its specificity suggest that the flavonols act as signal molecules, similar to hormones.

A male-sterile mutation (*pop1*) that affects pollen–pistil interaction by eliminating the extracellular pollen coat has been isolated in *Arabidopsis* (Preuss et al. 1993); the mutant pollen, even if viable, fails to germinate, because it does not absorb water from the stigma. Moreover, stigma cells entering into contact with the mutant pollen are stimulated to produce callose, normally synthesized in response to foreign pollen. Because callose does not occur when killed wild-type pollen is applied to the stigma, callose production cannot be due merely to a defect in pollen germination. Analysis of mature mutant pollen indicates that it is deficient in long-chain lipid and has none of the lipoidal tryphine normally present on its surface. Moreover, pollen of other wax-defective mutants with reduced fertility have some of the lipid droplets normally present in tryphine. Thus, tryphine components appear to be critical in the first phases of pollen–stigma interaction.

A category of genes directly controlling pollen–style interaction is represented by the so-called gametophytic factors reported in many species, such as maize (Nelson 1952, 1994), lima bean (Allard 1963), barley (Tabata 1961), rice (Iwata et al. 1964), and *Oenothera* (Harte 1969). These genes strongly affect pollen tube growth; pollen bearing the recessive *ga* allele has a disadvantage when competing with pollen carrying the dominant *Ga* allele in stylar tissue with the same dominant allele, but not in *ga/ga* styles. In maize, some *Ga* factors have been extensively studied using both *Ga* and *ga* pollen labeled with [32]P (House and Nelson 1964); *ga* pollen germinated and pollen tubes grew into the stylar tissues, but *ga* pollen growth became progressively slower than that of *Ga* pollen and eventually ceased before reaching the ovules.

A similar case of unidirectional cross-incompatibility has recently been described in maize (Rashid and Peterson 1992). When the female parent is homozygous recessive for one gene (*cif* locus) and the male parent is homozygous recessive for two different genes (*cim1* and *cim2* loci), there is reduced seed set (RSS); the reciprocal cross, however, results in normal seed set. This may provide a suitable system for identifying the gene products associated with such incompatibility reactions.

Applications in plant breeding

In a broad sense, pollen biotechnology refers both to direct manipulation of the male gametophyte genome and to manipulation of normal pollen development and function. Both these technologies have great potential in plant breeding.

Direct genetic manipulation of the pollen genome, that is, the use of pollen as a natural vector for DNA in order to obtain transgenic plants, is being attempted in many laboratories (for a thorough treatment of the subject, see Chapter 21). Manipulation of pollen development and function are based on (a) sporophyte–gametophyte correlation for specific traits – in particular, tolerance to a wide range of environmental stresses; (b) the large extent of gametophytic gene expression and of haplo-diploid genetic overlap observed in most plant species; and (c) mentor or pioneer pollen effects. Mentor pollen effects are not covered here because they are discussed in Chapter 15.

Pollen assay for the identification of plants with desirable traits

Association of gametophytic and sporophytic characters does not necessarily imply that the same genes are expressed in both phases. It simply indicates, irrespective of the underlying mechanisms, a parallel behavior of the plant and pollen. This allows pollen to be used to test the ability of different genotypes or species to withstand environmental stresses. The advantages of pollen assays stem from the large populations that can be tested in a small space and under controlled conditions by means of simple and inexpensive technologies. Thus, the use of pollen assay is more effective, much quicker, and less expensive when compared to the use of plants for identifying those tolerant to environmental stresses.

Correlations between pollen and sporophytic characters have been reported in a wide range of species for tolerance to temperature stress, salinity, heavy metals, fungal toxins, herbicides, and so forth (Table 16.1, part A). In all these cases, there is a positive correlation between the response of pollen (in terms of germination and tube growth rate) in the presence of stresses and the response of the sporophyte; therefore, such a positive correlation is a good parameter for predictive purposes.

Application of pollen assays is not restricted to stress tolerance. In maize, for instance, it has been shown that gametophytic performance can be used to predict combining ability (Ottaviano et al. 1980), and a positive correlation between PCA (in terms of pollen tube growth rate) and sporophytic growth and vigor has been demonstrated (D. L. Mulcahy 1971, 1974; Landi et al. 1989). Similar results have been reported in other species with regard to a number of agronomically important traits (Table 16.1, part B). However, generalization is not possible, because cases have been reported (Maisonneuve and den Njis 1984) in which no correlation was found between the responses at the plant and the pollen levels.

Table 16.1. *Sporophytic–gametophytic expression of agronomically important traits*

Characters	Species	Authors
A. Tolerance to:		
Low temperature	Tomato	Zamir et al. 1982*, Zamir and Gadish 1987*
	Maize	Kovacs & Barnabas 1992* Bocsi et al. 1990* Lyakh & Soroka 1993*
	Potato	Kristjansdottir 1990
	Prunus	Weinbaum et al. 1984
	Pistacia	Polito et al. 1988
High temperature	Cotton	Rodriguez-Garay & Barrow 1988*
	Maize	Frova et al. 1995* Petolino et al. 1990* Herrero & Johnson 1980
	Pearl millet	Robert et al. 1989*
Salinity	Tomato	Sacher et al. 1983* Sari-Gorla et al. 1988*
	Rosa	Weber & Reimann-Philip 1989
Metals	*Silene dioica* *Mimulus guttatus*	Searcy & Mulcahy 1985a,* 1985b
	Tomato	Searcy & Mulcahy 1990
	Pinus	Holub & Zelenakova 1986
Acidity	Forest species	Cox 1986
Phytotoxins	Tomato Maize	Rabinowich et al. 1978* Laughnan & Gabay 1973
	Sugarbeet	Smith 1986
	Brassica	Hodgkin 1990* Shivanna & Sawhney 1993
	Lycopersicon	Bino et al. 1988
Herbicides	Maize	Sari-Gorla et al. 1989* Sari-Gorla et al. 1992* Frascaroli et al. 1992 Sari-Gorla et al. 1994*
	Sugarbeet	Smith & Mozer 1985
Ozone	many	Feder 1986

Table 16.1 (cont.)

Characters	Species	Authors
B. Pollen and plant quality		
Kernel and seedling weight, root growth, plant vigor	Maize	Mulcahy 1971, 1974 Ottaviano et al. 1980, 1982*, 1988* Landi et al. 1989*
	Cotton	Ter Avanesian 1978*
	Vigna	
	Wheat	
	Dianthus	Mulcahy & Mulcahy 1975* McKenna & Mulcahy 1983*
	Turnera ulmifolia	McKenna 1986*
	Cucurbita pepo	Stephenson et al. 1986* Winsor et al. 1987* Schlichting et al. 1990*
	Cassia fasciculata	Lee & Hartgerink 1986*
Fertility	*Vigna*	Ter Avanesian 1978*
	Petunia	Mulcahy et al. 1978*
	C. pepo	Davis et al. 1987*
Stem length, leaf number	*Lotus corniculatus*	Schlichting et al. 1987*
Fatty acid composition	*Brassica*	Evans et al. 1988

* sporophytic effect of gametophytic selection

References for Table 16.1
Bino RJ, Franken J, Witsenboer HMA, Hille J, Dons JJM (1988) *Theor Appl Genet* 76:204–208; Bocsi J, Kovacs G, Barnabas B, (1990) *Cer Res Comm* 18(4):347–354; Cox RM, (1986) In: Mulcahy DL, Mulcahy GB, Ottaviano E (eds) *Biotechnology and Ecology of Pollen*. Springer-Verlag, New York, pp 95–100; Davis LE, Stephenson AG, Winsor JA (1987) *Proc Am Soc Hortic Sci* 112:712–716; Evans DE, Rothnie NE, Sang JP, Palmer MV, Mulcahy DL, Sing MB, Knox RB (1988) *Theor Appl Genet* 76:411–419; Feder WA (1986) In: Mulcahy DL, Mulcahy BG, Ottaviano E (eds) *Biotechnology and Ecology of Pollen*. Springer-Verlag, New York, Berlin, pp 89–94; Frascaroli E, Landi P, Sari-Gorla M, Ottaviano E (1992) *J Genet & Breed* 46:49–56; Frova C, Portaluppi P, Villa M, Sari-Gorla M (1995) *J Hered* 86:50–54; Herrero MP, Johnson RR (1980) *Crop Sci* 20:796–800; Hodgkin T (1990) *Sex Plant Reprod* 3:116–120; Holub Z, Zelenakova E (1986) *Ekologia* 5(1):81–90; Kovacs G, Barnabas B (1992) In: Ottaviano E, Mulcahy DL, Sari-Gorla M, Bergamini Mulcahy G (eds) *Angiosperm Pollen and Ovules*. Springer-Verlag, New York, Berlin, pp 359–363; Kristjandottir IS, (1990) *Theor Appl Genet* 80:139–142; Landi P, Frascaroli E, Tuberosa R, Conti S (1989) *Theor Appl Genet* 77:761–767; Laughnan JR, Gabay SJ (1973) *Crop Sci* 43:681–684; Lee TD, Hartgerink AP (1986) In: Mulcahy DL, Bergamini Mulcahy G, Ottaviano E (eds) *Biotechnology and Ecology of Pollen*. Springer-Verlag, New York, Berlin, pp 417–422; Lyakh VA, Soroka AV (1993) *Mydica* 38:67–71; McKenna M (1986) In: Mulcahy DL, Bergamini Mulcahy G, Ottaviano E (eds) *Biotechnology and Ecology of Pollen*. Springer-Verlag, New York, Berlin, pp 443–448; McKenna M, Mulcahy DL (1983) In: Mulcahy DL, Ottaviano E (eds) *Pollen: Biology and Implications for Plant Breeding*. Elsevier, New York, Berlin, Amsterdam, pp 419–424; Mulcahy DL (1971) *Science* 171:1155–1156; Mulcahy DL (1974) *Nature* 249:491–492; Mulcahy DL, Mulcahy BG, Ottaviano E (1978) *Soc Bot Fr Actual* Bot 1–2:57–60; Mulcahy DL, Mulcahy BG (1975) *Theor Appl Genet* 46:277–280; Ottaviano E, Sari Gorla M, Mulcahy DL (1980) *Science* 210:437–438; Ottaviano E, Sari-Gorla M, Pe' ME (1982) *Theor Appl Genet* 63: 249–254; Ottaviano E, Sari-Gorla M, Villa M (1988) *Theor Appl Genet* 76:601–608; Petolino JF, Cowen NM, Thompson SA, Mitchell JC (1990) *J Plant Physiol* 136:219–224; Polito VS, Luza JG, Weinbaum SA, (1988) *Sci Hortic* 35:269–274; Rabinowich HD, Reting N, Kedar N (1978) *Euphytica* 27: 219–224; Robert T, Sarr A, Pernes J (1989) *Genome* 32:946–952; Rodriguez-Garay B, Barrow

Use of pollen assay to identify suitable plants in given populations is particularly useful in woody perennials, such as fruit and forest trees, which have a long generation time and for which the setting up of large nurseries is expensive. Some of the examples of positive pollen–plant correlations are cold response in *Prunus* species (Weinbaum et al. 1984) and *Pistacia* (Polito et al. 1988), metal tolerance in *Pinus* species (Holub and Zelenakova 1986), pH in forest plant species (Cox 1986), and ozone in several species (reported in Feder 1986). However, fungal infection in apples (Visser and van der Meys 1986) was not correlated with the response of the pollen.

An extension of pollen assays is the monitoring of polluting agents, such as SO_2, in the environment. Since pollen is usually more sensitive to the pollutants than are sporophytic tissues, the presence of pollutants, even in trace levels, can be detected by pollen assays (Wolters and Martens 1987).

Pollen selection

Pollen can be the target of selective pressures with the aim of improving the progeny. The rationale of male gametophytic selection (MGS) is based on the assumption that adaptive traits are controlled by genes expressed both in the gametophytic and in the sporophytic phase. Although this cannot be simply inferred from plant–pollen correlated behavior (compare simple correlation with true pollen selection cases in Table 16.1), 60%–80% sporophytic–gametophytic genetic overlap has been demonstrated in numerous plant species (for review, see Ottaviano and Mulcahy 1989). Selective pressures applied during the male gametophytic phase for characters controlled by genes showing haplo-diploid transcription would result in nonrandom fertilization; pollen grains carrying genes that confer better adaptability sire a larger proportion of seeds. This would increase the frequency of favorable alleles in the resulting sporophytic generation.

Because of haploidy and the very large size of pollen populations (often millions from a single plant), MGS is expected to be extremely efficient. In

JR (1988) *Crop Sci* 28:857–859; Sacher RF, Mulcahy DL, Staples RC (1983) In: Mulcahy DL, Ottaviano E (eds) *Pollen: Biology and Implications for Plant Breeding*. Elsevier, New York, Berlin, Amsterdam, pp 329–334; Sari-Gorla M, Mulcahy DL, Gianfranceschi L, Ottaviano E (1988) *Genet Agr* 42:92–93; Sari-Gorla M, Ottaviano E, Frascaroli E, Landi P (1989) *Sex Plant Reprod* 2:65–69; Sari-Gorla M, Ferrario S, Gianfranceschi L, Villa M (1992) In: Ottaviano E, Mulcahy DL, Sari-Gorla M, Bergamini Mulcahy G (eds) *Angiosperm Pollen and Ovules*. Springer-Verlag, New York, Berlin, pp 364–369; Sari-Gorla M, Ferrario S, Frascaroli E, Frova C, Landi P, Villa M (1994) *Theor Appl Genet*; Schlichting CD, Stephenson AG, Davis LE, Winsor JA (1987) *Evol Trends Plants* 1:35–39; Schlichting CD, Stephenson AG, Small LE, Winsor JA, (1990) *Evolution* 44(5):1358–1372; Searcy KB, Mulcahy DL (1985a) *Am J of Botany* 72:1700–1706; Searcy KB, Mulcahy DL (1985b) *Theor Appl Genet* 69:597–602; Searcy KB, Mulcahy DL (1990) *Theor Appl Genet* 80:289–295; Shivanna KR, Sawhney VK (1993) *Theor Appl Genet* 86:339–344; Smith GA, Mozer HS (1985) *Theor Appl Genet* 71:231–237; Smith GA (1986) In: Mulcahy DL, Bergamini-Mulcahy G, Ottaviano E (eds) *Biotechnology and Ecology of Pollen*. Springer-Verlag, New York, pp 83–88; Stephenson AG, Winsor JA, Davis LE (1986) In: Mulcahy DL, Bergamini-Mulcahy G, Ottaviano E (eds) *Biotechnology and Ecology of Pollen*. Springer, New York, Berlin, pp 429–434; Ter Avanesian DV (1978) *Theor Appl Genet* 52:77–79; Weber J, Reimann-Philipp R, (1989) *Acta Hort* 246:353–354; Weinbaum SA, Parfitt DE, Polito VS, (1984) *Euphytica* 33:419–426; Winsor JA, Davis LE, Stephenson AG (1987) *Am Nat* 129:643–656; Zamir D, Tanksley SD, Jones RA (1982) *Genetics* 101: 129–137; Zamir D, Gadish I (1987) *Theor Appl Genet* 74: 545–548.

fact, a very high selection pressure can be sustained by pollen, since, even if most of the pollen grains are eliminated during selection pressure, the remaining small fraction is enough to ensure fertilization of most of the ovules. In particular, for quantitative traits controlled by complex gene combinations, the large amount of genetic load produced by recombination can be removed in the haploid phase without affecting the mean fitness of the sporophytic generation (Ottaviano et al. 1988a).

MGS can be effectively integrated to improve the efficiency of conventional breeding methods. One of the major advantages of MGS is that it is expected to be successful for characters, particularly quantitative traits, even when their genetic control is not known. In comparison to the selection applied to the sporophyte, MGS greatly increases the probability of selecting complex allele combinations, since the number of genetic combinations for n genes is much lower in haploids than in diploids (Pfahler 1983). This is especially advantageous if the favorable alleles are present at a low frequency in the population (Sari-Gorla 1992). Because most of the agronomically important traits are quantitatively controlled and/or the relative genes have not yet been identified, the potential of MGS, in comparison with other biotechnologies, is great. Although this is widely acknowledged, the impact of MGS on plant breeding has so far been small, probably for reasons that will be discussed later.

MGS efficiency has been demonstrated for characters related to plant growth and vigor, and for tolerance to a variety of biotic and abiotic stresses (see Table 16.1, references marked with asterisk). Extensive studies by Ottaviano et al. (1980, 1982, 1988a) have shown that the progeny derived from pollen selected for faster tube growth resulted in increased seedling vigor, kernel weight, and root growth in vitro. In several species, MGS resulted in an increase of sporophytic tolerance to environmental stresses (Table 16.1).

Successful integration of MGS into breeding programs is dependent on the following: (a) Correlation between the performance of the gametophyte and the sporophyte must be clearly established; (b) pollen selection must produce a response at the pollen level as well as at the sporophyte level; and (c) the selected trait has to be stably transmitted to the subsequent generations. All these points have been documented in the investigations listed in Table 16.1. In particular, a recent work on maize tolerance to the herbicide Alachlor (Sari-Gorla et al. 1994a) clearly demonstrated that (a) a pollen assay is able to predict plant performance in the field; (b) the positive response to pollen selection is independent of the pollen genetic source, thus allowing the use of material derived from commercial elite lines; and (c) the selected trait is transmitted and maintained through subsequent generations.

Two additional points deserve discussion. The first concerns the effective stage for application of selection pressure to the male gametophyte. The stage of pollen development at which the genes controlling the character under selection are expressed need to be established. Pollen selection will induce maximum effect only when applied at the stage in which genetic differences between pollen grains are fully expressed. Data on the expression of genes controlling specific traits is very scant, but indirect information can be obtained from the comparison of results of some of the experiments reported in Table 16.1. For instance, Chlorsulfuron (Sari-Gorla et al. 1989) or heavy metal (Searcy and Mulcahy 1985a, b) treatments were more effective in pre-pollination stages, whereas selection for tolerance to low temperature in tomato (Zamir and Vellejos 1983) was more effective when applied during pollen tube growth. In the case of high temperature tolerance in maize, both preshedding (Frova et al. 1995) and postshedding (Petolino et al. 1990) treatments were effective, although on different components of the character.

The second point concerns the evaluation of MGS efficiency. The main difficulty arises from the fact that MGS acts only on genetic differences expressed at cellular or subcellular levels, whereas the response is evaluated in the sporophyte, in which most of the characters are the result of complex interactions at different levels of the plant organization. Because of the simultaneous segregation of a large number of genes controlling different character components, the effect of selection on the components expressed at the cellular level is not easy to detect and evaluate. This difficulty can be overcome, however, with the aid of molecular markers. The approach is dependent on the identification of molecular markers (isozymes, RFLPs, RAPDs, microsatellites, and so forth) showing a significant genetic correlation with the character of interest in a segregating population. Besides allowing the detection of chromosome regions carrying the genes controlling the character, this approach is useful in evaluating the efficiency of selection in terms of changes of gene frequencies of the markers. So far, mainly isozymes have been used. The isozyme method was first applied by Zamir et al. (1982) for the evaluation of the response to selection for low temperature tolerance in tomato. MGS induced differential transmission of isozyme alleles linked to the genes controlling the character. Similar results were obtained for tolerance to salinity in the same species by Sari-Gorla et al. (1988). For this purpose, however, DNA-based markers proved to be much more powerful because they can be found in very large numbers, potentially sufficient to saturate the entire genome (Sari-Gorla et al. 1996).

Concluding remarks

The considerable information accumulated in the last 15 years concerning the physiology and genetics of the male gametophyte has opened the way for exploitation of the male gametophyte in plant breeding. Genes controlling PCA and involved in pollen–pistil interaction are being identified. Cloning of these genes is now made possible by the availability of a large number of molecular markers and of densely saturated maps. The molecular characterization of these genes is expected to clarify several aspects of pollen tube growth through the style. Improvement of complex traits can be achieved by means of male gametophytic selection. Because the biological bases of the phenomenon and its effectiveness have been demonstrated, its large-scale application is now possible.

References

Adair, W.S. & Mecham, R.P., eds. (1990). *Organization and Assembly of Plant and Animal Extracellular Matrix: Biology of Extracellular Matrix.* San Diego: Academic Press.

Allard, R.W. (1963). An additional gametophyte factor in the lima bean. *Züchter* 33:212–216.

Barnabás, B. & Fridvalszky, L. (1984). Adhesion and germination of differently treated maize pollen grains on the stigma. *Acta. Bot. Hungar.* 30:229–232.

Barnabás, B. & Kovács, G. (1992). In vitro pollen maturation and successful seed production in detached spikelet cultures in wheat (*Triticum aestivum* L.). *Sex. Plant Reprod.* 5:286–291.

Boudelier, K.A., Smith, A.G. & Gasser, C.S. (1990). Regulation of a stylar transmitting tissue–specific gene in wild-type and transgenic tomato and tobacco. *Mol. Gen. Genet.* 224:183–192.

Brewbaker, J.L. & Kwack, B.H. (1963). The essential role of calcium ion in pollen germination and pollen tube growth. *Amer. J. Bot.* 50:859–865.

Cheng, D.S. & Freeling, M. (1976). Methods of maize pollen germination in vitro, collection, storage and treatment with toxic chemicals; recovery of resistant mutants. *MGC Newslett.* 50:11–13.

Coe, E.H., McCormick, S.M. & Modena, S.A. (1981). White pollen in maize. *J. Hered.* 72:318–320.

Cox, R.M. (1986). *In vitro* and *in vivo* effects of acidity and trace elements on pollen function. In *Biotechnology and Ecology of Pollen,* eds. D.L. Mulcahy, G.B. Mulcahy, and E. Ottaviano, pp. 95–100. New York: Springer-Verlag.

Cruzan, M.B. (1986). Pollen tube distribution in *Nicotiana glauca*: evidence for density dependent growth. *Amer. J. Bot.* 73:902–907.

Dickinson, D.B. (1967). Permeability and respiratory properties of germinating pollen. *Physiol. Plant.* 20:118–127.

Dwyer, K.G., Lalonde, B.A., Nasrallah, J.B. & Nasrallah, M.E. (1992). Structure and expression of AtS1, an *Arabidopsis thaliana* gene homologous to the S-locus related genes of *Brassica. Mol. Gen. Genet.* 231:442–448.

Feder, W.A. (1986). Predicting species response to ozone using a pollen screen. In *Biotechnology and Ecology of Pollen,* eds. D.L. Mulcahy, B.G. Mulcahy, and E. Ottaviano, pp. 89–94. New York: Springer-Verlag.

Frova, C., Portaluppi, P., Villa, M. & Sari-Gorla, M. (1995). Sporophytic and gametophytic components of thermotolerance affected by pollen selection. *J. Hered.* 86:50–54

Ganeshaiah, K.N. & Uma Shaanker, R. (1988). Regulation of seed number and female incitation of mate competition by a pH-dependent proteinaceous inhibitor of pollen grain germination in *Leucaena leucocephala. Oecologia* 75:110–113.

Harte, C. (1969). Gonenkonkrenz bei *Oenothera* unter dem Einflus eines gametophitischwirksamen Gensin der ersten Koppelungsgruppen sowie ein Modell für die Untersuchung verzweigter Koppelungsgruppen. *Theor. Appl. Genet.* 39:163–178.

Heberle-Bors, E. (1989). Isolated pollen culture in tobacco: plant reproductive development in a nutshell. *Sex. Plant Reprod.* 2:1–10.

Hi Zhang, Y., Craker, L.E. & Mulcahy, D.L. (1993). A method to separate germinated from ungerminated pollen grains. *Environ. Experim. Bot.* 33:415–421.

Hodgkin, T. (1987). A procedure suitable for *in vitro* pollen selection in *Brassica oleracea. Euphytica* 36:153–159.

Holub, Z. & Zelenakova, E. (1986). Tolerance of reproduction processes of woods to the influence of heavy metals. *Ekologia* 5(1):81–90.

House, L.R. & Nelson, O.E. (1964). Tracer study of pollen tube growth in cross-sterile maize. *J. Hered.* 49:18–21.

Iwanami, Y. (1970). Researches of pollen. XX. Population effect and mixture effect in pollen culture. *Bot. Mag. (Tokyo)* 83:364–373.

Iwanami, Y. (1980). Stimulation of pollen tube growth *in vitro* by dicarboxylic acids. *Protoplasma* 102:111–115.

Iwata, N., Nagamatsu, T. & Omura, T (1964). Abnormal segregation of waxy and apiculus coloration by a gametophyte gene belonging to the first linkage group in rice. *Japan J. Breed.* 14:33–39.

Kearns, C.A. & Inouye, D.W. (1993). *Techniques for Pollination Biologists.* Niwot Ridge, CO: University Press.

Kranz, E. & Lorz, H. (1990). Micromanipulation and *in vitro* fertilization with single pollen grains of maize. *Sex. Plant Reprod.* 3:160–169.

Landi, P., Frascaroli, E., Tuberosa, R. & Conti, S. (1989). Comparison between responses to gametophytic and sporophytic recurrent selection in maize (*Zea mays* L.). *Theor. Appl. Genet.* 77:761–767.

Maisonneuve, B. & den Njis, A.P.M. (1984). *In vitro* pollen germination and tube growth of tomato (*Lycopersicon esculentum* Mill.) and its relation with plant growth. *Euphytica* 33:833–840.

McCormick, S. (1991). Molecular analysis of male gametogenesis in plants. *Trends Genet.* 7:289–303.

Meinke, D.W. (1982). Embryo-lethal mutants of *Arabidopsis thaliana:* evidence for gametophytic expression of the mutant genes. *Theor. Appl. Genet.* 63:381–386.

Meinke, D.W. (1985). Embryo-lethal mutants of *Arabidopsis thaliana*: analysis of mutants with a wide range of lethal phases. *Theor. Appl. Genet.* 69:543–552.

Mo, Y., Nagel, C. & Taylor, L.P. (1992). Biochemical complementation of chalcone synthase mutants defines a role for flavonols in functional pollen. *Proc. Nat. Acad. Sci.* 89:7213–7217.

Mulcahy, D.L. (1971). A correlation between gametophytic and sporophytic characteristics in *Zea mays* L. *Science* 171:1155–1156.

Mulcahy, D.L. (1974). Correlation between speed of pollen tube growth and seedling weight in *Zea mays* L. *Nature* 249:491–492.

Mulcahy, D.L. & Mulcahy, G.B. (1988). The effect of supplemented media *in vitro* on bi- and trinucleate pollen. *Plant Sci.* 55:213–216.

Mulcahy, G.B. & Mulcahy, D.L. (1985). Ovaries' influence on pollen tube growth as indicated by semivivo technique. *Amer. J. Bot.* 72:1078–1080.

Nelson, O.E. (1952). Non reciprocal cross sterility in maize. *Genetics* 37:101–124.

Nelson, O.E. (1994). The gametophyte factors of maize. In *The Maize Handbook*, eds. M. Freeling and V. Walbot, pp. 496–502. New York: Springer-Verlag.

Ori, N., Sessa, G., Lotan, T., Himmelhoch, S. & Fluhr, R. (1990). A major stylar matrix polypeptide (sp41) is a member of the pathogenesis-related superclass. *EMBO J.* 9:3429–3436.

Ottaviano, E. & Mulcahy, D.L. (1989). Genetics of angiosperm pollen. *Adv. Genet.* 26:1–64.

Ottaviano, E. & Sari-Gorla, M. (1993). Gametophytic and sporophytic selection. In *Plant Breeding: Principles and Prospects,* eds. M.D. Hayward, N.O. Bosemark, and I. Romagosa, pp. 332–352. London: Chapman Hall.

Ottaviano, E., Sari-Gorla, M. & Mulcahy, D.L. (1980). Pollen tube growth rates in *Zea mays*: implications for genetic improvement of crops. *Science* 210:437–438.

Ottaviano, E., Sari-Gorla, M. & Pe', M.E. (1982). Male gametophytic selection in maize. *Theor. Appl. Genet.* 63:249–254.

Ottaviano, E., Sari-Gorla, M. & Villa, M. (1988a). Pollen competitive ability in maize: within population variability and response to selection. *Theor. Appl. Genet.* 76:601–608.

Ottaviano, E., Petroni, D. & Pe', M.E. (1988b). Gametophytic expression of genes controlling endosperm development in maize. *Theor. Appl. Genet.* 75:252–258.

Pareddy, D.R. & Petolino, J.F. (1992). *In vitro* maturation of maize pollen. In *Angiosperm Pollen and Ovules,* eds. E. Ottaviano, D.L. Mulcahy, M. Sari-Gorla, and G. Bergamini Mulcahy, pp. 303–308. New York: Springer-Verlag.

Pedersen, S. (1988). Pollen competition in barley. *Hereditas* 109:75–81.

Petolino, J.F., Cowen, N.M., Thompson, S.A. & Mitchell, J.C. (1990). Gamete selection for heat stress tolerance in maize. *J. Plant Physiol.* 136:219–224.

Pfahler, P.L. (1967a). *In vitro* germination and pollen tube growth of maize (*Zea mays* L.) pollen. I. Calcium and boron effects. *Can. J. Bot.* 45:839–845.

Pfahler, P.L. (1967b). Fertilization ability of maize pollen grains. II. Pollen genotype, female sporophyte and pollen storage interaction. *Genetics* 57:513–521.

Pfahler, P.L. (1983). Comparative effectiveness of pollen genotype selection in higher plants. In *Biotechnology and Ecology of Pollen,* eds. D.L. Mulcahy and E. Ottaviano, pp. 361–367, New York: Elsevier Biomedical.

Pfahler, P.L. (1992). Analysis of ecotoxic agents using pollen tests. In *Modern Methods of Plant Analysis, Vol. 13: Plant Toxin Analysis,* eds. H.F. Linskens and J.F. Jackson, pp. 317–331. Berlin: Springer-Verlag.

Polito, V.S., Luza, J.G. & Weinbaum, S.A. (1988). Differential low temperature germination responses by pollen of *Pistacia vera* clones with different bloom dates. *Sci. Hortic.* 35:269–274.

Preuss, D., Lemieux, B., Yen, G. & Davis, R.W. (1993). A conditional sterile mutation eliminates surface components from *Arabidopsis* pollen and disrupts cell signalling during fertilization. *Genes. Dev.* 7:974–985.

Rashid, A. & Peterson, P.A. (1992). The RSS system of unidirectional cross-incompatibility in maize: I. *Genet. J. Hered.* 83:130–134.

Robert, T., Lespinasse, R., Pernes, J. & Sarr, A. (1991). Gametophytic competition as influencing gene flow between wide and cultivated form of pearl millet (*Pennisetum typhoides*). *Genome* 34:195–200.

Sanders, L.C. & Lord, E.M. (1989). Directed movement of latex particles in the gynoecia of three species of flowering plants. *Science* 243:1606–1608.

Sanders, L.C. & Lord, E.M. (1992). A dynamic role for the stylar matrix in pollen tube extension. *Int. Rev. Cytol.* 140:297–318.

Sari-Gorla, M. (1992). Effects of gametophytic selection on the genetic structure of populations. In *Sexual Plant Reproduction,* eds. M. Cresti and A. Tiezzi, pp. 151–159. Berlin: Springer-Veralg.

Sari-Gorla, M., Ottaviano, E. & Bellintani, R. (1976). Competitive ability of maize pollen: interaction between genotypes of pollen and stylar tissues. *Maydica* 21:77–80.

Sari-Gorla, M., Mulcahy, D.L., Gianfranceschi, L. & Ottaviano, E. (1988). Gametophytic selection for salt tolerance. *Genet. Agrar.* 42:92–93.

Sari-Gorla, M., Ottaviano, E., Frascaroli, E. & Landi, P. (1989). Herbicide-tolerant corn by pollen selection. *Sex. Plant Reprod.* 2:65–69.

Sari-Gorla, M., Pe', M.E., Mulcahy, D.L. & Ottaviano, E. (1992). Genetic dissection of pollen competitive ability in maize. *Heredity* 69:423–430.

Sari-Gorla, M., Ferrario, S., Frascaroli, E., Frova, C., Landi, P. & Villa, M. (1994a). Sporophytic response to pollen selection for Alachlor tolerance in maize. *Theor. Appl. Genet.* 88:812–817.

Sari-Gorla, M., Pe', M.E. & Rossini, L. (1994b). Detection of QTLs controlling pollen germination and growth in maize. *Heredity* 72:332–335.

Sari-Gorla, M., Binelli, G., Pe', M.E. & Villa, M. (1995). Detection of genetic factors controlling pollen–style interaction in maize. *Heredity* 74:62–69.

Sari-Gorla, M., Rampoldi, L., Binelli, G., Frova, C. & Pe' M.E. (1996). Identification of genetic factors for Alachlor tolerance in maize by molecular markers. *Mol. Genet.* 251:551–555.

Sarr, A., Sandmeier, M. & Pernes, J. (1988). Gametophytic competition in pearl millet, *Pennisetum typhoides* (Stapf and Hubb). *Genome* 30:924–928.

Schrauwen, J. & Linskens, M.F. (1967). Mass culture of pollen tubes. *Acta Bot. Neerl.* 16:177–179.

Searcy, K.B. & Mulcahy, D.L. (1985a). The parallel expression of metal tolerance in pollen and sporophytes of *Silene dioica* (l.) Clairv., *Silene alba* (Mill.) Krause and *Mimulus guttatus* DC. *Theor. Appl. Genet.* 69:597–602.

Searcy, K.B. & Mulcahy D.L. (1985b). Pollen selection and the gametophytic expression of metal tolerance in *Silene dioica* (Caryophyllaceae) and *Mimulus guttatus* (Scrophulariaceae). *Amer. J. Bot.* 72:1700–1706.

Shivanna, K.R., Xu, H., Taylor, P. & Knox, R.B. (1988). Isolation of sperm from the pollen tubes of flowering plants during fertilization. *Plant Physiol.* 87:647–650.

Shivanna, K.R., Linskens, H.F. & Cresti, M. (1991). Responses of tobacco pollen to high humidity and heat stress: viability and germinability *in vitro* and *in vivo*. *Sex. Plant Reprod.* 4:104–109.

Shivanna, K.R. & Rangaswamy, N.S. (1992). *Pollen Biology: A Laboratory Manual.* Heidelberg: Springer-Verlag.

Snow, A.A. (1982). Pollination intensity and potential seed set in *Passiflora vitifolia.* *Oecologia* 55:231–237.

Snow, A.A. & Spira, T.P. (1991). Differential pollen tube growth rates and nonrandom fertilization in *Hibiscus moscheutos* (Malvaceae). *Amer. J. Bot.* 78:1419–1426.

Stauffer C., Benito-Moreno, R.M. & Heberle-Bors, E. (1991). Seed set after pollination with *in-vitro*-matured isolated pollen on *Triticum aestivum.* *Theor. Appl. Genet.* 81:576–580.

Tabata, M. (1961). Studies of a gametophyte factor in barley. *Japan. J. Genet.* 36:157–167.

Taylor, L.P. & Jorgensen, R. (1992). Conditional male fertility in chalcone synthase–deficient *Petunia. J. Hered.* 83:11–17.

van der Meer, I.M., Stam, M.E., van Tunen, A.J., Mol, J.N.M. & Stultje, A.R. (1992). Antisense inhibition of flavonoid biosynthesis in petunia anthers results in male sterility. *Plant Cell* 4:253–262.

Vasil, I.K. (1987). Physiology and culture of pollen. *Int. Rev. Cytol.* 107:127–171.

Visser, T. & van der Meys, Q. (1986). *In vitro* apple pollen and apple scab fungus (*Venturia inaequalis* Cke Wint). In *Biotechnology and Ecology of Pollen,* eds. D.L. Mulcahy, G.B. Mulcahy, and E. Ottaviano, pp. 119–124. New York: Springer-Verlag.

Visser, T., Sniezko, R. & Marcucci, C.M. (1988). The effect of pollen load on pollen tube performance in apple, pear and rose styles. In *Sexual Reproduction in Higher Plants,* eds. M. Cresti, P. Gori, and E. Pacini, pp. 75–80. Berlin: Springer-Verlag.

Wang, H., Wu, H.M. & Cheung, A.Y. (1993). Development and pollination regulated accumulation and glycosilation of a stylar transmitting tissue–specific prolin-rich protein. *Plant Cell* 5:1639–1650.

Weinbaum, S.A., Parfitt, D.E. & Polito, V.S. (1984). Differential cold sensitivity of pollen grain germination in two *Prunus* species. *Euphytica* 33:419–426.

Wing, R.A., Yamaguchi, J., Larabell, S.K., Ursin, V.M. & McCormik, S. (1989). Molecular and genetic characterization of two pollen-expressed genes that have sequence similarity to pectate lyases of the plant pathogen Erwinia. *Plant. Mol. Biol.* 14:17–28.

Wolters, J.H.B. & Martens, M.J.M. (1987). Effects of air pollutants on pollen. *Bot. Rev.* 53(3):372–414.

Ylstra, B., Touraev, A., Benito Moreno, RM., Stoger, E., van Tunen, A., Viciente, A., Mol, J.N.M. & Heberle-Bors, E. (1992). Flavonols stimulate development, germination and tube growth of tobacco pollen. *Plant Physiol.* 100:902–907.

Zamir, D. & Vellejos, E.C. (1983). Temperature effects on haploid selection of tomato microspores and pollen grains. In *Pollen: Biology and Implications for Plant Breeding,* eds. D.L. Mulcahy and E. Ottaviano, pp. 335–342. New York: Elsevier.

Zamir, D., Tanskley, S.D. & Jones, R.A. (1982). Haploid selection for low temperature tolerance in tomato pollen. *Genetics* 101:129–137.

Zhang, Hong-Qi & Croes, A.F. (1982). A new medium for pollen germination *in vitro. Acta Bot. Neerl.* 31:113–119.

17

Isolation and manipulation of sperm cells

DAVID D. CASS

Summary   352
Introduction   353
Isolation of sperm cells   353
    Pollen pretreatment is important   354
    Extrusion of pollen cytoplasm and sperm cells   354
Purification of sperm cells   354
Microbiological sterility of pollen   355
Assessment of sperm cell numbers and viability   355
    Microscopical methods   355
    Dye exclusion and uptake   356
    Fluorescent procedures   356
    Flow cytometry   356
    Further analysis of Brewbaker–Kwack salts   358
    Cell number determinations   358
Manipulation of isolated sperm cells   359
    Manipulation of sperm cells leading to fusions with isolated egg cells   359
    Other types of manipulations   360
    A new way to look at fusion using isolated sperm cells   360
Prospects for sperm cell research   361
References   361

Summary

The now nearly routine isolation of large numbers of viable sperm cells from pollen or pollen tubes of both dicot and monocot flowering plants has opened up a new field of research in plant biology. These unique plant cells are being studied using techniques that were previously more amenable to cultured animal cells. Investigations of synthetic processes, membrane structure, transport, and fusion and recognition properties are being carried out. When the ability to isolate sperms is equaled by the ability to isolate eggs and/or embryo sacs, biologists studying flowering plants will have powerful tools to probe the molecular aspects of fertilization and early embryogenesis, and the opportunity of attempting to achieve genetic transformation during fertilization.

Introduction

The successful isolation and manipulation of sperm and egg cells of flowering plants has changed the approach to studies of fertilization and embryogenesis from a rather descriptive one to a highly experimental one. The history of flowering plant sperm cell isolation goes back to the 1920s, but there was a considerable gap between the work of Finn (1925) on isolation of sperm cells of *Asclepias* and that of Cass (1973) on isolation of sperms from pollen of barley for differential interference contrast (DIC) microscopy. The latter work combined DIC microscopy with transmission electron microscopy of sperms in pollen grains. The work of Russell (1984, 1985, 1986) on fertilization and sperm cell dimorphism, including isolation, in *Plumbago* provided further impetus to others (Dupuis et al. 1987; Matthys-Rochon et al. 1987; Mogensen et al. 1990; Russell et al. 1990; Theunis et al. 1991; Yang and Zhou 1989) to reexamine not only the morphology but also the function of sperms and to consider using them as fusion protoplasts in experimental fusions. The achievement by Kranz and Lorz (1993), who fused maize sperms and eggs and obtained normal embryos and adult plants, is a milestone in the developmental biology of flowering plants. If large numbers of such fusions and developing embryos can be obtained, there will be an unprecedented opportunity to study zygotic plant embryogenesis in novel ways. In terms of crop improvement, the in vitro approach to fertilization may allow the insertion of new genes at the moment of fusion of sperm and egg. The possibility of doing genetic transformation at such an early stage has considerable potential, particularly in cereal crops. This chapter will discuss the isolation of sperm cells from pollen grains and tubes of flowering plants and some of the things that can be done with these cells once their viability and numbers have been determined. The maize sperm cell will be central to much of this discussion.

Isolation of sperm cells

There are two broad categories of isolation, depending on the condition of pollen grains at dehiscence. One, in the case of pollen grains that are two-celled (hence, generative cell division is delayed until after tube growth has begun), and the other in the case of three-celled pollen. In the former, isolation of sperm cells must be done either by severing the pollen tubes (Shivanna et al. 1988) or by waiting until the pollen tubes have discharged. Practically, the former is preferred because the tubes can be severed at the same time and the largest possible number of sperm cells harvested. In terms of cell numbers, harvesting sperm cells from growing pollen tubes is not optimal. However, the issue of cell number is less important than sperm quality in virtually all experimental protocols. In terms of sperm cell physiology, severing growing tubes may have the advantage of exposing sperm cells, even for short periods of time, to the active cytoplasm of growing pollen tubes. Harvesting sperm cells from growing pollen tubes, either by severance or by discharge, is the only option available for taxa with two-celled grains.

Examples of taxa with two-celled grains include many dicot crop plants, such as members of Solanaceae (for example, *Nicotiana*, *Datura*, *Capsicum*). Taxa having three-celled grains include the monocot crop plants, such as all members of Poaceae. Some of these, most dramatically *Zea mays* L., produce prodigious quantities of pollen and, hence, very large numbers of sperm cells. Sperm cells from three-celled pollen are isolated through osmotic shock treatment or through gentle mechanical grinding.

Pollen pretreatment is important

The usual approach to harvesting sperm from three-celled pollen has been to place the pollen in an aqueous medium that causes osmotic grain bursting. The assessment of pollen grain samples by the method of Kerhoas et al. (1987) could be made a standard first step in the harvesting procedure; this technique would identify marginal or poor pollen batches before going through the entire sperm isolation procedure. The second pretreatment involves placing mature pollen grains in a humidifying chamber for several minutes before placing the grains in the bursting solution (Zhang et al. 1992a).

Extrusion of pollen cytoplasm and sperm cells

Hydrated pollen grains are placed in the pollen burst medium (in the author's laboratory, the pollen burst medium is similar to the sperm cell maintenance medium). The semiviscous pollen cytoplasm flows out through the aperture; the sperm cells are normally within this viscous cytoplasmic material, which includes large numbers of starch grains and various cytoplasmic particles from the grains. Sperms from three-celled grains can also be harvested by severing growing pollen tubes.

Purification of sperm cells

Whatever the isolation technique used, the sperm cells are normally subjected to purification procedures, usually Percoll density centrifugation, to remove unwanted materials from the pollen or pollen tube cytoplasm. Starch grains constitute a major part of these materials. The quantities of unwanted materials are higher in the case of sperm cell isolation from the grain itself. The bursting medium also contains exine fragments, which can damage sperm cell membranes. We normally sieve such fragments soon after bursting but before Percoll centrifugation to minimize damage (Zhang et al. 1992b). A discontinuous Percoll gradient is prepared by layering 30%, 15%, and 10% Percoll (Sigma Chemical Co.) solutions in an isolation buffer in 15 ml Corex centrifuge tubes (Zhang et al. 1992b); 5–10 ml of filtered sperm suspension, also containing the isolation buffer, is layered on top of the 10% Percoll.

Centrifugation is at 3,000 g for 60 minutes at 4°C. Sperm cells collect at the interface between 15% and 30% Percoll (Zhang et al. 1992b). Washing steps may be employed after Percoll separation if enhanced sperm cell purity is desired, but with increased purity comes higher probability of cellular damage. To optimize sperm cell physiology, care needs to be taken in determining the stage at which pollen is collected from anthers; too young or too old pollen can seriously affect the results of both the harvesting step and the quality of sperm cells after separation procedures.

Microbiological sterility of pollen

Another issue that has not, as far as the author is aware, been adequately dealt with is the possibility of bacterial contamination of the pollen sample. For those wishing to use sperm cells in experimental fusions, possibly including in vitro fertilization, it is vital that the sperm cells and their suspending media be sterile. Whether approaches that might include surface sterilization of anther tissue prior to dehiscence (Zenkteler 1990; Shivanna and Rangaswamy 1992) and bagging of anthers with sterile bags would ensure sterility without interfering with pollen release is not known at present. It may be necessary to utilize both of these measures and one or more antibiotics in the isolation medium.

Assessment of sperm cell numbers and viability

Determination of pollen quality and its effect on the viability of isolated sperm cells is an important component. The laboratory of Christian Dumas has pioneered several techniques for this evaluation, including one that attempts to relate pollen quality to the quality of sperm cells that can be harvested from the pollen. Using nuclear magnetic resonance technique, Kerhoas et al. (1987) found a strict correlation between pollen water content and the functionality of the pollen plasmalemma. This is important as it allows decisions about probable sperm quality to be made before the isolation procedures. In addition to evaluations of pollen quality, both the Dumas group and the author's group have used various approaches for the analysis of sperm viability after isolation. Because isolated flowering plant sperms are nondividing, assessment techniques depend on morphological and/or functional tests.

Microscopical methods

The easiest way to observe isolated flowering plant sperms is with phase-contrast microscopy; this can be done reliably with 25x or 40x objectives. We can usually get some idea about cell quality based on morphology. The sperm cells of maize are about 7 micrometers in diameter and are spherical after isolation. The phase-contrast (PC) image is a fair indicator of sperm cell condition. The PC image of maize sperms shortly after isolation is that of quite dark, relatively uniform, spherical cells. The author's group has performed flow cytometry on such cells and usually found them to be quite viable. Impaired cells viewed under PC image are not uniformly dark or even spherical (Cass and Fabi 1988). These cells can be recognized as impaired by using flow cytometry. Batches of sperm cells exhibiting this PC morphology are generally discarded. DIC microscopy is also useful for studying the morphology of these isolated protoplasts (Figures 17.1, 17.2), but for routine purposes PC is quite adequate. Cell morphology is, however, not a good indicator of viability, so additional steps are necessary. The additional steps can identify marginal cells that do not exhibit changes in their PC image.

Dye exclusion and uptake

The dye, Evans blue, is excluded from cells whose membranes are intact. Maize sperm cells whose membranes are intact will appear white in a dark blue background (Cass and Fabi 1988). Neutral red accumulates in living cells. Maize sperm cells that are living will appear bright red in a pale environment after staining with neutral red (Cass and Fabi 1988).

Fluorescent procedures

To date, one of the most sensitive methods the author and others have used is the fluorescein diacetate (FDA) test (Heslop-Harrison and Heslop-Harrison 1970). This hydrophobic molecule is internalized in maize sperm cells. If the cells are viable, the diacetate portion is cleaved by cytoplasmic esterases, releasing fluorescein into the cytoplasm. If the cell membrane is intact, the fluorescein will remain inside and will fluoresce (Dupuis et al. 1987; also see Chapter 14).

A DNA fluorochrome, dihydroethidium (DHE) (Molecular Probes, Eugene, Oregon), offers considerable promise for sperm cell technology. This is a cell permeant reagent that is oxidized to ethidium after entering the cell cytoplasm. The ethidium then enters the nuclei and intercalates with DNA. As a result, the nuclei become brightly fluorescent. This compound stains the nuclei of living sperm cells, indicating the physiological state of the cells; cells that are damaged or nonliving are not likely to carry out the oxidation of DHE and will not be fluorescent after use of this reagent. An additional fluorescent procedure that positively identifies cells whose membranes have been damaged is propidium iodide (PI) (Ross et al. 1989). As a nonpermeant DNA fluorochrome, it enters only injured cells. A membrane fluorochrome, PKH2 (Sigma Chemical Co.), provides a means to stably label sperm cell membranes. This reagent, used in combination with DHE, provides 2-wavelength fluorescence to isolated, living maize sperm cells and the possibility of carrying out confocal analyses.

Flow cytometry

The fluorescent methods just discussed can be used with any fluorescent microscope that delivers the appropriate excitation and emission wavelengths. The

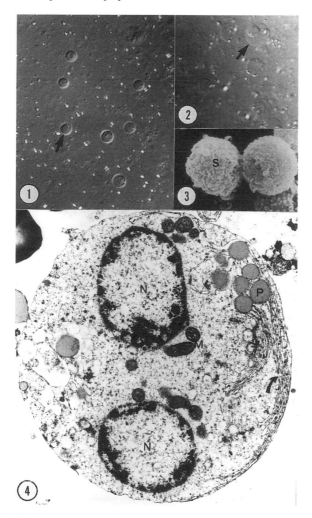

Figures 17.1–17.4

17.1. A number of isolated sperm cells of maize under DIC microscopy (arrow). The suspension medium here is the one normally used for isolation and purification. *17.2.* Isolated maize sperms under DIC microscopy after dilution of the normal medium with water to induce osmotic swelling of the sperms (arrow). (17.1) and (17.2) were photographed at the same magnification using a 40× planapochromatic lens with oil. *17.3.* Two maize sperms (S) in contact prior to fusion under scanning electron microscopy. The medium used to induce fusion can be either Brewbaker–Kwack salts (original strength) or the normal isolation/maintenance medium containing 1 mM or higher calcium nitrate. *17.4.* A transmission electron micrograph of an isolated maize sperm cell showing that it is a protoplast. Endoplasmic reticulum (arrow) is visible in the outer periphery of the cell. N = nucleus, P = plastid.

problem with any of these techniques is that the numbers of cells observed will be limited to a few microscopic fields. Further, marginal fluorescence is difficult to score. Accordingly, the author's lab has coupled FDA, PI, and DHE use in fluorescence microscopy with flow cytometry (Zhang et al. 1992a). Flow cytometry also allows us to examine light scattering in isolated maize sperm cells, which gives us an idea about cell morphology. Changes in light scattering indicate changes in density of the cytoplasm as well as morphological changes. We can measure fluorescence intensity after staining the cells with FDA. We can assess viability of thousands of maize sperm cells in a very short period of time. We can also use the flow cytometer to identify cells that are not in a healthy state, particularly relating to membrane integrity. Any batches of maize sperm cells that we are concerned about can be stained with PI. Cells with damaged membranes will allow PI to enter; it then stains nuclei bright red (Zhang et al. 1992b). The author's group and others originally used the Brewbaker–Kwack salts + sucrose (BKS) (1963) medium for sperm cell isolation and maintenance. This medium was designed for pollen tube growth and was presumed to be appropriate for isolation of sperm cells. However, a large percentage of maize sperms were impaired after a relatively short period of time in BKS, which contains boric acid (1.62 mM), calcium nitrate (1.27 mM), magnesium sulfate (0.81 mM), and potassium nitrate (0.99 mM); sucrose is normally added as an osmoticum. The next step was to design an isolation medium that did not induce the deleterious changes in sperm cells attributed to BKS. The new medium contains the cell culture buffer 2-(N-morpholino) ethanesulfonic acid (MES), galactose, and bovine serum albumin (BSA) at a pH of 6.7 (Zhang et al. 1992b).

Further analysis of Brewbaker–Kwack salts

In subsequent detailed studies of BKS, the author's research group learned that calcium salts alone, regardless of anion, at 1 mM or higher are strongly damaging to maize sperm cells (Zhang et al. 1995). In their opinion, the use of BKS at the concentrations given is not appropriate for isolated maize sperm cells. Calcium at 1 mM or higher induces membrane changes leading to premature fusion of sperm cells, even with other sperm cells in the same medium (Figure 17.3). There appears to be no tendency for sperm–sperm fusions in media lacking calcium. The maintenance medium that the author's lab has developed for maize sperm cells maintains viability for up to 96 hours, allowing a number of additional experiments to be performed on these cells. It is unlikely that this MES, galactose, and BSA medium at pH 6.7 is totally appropriate for maize sperms; additional changes to this medium are contemplated.

Cell number determinations

The author's lab normally performs hemacytometry using phase-contrast microscopy with a 25× objective. This is tedious, but it can be done. The use of electronic counting techniques, such as the Coulter Counter, may not be possible because of low electrolytes in the original Brewbaker–Kwack salt solution.

Manipulation of isolated sperm cells

Working with specialized cells like the sperms of flowering plants is a challenge, especially in considering the issue of which isolation and/or maintenance media are appropriate. Sperm cells normally live in an isolated environment about which we know relatively little. One must try to develop an appropriate medium by using several viability tests and then put the medium together, using the viability tests at every critical juncture.

Manipulations of sperm cells leading to fusions with isolated egg cells

Fusion of sperm cell with isolated egg cell has great importance to plant developmental biology and technologies that can lead to monocot crop improvement. The first demonstration that isolated sperm cells from flowering plants could be induced to participate in an in vitro fusion was reported by Kranz et al. (1991; also see Chapter 19). They accomplished this by electrofusing isolated sperms and eggs. The putative zygotes underwent some divisions but did not undergo normal embryogenesis. In 1993, Kranz and Lorz obtained normal embryogenesis in zygotes resulting from electrofusion after developing a feeder layer derived from nucellar tissue from a fertilized kernel that provided the proper conditions for embryogenesis. Faure et al. (1994) studied experimental fusions between isolated maize sperm cells and target protoplasts from various tissues; their experiments also included isolated egg cells. They discovered that there is a degree of fusion specificity and that high concentrations of calcium (0.05 M) can induce fusion without an electric current. They also made the interesting observation that once an egg of maize has fused with a sperm cell, additional sperm cells do not fuse with the zygote, suggesting some change in the zygote membrane that reduces its receptivity, analogous to prevention of polyspermy in animal fertilization. Kranz and Lorz (1994) have also attempted fusions between single maize egg and sperm cells using high calcium (again, 0.05 M) at pH 11. The fusions appeared to work but, to date, have not resulted in developing embryos but rather in microcalli of about 30–50 cells (see Chapter 19). Matthys-Rochon et al. (1994; also see Chapter 18) are attempting to fuse sperm and egg nuclei by microinjecting sperm nuclei into enzymatically isolated maize embryo sacs (Mòl et al. 1993). The microinjection technique, which Matthys-Rochon et al. (1994) and the author's laboratory are working to perfect, combined with a completely surgical isolation of unfertilized maize embryo sacs in small slabs of nucellar tissue (Campenot et al. 1992) may become one of the most simple and rapid methods for achieving in vitro fusions of this sort. The intact embryo sac represents the ideal environment for early embryogenesis. The additional advantage of the microinjection technique is that it provides access to the central cell for endosperm initiation and possible transformation.

Other types of manipulations

The isolated sperm cell of flowering plants represents a unique plant cell, one that offers considerable potential for other experiments that have no direct connection to in vitro fertilization. Pennell et al. (1987) injected purified sperm cells of *Plumbago* into mice in an attempt to obtain antibodies specific to the sperms. After fusing spleens from immunized animals with myeloma cells, they obtained 195 stable hybridomas, 91% of which produced antibodies; 46 of these proved to be specific to the sperms. This approach may ultimately produce useful information about the types of interactions occurring during fusion between sperms and target protoplasts. Geltz and Russell (1988) extracted proteins from the male germ unit (MGU) fraction (two sperms and the vegetative nucleus) of *Plumbago*, obtaining 18 unique polypeptides from the MGU-rich fraction. This information will also contribute to our understanding of some interactions occurring during fusion between sperms and target protoplasts, including eggs and central cells.

The author's research group has studied macromolecular synthesis in isolated, living maize sperm cells (Zhang et al. 1993) and learned that they are synthetically active, but that the rates of synthesis increase within 24 hours of isolation. These cells synthesize a considerable amount of protein and some RNA. In fact, there may be qualitative changes in proteins formed from freshly isolated cells and those that are 24 hours old. Isolated sperm cells are also suitable for studies of transport. For this purpose, the absence of cell walls makes them particularly well suited (Figure 17.4). The author's lab is currently examining uptake of labeled amino acids into isolated sperms and has found that the sperm cells are completely amenable to standard uptake assay techniques used in membrane transport research (Williams et al., in press). The damage to isolated maize sperms caused by calcium is also being investigated (Williams et al., in press). Use of ^{45}Ca has made it clear that calcium is internalized by isolated maize sperms and that anything that increases the rate of internalization contributes to more rapid cell lysis. The author's lab is currently attempting to characterize the path by which calcium internalization occurs. An interpretation of the calcium-induced damage is that calcium may promote sperm membrane changes that are related to ultimate fusion with target protoplasts. This interpretation would explain several observations that this lab and others have made.

A new way to look at fusion using isolated sperm cells

The confocal microscope provides a new way for observing the fusion process, whatever fusion technique is being employed. The use of the DNA fluorochrome DHE, combined with the cell membrane fluorochrome PKH2, makes it possible to perform confocal microscopy on isolated sperms. This seems to be an excellent way to do the microscopic assessment of fusions using isolated sperms of flowering plants (L. Sanders et al., unpublished results).

Prospects for sperm cell research

The isolated sperm cell of flowering plants offers several advantages over many other plant cell types for experiments in cell biology and physiology. Virtually all are natural protoplasts, so that no cell wall removal procedures are required to expose cell membranes; they are relatively durable as long as osmotic and ion requirements are met; and they are available in very large numbers, using relatively simple isolation techniques compared with other plant protoplasts. The isolated flowering plant sperm represents a differentiated cell type; although it does not divide, it carries out numerous physiological functions, such as transport and protein synthesis. The possibility of looking at kinetics of uptake of various permeants opens the door to the study of membranes in a minimally altered plant cell. The role of flowering plant sperms in fertilization may well be mediated by recognition events between the sperm and the target cell (that is, egg and central cell). This has already been alluded to in the work of Russell (1985). The fact that these cells are amenable to large-scale, rapid screening using such techniques as flow cytometry means that researchers could efficiently look for transport and/or recognition molecules using a variety of molecular biology techniques.

References

Brewbaker, J.L. & Kwack, B.H. (1963). The essential role of calcium in pollen germination and pollen tube growth. *Amer. J. Bot.* 50:859–865.

Campenot, M.K., Zhang, G., Cutler, A.J. & Cass, D.D. (1992). *Zea mays* embryo sacs in culture. I. Plant regeneration from 1 day after pollination embryos. *Amer. J. Botany* 79:1368–1373.

Cass, D.D. (1973). An ultrastructural and Nomarski-interference study of the sperms of barley. *Can. J. Bot.* 51:601–605.

Cass, D.D. & Fabi, G.C. (1988). Structure and properties of sperm cells isolated from the pollen of *Zea mays. Can. J. Bot.* 66:819–825.

Dupuis, I., Roeckel, P., Matthys-Rochon, E. & Dumas, C. (1987). Procedure to isolate sperm cells from corn (*Zea mays* L.) pollen grains. *Plant Physiol.* 85:876–878.

Faure, J.-E., Digonnet, C. & Dumas, C. (1994). An *in vitro* system for adhesion and fusion of maize gametes. *Science* 263:1598–1600.

Finn, W.W. (1925). Male cells in angiosperms. I. Spermatogenesis and fertilization in *Asclepias cornuti. Bot. Gazette* 80:1–25.

Geltz, N.R. & Russell, S.D. (1988). Two-dimensional electrophoretic studies of the proteins and polypeptides in mature pollen grains and the male germ unit of *Plumbago zeylanica. Plant Physiol.* 88:764–769.

Heslop-Harrison, J. & Heslop-Harrison, Y. (1970). Evaluation of pollen viability by enzymatically induced fluorescence: intracellular hydrolysis of fluorescein diacetate. *Stain Tech.* 45:115–120.

Kerhoas, C., Gay, G. & Dumas, C. (1987). A multidisciplinary approach to the study of the plasma membrane of *Zea mays* pollen during controlled dehydration. *Planta* 171:1–10.

Kranz, E. & Lorz, H. (1993). In vitro fertilization with isolated, single gametes results in zygotic embryogenesis and fertile maize plants. *Plant Cell* 5:739–746.

Kranz, E. & Lorz, H. (1994). *In vitro* fertilization of maize by single egg and sperm cell protoplast fusion mediated by high calcium and high pH. *Zygote* 2:125–128.

Kranz, E., Bautor, J. & Lorz, H. (1991). In vitro fertilization of single, isolated gametes of maize mediated by electrofusion. *Sex. Plant Reprod.* 4:12–16.

Matthys-Rochon, E., Vergne, P., Detchepare, S. & Dumas, C. (1987). Male germ unit isolation from three tricellular species: *Brassica oleracea, Zea mays*, and *Triticum aestivum. Plant Physiol.* 83:464–466.

Matthys-Rochon, E., Mòl, R., Heizmann, P. & Dumas, C. (1994). Isolation and microinjection of active sperm nuclei into egg cells and central cells of isolated maize embryo sacs. *Zygote* 2:29–35.

Mogensen, H.L., Wagner, V.T. & Dumas, C. (1990). Quantitative three-dimensional ultrastructure of isolated corn (*Zea mays*) sperm cells. *Protoplasma* 153:136–140.

Mòl, R., Matthys-Rochon, E. & Dumas, C. (1993). *In vitro* culture of fertilised embryo sacs of maize: zygotes and two-celled proembryos can develop into plants. *Planta* 189:213–217.

Pennell, R.I., Geltz, N.R., Koren, E. & Russell, S.D. (1987). Production and partial characterization of hybridoma antibodies elicited to the sperm of *Plumbago zeylanica. Bot. Gazette* 148:401–406.

Ross, D.D., Joneckis, C.C., Ordonez, J.V., Sisk, A.M., Wu, R.K., Hamburger, A.W. & Nora, R.E. (1989). Estimation of cell survival by flow cytometric quantification of fluorescein diacetate/propidium iodide viable cell number. *Cancer Res.* 49:3776–3782.

Russell, S.D. (1984). Ultrastructure of the sperm of *Plumbago zeylanica*. 2. Quantitative cytology and three-dimensional reconstruction. *Planta* 162:385–391.

Russell, S.D. (1985). Preferential fertilization in *Plumbago*: ultrastructural evidence for gamete-level recognition in an angiosperm. *Proc. Nat. Acad. Sci. USA* 82:6129–6132.

Russell, S.D. (1986). Isolation and characterization of sperm cells from the pollen of *Plumbago zeylanica. Plant Physiol.* 81:317–319.

Russell, S.D., Cresti, M. & Dumas, C. (1990). Recent progress on sperm characterization in flowering plants. *Physiologia Plantarum* 80:669–676.

Shivanna, K.R. & Rangaswamy, N.S. (1992). *Pollen Biology: A Laboratory Manual.* Heidelberg: Springer-Verlag.

Shivanna, K.R., Xu, H., Taylor, P. & Knox, R.B. (1988). Isolation of the sperms from the pollen tubes of flowering plants during fertilization. *Plant Physiol.* 87:647–650.

Theunis, C.H., Pierson, E.S. & Cresti, M. (1991). Isolation of male and female gametes in higher plants. *Sex. Plant Reprod.* 4:145–154.

Williams, C.M., Zhang, G. & Cass, D.D. (in press). Characterization of calcium uptake in isolated maize sperm cells. *Plant Physiol.*

Yang, H.Y. & Zhou, C. (1989). Isolation of viable sperms from pollen of *Brassica napus, Zea mays* and *Secale cereale. Chinese J. Bot.* 1:80–84.

Zenkteler, M. (1990). *In vitro* fertilization and wide hybridization in flowering plants. *CRC Critical Reviews in Plant Science* 9:267–279.

Zhang, G., Campenot, M.K., McGann, L.E. & Cass, D.D. (1992a). Flow cytometric characteristics of sperm cells isolated from pollen of *Zea mays* L. *Plant Physiol.* 99:54–59.

Zhang, G., Williams, C.M., Campenot, M.K., McGann, L.E. & Cass, D.D. (1992b). Improvement of longevity and viability of sperm cells isolated from pollen of *Zea mays* L. *Plant Physiol.* 100:47–53.

Zhang, G., Gifford, D.J. & Cass, D.D. (1993). RNA and protein synthesis in sperm cells isolated from *Zea mays* L. pollen. *Sex. Plant Reprod.* 6:239–243.

Zhang, G., Williams, C.M., Campenot, M.K., McGann, L.E., Cutler, A.I. & Cass, D.D. (1995). Effects of calcium, magnesium, potassium, and boron sperm cells isolated from pollen of *Zea mays* L. *Sex. Plant Reprod.* 8:113–122.

18

Isolation and micromanipulation of the embryo sac and egg cell in maize

E. MATTHYS-ROCHON, R. MÒL, J. E. FAURE,
C. DIGONNET, and C. DUMAS

Summary   363

Introduction   364

Progress in the cell biology of the embryo sac and egg cell in maize   364

    Maize as a model system   364

    The embryo sac and female germ unit   365

    Manipulation of the embryo sac   368

    Isolation and characterization of egg cells   370

    In vitro manipulation of the egg and sperm cells   371

Conclusions   372

References   373

Summary

Maize is a flowering plant well suited for biological research on reproduction processes. In addition to interesting sexual traits, there is in this species a large amount of data regarding genetics and cytogenetics. The embryo sac and the female germ unit have been characterized, and precise studies have been developed on male gametes. Finally, the time course of double fertilization has recently been determined. Nevertheless, the intimate mechanisms of double fertilization are still unknown. A new way to approach this phenomenon is to develop systems of "in vitro fertilization." In this chapter, the technical procedures to isolate and manipulate male and female gametes in maize are described. The embryo sac and egg cells are prepared with the help of an enzymatic treatment. Sperm cells or male nuclei from viable pollen grains are released in an acidic medium and selected from a specific layer of a Percoll gradient after centrifugation steps. With these cellular tools, two in vitro methods of fertilization have been developed: (1) spontaneous fusion of male and female gametes in the presence of 5 mM of calcium and (2) microinjection of male nuclei either into the egg or into the central cell of a mature embryo sac. Culture of the artificial zygotes and microinjected embryo sacs is in progress. All these techniques provide a new window for investigating the first steps of fertilization and early embryogenesis at the cellular and molecular levels.

Introduction

Camerarius (1694, reviewed in Ducker and Knox 1985) was the pioneer in plant reproductive biology. He observed that in several plants, including *Zea mays*, the anthers are essential for seed production by the ovary, although he was not aware of the exact role played by the anthers. Amici (1824) has been credited with the first report of pollen tube growth. A special tribute has to be made to Wilhelm Hofmeister (1824–1877), who provided the first unequivocal evidence to document the actual roles of pollen tubes and egg in angiosperm embryogenesis. This author proved – by presenting an exhaustive survey of the entire embryological stages among 40 flowering plants – that the fertilized egg gave rise to the embryo (reviewed in Ducker and Knox 1985). The female gametophyte was first described accurately by Strasburger (cited in Ducker and Knox 1985). Finally, pollen tube involvement in double fertilization was generalized by Navashin (1898) and Guignard (1899), and confirmed in maize by Guignard (1901) and Miller (1919).

Considerable progress has been made in establishing the female gametophyte as a developmental model mainly in maize and *Plumbago zeylanica* (see Russell 1992). In particular, the isolation of living and fixed embryo sacs and female gametophytic cells (Mòl 1986; review in Theunis et al. 1991), the determination of their precise localization within the ovule (Wagner et al. 1990), the isolation of male gametes (see reviews, Russell 1991; Chaboud and Perez 1992; also see Chapter 17), and the recent research on an in vitro fusion of gametes (Kranz et al. 1991a, b; Faure et al. 1993, 1994; Kranz and Lörz 1993; also see Chapter 19) represent some of the most important accomplishments in exploiting the gametophyte as a source of haploid cells for experimental manipulation. Moreover, the visualization of the cytoskeleton (Huang et al. 1993) and the micromanipulation and in vitro culture of the female gametophyte (Campenot et al. 1992; Mòl et al. 1993; Matthys-Rochon et al. 1994) open new research avenues.

In this chapter, attention is given to *Zea mays* because most progress has been realized in this species. It is thus considered as a model system to analyze fertilization as well as natural or experimental embryogenesis (Dumas and Mogensen 1993).

Progress in the cell biology of the embryo sac and egg cell in maize

Maize as a model system

In many ways, maize is the flowering plant best suited for biological research on gametes (Freeling and Walbot 1993), fertilization, and embryogenesis (Dumas and Mogensen 1993). Maize has unisexual, rather than bisexual, flowers, and male and female organs are well separated on the stem. A tassel grouping staminate flowers is located at the top of the main stem and one or more female spikes are found at the ends of short branches near the middle of the stem. Several million pollen grains are released from the tassel and four to five hundred ovules are available per spike. Controlled pollination in maize is much easier than in other cereals, in which emasculation of each flower is required before it can be used as a female parent.

In addition to these interesting sexual traits, there is a large amount of data regarding genetics and cytogenetics of maize (Sheridan 1988). More recently, considerable attention has been focused on molecular analysis including the use of restriction fragment length polymorphism (RFLP) markers (more than 1,200 are known today), developmental mutants, and gene sequencing. Good examples of homeotic genes are the *knotted* locus, which alters leaf development (Vollbrecht et al. 1991), and *ZAG1*, which controls flower morphogenesis (Schmidt et al. 1993). The *Zmhox1* gene encodes a nuclear protein with a central acidic domain, indicating that it may have a transcriptional activator function (Bellmann and Werr 1992). Finally, a large collection of developmental mutants is available for embryogenesis (Clark and Sheridan 1991; Meinke 1991).

The embryo sac and female germ unit

The embryo sac: localization and characterization

The embryo sac and the constituent cells of an embryo sac inside the sporophytic tissues of maize have been localized by scanning electron microscopy, serial sectioning, and three-dimensional reconstruction. In most angiosperms, including maize, the embryo sac is located deep within the ovule, which, in turn, is located within the ovary (Figure 18.1; see also Van Lammeren 1986; Wagner et al. 1990). This location, which is a general trait of all angiosperms, constitutes a real problem in accessing the female cells (see later section on embryo sac isolation procedure). The mature embryo sac is composed of an egg cell, two synergids, the central cell with two polar nuclei, and about 20 antipodals (Figure 18.2a). Dow and Mascarenhas (1991a, b) have employed in situ hybridization to determine the relative number of ribosomes within the cells of the embryo sac of maize. During embryo sac maturation, the accumulation of ribosomes appears to proceed at a constant and high rate, with the highest rate in the developing central cell.

As in other angiosperms, synergids and two female gametes are integrally involved in the process of double fertilization and are recognized as the female germ unit.

The female germ unit: concept and fate

The female germ unit concept was recently reviewed by Huang and Russell (1992) in an issue devoted entirely to plant reproductive biology (Russell and Dumas 1992). In his pioneering studies of the embryo sac in cotton, Jensen (1974) proposed that the egg apparatus, composed of the egg and the two synergids, serves as the functional female unit during fertilization. The authors have proposed the concept of the female germ unit to encompass both of the female reproductive cells: the egg cell (with the synergids) and the central cell

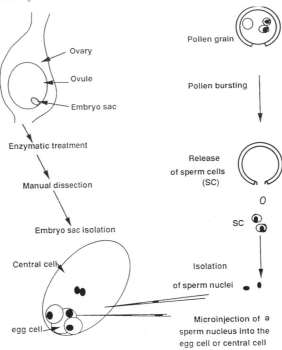

Figure 18.1 A general protocol of microinjection of sperm nuclei into the female gametophyte of maize.

(Dumas et al. 1984). This new proposal is supported by the fact that no wall is present at the target zone where the two sperm cells must fuse and enter. There is a virtual window in the egg cell wall, whereas the central cell has no wall, only bounding protoplasmic membranes (Russell 1992).

The intimate relations between the cells of the female germ unit are of special importance for understanding the mechanisms of male gamete delivery and syngamy. As was shown recently in *Plumbago* and *Nicotiana*, actin elements are abundant at the egg–central cell boundary and may be involved in sperm migration (Huang et al. 1993; Russell 1993). Many actin microfilaments also appear in the pollen tube (Pierson and Cresti 1992, and references therein). An ultrastructural and histochemical study suggests that pollen tube cytoplasm, rather than the cytoplasm of the penetrated synergid, enters between the egg cell and central cell in maize (Diboll 1968). So it is likely that the pollen tube residue in maize embryo sacs also contains actin. Maize embryo sacs cleared or stained for light microscopy showed granular material similar to actin "coronas" of tobacco and *Plumbago* (R. Mòl and E. Matthys-Rochon, unpublished results).

The synergids play an essential role in receiving the pollen tube and discharging male gametes into the embryo sac. The egg cell and the central cell fuse with the sperm cells and form the zygote and the endosperm, respectively. Because no cell walls appear in the sperm cells and at the egg–cen-

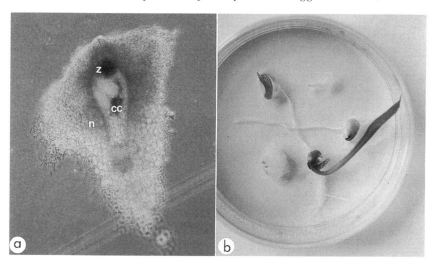

Figure 18.2 (**a**) An isolated fertilized embryo sac; z = zygote; cc = central cell; n = nucellar cells. Bar = 50μm. (**b**) After 2–3 weeks of culture, small plantlets develop in Petri dishes (diameter = 5cm).

tral cell boundary, the models of plasmogamy are based on membrane fusion (Wilms 1981; Russell 1983; Mogensen 1990). In situ fertilization in *Zea mays* was analyzed only at the light-microscopic level. The first observations of Guignard (1901) were confirmed by several authors 40–60 years ago, and a detailed study on the kinetics of this process appeared recently (Mòl et al. 1994). The period from pollen tube arrival to gamete fusion is short, possibly less than 1 hour. Both sperm cells and dense cytoplasm are released from the degenerated synergid into a narrow space between the egg cell and central cell and then fuse with the female gametes. It is difficult to judge the precise time required for plasmogamy in situ. It must be in the range of minutes. The fate of male cytoplasm is unknown, but the sperm nuclei must move quickly toward the female nuclei because the sperm nuclei are usually found in cytological preparations very close to the egg nucleus and polar nuclei or the secondary nucleus (Vazart 1955; Mòl et al. 1994). The karyogamy is of the premitotic type. Male and female nucleoli fuse within 5 hours in the egg cell and 3 hours in the central cell. The time course of in situ and in vitro karyogamy in the egg cell is shown by Mòl et al. (1994) and Faure et al. (1993). The polarity of maize zygotes is opposite to that of the egg cell (Van Lammeren 1986). This shift in cell polarity starts during karyogamy and continues during zygote maturation (Mòl et al. 1994). In parallel with fusion of the sperm and egg nuclei, the cytoplasm displaces from the basal region to the perinuclear area in the middle part of the cell. The young zygote does not divide for about 16 hours after karyogamy, and its nucleus and most of the cytoplasm occupy the apical position (Mòl et al. 1994).

Manipulation of the embryo sac

Female receptivity: a prerequisite for embryo sac manipulation

An important property of the pistil is the period during which the stigma is receptive and can be actually pollinated. This period more or less corresponds to the embryo sac maturation period and has most often been determined by pollination experiments. It has been shown that both stigma and ovule receptivity are quite synchronous. Several methods are currently used for assessing stigma receptivity: determination of seed set after pollination at different times relative to anthesis (pre-and postanthesis); detection of the presence of stigmatic secretion on the papilla surface (for example, Knox et al. 1986); and cytochemical detection of some nonspecific enzymes, for example ATPase and peroxidase (for example, Gaude and Dumas 1987). In maize, the silk length can be correlated to the maturation of the embryo sac (Dupuis and Dumas 1990). Ovule receptivity may be assessed by using aniline blue fluorescence (ABF). Nonfunctional and senescent ovules accumulate callose and consequently show bright yellow fluorescence. By contrast, functional ovules exhibit low fluorescence or only autofluorescence (Dumas and Knox 1983). It is clear that precise knowledge of the receptivity period aids in the determination of the developmental stage of the embryo sac and for subsequent manipulations.

Embryo sac isolation procedure

Wagner et al. (1989b) developed a procedure for the isolation of embryo sacs in maize. The routine technique consists of collection of unpollinated ears, which are surface sterilized with 70% ethanol. Then the emerged silks and husks are removed. In the middle portion of the ears the upper halves of the ovaries are trimmed row by row, and ovule halves (about 20 per row) are collected in a convenient medium (osmotic pressure: 580–600 mOsm) (Matthys-Rochon et al. 1993, 1994). The ovule halves are transferred to an enzymatic solution modified from Wagner et al. (1989b) and composed of 0.5% macerozyme, 0.5% cellulase, 0.75% pectinase, and 0.25% pectolyase, for 30 minutes at 4°C. Ovules are washed in a fresh medium and embryo sacs are then dissected with insect needles under a stereo microscope (SMZ 10, Nikon, Japan). Generally, the isolated embryo sacs are still surrounded with nucellar cells. At low magnification, only the position of the central cell and the egg cell can be distinguished (Figure 18.2a). With this method, viable embryo sacs can be prepared (Matthys-Rochon et al. 1993).

Microinjection of lucifer yellow and sperm nuclei into the female cells.

Before micromanipulation, it is necessary to characterize the embryo sac, particularly to determine the composition of the wall. With that in mind, the authors used aniline blue and calcofluor white, which respectively demonstrated that callose and cellulose are present in the wall of the embryo sac. This envelope is in fact a barrier to microinjection. The first assays showed that a microcapillary cannot penetrate into the embryo sac, which bends when held with a holding pipette in a liquid medium. For this reason, the authors' procedure is to place the embryo sacs on a coverslip, aspire the liquid medium, and deposit a drop of low melting point agarose (1%). The fluorescein diacetate test showed that cells are still alive after embedding.

With embedded embryo sacs, the authors have succeeded in injecting lucifer yellow into the central cell or into the egg cell (Matthys-Rochon et al. 1993). No diffusion of the fluorescent dye was observed either between cells or through the embryo sac envelope. In this type of experiment, the survival of the gametophyte is dependent on the environmental conditions and also on the accuracy of manipulation. The microcapillary (diameter 0.5 μm) must be firmly inserted through the wall and then delicately pushed onto the target cell. The injection pressure is then activated and the fluorescent dye spreads into the cell. Finally, to avoid cell injury, the microcapillary must be removed very gently.

The success of these first attempts led to a more complex manipulation: the microinjection of sperm nuclei into female cells of *Zea mays* (Figure 18.1). The first step is to isolate active male nuclei from pollen grains. A pollen suspension is prepared in a sucrose solution (350 mM) and 0.5% Triton x100 is added to disrupt the membranes of the released sperm cells. After filtration, the mixture is centrifuged and the pellet containing the nuclei is resuspended in a protective medium modified from de Paepe et al. (1990). The nuclei are then purified with Percoll gradient centrifugation. The pure and morphologically intact nuclei are checked for their transcriptional activity; 10^5 nuclei were labeled with ^{32}P UTP (uridine triphosphate) and the kinetics of incorporation of the labeled nucleotides was followed. Freshly isolated nuclei are able to incorporate precursors of RNA. Autoradiography confirmed that the sperm nuclei themselves are transcriptionally active under in vitro conditions (Matthys-Rochon et al. 1994).

In parallel to sperm nuclei isolation, the embryo sacs are embedded in agarose as described previously. Microinjection of nuclei involves four steps: (1) With the help of a micromanipulator, the micropipette containing the single nucleus is brought laterally to the targeted female cell. (2) The microcapillary is pushed firmly until it penetrates into the female cell and it is then slowly inserted a fraction more. (3) With the help of an oil pressure in a screw syringe microinjector, the selected nucleus is injected into the cell. (4) Finally, the micropipette is withdrawn very slowly in order to avoid cell damage.

To follow the injection of sperm nuclei, the authors stained them with the fluorochrome DAPI. To ascertain the presence of sperm nuclei within the female cells, a cytological analysis was performed. The examination of stained microtome sections showed that the microinjected nuclei appeared in the cytoplasm of the egg or central cell with a success rate of 14% for each cell type (Matthys-Rochon et al. 1994).

Such experiments help us understand the behavior of the nucleus within the female cells during the fertilization process and, in turn, may constitute a new approach to the study of sexual plant reproduction.

In vitro regeneration of plants from isolated fertilized embryo sac

Fertilized embryo sacs of maize can also be isolated in a manner similar to that used for the mature, unfertilized embryo sacs (E. Matthys-Rochon and R. Mòl, unpublished results). In maize, another technique – microdissection without any enzymatic treatment – yields embryo sacs containing zygotes surrounded only by some nucellar cells (Figure 18.2a) (Mòl et al. 1993). Such gametophytes cultured in vitro develop zygotic embryos and cellular endosperm. On the best medium supplemented with zeatin, 58% of isolated embryos are well differentiated and similar to maize embryos in vivo, and they develop coleoptile or one or two leaf primordia (Mòl et al. 1995). Isolated embryos germinate precociously on a hormone-free medium (Figure 18.2b) and up to 62% of embryos give fertile plants in 8 weeks. The main advantages of this culture system are the high regeneration capacity of maize zygotes cultured within fertilized embryo sacs, the normal structure of many embryos, direct plant regeneration, and access to the zygote for micromanipulation. However, two other approaches for plant regeneration from maize zygotes have been recently developed: the feeder cell system for zygotes resulting from electrofusion of isolated gametes (Kranz and Lörz 1993; also see Chapter 19), and the culture of nucellar tissue containing fertilized embryo sacs (Campenot et al. 1992).

Isolation and characterization of egg cells

Egg cell isolation procedure

Ears bearing female flowers were bagged before silk emergence and cut off at the mature stage as indicated by the silk length (Dupuis and Dumas 1990). Ovule pieces containing embryo sacs were dissected from ovaries of the middle part of the ear, temporarily collected in 0.5 M mannitol, digested for 10 minutes at 24°C with an enzyme solution containing pectinase, pectolyase, hemicellulase, and cellulase and adjusted to 570 mOsm with mannitol (Wagner et al. 1989b; Kranz and Lörz 1990; Kranz et al. 1991a, b). The egg cells were mechanically isolated from the digested ovular tissues by means of microneedles under an inverted microscope.

An important step in single gamete manipulation is the transfer of cells without damage. A suitable system is that of Koop and Schweiger (1985) using a computer-controlled dispenser–dilutor, a hydraulic system, and microcapillaries connected by means of Teflon tubing filled with mineral oil (Kranz and Lörz 1990; Kranz et al. 1991a, 1992). A simpler system uses manual tube clamps (Faure et al. 1994).

The isolation of viable egg cells is difficult, and, routinely, 5 egg cells per 20 ovule pieces can be isolated (Kranz et al. 1991a, b).

Characterization of isolated eggs (ultrastructure, quantification), and comparison with sperm cells

Transmission electron microscopy allows the study of the ultrastructure of the isolated egg cell, which can be compared to that of the isolated male gametes.

Isolated egg cell protoplasts are highly vacuolated with numerous peripheral vacuoles. Most of the organelles are aggregated. The center of this aggregate is occupied by polymorphic mitochondria that are surrounded by large plastids containing starch grains (Faure et al. 1992). Isolated egg cells are about three hundred times larger than isolated sperm cell protoplasts. The spherical nucleus is euchromatic and contains a dense nucleolus, whereas the nucleus is heterochromatic in the mature sperm cells (Wagner et al. 1989a; Mogensen et al. 1990; Faure et al. 1992).

In vitro manipulation of the egg and sperm cells

Procedures for producing fusion of isolated gametes

Two procedures have been successful in producing the fusion of isolated gametes;

1. ELECTROFUSION. A method of in vitro fertilization using electrofusion was first described by Kranz and Lörz (1990) (see Chapter 19). By using electron microscopy combined with serial sectioning and computer-aided reconstruction, it was demonstrated that karyogamy occurs 1 hour after electric membrane fusion (Faure et al. 1993). Electrofusion might be useful in studies of cytoplasmic inheritance and fusion of individual sperm cells with nongametic cells of the embryo sac (Kranz et al. 1991a, b, 1992).

2. IN VITRO "SPONTANEOUS" FUSION. In order to get a better understanding of the double fertilization process and to study gamete recognition and the molecular mechanism of adhesion and fusion, a nonelectric alternative has been developed by Faure et al. (1994). Fusion of male and female isolated gametes occurs in 0.5 M mannitol supplemented with 5 mM calcium at a frequency of 80% (Figure 18.3). In the fusion products, two nucleoli were visible in the nucleus when maintained for 24 hours in the same medium. Through comparison to in vivo studies (Mòl et al. 1994), the presence of the two nucleoli indicates that karyogamy has occurred. No fusion was observed between additional gametes and the fusion product, or between mesophyll protoplasts and egg cells. Therefore this in vitro system allows fusion in conditions close to the natural process and will permit the study of polyspermy barriers and gamete recognition.

Zygote regeneration from in vitro fertilization and from in vivo fertilization

Fertile plants were regenerated from zygotes produced by electrofusion. Regeneration occurs via zygotic embryogenesis using a feeder cell system

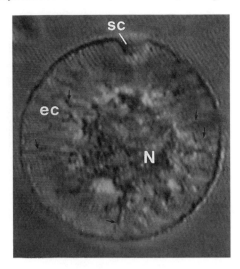

Figure 18.3 In vitro "spontaneous" fusion of a sperm cell (sc) and egg cell (ec) from maize. Note the nucleus area (N) and the cytoplasmic tracts (arrows).

with a nonmorphogenic maize cell suspension. After 10–12 days, proembryos developed from fusion products and transition-phase embryos were transferred onto modified MS (Murashige and Skoog) media. Plants were transferred into soil 45–109 days after gamete electrofusion. The hybrid nature of the plants was confirmed (Kranz and Lörz 1993). A Danish team has recently succeeded in developing fertile plants from isolated barley and wheat zygotes when co-cultivated with embryogenic microspores (Holm et al. 1994). In the authors' laboratory, isolated maize zygotes have also been grown into fertile plants (Leduc et al. 1996).

Conclusions

The obtaining of female and male reproductive cells (embryo sacs, egg and central cells, sperm cells) allows biochemical and molecular studies and dissection of the fertilization process. Preliminary biochemical studies have been performed and have revealed the presence of specific proteins in the female gametophyte (Y. C. Song, P. Vergne, and E. Matthys-Rochon, unpublished results). cDNA libraries from egg cells and electrofused zygotes have been already constructed (Kranz et al. 1994) and could lead to the discovery of specific genes acting during fertilization. The spontaneous fusion system developed by Faure et al. (1994) opens new horizons for the study of a possible gametic determinant that targets the double fertilization process. In fact, Russell (1985) has shown in *Plumbago* that the two male gametes differ by their organelle content: One is plastid rich and the other is mitochondria rich. Moreover, the same author has demonstrated that the plastid-rich male gamete preferentially fuses with the egg cell. These cytological results suggest that recognition events occur at the gamete level. In maize, no clear morphological heterospermy has been observed, but nuclear differences have been reported in the Black Mexican Sweet (BMS) line, in which the sperm bearing the nondisjuncted B-chromosome preferentially fuses with the egg cell (Roman 1948). It is thus reasonable to suggest that higher plant gametes might have specific cell surface determinants that intervene during mating. One strategy to analyze these phenomena would consist of the use of monoclonal antibodies raised against gametes, which could potentially inhibit the fusion process and reveal the role of gamete membrane molecules.

In addition, the development of fertilized embryo sacs (zygote stage) and the obtaining of fertile plants from fusion products and from natural zygotes in in vitro conditions may provide an effective approach to achieving stable transformation in maize via microinjection or electroporation. Thus, maize seems to be one of the most appropriate plant models for studying gametes and early embryogenesis (Dumas and Mogensen 1993).

Although double fertilization was first described by Navashin in 1898 and significant cytological and physiological work has been done on this unique feature of flowering plants, little progress has been made in terms of molecular analysis. Gamete isolation procedures, gamete manipulation experiments, and in vitro development of fertilized embryo sacs and zygotes open up new possibilities for study in this field.

References

Amici, G. (1824). Observations microscopiques sur diverses espèces de plantes. *Ann. Sci. Nat.* 2:41–70.

Bellman, R. & Werr, W. (1992). *ZmHox1a*, the product of a novel maize homeobox gene, interacts with the shrunken 26-bp feedback control element. *EMBO J.* 11:3367–3374.

Campenot, M.K., Zhang, G., Cutler, A.J. & Cass, D.D. (1992). *Zea mays* embryo sacs in culture. I. Plant regeneration from 1 day after pollination embryos. *Amer. J. Bot.* 79:1368–1373.

Chaboud, A. & Perez, R. (1992). Generative cells and male gametes: isolation, physiology, and biochemistry. *Int. Rev. Cytol.* 140:205–232.

Clark, J.K. & Sheridan, W.F. (1991). Isolation and characterization of 51 embryo-specific mutations of maize. *Plant Cell* 3:935–951.

Diboll, A.G. (1968). Fine structural development of the megagametophyte of *Zea mays* following fertilization. *Amer. J. Bot.* 55:787–806.

Dow, D.A. & Mascarenhas, J.P. (1991a). Optimization of conditions for *in situ* hybridization and determination of the relative number of ribosomes in the cells of the mature embryo sac of maize. *Sex. Plant Reprod.* 4:244–249.

Dow, D.A. & Mascarenhas, J.P. (1991b). Synthesis and accumulation of ribosomes in individual cells of the female gametophyte of maize during its development. *Sex. Plant Reprod.* 4:250–253.

Ducker S.C. & Knox, R.B. (1985). Pollen and pollinisation: a historical review. *Taxon* 34:401–419.

Dumas, C. & Knox, R.B. (1983). Callose and determination of pistil viability and incompatibility. *Theor. Appl. Genet.* 67:1–10.

Dumas, C. & Mogensen, H.L. (1993). Gametes and fertilization: maize as a model system for experimental embryogenesis in flowering plants. *Plant Cell* 5:1337–1348.

Dumas, C., Knox, R.B., McConchie, C.A. & Russell, S.D. (1984). Emerging physiological concepts in fertilization. *What's New Plant Physiol.* 15:168–174.

Dupuis, I. & Dumas, C. (1990). Biochemical markers of female receptivity in maize (*Zea mays* L.) assessed using *in vitro* fertilization. *Plant Sci.* 70:11–19.

Faure, J.E., Mogensen, H.L., Kranz, E., Digonnet, C. & Dumas, C. (1992). Ultrastructural characterization and three-dimensional reconstruction of isolated maize (*Zea mays* L.) egg cell protoplasts. *Protoplasma* 170:97–103.

Faure, J.E., Mogensen, H.L., Dumas, C., Lörz, H. & Kranz, E. (1993). Karyogamy after electrofusion of single egg and sperm cell protoplasts from maize: cytological evidence and time course. *Plant Cell* 5:747–755.

Faure J.E., Digonnet C. & Dumas C. (1994). An *in vitro* system for adhesion and fusion of maize gametes. *Science* 263:1598–1600.

Freeling, M. & Walbot, V., eds. (1993). *The Maize Handbook*. New York: Springer-Verlag.

Gaude, T. & Dumas, C. (1987). Molecular and cellular events of self incompatibility. *Int. Rev. Cytol.* 107:333–366.

Guignard, L. (1899). Sur les anthérozoïdes et la double copulation sexuelle chez les végétaux angiospermes. *Rev. Gén. de Bot.* 11:129–135.

Guignard, L. (1901). La double fécondation dans le maïs. *J. Bot.* 15:37–50.

Holm, P.B., Knudsen, S., Mouritzen, P., Negri, D., Olsen, F.L. & Roué, C. (1994). Regeneration of fertile barley plants from mechanically isolated protoplasts of the fertilized egg cell. *Plant Cell* 6:531–543.

Huang, B.Q. & Russell, S.D. (1992). Female germ unit: organization, isolation, and function. *Int. Rev. Cytol.* 140:233–293.

Huang, B.Q., Pierson, E.S., Russell, S.D., Tiezzi, A. & Cresti, M. (1993). Cytoskeletal organization and modification during pollen tube arrival, gamete delivery and fertilization in *Plumbago zeylanica*. *Zygote* 1:143–154.

Jensen, W.A. (1974). Reproduction in flowering plants. In *Dynamic Aspects of Plant Ultrastructure*, ed. A.W. Robards, pp. 481–503. New York: McGraw Hill.

Knox, R.B., Williams, E.G. & Dumas, C. (1986). Pollen, pistil and reproductive function in crop plants. *Plant Breeding Rev.* 4:9–79.

Koop, H.U. & Schweiger, H.G. (1985). Regeneration of plants from individually cultivated protoplasts using an improved microculture system. *J. Plant Physiol.* 121:245–257.

Kranz, E. & Lörz, H. (1990). Micromanipulation and *in vitro* fertilization with single pollen grains of maize. *Sex. Plant Reprod.* 4:160–169.

Kranz, E. & Lörz, H. (1993). *In vitro* fertilization with isolated, single gametes results in zygotic embryogenesis and fertile maize plants. *Plant Cell* 5:739–746.

Kranz, E., Bautor, J. & Lörz, H. (1991a). *In vitro* fertilization of single, isolated gametes of maize by electrofusion. *Sex. Plant Reprod.* 4:12–16.

Kranz, E., Bautor, J. & Lörz, H. (1991b). Electrofusion-mediated transmission of cytoplasmic organelles through the *in vitro* fertilization process, fusion of sperm cells with synergids and central cells, and cell construction in maize. *Sex. Plant Reprod.* 4:17–21.

Kranz, E., Lörz, H., Digonnet, C. & Faure, J.E. (1992). *In vitro* fusion of gametes and production of zygotes. *Int. Rev. Cytol.* 140:407–423.

Kranz, E., Dresselhaus T. & Lörz, H. (1994). *In vitro* fertilization with single maize gametes. In *Frontiers in Sexual Plant Reproduction Research*, eds. E. Heberle-Bors, M. Hesse, and O. Vicente, p. 64. Vienna: University of Vienna Press.

Leduc, N., Matthys-Rochon, M., Rougier, M., Mogensen, L., Holm, P., Magnard, J-

L. & Dumas, C. (1996). Isolated maize zygotes mimic in vivo embryonic development and express microinjected genes when cultured in vitro. *Develop. Biol.* 177:190–203.

Matthys-Rochon, E., Digonnet C. & Dumas, C. (1993). Characterisation of *Zea mays* embryo sac using fluorescent probes and microinjection of lucifer yellow into the female cells. In *Biotechnology Applications of Microinjection, Microscopic Imaging and Fluorescence,* eds. P.H. Bach, C.H. Reynolds, J.M. Clark, J. Mottley, and P.L. Poole, pp. 53–60. New York: Plenum Press.

Matthys-Rochon, E., Mòl, R., Heizmann, P. & Dumas, C. (1994). Isolation and microinjection of active sperm nuclei into egg cells of isolated maize embryo sacs. *Zygote* 2:29–35.

Meinke, D.W. (1991). Perspectives on genetic analysis of plant embryogenesis. *Plant Cell* 3:857–866.

Miller, E.C. (1919). Development of the pistillate spikelet and fertilization in *Zea mays*. *J. Agric. Res.* 18:255–257.

Mogensen, H.L. (1990). Fertilization and early embryogenesis. In *Reproductive Versatility in the Grasses,* ed. G.P. Chapman, pp. 76–99. Cambridge, New York, Port Chester, Melbourne, Sydney: Cambridge University Press.

Mogensen, H.L., Wagner, V.T. & Dumas, C. (1990). Quantitative three-dimensional ultrastructure of isolate corn (*Zea mays*) sperm cells. *Protoplasma* 153:136–140.

Mòl, R., (1986). Isolation of protoplasts from female gametophytes of *Torenia fournieri*. *Plant Cell Reprod.* 3:202–206.

Mòl, R., Matthys-Rochon, E. & Dumas, C. (1993). *In vitro* culture of fertilized embryo sacs of maize: zygotes and two-celled proembryos can develop into plants. *Planta* 189:213–217.

Mòl, R., Matthys-Rochon, E. & Dumas, C. (1994). The kinetics of cytological events during double fertilization in *Zea mays* L. *Plant J.* 5:197–206.

Mól R., Matthys-Rochon, E. & Dumas C. (1995). Embryogenesis and plant regeneration from maize zygotes by *in vitro* culture of fertilized embryo sacs. *Plant Cell Reports* 14:743–747.

Navascin, S.G. (1898). Über das Verhalten des Pollenschlauches bei der Ulme. *Bul. Acad. Imp. des Sci. St. Petersburg* 8:345–357.

de Paepe, R., Koulou, A., Pham, J.L. & Brown, S.C. (1990). Nuclear DNA content and separation of *Nicotiana sylvestris* vegetative and generative nuclei at various stages of male gametogenesis. *Plant Sci.* 70:255–65.

Pierson, E.S. & Cresti, M. (1992). Cytoskeleton and cytoplasmic organization of pollen and pollen tubes. *Int. Rev. Cytol.* 140.73–125.

Roman, H. (1948). Directed fertilisation in maize. *Proc. Nat. Acad. Sci. USA* 34:36–42.

Russell, S.D. (1983). Fertilization in *Plumbago zeylanica*: gametic fusion and fate of the male cytoplasm. *Amer. J. Bot.* 70:416–434.

Russell, S.D. (1985). Preferential fertilization in *Plumbago*: ultrastructural evidence for gamete level recognition in an angiosperm. *Proc. Nat. Acad. Sci. USA* 34:36–42.

Russell, S.D. (1991). Isolation and characterization of sperm cells in flowering plants. *Ann. Rev. Plant. Physiol. Plant Mol. Biol.* 42:189–204.

Russell, S.D. (1992). Double fertilization. *Int. Rev. Cytol.* 140:357–388.

Russell, S.D. (1993). The egg cell: development and role in fertilization and early embryogenesis. *Plant Cell* 5:1349–1359.

Russell, S.D. & Dumas, C., eds. (1992). Sexual reproduction in flowering plants. In *Int. Rev. Cytol.,* 140:565–592. San Diego: Academic Press.

Schmidt, R.J., Velt, B., Alejandra Menel, M., Mena, M., Hake, S. & Yanofsky, M.F.

(1993). Identification and molecular characterization of *ZAG1*, the maize homolog of the *Arabidopsis* floral homeotic gene AGAMOUS. *Plant Cell* 5:729–737.

Sheridan, W.F. (1988). Maize developmental genetics: genes of morphogenesis. *Ann. Rev. Genet.* 22:353–385.

Theunis, C.H., Pierson, E.S. & Cresti, M. (1991). Isolation of male and female gametes in higher plants. *Sex. Plant Reprod.* 4:145–154.

Van Lammeren, A.A.M. (1986). A comparative ultrastructural study of the megagametophytes in two strains of *Zea mays* L. before and after fertilization. *Agric. Univ. Wageningen Papers* 86:1–37.

Vazart, B. (1955). Contribution à l'étude caryologique des éléments reproducteurs et de la fécondation chez les végétaux angiospermiens. *Rev. Cytol. Biol. Végét.* 16:209–407.

Vollbrecht, E., Veit, B., Sinha, N. & Hake, S. (1991). The developmental gene *Knotted-1* is a member of a maize homeobox gene family. *Nature* 350:241–243.

Wagner, V.T., Dumas, C. & Mogensen, H.L. (1989a). Morphometric analysis of isolated sperm. *J. Cell Sci.* 93:179–184.

Wagner, V.T., Song, Y.C., Matthys-Rochon E. & Dumas, C. (1989b). Observations on the isolated embryo sac of *Zea mays* L. *Plant Sci.* 59:127–132.

Wagner, V.T., Dumas, C. & Mogensen, H.L. (1990). Quantitative three-dimensional study on the position of the female gametophyte and its constituent cells as a prerequisite for corn (*Zea mays*) transformation. *Theor. Appl. Genet.* 79:72–76.

Wilms, H.J. (1981). Pollen tube penetration and fertilization in spinach. *Acta Bot. Neerl.* 30:101–122.

19

In vitro fertilization with single isolated gametes

ERHARD KRANZ

Summary 377
Introduction 377
Isolation and selection of male and female gametes, synergids, and central cells 378
In vitro fusion of male and female gametes 381
Culture of in vitro–produced zygotes and plant regeneration 383
Concluding remarks and prospects 385
References 387

Summary

The ability to isolate angiosperm gametes has opened new experimental avenues, including in vitro fertilization. The use of biotechnological methods such as micromanipulation and single cell culture has led to the technique of in vitro fertilization at the single cell level. In this chapter, a description is given of (a) in vitro fusion techniques using single isolated egg and sperm cell protoplasts of maize and (b) the subsequent development of the fusion product, the zygote, in individual culture. The electrofusion of the gametic protoplasts leads to zygotic embryogenesis and fertile hybrid plants. Furthermore, a non-electrical alternative technique to fuse isolated higher plant gametes using a fusiogenic medium is given and its relevance to studies of adhesion, recognition, and fusion of these gametes is discussed. The experimental experiences obtained are so far limited to maize. These micromanipulation techniques and their possible application for fundamental and applied studies are described.

Introduction

Biotechnological methods have been applied in addition to sexual crossings in breeding programs for a number of years. For example, cell and tissue culture techniques such as somatic cell genetics, anther and microspore culture, and methods of ovule culture are extensively used. Since the first successful in vitro pollination/fertilization of excised ovules with mature pollen of *Papaver somniferum* L. was performed (Kanta et al. 1962), techniques of embryo rescue (Stewart 1981) and in vitro pollination/fertilization of flower explants, ovaries, and ovules have been used to overcome cases of self- and cross-incompatibility (for example, Rangaswamy and Shivanna 1967, 1971;

Rangaswamy 1977; Zenkteler 1990, 1992). Meiotic and pollen tetrad protoplasts were used in fusion experiments (Ito and Maeda 1973; Deka et al.
1977). Gametosomatic hybrid plants were regenerated by fusion of somatic
protoplasts with tetrad protoplasts (Pirrie and Power 1986; Lee and Power
1988; Pental et al. 1988). Progress has been made in generative cell manipulation by the use of pollen protoplasts and generative cell protoplasts for
fusion experiments (Ueda et al. 1990; Yang and Zhou 1992).

An extension of these approaches is the technique of in vitro fertilization
(IVF) with isolated single gametes (Kranz et al. 1990, 1991a). Maize (*Zea
mays* L.) was used for the development of this method. Pairs of single sperm
and egg cell protoplasts were fused through electrofusion. The individually
cultured fusion products developed via direct embryogenesis to globular
structures, proembryos, transition-phase embryos, and finally to phenotypically normal and fertile maize plants. These plants were examined karyologi-
cally and morphologically, and the segregation ratios of the kernel color of the
selfed fusion plants confirmed their hybrid nature (Kranz and Lörz 1993).
Additionally, karyogamy was observed in the fusion products at a high frequency (Faure et al. 1993). The fusion of the female and male gametic protoplasts of maize is also mediated by calcium (Kranz 1993; Faure et al. 1994),
leading to multicellular structure formation (Kranz and Lörz 1994). These
experiments parallel those performed with somatic protoplasts more than 20
years ago (Keller and Melchers 1973).

To date there are no other reports on the technique of IVF with single
gametes leading to zygotic embryogenesis and plant regeneration. Therefore,
details of this new and powerful method are described. Its relevance for
research and breeding is also discussed.

Isolation and selection of male and female gametes, synergids, and central cells

In recent years, protocols have been developed for the isolation of pollen protoplasts and generative and sperm cells in a wide range of plant species.
Procedures have also been developed to isolate angiosperm embryo sacs and
egg cells from many species. The potentials of isolation and manipulation of
these cells for experimental plant reproductive biology have been reviewed in
several publications; for example, by Theunis et al. (1991), Chaboud and
Perez (1992), Huang and Russell (1992), Kranz et al. (1992), Yang and Zhou
(1992), Dumas and Mogensen (1993); also see Chapters 17 and 18.

A prerequisite for IVF experiments is the use of viable sperm cells. They
are isolated from viable pollen, collected from freshly dehisced anthers, and
used immediately. Maize sperm cells can also be used from pollen that was
stored in a moistened atmosphere for some hours (Kranz and Lörz 1990;
Kranz et al. 1991a). The long-time storage of pollen (Barnabás and Rajki
1976, 1981; also see Chapter 14) and the improvement of the longevity and
viability of isolated sperm cells (Zhang et al. 1992; Roeckel and Dumas 1993;

also see Chapter 17) can be important factors in the development of IVF with other plant species. Maize sperm cells can be obtained from tricellular pollen grains by osmotic shock (Dupuis et al. 1987; Matthys-Rochon et al. 1987; Cass and Fabi 1988). Sperm cells from binucleate pollen have to be isolated from in vitro–grown pollen tubes by osmotic bursting, grinding, or squashing (reviewed in Theunis et al. 1991; also see Chapter 17).

Viable embryo sacs including egg cells are isolated primarily through methods of enzymatic degradation of cell walls or in combination with microdissection (reviewed in Theunis et al. 1991; also see Chapter 18). Recently, living egg cells of barley and wheat were obtained by a mechanical procedure without any enzymatic treatment (Holm et al. 1994; Kovács et al. 1994). This method might be useful for studies of adhesion, recognition, and fusion of isolated gametes. The isolation of viable embryo sacs of maize was pursued by Wagner et al. (1988, 1989) using methods of enzymatic macera-tion (Zhou and Yang 1984, 1985). The embryo sacs containing ovular tissues of maize are generally treated with an enzyme mixture (0.75% pectinase, 0.25% pectolyase Y23, 0.5% hemicellulase, and 0.5% cellulase Onozuka RS) for about 30 minutes at 24°C (Kranz and Lörz 1993). As reported by Wagner et al. (1989), Huang and Russel (1989), and Kranz et al. (1990), the timing of exposure of the ovular tissue to the enzyme solution was important in pre-venting unwanted fusion of cells of the embryo sac. The enzyme concentra-tions and the incubation time must be determined exactly. In maize, the yield of isolated embryo sacs and egg cells could be improved by reducing the enzyme concentrations and by shortening the time of incubation of ovule pieces in the enzyme mixture and especially by manual dissection using microneedles (Kranz 1992). Thin glass needles, prepared with a microforge, may be used for the dissection of the material. The yield is finally determined by the manual isolation step. Using maize, about 5 egg cells per 20 ovule pieces can be obtained routinely. Depending on the "quality" of the plant material, up to 40 egg cells can be isolated by microdissection within 2 to 3 hours. The "bottleneck" is getting a relatively high number of isolated egg cells for performing IVF by mass fusion. This problem can be circumvented by the use of single cell techniques, which include individual selection, trans-fer, and fusion of gametes (Figure 19.1). Therefore, IVF so far is performed only with single gametes.

Egg cell protoplasts are taken up by a microcapillary with a tip opening of about 200 μm. The diameter of the tip opening of the capillary used for the selection of the sperm cells is about 20 μm. The two sperm cell protoplasts can be differentiated from the smaller vegetative nucleus. Sperm, egg, and central cell protoplasts are selected by the microcapillaries and transferred by use of a hydraulic system and a computer-controlled dispenser–dilutor (Koop and Schweiger 1985a; Kranz 1992; Figure 19.2). Microcapillaries are pre-pared by hand or by use of a micropipette puller and a microforge. The use of donor plants, growing under controlled growth chamber conditions, is preferable.

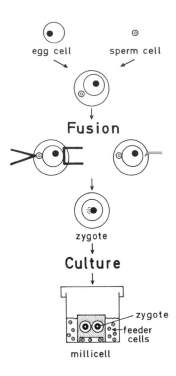

egg cell   sperm cell

Fusion

zygote

Culture

zygote
feeder
cells

millicell

Figure 19.1 Method of in vitro fertilization with isolated single gametes. Sperm and egg cell protoplasts are isolated and transferred into the fusion droplet. Pairs of the gametes are fused, either by the application of an electrical pulse or in a fusiogenic medium. The fusion products are transferred into "Millicell" inserts surrounded by feeder cells for growth of the zygotes. (*With permission from Kranz et al. 1991a.*)

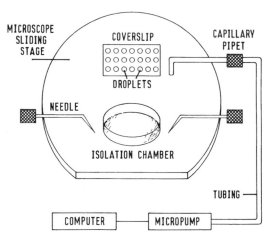

Figure 19.2 Set-up for individual selection and transfer of the protoplasts of the embryo sac and sperm cells. The procedures are performed on a sliding stage of an inverted microscope. Egg cell, synergid, and central cell protoplasts are isolated mechanically by needles from embryo sacs of enzymatically softened ovule pieces in an isolation chamber (plastic dish). The protoplasts are transferred by a microcapillary, connected with a hydraulic system, to a computer-directed micropump into microdroplets on a coverslip. This is followed by the selection of sperm cell protoplasts, released after bursting of the pollen grains in another plastic dish. The sperm cell protoplasts are transferred into droplets containing the egg cell protoplasts for fusion. After fusion, the fusion products are transferred into "Millicell" inserts in a plastic dish for culture.

In vitro fusion of male and female gametes

Fertilization, the union of male and female gametes, can be performed in vitro either electrically by applying an electrical pulse, or chemically in a fusiogenic medium, for example, mediated by calcium (Kranz et al. 1990, 1991a; Kranz and Lörz 1994; Faure et al. 1994; Figures 19.3A, B). Because gametes are used as protoplasts, IVF might also be performed by other methods available for the fusion of somatic protoplasts. In maize, sperm cells are able to fuse after osmotic-shock–mediated isolation without any further treatment with other sperm cells, egg cells, and somatic protoplasts (Kranz et al. 1991a, b; Kranz and Lörz 1994). Further indication that these isolated sperm cells are true protoplasts, lacking a cell wall, is demonstrated by electron microscopy (Dupuis et al. 1987; McConchie et al. 1987; Cass and Fabi 1988). Before fusion, the isolated protoplasts of the embryo sacs are washed by transferring those from one microdroplet into another droplet. Comparable with somatic protoplast fusion, the osmolarity of the fusion medium is an important factor for the fusion of the protoplasts of the egg cell, the sperm cell, and the nongametic cells of the embryo sac. This is true for electrical fusion as well as for the fusion in a fusiogenic medium. For the electrical fusion, 540–570 mOsmol/kg

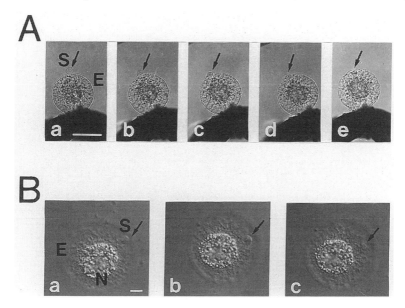

Figure 19.3 Fusion sequences of selected pairs of egg (E) and sperm cell (S) protoplasts of maize (*Zea mays* L.). Arrows indicate the fusion sites. (**A**) Electrofusion sequence (a–e). The time interval from (a) to (e) was 4 seconds, but generally it was less than 1 second. Scale bar = 50 μm. (*With permission from Kranz et al. 1991a.*) (**B**) Sequence of fusion mediated by high calcium and high pH (a–c). Following an adhesion time of about 10–20 minutes, the time interval from (a) to (c) was 2 seconds, but generally it was less than 1 second. N indicates the egg nucleus and nucleolus. Scale bar = 10 μm. (*From Kranz & Lörz 1994.*)

H_2O, and, for the high calcium/high pH–mediated fusion, 400–430 mOsmol/kg H_2O, were found to be useful for maize.

Pairs of one egg and one sperm cell protoplast, as well as pairs of one sperm cell and one central cell protoplast of maize, were fused electrically, as described in Kranz et al. (1991a, b) and Kranz (1992). The conditions for the electrical fusion of somatic cell protoplasts could also be used for the gametic protoplasts of maize. The fusions are performed on a coverslip in micro-droplets that are overlayered by mineral oil. Each microdroplet consists of 2,000 nl mannitol solution. The individual electrical fusion is performed under microscopic observation with electrodes fixed to a support mounted under the condenser of an inverted microscope. Cell fusion is induced by single or multiple negative DC pulses after dielectrophoretic alignment on one of the electrodes (Zimmermann and Scheurich 1981; Koop et al. 1983; Koop and Schweiger 1985b; Schweiger et al. 1987; Spangenberg and Koop 1992). The frequencies of electrofusion of egg and sperm cell protoplasts are high (mean frequency 85% of about 2,000 fusion products). High fusion frequencies were also obtained using combinations of sperm cell protoplasts with sperm cell protoplasts, synergid protoplasts, central cell protoplasts, and somatic cyto-plasts (Kranz et al. 1991b).

The individual gamete fusion can also be performed in a microdroplet of a fusiogenic medium, for example, one containing a high calcium concentration and at a high pH (0.01–0.05 M $CaCl_2$; pH 11.0) (Kranz and Lörz 1994; Figure 19.3B). This medium was developed for the fusion of somatic tobacco proto-plasts (Keller and Melchers 1973). Microneedles can be used to accomplish the alignment of the two gametic protoplasts (Figure 19.2). The mechanical alignment leading to adhesion of the gametic protoplasts, however, was more time consuming than the dielectrophoretic alignment procedure. Compared with somatic protoplasts, no deleterious effects in the gametic cells could be observed after the treatment. Furthermore, fusion products of the gametic pro-toplasts were able to develop to microcalli, as was previously observed with somatic protoplasts.

Whether gamete-specific receptors are involved in the adhesion and fusion, or whether these processes are occurring in a non–gamete-specific manner, comparable with the fusion of somatic protoplasts, needs to be inves-tigated. The likelihood of non–gamete-specific fusion is indicated by the observation that the high calcium/high pH media mediate the fusion of two egg cell protoplasts as well as the fusion of two sperm cell protoplasts (Kranz and Lörz 1994). IVF methods performed with somatic protoplast fusion media, for example, containing sodium nitrate (Power et al. 1970) or calcium (Keller and Melchers 1973; Ward et al. 1979; Boss et al. 1984; Grimes and Boss 1985), or even spontaneously induced fusion (Ito and Maeda 1973), may allow studies of adhesion, fusion, and possible mechanisms to prevent polyspermy. However, for obtaining artificial zygotes, the electrically per-formed IVF technique was much more efficient compared to the procedure using a fusiogenic medium.

Culture of in vitro–produced zygotes and plant regeneration

Maize zygotic embryogenesis and plant regeneration from in vitro–produced zygotes was achieved by electrofusion-mediated IVF (Kranz and Lörz 1993; Figures 19.4 and 19.5). High frequency multicellular structure formation was obtained using a wide range of different maize lines as donor plants for sperm and egg cells. A prerequisite for the successful culture of the zygotes, which can be obtained only in a limited amount, is the use of a suitable microculture system. A "Millicell/feeder suspension" system, which involved the use of

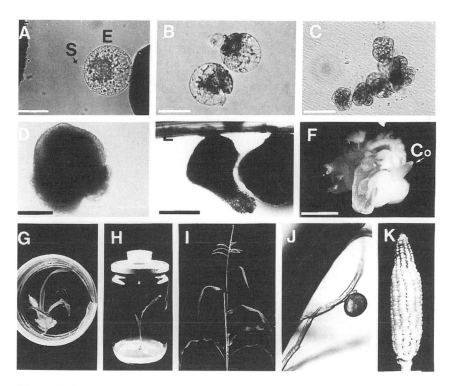

Figure 19.4 In vitro zygotic embryogenesis and plant regeneration after electrofusion of single maize (*Zea mays* L.) sperm and egg cell protoplasts. (**A**) Egg cell (E) and sperm cell (S) protoplast alignment on one electrode before fusion. The arrow indicates the sperm cell protoplast. Bar = 50 μm. (**B**) First unequal cell divisions 42 hours after fusion. Bar = 50 μm. (**C**) Multicellular structures 5 days after fusion. Bar = 100 μm. (**D**) Embryo structure with a protoderm-like outer cell layer at the meristematic region and vacuolated cells at the other pole 12 days after fusion. Bar = 200 μm. (**E**) Transition-phase embryo 14 days after fusion. Bar = 200 μm. (**F**) Structure with compact green and white tissue with a coleoptile (Co, arrow) 30 days after fusion. Bar = 4 mm. (**G**) Plantlet 35 days after fusion. Bar = 2 cm. (**H**) Plant 39 days after fusion. Bar = 6 cm. (**I**) Flowering plant 99 days after fusion. Bar = 50 cm. (**J**) Self-pollination. The pollen tube penetrated the trichome and the grain is half emptied of its content. Bar = 100 μm. (**K**) Mature cob 148 days after fusion. Bar = 4 cm. (*With permission from Kranz and Lörz 1993.*)

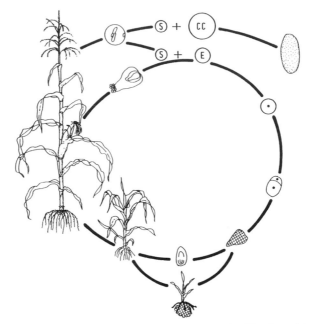

Figure 19.5 Growth cycle of maize following in vitro fertilization with isolated single gametes and zygotic embryogenesis. S = sperm cell protoplast, E = egg cell protoplast, CC = central cell protoplast.

suspension cells as nurse cultures, proved to be capable of achieving sustained growth of the zygotes. The nonmorphogenic feeder suspension cells were derived from excised zygotic maize embryos (Kranz et al. 1991a). The fusion products were cultivated in a liquid medium on a semipermeable membrane of "Millicell-CM" dishes, which were placed in a feeder suspension (Figure 19.1). The feeder suspension consisted of actively growing cells. Sustained growth of the multicellular structures could be achieved only with feeder cells that were cytoplasmically rich. This observation suggests that the choice of the culture conditions, especially the growth medium composition, for optimal growth of the in vitro–produced zygotes depends largely on those conditions that are optimal for feeder cell growth. Depending on the feeder cell line, modified MS medium (Murashige and Skoog 1962) containing 1.0 mg/l 2,4-D and 0.02 mg/l kinetin, or 2.0 mg/l 2,4-D without kinetin, as well as modified N_{6ap} medium (Rhodes et al. 1988) with 1.0 mg/l 2,4-D, was found to be suitable. In addition, the feeder suspension lines have to be adapted to a higher osmolarity than that normally used for culture. To prevent bursting of the fusion products and zygotes, the suspension cultures were generally adapted to 600 mOsmol/kg H_2O before they were used as feeder.

Karyogamy was demonstrated within 1 hour after electro-mediated gamete fusion (Faure et al. 1993). The first cell division occurred between 40 and 60 hours after cell fusion. The early development of the zygotes – for example, the unequal first division of most zygotes, the formation of multicellular structures, proembryos, and transition-phase embryos – resembles that observed in vivo (Van Lammeren 1981, 1986; Schel et al. 1984; Clark and Sheridan 1991). Interestingly, whereas the maize zygotes develop under comparable conditions with a high frequency of cell division, egg cells never divide in culture (Kranz et al. 1991a). Development of fusion products of a sperm cell with a central cell was achieved in culture; thus, endosperm can be produced in vitro by this fusion (Kranz et al. 1991b; E. Kranz, unpublished data).

Embryo structures were formed with a pronounced axis, consisting of a meristematic region at the top and a suspensor at the base (Figure 19.4E). They were transferred onto solid regeneration medium, usually 10–12 days after gamete fusion, when they reached a size of about 0.4 x 0.5 mm. Coleoptiles and plantlets developed from enlarged structures about 22–34 days after electrofusion. Further development of in vitro–produced early embryos was accompanied by an overgrowth and occasionally by polyembryony. These structures were comparable to those obtained by somatic embryogenesis (for example, Lu et al. 1982). Phenotypically normal and fertile maize plants were obtained within 99–171 days after gamete fusion (Figures 19.4I, J, K). In four independent experiments, 11 fertile plants were regenerated from 28 fusion products. For rapid plant development, regeneration media should be used that are generally suitable for tissue culture of the species of choice. In order to optimize the zygotic embryogenesis in vitro, especially the later embryo stages, modified regeneration media (for example, special maturation media) may have to be developed. The development of the few-celled embryo following the first unequal cell division and a cell size decrease was found to occur without any detectable orientation pattern, comparable with the observations made in vivo in maize (Randolph 1936). For studies of the early development of the embryo in vitro, it may be interesting to perform IVF with dicotyledonous plants, where the morphological stages of the early embryo are more pronounced.

Concluding remarks and prospects

With access to isolated gametes, experiments comparable to those with animal and lower plant systems can be carried out to study the phenomena of adhesion, fusion, and recognition, and the prevention of polyspermy in higher plants. Furthermore, the individual culture of single zygotes offers new possibilities of investigations of zygotic embryogenesis in vitro, especially of very early events (Kranz and Dresselhaus 1996).

IVF was first developed using different maize lines. This model may also be applicable to other plant species. Tissue culture methods are available for fusion and plant regeneration of many different androgenic and somatic protoplasts from cells and tissues of various species (for example, see Lindsey 1992). Although difficulties still exist, the regeneration of fertile plants is also possible from protoplasts of all the important cereal crops, for example, rice, maize, barley, and wheat (reviewed in Vasil and Vasil 1992; Jähne et al. 1995).

Such tissue and protoplast techniques are very important, especially for the development of single cell culture methods, which are necessary to allow sustained growth of the artificially produced zygotes.

The usefulness of this technique in plant breeding, especially in combinations with remote species, has still to be demonstrated. Intra- or interspecies incompatibilities, located on the stigma or in the style or ovary, might be overcome by this method. Zygotic incompatibility might be circumvented by the use of special tissue culture conditions. In preliminary experiments, high rates of microcallus formation were found after electrofusion of sperm cell protoplasts of *Coix lacryma-jobi* and egg cell protoplasts of *Zea mays*, and sperm cell protoplasts of *Sorghum bicolor* and egg cell protoplasts of *Zea mays* (E. Kranz and A. Jahnke, unpublished results). The technology of somatic protoplast fusion holds considerable promise for new combinations of genomes and plastomes that cannot be produced by other methods. Since the first parasexual hybrids were produced among *Nicotiana* species (Carlson et al. 1972), numerous interspecific, intergeneric, and intertribal somatic hybrids and cybrid plants have been obtained through somatic protoplast fusion (for example, Bajaj 1989). Any somatic protoplasts, irrespective of the taxonomic distances, can be fused. This may also be true for the gametic protoplasts and protoplasts of the central cell. Fertile triploid hybrids via gametosomatic hybridization have been produced (Pirrie and Power 1986; Giddings and Rees 1992). Asymmetric somatic hybrids were obtained and used for sexual hybridization. However, in wide distance somatic hybridizations, the regeneration capacity often is absent or reduced or results in sterility, caused by polyploidization, aneuploidy, and the elimination of parts of, or entire, genomes (Schoenmakers et al. 1994). The karyotypical instability has been extensively studied and also observed in hybrids after conventional crossing (for example, Laurie et al. 1990; O'Donoughue and Bennett 1994). In asymmetric somatic hybridization, cases of incongruity between donor chloroplasts and the recipient nucleus of remote species also need to be considered (Derks et al. 1992; Wolters et al. 1993). Although IVF may not be applicable to the generation of hybrids between highly diverse species, it may be useful for more closely related species.

Genetic transformation by transfer of isolated genes might be independent of incompatibility barriers (Schell 1987; Lörz et al. 1988). Direct gene transfer by electroporation of the male and female gametes and its fusion products, as well as by microinjection of the zygotes, seems attractive for detailed studies of genetic transformation, because the possibility of regeneration to normal and fertile plants from single fused gametes has been demonstrated (Kranz and Lörz 1993; Kranz and Dresselhaus 1996). Despite the limited yield of isolated egg cells, the laborious individual cell fusion and other micromanipulation methods are acceptable because of high fusion and cell division rates, short in vitro culture time, and rapid regeneration.

A further application of this technique is the transfer of alien cytoplasms through the fertilization process for studies of cytoplasmic inheritance. For example, the fate of chloroplasts and mitochondria transferred by the sperm cell protoplast or by an additional cytoplast might be studied during the development of the in vitro zygote (Kranz et al. 1991b; Faure et al. 1993). Another promising application of IVF technique might be the injection of sperm cells (Keijzer et al. 1988) or isolated sperm cell nuclei (Matthys-Rochon et al. 1994) into the embryo sac to perform double fertilization in vitro.

Recently, partially isolated embryo sacs containing zygotes, several-celled zygotic proembryos (Campenot et al. 1992; Liu et al. 1993; Mòl et al. 1993), and egg cells (Holm et al. 1994) from in vivo fertilized ovules were cultured, and normal fertile plants were regenerated. This allows access to experimentation with zygotes and very early embryos for studies of early events of embryogenesis and for genetic manipulation. The versatile technique of IVF widens such studies with the possibilities of gamete manipulation before fertilization as well as providing experimental access to very early events (for example, egg cell activation and zygote polarity) without any influence of surrounding nucellar cells. It is expected that the experimental system described herein will be used widely for future developmental studies at the molecular level, such as the detection of zygote-specific genes from very early stages of zygote development. As a first step toward this direction, a representative cDNA library of as few as 128 isolated maize egg cells was constructed (Dresselhaus et al. 1994).

The gamete fusion system fills a gap between sexual crossing and somatic hybridization. The success of future plant breeding programs depends on new biotechnological methods in DNA transfer or genetic engineering, chromosomal engineering, and cellular handling, including somatic cell fusion and clonal proliferation at the tissue level; these methods have been categorized by Adachi (1991). Methods of in vitro fertilization with isolated gametes will certainly contribute to these advances in the near future.

Acknowledgments

The critical reading of the manuscript by Drs. H. Lörz and P. Pechan, the preparation of the photographs by Mrs. C. Adami, the graphical work by Mr. A. Braeutigam, the manuscript preparation by Mrs. R. Schmidt, and the financial support from the Deutsche Forschungsgemeinschaft (grant No. Kr 1256/1-2) are gratefully acknowledged.

Note added in proof

Recently several reports on in vitro fertilization and related topics have been published. These have been reviewed by Kranz and Dresselhaus in *Trends in Plant Science* 1:82–89 (1996).

References

Adachi, T. (1991). How to overcome breeding barriers by means of plant biotechnology. Past, present and future. In *Proceedings of International Coll. on Overcoming Breeding Barriers by Means of Plant Biotechnology*, Miyazaki, March 11–15, 1991, ed. T. Adachi, pp. 1–4.

Bajaj, Y.P.S. (1989). Genetic engineering and in vitro manipulation of plant cells: technical advances. In *Biotechnology in Agriculture and Forestry 9: Plant Protoplasts and Genetic Engineering II*, ed. Y.P.S. Bajaj, pp. 1–25. Berlin, Heidelberg, New York: Springer-Verlag.

Barnabás, B. & Rajki, E. (1976). Storage of maize (*Zea mays* L.) pollen at −196°C in liquid nitrogen. *Euphytica* 25:747–752.

Barnabás, B. & Rajki, E. (1981). Fertility of deep-frozen maize (*Zea mays* L.) pollen. *Ann. Bot.* 48:861–864.

Boss, W.F., Grimes, H.D. & Brightman, A. (1984). Calcium-induced fusion of fusogenic wild carrot protoplasts. *Protoplasma* 120:209–215.

Campenot, M.K., Zhang, G., Cutler, A.J. & Cass, D.D. (1992). *Zea mays* embryo sacs in culture. I. Plant regeneration from 1 day after pollination embryos. *Amer. J. Bot.* 79:1368–1373.

Carlson, P. S., Smith, H.H. & Dearing, R.D. (1972). Parasexual interspecific plant hybridization. *Proc. Nat. Acad. Sci. USA* 69:2292–2294.

Cass, D.D. & Fabi, G.C. (1988). Structure and properties of sperm cells isolated from pollen of *Zea mays*. *Can. J. Bot.* 66:819–825.

Chaboud, A. & Perez, R. (1992). Generative cells and male gametes: isolation, physiology, and biochemistry. *Int. Rev. Cytol.* 140:205–232.

Clark, J.K. & Sheridan, W.F. (1991). Isolation and characterization of 51 embryospecific mutations of maize. *Plant Cell* 3:935–951.

Deka, P.C., Mehra, A.K., Pathak, N.N. & Sen, S.K. (1977). Isolation and fusion studies on protoplasts from pollen tetrads. *Experimentia* 33:182–184.

Derks, F.H.M., Hakkert, J.C., Verbeek, W.H.J. & Colijn-Hooymans, C.M. (1992). Genome composition of asymmetric hybrids in relation to the phylogenetic distance between the parents: nucleus–chloroplast interaction. *Theor. Appl. Genet.* 84:930–940.

Dresselhaus, T., Lörz, H. & Kranz, E. (1994). Representative cDNA libraries from few plant cells. *Plant J.* 5:605–610.

Dumas, C. & Mogensen, H.L. (1993). Gametes and fertilization: maize as a model system for experimental embryogenesis in flowering plants. *Plant Cell* 5:1337–1348.

Dupuis, I., Roeckel, P., Matthys-Rochon, E. & Dumas, C. (1987). Procedure to isolate viable sperm cells from corn (*Zea mays* L.) pollen grains. *Plant Physiol.* 85:876–878.

Faure, J.-E., Mogensen, H.L., Dumas, C., Lörz, H. & Kranz, E. (1993). Karyogamy after electrofusion of single egg and sperm cell protoplasts from maize: cytological evidence and time course. *Plant Cell* 5:747–755.

Faure, J.-E., Digonnet, C. & Dumas, C. (1994). An in vitro system for adhesion and fusion of maize gametes. *Science* 263:1598–1600.

Giddings, G.D. & Rees, H. (1992). A *Nicotiana* gametosomatic hybrid and its progenies. *J. Exper. Bot.* 43:419–425.

Grimes, H.D. & Boss, W.F. (1985). Intracellular calcium and calmodulin involvement in protoplast fusion. *Plant Physiol.* 79:253–258.

Holm, P.B., Knudson, S., Mouritzen, P., Negri, D., Olsen, F.L. & Roué, C. (1994). Regeneration of fertile barley plants from mechanically isolated protoplasts of the fertilized egg cell. *Plant Cell* 6:531–543.

Huang, B.Q. & Russell, S.D. (1989). Isolation of fixed and viable eggs, central cells and embryo sacs from ovules of *Plumbago zeylanica*. *Plant. Physiol.* 90:9–12.

Huang, B.Q. & Russell, S.D. (1992). Female germ unit: organization, isolation, and function. *Int. Rev. Cytol.* 140:233–293.

Ito, M. & Maeda, M. (1973). Fusion of meiotic protoplasts in liliaceous plants. *Exper. Cell Res.* 80:453–456.

Jähne, A., Becker, D. & Lörz, H. (1995). Genetic engineering of cereal crop plants: a review. *Euphytica* 85:35–44.

Kanta, K., Rangaswamy, N.S. & Maheshwari, P. (1962). Test-tube fertilization in a flowering plant. *Nature* 194:1214–1217.

Keijzer, C.J., Reinders, M.C. & Leferink-ten Klooster, H.B. (1988). A micromanipulation method for artificial fertilization in *Torenia*. In *Sexual Reproduction in Higher Plants*, eds. M. Cresti, P. Gori, and E. Pacini. Proceedings of the Tenth International Symposium on the Sexual Reproduction in Higher Plants, 30 May–4 June 1988, Siena, Italy, pp.119–124. Berlin, Heidelberg, New York: Springer-Verlag.

Keller, W.A. & Melchers, G. (1973). The effect of high pH and calcium on tobacco leaf protoplast fusion. *Z. Naturforsch.* 28c:737–741.

Koop, H.-U. & Schweiger H.-G. (1985a). Regeneration of plants from individually cultivated protoplasts using an improved microculture system. *J. Plant Physiol.* 121: 245-257.

Koop, H.-U. & Schweiger H.-G. (1985b). Regeneration of plants after electrofusion of selected pairs of protoplasts. *Eur. J. Cell Biol.* 39:46–49.

Koop, H.U., Dirk, J., Wolff, D. & Schweiger, H.G. (1983). Somatic hybridization of two selected single cells. *Cell. Biol. Int. Rep.* 7:1123–1128.

Kovács, M., Barnabás, B. & Kranz, E. (1994). The isolation of viable egg cells of wheat (*Triticum aestivum* L.). *Sex. Plant Reprod.* 7:311–312.

Kranz, E. (1992). *In vitro* fertilization of maize mediated by electrofusion of single gametes. In *Plant Tissue Culture Manual*, ed. K. Lindsey, Supplement 2, E1, pp. 1–12. Dortrecht, Boston: Kluwer Academic Publishers.

Kranz, E. (1993). In vitro Befruchtung mit einzelnen isolierten Gameten von Mais (*Zea mays* L.). *Habilitationsschrift,* Universität Hamburg.

Kranz, E. & Dresselhaus, T. (1996). In vitro fertilization with isolated higher plant gametes. *Trends in Plant Sci.* 1:82–89.

Kranz, E. & Lörz H. (1990). Micromanipulation and in vitro fertilization with single pollen grains of maize. *Sex. Plant Reprod.* 3:160–169.

Kranz, E. & Lörz, H. (1993). In vitro fertilization with isolated, single gametes results in zygotic embryogenesis and fertile maize plants. *Plant Cell* 5:739–746.

Kranz, E. & Lörz, H. (1994). *In vitro* fertilization of maize by single egg and sperm cell protoplast fusion mediated by high calcium and high pH. *Zygote* 2:125–128.

Kranz, E., Bautor, J. & Lörz, H. (1990). *In vitro* fertilization of single, isolated gametes, transmission of cytoplasmic organelles and cell reconstitution of maize (*Zea mays* L.). In *Progress in Plant Cellular and Molecular Biology*, eds. H.J.J. Nijkamp, L.H.W. Van der Plas, and J. Van Aartrijk. Proc. of the VIIth International Congress on Plant Tissue and Cell Culture, Amsterdam, The Netherlands, 24–29 June 1990, pp. 252–257. Dordrecht, Boston, London: Kluwer Academic Publishers.

Kranz, E., Bautor, J. & Lörz, H. (1991a). *In vitro* fertilization of single, isolated gametes of maize mediated by electrofusion. *Sex. Plant Reprod.* 4:12–16.

Kranz, E., Bautor, J. & Lörz, H. (1991b). Electrofusion-mediated transmission of cytoplasmic organelles through the in vitro fertilization process, fusion of sperm cells with synergids and central cells, and cell reconstitution in maize. *Sex. Plant Reprod.* 4:17–21.

Kranz, E., Lörz, H., Digonnet, C. & Faure, J.-E. (1992). *In vitro* fusion of gametes and production of zygotes. *Int. Rev. Cytol.* 140:407–423.

Laurie, D.A., O'Donoughue, L.S. & Bennett, M.D. (1990). Wheat x maize and other wide sexual hybrids: their potential for genetic manipulation and crop improvement. In *Gene Manipulation in Plant Improvement*. II. *19th Stadler Genetics Symposium,* ed. J.P. Gustafson, pp. 95–126. New York, London: Plenum Press.

Lee, C.H. & Power, J.P. (1988). Intraspecific gametosomatic hybridization in *Petunia hybrida*. *Plant Cell Rep.* 7:17–18.

Lindsey, K. (1992). *Plant Tissue Culture Manual*. Dortrecht, Boston: Kluwer Academic Publishers.

Liu, C-M., Xu, Z.-H. & Chua, N.H. (1993). Proembryo culture: *in vitro* development of early globular-stage zygotic embryos from *Brassica juncea*. *Plant J.* 3:291–300.

Lörz, H., Göbel, E. & Brown, P.T.H. (1988). Advances in tissue culture and progress towards genetic transformation of cereals. *Plant Breeding* 100:1–25.

Lu, C., Vasil, I.K. & Ozias-Akins P. (1982). Somatic embryogenesis in *Zea mays* L. *Theor. Appl. Genet.* 62:109–112.

Matthys-Rochon, E., Vergne, P., Detchepare, S. & Dumas, C. (1987). Male germ unit isolation from three tricellular pollen species: *Brassica oleracea, Zea mays,* and *Triticum aestivum*. *Plant Physiol.* 83:464–466.

Matthys-Rochon, E., Mòl, R., Heizmann, P. & Dumas, C. (1994). Isolation and microinjection of active sperm nuclei into egg cells and central cells of isolated maize embryo sacs. *Zygote* 2:29–35.

McConchie, C.A., Hough, T. & Knox, R.B. (1987). Ultrastructural analysis of the sperm cells of mature pollen of maize, *Zea mays*. *Protoplasma* 139:9–19.

Mòl, R., Matthys-Rochon, E. & Dumas, C. (1993). *In vitro* culture of fertilized embryo sacs of maize: zygotes and two-celled proembryos can develop into plants. *Planta* 189:213–217.

Murashige, T. & Skoog, F. (1962). A revised medium for rapid growth and bioassays with tobacco tissue cultures. *Physiol. Plant.* 15:473–497.

O'Donoughue, L.S. & Bennett, M.D. (1994). Comparative responses of tetraploid wheats pollinated with *Zea mays* L. and *Hordeum bulbosum* L. *Theor. Appl. Genet.* 87:673–680.

Pental, D., Mukhopadhyay, A., Grover, A. & Pradhan, A.K. (1988). A selection method for the synthesis of triploid hybrids by fusion of microspore protoplasts (n) with somatic cell protoplasts (2n). *Theor. Appl. Genet.* 76:237–243.

Pirrie, A. & Power, J.B. (1986). The production of fertile, triploid somatic hybrid plants (*Nicotiana glutinosa* (n) + *N. tabacum* (2n) via gametic–somatic protoplast fusion. *Theor. Appl. Genet.* 72:48–52.

Randolph, L.F. (1936). Developmental morphology of the caryopsis in maize. *J. Agric. Res.* 53:881–916.

Rangaswamy, N.S. (1977). Applications of *in vitro* pollination and *in vitro* fertilization. In *Applied and Fundamental Aspects of Plant, Cell Tissue and Organ Culture*, eds. J. Reinert and Y. Baja, pp. 412–425. Berlin: Springer-Verlag.

Rangaswamy, N.S. & Shivanna, K.R. (1967). Induction of gametophytic compatibility and seed formation in axenic cultures of a diploid self-incompatible species of *Petunia*. *Nature* 216:937–939.

Rangaswamy, N.S. & Shivanna, K.R. (1971). Overcoming self-incompatibility in *Petunia axillaris* (Lam.) B.S.P. II. Placenta pollination *in vitro*. *J. Indian Bot. Soc.* 50A:286–296.

Rhodes, C.A., Lowe, K.S. & Ruby, K.L. (1988). Plant regeneration from protoplasts isolated from embryogenic maize cell cultures. *Bio/Technol.* 6:56–60.

Roeckel, P. & Dumas, C. (1993). Survival at 20°C and cryopreservation of isolated sperm cells from *Zea mays* pollen grains. *Sex. Plant Reprod.* 6:212–216.

Schel, J.H.N., Kieft, H. & Van Lammeren A.A.M. (1984). Interactions between embryo and endosperm during early developmental stages of maize caryopses (*Zea mays*). *Can. J. Bot.* 62:2842–2853.

Schell, J.S. (1987). Transgenic plants as tools to study the molecular organization of plant genes. *Science* 237:1176–1183.

Schoenmakers, H.C.H., Wolters, A.-M.A., de Haan, A., Saiedi, A.K. & Koorneef, M. (1994). Asymmetric somatic hybridization between tomato (*Lycopersicon esculentum* Mill) and gamma-irradiated potato (*Solanum tuberosum* L.): a quantitative analysis. *Theor. Appl. Genet.* 87:713–720.

Schweiger, H.-G., Dirk, J., Koop, H.-U., Kranz, E., Neuhaus, G., Spangenberg, G. & Wolff, D. (1987). Individual selection, culture and manipulation of higher plant cells. *Theor. Appl. Genet.* 73:769–783.

Spangenberg, G. & Koop, H.-U. (1992). Low density cultures: microdroplets and single cell nurse cultures. In *Plant Tissue Culture Manual,* ed. K. Lindsey, Supplement 2, A10, pp. 1–28. Dordrecht, Boston: Kluwer Academic Publishers.

Stewart, J.M.D. (1981). *In vitro* fertilization and embryo rescue. *Env. Exp. Bot.* 21:301–315.

Theunis, C.H., Pierson, E.S. & Cresti, M. (1991). Isolation of male and female gametes in higher plants. *Sex. Plant Reprod.* 4:145–154.

Ueda, K., Miyamoto, Y. & Tanaka, I. (1990). Fusion studies of pollen protoplasts and generative cell protoplasts in *Lilium longiflorum. Plant Sci.* 72:259–266.

Van Lammeren, A.A.M. (1981). Early events during embryogenesis in *Zea mays* L. *Acta Soc. Bot. Pol.* 50:289–290.

Van Lammeren, A.A.M. (1986). Developmental morphology and cytology of the young maize embryo (*Zea mays* L.). *Acta Bot. Neerl.* 35:169–188.

Vasil, I. & Vasil, V. (1992). Advances in cereal protoplast research. *Physiol. Plant. Copenhagen* 85:279–283.

Wagner, V.T., Song, Y., Matthys-Rochon, E. & Dumas, C. (1988). The isolated embryo sac of *Zea mays*: structural and ultrastructural observations. In *Sexual Reproduction in Higher Plants,* eds. M. Cresti, P. Gori, and E. Pacini. Proc. of The Tenth International Symposium on the Sexual Reproduction in Higher Plants, 30 May–4 June, 1988, Siena, Italy, pp. 125–130. Berlin, Heidelberg, New York: Springer-Verlag.

Wagner, V.T., Song, Y.C., Matthys-Rochon, E. & Dumas, C. (1989). Observations on the isolated embryo sac of *Zea mays* L. *Plant Sci.* 59:127–132.

Ward, M., Davey, M.R., Mathias, R.J., Cocking, E.C., Clothier, R.H., Balls, M. & Lucy, J.A. (1979). Effects of pH, Ca^{2+}, temperature, and protease pretreatment in interkingdom fusion. *Somatic Cell Genet.* 5:529–536.

Wolters, A.M.A., Koornneef, M. & Gilissen, L.J.W. (1993). The chloroplast and the mitochondrial DNA type are correlated with the nuclear composition of somatic hybrid calli of *Solanum tuberosum* and *Nicotiana plumbaginifolia. Curr. Genet.* 24:260–267.

Yang, H.-Y. & Zhou, C. (1992). Experimental plant reproductive biology and reproductive cell manipulation in higher plants: now and the future. *Amer. J. Bot.* 79:354–363.

Zenkteler, M. (1990). *In vitro* fertilization and wide hybridization in higher plants. *Crit. Rev. Plant Sci.* 9:267–279.

Zenkteler, M. (1992). Wide hybridization in higher plants by applying the method of test tube pollination of ovules. In *Reproductive Biology and Plant Breeding,* eds. Y. Datte'e, C. Dumas, and A. Gallais. The XIIIth Eucarpia Congress from the 6th to the 11th July 1992, Angers, France, pp. 205–214. Berlin, Heidelberg, New York: Springer-Verlag.

Zhang, G., Williams, C.M., Campenot, M. K., McGann, L.E. & Cass, D.D. (1992). Improvement of longevity and viability of sperm cells isolated from pollen of *Zea mays* L. *Plant Physiol.* 100:47–53.

Zhou, C. & Yang, H.Y. (1984). The enzymatic isolation of embryo sacs from fixed and fresh ovules of *Antirrhinum majus* L. *Acta Biol. Exp. Sin.* 17:141–147.

Zhou, C. & Yang, H.Y. (1985). Observations on enzymatically isolated, living and fixed embryo sacs in several angiosperm species. *Planta* 165: 225-231.

Zimmermann, U. & Scheurich, P. (1981). High frequency fusion of plant protoplasts by electric fields. *Planta* 151: 26-32.

20

Pollen embryos

C. E. PALMER and W. A. KELLER

Summary 392
Introduction 393
Anther and microspore culture 395
Factors affecting pollen embryogenesis 396
    Genotype 396
    Donor plant physiology 397
    Stage of microspore development 397
    Medium composition 398
    Culture conditions 399
Developmental aspects of pollen embryogenesis 400
Comparison of zygotic and pollen embryos 403
Pollen embryo maturation 405
Plant regeneration and doubled haploid production 406
Utilization of pollen embryos 407
    Breeding and genetic studies 407
    Mutation and selection 407
    Gene transfer 408
    Biochemical and physiological studies 409
    Artificial seeds and germplasm storage technology 409
Conclusions and future potentials 410
References 411

Summary

Under the appropriate in vitro culture conditions, anthers and isolated microspores of higher plants develop haploid embryos by a process referred to as androgenesis. Embryo development can be a direct recapitulation of the developmental stages characteristic of zygotic embryos, or it can be preceded by a callus stage.

There are a number of factors governing pollen embryogenesis, but genotype, donor plant physiology, stage of microspore development, and in vitro culture conditions are the most important. Some characteristics of embryogenic microspores have been identified, and late uninucleate to early binucleate cells are the most responsive. Under inductive culture conditions, uninucleate microspores divide symmetrically to initiate embryogenesis, whereas binucleate pollen exhibits sustained cell division of the vegetative or generative cell. High carbohydrate levels and an initial period at high temperature are conducive to pollen embryogenesis in some species.

Pollen embryos are developmentally similar to zygotic embryos and, under the appropriate culture conditions, they can mature and accumulate seed-specific storage products in a comparable manner. Such embryos develop directly into plants, although plants may arise from secondary structures. Double haploid plants are produced through chromosome doubling techniques. These homozygous plants are useful in plant breeding and genetic studies. In addition, haploid embryos are used in mutant isolation, gene transfer, studies of storage product biochemistry, and physiological aspects of embryo maturation.

Introduction

The occurrence of haploid embryos in plants was first reported by Blakeslee et al. in 1922. Other reports indicated the recovery of haploids, probably pollen derived, as a consequence of interspecific hybridization and embryo development without fertilization (Kostoff 1934). In early reviews of haploid plant production (Kimber and Riley 1963) there was no documented evidence of tissue culture–derived haploids, and interspecific crossing was regarded as the main route to haploidy.

The first definitive demonstration of microspore-derived embryos from anther culture was by Guha and Maheshwari (1964, 1966), who succeeded in recovering pollen embryos from anthers of *Datura innoxia* cultured on a simple agarized medium. Other reports appeared (Bourgin and Nitsch 1967; Kameya and Hinata 1970) and, to date, such embryos have been recorded from anther and microspore culture of over two hundred species in a large number of families (Maheshwari et al. 1982; Heberle-Bors 1985; Hu and Huang 1987; Keller et al. 1987). Table 20.1 shows some species in which pollen embryos have been reported since 1990. For other listings, see Ferrie et al. (1995).

In view of the potential usefulness of haploids in crop improvement and basic studies of embryogenesis, the findings of Guha and Maheshwari (1964) provided the stimulus for experimentation leading to the development of protocols now adaptable to a number of agronomically important crop species. The importance of genotype, stage of microspore development, donor plant physiology, and media factors to successful pollen embryogenesis is now well recognized. This is by no means indicative of an understanding of the physiological and biochemical basis for this process. An explanation of the sporophytic development from a potential gamete is yet to be advanced. However, empirical analyses have allowed researchers to manipulate pollen embryogenesis with a reasonable degree of success, at least in some species.

The purpose of this chapter is to examine the experimental aspects of pollen embryogenesis in vitro. The treatment, in some cases, will be brief, but

Table 20.1. *Some important crop species in which pollen-derived embryos have been reported since 1990**

Species	Response	Reference
Asparagus officinalis	E, P	Feng & Wolyn 1991, 1993
Brassica campestris	E, P	Baillie et al. 1992 Burnett et al. 1992
Brassica napus ssp. *rapifera*	E	Hansen & Svinnset 1993
Brassica oleracea	E, P	Duijs et al. 1992
Brassica oleracea convar. *capitata* (L.) Alef.	E, P	Roulund et al.1991
Brassica oleracea var. *botrytis*	E, P	Yang et al. 1992
Brassica oleracea var. *capitata*	E, P	Osolnik et al. 1993
Brassica oleracea convar. acephala	E, P	Kieffer et al. 1993
Brassica juncea	E, P	Agarwal & Bhojwani 1993
Camellia japonica	E, P	Pedroso & Pais 1993
Fagopyrum esculentum	E, P	Bohanec et al. 1993
Guizotia abyssinica (Coss)	E, P	Sarvesh et al. 1993
Helianthus annuus	E, P	Gürel et al. 1991 Thengane et al. 1994
Hordeum vulgare	E, P	Ziauddin et al. 1990 Olsen 1991 Hoekstra et al. 1992 Mouritzen & Holm 1992 Mordhorst & Lörz 1993 Scott and Lyne 1994
Hordeum bulbosum	E, P	Gadu et al. 1993
Hordeum spontaneum	E, P	Piccirilli & Arcioni 1991
Lupinus albus	E	Ormerod & Caligari 1994
Oryza sativa	E, P	Zhang & Qifeng 1993
Populus maximowiczii	E, P	Stoehr & Zsuffa 1990
Secale cereale	E, P	Fehlinghaus et al. 1991
Sinocalamus latiflora	E, P	Tsay et al. 1990
Solanum tuberosum	E, P	Tiainen 1992
Triticale	E, P	Hassawi & Laing 1990

Species	Response	Reference
Triticum aestivum	E, P	Hassawi & Laing 1990 Simonsen et al. 1992 Tuvesson & Ohuland 1993
Triticum Turgidum	E, P	Ghaemi et al. 1993
Zea mays	E, P	Afele et al. 1992 Buter et al. 1993
Tritordeum	E, P	Barcelo et al. 1994

E = embryos, P = plants
*This list includes species where improved methodology has increased the frequency of embryo production.

the reader can consult a number of reviews on various aspects of pollen embryogenesis (Maheshwari et al. 1982; Dunwell 1986; Hu and Yang 1986; Bajaj 1990; Ferrie et al. 1995). Because of the usefulness of pollen-derived embryos in studies of storage product biosynthesis and accumulation, this aspect will be dealt with at some length and the literature updated.

Anther and microspore culture

General methods for pollen embryo production involve the culture of anthers, isolated microspores, and sometimes whole flower buds (Dore 1989). Procedures have been developed that now permit a high frequency of embryogenesis from a wide variety of plant species (Dunwell 1986; Hu and Yang 1986; Keller et al. 1987; Bajaj 1990). In some species, especially graminaceous species, anther culture is the most frequent method for pollen embryogenesis and may be superior to the culture of isolated microspores (Kasha et al. 1990; Tuvesson and Öhuland 1993). Either embryos develop within the anthers and are released by degeneration of the anther walls, or pollen grains are shed into the culture medium before embryo development. However, there are some distinct advantages to the use of isolated microspores for embryo production and even though anther culture is very successful in Gramineae, increasing emphasis is now being placed on microspore culture (Olsen 1991; Hoesktra et al. 1992; Mordhorst and Lörz 1993).

The number of microspores, and consequently the number of potential embryogenic units, vary with species but may be in excess of 20,000 in each anther (Pan et al. 1991). In the most efficient anther culture systems, embryo yields are in the order of 100–200 per anther (Kasha et al. 1990). This is considerably lower than the yield with microspore culture. In *Brassica napus,* a comparison of anther and microspore culture revealed the latter to be about 10 times more efficient in embryogenesis (Lichter 1989; Siebel and Pauls 1989a; Arnison and Keller 1990; Kieffer et al. 1993). These findings suggest that the anther wall environment may be restricting embryogenesis, although there may not be selective regeneration from the microspores (Lichter 1989). Also, the fact that efficient embryogenesis can be achieved with isolated microspores indicates that there is no specific nutritive or hormonal function for the anther walls, which cannot be replaced by a defined medium.

The advantages of microspore culture compared to anther culture are as follows:

(a) It is easy to isolate microspores. Whole buds can be macerated and microspores recovered by filtration and centrifugation, which is less time consuming compared to isolation of individual anthers. In some species, the shed pollen technique is employed, in which anthers are isolated, and then floated on culture medium with high osmoticum to allow anther wall dehiscence and release of pollen.

(b) The culture density for optimum response can be conveniently adjusted with isolated microspores.

(c) Microspore populations can be enriched for potentially embryogenic microspores by cell sorting or gradient centrifugation, which is not possible with anthers.

(d) The stages of embryogenesis can be easily monitored with the isolated microspore system, and the importance of factors affecting each stage of embryogenesis can be identified.

(e) The isolated microspore system is more adaptable to biochemical and physiological studies of embryogenesis.

(f) Where the frequency of embryogenesis is low, callusing of the anther walls can overgrow microspore-derived embryos and obscure their development. This problem is avoided with isolated microspore culture.

(g) The isolated microspore culture system provides a much more convenient target for gene transfer techniques and mutagenesis compared to anther culture.

With these advantages, it is likely that isolated microspore culture will replace anther culture as the route to pollen embryos.

Factors affecting pollen embryogenesis

Genotype

Pollen embryogenesis is influenced by the genotype of the donor plant, and both the frequency of embryogenesis and the quality of embryos are affected (Foroughi-Wehr and Friedt 1984; Phippen and Ockendon 1990; Pickering and Devaux 1992; Zhang and Qifeng 1993). Exhaustive studies on genotypic influence on pollen embryogenesis have been reported for a number of species, and it is possible to transfer embryogenic responsiveness from high-responding genotypes to low-responding ones (Singsit and Veilleux 1989;

Arnison et al. 1990a, b; Zhang and Qifeng 1993). In wheat, *Triticum aestivum*, pollen embryogenesis is controlled by at least three genes (Agache et al. 1989), which appear to be dominant (Marsolais et al. 1986). This is an important aspect of pollen embryo technology as such traits can be transferred from highly responsive genotypes, of little agronomic value, to elite germplasm.

The basis of such genotypic control remains unexplained, although in *Brassica oleracea*, genotypic differences in ethylene production by cultured anthers is a controlling factor in embryogenesis (Biddington and Robinson 1991). Undoubtedly, genetic factors interact with other factors to control pollen embryogenesis, because donor plant growth conditions and manipulation of in vitro culture conditions can neutralize genotypic differences to some extent (Dunwell and Thurling 1985; Gland et al. 1988; Lichter 1989; Baillie et al. 1992).

Donor plant physiology

An important aspect of pollen embryo development from cultured anthers and microspores is the physiological status of the donor plant, which is determined by the growth conditions. For most species, plants grown under low temperature regimes yield more responsive microspores (Dunwell and Cornish 1985; Keller et al. 1987; Roulund et al. 1991; Takahata and Keller 1991; Baillie et al. 1992; Lo and Pauls 1992). Even where donor plants are not initially grown at low temperatures, a low temperature pretreatment of buds or isolated anthers enhanced embryogenesis (Lichter 1982; Heberle-Bors 1989; Huang and Keller 1989; Mordhorst and Lörz 1993; Osolnik et al. 1993). Plant age appears to be important, but this influence may be dependent on genotype and species. In some species, higher frequencies of embryogenesis occurred with microspores from young plants (Maheshwari et al. 1980; Jing et al. 1982), whereas in others, notably *Brassica oleracea*, older plants yielded more responsive microspores (Takahata et al. 1991).

Other factors, such as light intensity and nutritional status of the donor plant, affect the frequency of pollen embryogenesis and may interact with growth temperature. The mechanisms by which plant growth conditions affect the microspore response are unexplained. However, cytological modifications have been noted in embryogenic microspores from plants maintained at low temperatures (Lo and Pauls 1992). Low temperature pretreatments are probably important in arresting gametophytic development of microspores (Foroughi-Wehr and Wenzel 1989) and may be related to the emergence of dimorphic microspores (Heberle-Bors 1985, 1989).

Stage of microspore development

For most species there is an optimum stage of microspore development at which embryos can be induced. This covers the late uninucleate to early binucleate stages (Fan et al. 1988; Hunter 1988; Huang and Keller 1989; Telmer et al. 1992a, b, 1993; Mordhorst and Lörz 1993). Before or after these stages, microspores or pollen are not generally responsive to inductive treatment. However, in *Nicotiana tabacum* the optimum stage for pollen embryogenesis is early to mid binucleate stage (Rashid and Reinert 1980; Kyo and Harada 1986), and in *Helianthus annuus* and *Camellia japonica* the tetrad stage is optimal (Gürel et al. 1991; Pedroso and Pais 1993).

It appears that at certain stages of microsporogenesis, microspores are not fully committed to gametophytic development and embryogenesis is inducible (Sunderland and Huang 1987). Some microspores may also be arrested at these stages and are destined to become embryos under inductive conditions. This results in the occurrence of distinct embryogenic and game-tophytic pollen (Heberle-Bors 1985, 1989). Although it is possible that all microspores are potentially embryogenic, there are indications that male sterility, and conditions that affect pollen development, favor pollen embryo-genesis in some species (Heberle-Bors 1985).

Medium composition

In early studies of pollen embryogenesis in cultured anthers, a relatively simple hormone-free medium with low sucrose concentration was effective (Bourgin and Nitsch 1967). With isolated microspores, the medium composition is likely to be different to compensate for any nutritive function of the anther walls. The carbohydrate content of the medium is now recognized as a critical component for pollen embryogenesis (Dunwell and Thurling 1985; Gland et al. 1988; Huang and Keller 1989; Lichter 1989; Kasha et al. 1990). This serves both as a carbon source and as an osmoticum (Dunwell and Thurling 1985). In some species, the frequency of embryogenesis increases with increasing sucrose con-centration (Dunwell and Thurling 1985; Roulund et al. 1991). In the case of *Brassica* spp., sucrose levels ranging from 8% to 13% have been employed (Huang and Keller 1989; Roulund et al. 1991).

Sucrose is the most effective carbohydrate in anther and microspore culture (Hamoaka et al. 1991; Roulund et al. 1991). However, in some species, sucrose and glucose plus fructose inhibit pollen embryogenesis, whereas maltose is effective (Hunter 1988; Batty and Dunwell 1989; Kasha et al. 1990; Scott and Lyne 1994). The use of maltose also improves the quality of embryos and the recovery of green plants, especially in cereal anther and microspore culture (Kasha et al. 1990; Hoekstra et al. 1992). Even among cereals, pollen embryo-genic response to maltose and sucrose may be genotype dependent, because maltose inhibits embryogenesis in responsive wheat genotypes but promotes it in nonresponsive ones (Trottier et al. 1993). There is no definitive explanation for the differential response to these carbohydrates. Indications are that metab-olism of sucrose leads to the accumulation of toxic metabolites that are not pro-duced when maltose is the carbon source (Kasha et al. 1990; Scott and Lyne 1994). Although embryogenesis from cultured barley anthers occurs in the pres-ence of sucrose or fructose, isolated microspores show 100% mortality with these carbohydrates (Finnie et al. 1989; Scott and Lyne 1994). This suggests a buffering effect of the anther walls on microspore response to carbohydrates.

Manipulation of other media components, such as nitrogen source, pH, gelling agent, and growth regulators, improves the efficiency of embryogenesis (Kyo and Harada 1985; Lichter 1989; Baillie et al. 1992; Simonsen et al. 1992; Mordhorst and Lörz 1993). The requirement for exogenous growth regulators varies with species (Charne and Beversdorf 1988). With cereals where callusing usually precedes embryogenesis, exogenous auxins are required (Kasha et al. 1990; Kao et al. 1991; Ziauddin et al. 1992). In some dicot species, exogenous growth regulators are not required, and inhibition of endogenous plant growth regulator action may be necessary to elicit embryogenesis (Keller et al. 1987; Biddington and Robinson 1991). Since auxin is critical to embryogenesis, absence of exogenous requirements reflects adequate endogenous levels to permit normal embryo development (Liu et al. 1993).

Culture conditions

The quality and frequency of pollen embryo formation are affected by anther and microspore culture density and other features, such as the use of activated charcoal to remove toxic metabolites (Kott et al. 1988; Buter et al. 1993). Medium exchange, aeration, and the use of feeder layers all have a positive effect on embryogenesis (Lichter 1989; Simmonds et al. 1991, Hansen and Svinnset 1993). The initial culture temperature is a critical factor for pollen embryogenesis. In most protocols, isolated anthers and microspores require an initial period of 12–72 hours at 30°–35°C in order to elicit embryo development (Keller and Armstrong 1977; Dunwell et al. 1985; Arnison et al. 1990a; Pechan et al. 1991; Baillie et al. 1992; Duijs et al. 1992). This period is primarily inductive, and embryo development is completed by subsequent incubation at lower temperatures. The need for this split temperature regime is particularly important in species such as *Brassica* (Keller and Armstrong 1977; Baillie et al. 1992), and the longer the microspores and anthers are held at 25°C before the high temperature treatment, the lower is the embryogenic response (Pechan et al. 1991). Temperature requirement appears to be species dependent: With *Hordeum vulgare* and *Oryza sativa*, embryogenesis occurred at 26°–28°C, whereas in wheat (*Triticum aestivum*), elevated temperatures enhanced embryogenesis (Kasha et al. 1990; Mordhorst and Lörz 1993; Zhang and Qifeng 1993). In a few cases, initial culture at low temperature, 14°–15°C, enhanced embryogenesis (Pescitelli et al. 1990; Buter et al. 1991).

There is little information on the mechanism of the high temperature response. However, there are indications that heat shock proteins may be involved (Arnison et al. 1990a; Pechan et al. 1991; Fabijanski et al. 1992). Reports also indicate a change in microtubule distribution leading to a more symmetric pattern of cell division in microspore embryogenesis, as opposed to the asymmetric division characteristic of gametophytic development (Hause et al. 1993; Telmer et al. 1993). The modification of cell division leading to anomalous pollen development was observed with intact plants exposed to high temperature (Sax 1935; LaCour 1949; Telmer et al. 1993) and can be regarded as a trigger for pollen embryogenesis. With *B. napus* microspores, culture temperature determined the pattern of development, which was gametophytic at low temperature and embryogenic at high temperature (Custers et al. 1994). It is unresolved whether high temperature acts as a stress factor or is required for specific metabolism unrelated to stress. The importance of sucrose to the high temperature response in *B. campestris* suggests a specific effect (Hamoaka et al. 1991). The inductive effect of nutrient starvation in *Nicotiana* pollen embryogenesis (Zarsky et al. 1992) argues for a stress-related response, although the replacement of nutrient stress by high temperature has not been reported.

Developmental aspects of pollen embryogenesis

In higher plants, the function of the male gametophyte is the production of sperm cells and a pollen tube. The differentiation of this structure has been the subject of several reviews in which both developmental and biochemical aspects have been addressed (Mascarenhas 1992; Tanaka 1993; see also Chapters 2 and 3). Under certain conditions, the normal ontogenetic development of the gametophyte can be disrupted, leading to a sporophytic pattern of development and the emergence of a pollen embryo or tissues able to undergo embryogenesis. There is still debate concerning the origin of pollen embryos. One view is that embryogenic pollen is predetermined following meiosis and exists along with gametophytic pollen (Horner and Street 1978; Heberle-Bors 1985). This view presupposes that under permissive culture conditions embryos develop from such pollen grains that normally degenerate in vivo and are never gametophytic. The occurrence of anomalous pollen development supports this view (Huang 1986).

Contrary to this argument for pollen dimorphism, there is growing evidence that all pollen or microspores are potentially embryogenic and embryos are inducible under specific stimuli in culture (Kyo and Harada 1985, 1986; Sunderland and Huang 1987; Benito-Moreno et al. 1988; Telmer et al. 1992a, b, 1993). Failure to undergo embryogenesis is therefore due to progression beyond the inducible stage during the process of microsporogenesis. In the majority of species in which pollen embryos have been observed, the optimum stage for induction was the late uninucleate to early binucleate stages. Therefore the embryo can be derived from a uninucleate or a binucleate structure.

Cytological studies on both monocot and dicot microspores during the initial stages of culture indicate that a centrally located nucleus, prominent nucleoli, and reduced vacuolation are characteristic of embryogenic microspores (Figure 20.1a), (Telmer 1992; Reynolds 1993). The presence of starch grains in microspores, prior to culture, is usually indicative of non-embryogenic microspores (Dunwell and Sunderland 1975; Rashid et al. 1982; Sangwan and Sangwan-Norreel 1987; Telmer 1992).

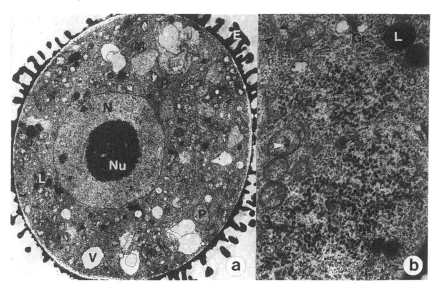

Figure 20.1 (a) Transmission election micrograph of embryogenic microspore of *Brassica napus* after 6 hours at 32.5°C. The nucleus and nucleolus are prominent; other organelles are present. N = nucleus; Nu = nucleolus; V = vacuole; M = mitochondria; P = plastid, L = lipid bodies. Magnification x 5,800. (b) Symmetric cell division of embryogenic microspore after 18 hours at 32.5°C. Arrowheads indicate the new cell wall. Magnification x6,900. (*Both figures are reproduced with permission of Dr. C. A. Telmer, Ph.D. dissertation 1992, Queen's University, Kingston, Ontario, Canada.*)

A critical cytological aspect of early microspore embryogenesis of uninucleate microspores is the first cell division. This is symmetrical (Figure 20.1b), giving two roughly equal cells (Fan et al. 1988; Zaki and Dickinson 1990, 1991; Reynolds 1993; Telmer et al. 1993), unlike the asymmetrical division in pollen development. Short-term exposure of microspores to antimicrotubule agents such as colchicine enhanced symmetrical cell division and the frequency of pollen embryos (Zaki and Dickinson 1991). This underscores the role of microtubules in controlling the plane of cell division (Simmonds et al. 1991; Zaki and Dickinson 1991; Hause et al. 1993). Evidence indicates that microspores held under noninductive conditions exhibit asymmetric cell divisions (Telmer 1992; Telmer et al. 1993). Following the first pollen mitosis, embryogenesis can still occur through cell division of the vegetative and/or generative cell (Raghavan 1990; Reynolds 1990). In anther culture of *Triticum aestivum* L., the initial cell division was asymmetric, defining a vegetative and a generative cell, and continued cell division of the former produced the embryo (Reynolds 1993). The various pathways of pollen embryogenesis are outlined in Figure 20.2. It is likely that from binucleate pollen, the embryo originates from the vegetative nucleus (Reynolds 1990).

Undoubtedly the commitment to embryogenesis requires the synthesis of specific proteins and other macromolecules characteristic of embryogenic

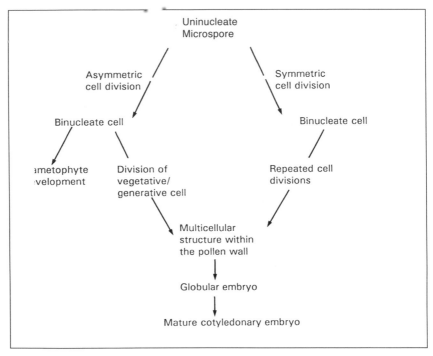

Figure 20.2 Developmental pathways to pollen embryos.

cells (Raghavan 1990; Reynolds 1990). Some of these proteins may be regulatory, functioning to redirect the microspore toward embryogenic development (Raghavan 1990). A number of embryoid-specific genes were isolated during pollen embryogenesis of wheat, and some of them were shown to be associated with the early stages of embryogenesis (Reynolds and Kitto 1992). Studies with *Nicotiana tabacum* revealed the appearance of specific mRNAs and phosphoproteins related to the induction of embryogenesis in microspores (Kyo and Harada 1990; Garrido et al. 1993). The induction of embryogenesis from binucleate pollen involves derepression of the G_1 arrest in the cell cycle of the vegetative cell (Zarsky et al. 1992). Conditions such as nutrient starvation and temperature may be responsible for this derepression.

Studies with *B. napus* binucleate pollen culture revealed extensive DNA replication of the vegetative cell under inductive temperatures (Binarova et al. 1993). Under noninductive temperatures, DNA replication occurred only in the generative cell. The conclusion by these authors was that pollen embryos are inducible at the G_1, S, or G_2 phases of the cell cycle. Inductive temperatures also elicited a number of mRNAs and proteins in *Brassica napus* microspores (Pechan et al. 1991). These results point to the importance of inductive conditions in triggering biochemical and physiological activities associated with pollen embryogenesis. The participation of heat shock proteins in this process is still unresolved (Telmer et al. 1993; Pechan et al. 1991).

Comparison of zygotic and pollen embryos

All the stages of embryogenesis characteristic of the zygote are recapitulated during microspore embryogenesis (Figure 20.3). However, development of zygotic and microspore-derived embryos occurs in completely different environments and this may account for some of the developmental, physiological, and biochemical differences that have been observed.

In terms of development, the initial difference appears to be at the early stages of cell division. In most angiosperms, the zygote divides asymmetrically to define an apical cell that gives rise to the embryo proper, and a basal cell that forms the suspensor (de Jong et al. 1993; Mayer et al. 1993; Rahman 1993). In contrast, the first division of an embryogenic microspore is usually symmetrical, delineating two equal cells (Fan et al. 1988; Zaki and Dickinson 1991; Reynolds 1993; Telmer et al. 1993). This suggests that the participation of a suspensor in microspore embryogenesis may not be important, although this structure has been observed (Rahman 1993).

The early stages of microspore embryogenesis occur within the pollen or microspore wall, and globular or bipolar structures are released by rupture of the wall (Rahman 1993; Reynolds 1993). In detailed comparative histological studies of microspore-derived and zygotic embryos of *Brassica napus* (Rahman 1993), differences were evident between the protodermal cell layers of the two types of embryos. Protodermal cells of microspore-derived embryos were more cytoplasmic and were defined much later in embryo development compared to the zygotic embryo. The shoot meristem of microspore-derived embryos contained an abundance of starch grains in com-

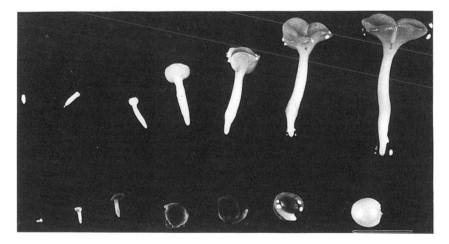

Figure 20.3 Comparison of the developmental stages of microspore-derived embryos (top) and zygotic embryos (bottom) of *Brassica napus*. Bar = 5mm. (*Reproduced with the permission of Dr. M. H. Rahman, Ph.D. dissertation, University of Calgary, Calgary, Alberta, Canada.*)

parison to the starch free meristem cells of the zygote. At a comparable stage of development, the shoot meristem of microspore-derived embryos was less defined compared to that of the zygote. The root axis of microspore-derived embryos accumulated starch granules and had an ill-defined epidermis compared to the zygotic embryo with few starch granules and a well-defined epidermis. Another feature of microspore-derived embryos was the development of intercellular spaces between the shoot meristem and the vascular tissues. This abnormality may be responsible for the low frequency of embryos that develop into mature plants.

The physiology of microspore-derived embryos is presumably different from that of the zygotic embryo because the latter is influenced in general by the intact plant and, more specifically, by the surrounding maternal tissues. In contrast, microspore-derived embryos are influenced by the anther wall and the culture medium. The normal pattern of embryogenesis is dictated by auxin, and interruption of its functions leads to abnormalities (Liu et al. 1993). In zygotic embryos, the source of auxin can be endogenous or from the various maternal tissues. With microspore-derived embryos, the culture medium or endogenous synthesis are the only sources. In cases where exogenous auxins are required for embryogenesis, there may be less regulatory control over tissue levels and this may be responsible for abnormalities observed in these embryos.

In addition to auxin, ABA (abscisic acid) plays a critical role in embryo development. It is important in embryo maturation and storage product accumulation during seed development (Finkelstein and Crouch 1986; Quatrano 1987; Johnson-Flanagan et al. 1992). Increased endogenous levels of ABA are correlated with normal desiccation tolerance and the acquisition of dormancy during seed development. Mutants impaired in ABA biosynthesis exhibit precocious germination through lack of dormancy (Niell et al. 1987). Microspore-derived embryos synthesize very little ABA and do not undergo the same maturation and desiccation tolerance phase as zygotic embryos (Mandel 1991). Exogenous ABA or culture conditions that enhance ABA synthesis allow maturation and development of desiccation tolerance in microspore-derived embryos (Senaratna et al. 1991; Brown et al. 1993).

The storage product profiles of zygotic and of microspore-derived embryos are comparable (Taylor et al. 1990; Pomeroy et al. 1991; Wiberg et al. 1991). In *Brassica napus*, lipids of the zygotic and microspore-derived embryos are similar, and culture manipulations result in a higher rate of lipid synthesis compared to zygotic embryos (Taylor et al. 1990; Pomeroy et al. 1991; Wiberg et al. 1991). With the application of ABA, the arrangement of oil bodies in microspore-derived embryos is comparable to that of the zygotic embryo (Rahman 1993). Protein bodies are not observed in microspore-derived embryos in the absence of ABA (Rahman 1993). This underscores the role of this hormone in storage protein accumulation (Bray and Beachy 1985; Wilen et al. 1990).

The increased starch accumulation detected in microspore-derived embryos of *Brassica napus* is most likely a consequence of the culture conditions where excess sucrose is converted to starch (Rahman 1993). Therefore, the accumulation of this storage product by these embryos, in contrast to the zygotic embryo, is a function of the culture environment.

Storage product synthesis and accumulation have been extensively examined in *Brassica* species but there are hardly any data for other species, such as cereals. In terms of development, a microspore-derived embryo of wheat exhibits well-developed scutellum, coleoptile, coleorhiza, and epiblast similar to the zygotic embryo (Reynolds 1993). It can therefore be assumed that the pattern of carbohydrate and protein accumulation is comparable.

In general, there is a close parallel in developmental pattern between microspore embryos and zygotic embryos. The physiological, biochemical, and morphological variations that occur are likely to be the result of the culture environment and not an inherent feature of the microspores.

Pollen embryo maturation

The physiological and biochemical differences between the microspore-derived and zygotic embryos are evidenced by the maturation process in which the former do not proceed through the normal dehydration, desiccation tolerance, and dormancy phases as seen in the latter. As a consequence, microspore-derived embryos tend to show precocious germination and poor conversion to whole plants. This is a characteristic of somatic embryos, and culture manipulation is required to allow normal maturation (Senaratna et al. 1990; Attree et al. 1992).

The critical factor in zygotic embryo maturation is the accumulation of ABA, which regulates the deposition of storage products and dormancy and desiccation tolerance (for example, Finkelstein and Somerville 1989; Rivin and Grudt 1991; Maquoi et al. 1993). Mutants impaired in ABA biosynthesis were unable to accumulate storage products and germinated precociously (Rivin and Grudt 1991). The low level of endogenous ABA may account for the abnormal maturation (Mandel 1991). However, the absence of the endosperm and the culture conditions may contribute to this behavior.

Application of exogenous ABA and the use of low osmotic potential in the culture medium enhanced maturation of microspore-derived embryos and resulted in storage product accumulation in a manner similar to that of the zygotic embryo (Eikenberry et al. 1991; Gruber and Röbbelen 1991; Senaratna et al. 1991; Brown et al. 1993). In these and other studies, desiccation tolerance of microspore-derived embryos was achieved through ABA treatment of embryos at the late torpedo or cotyledonary stage. Embryos not treated with ABA did not develop desiccation tolerance.

An important aspect of the embryo maturation process is the accumulation of storage products. ABA enhanced this process in microspore-derived embryos of *Brassica napus* by eliciting the transcription of genes for protein and lipid biosynthesis (Taylor et al. 1990; Wilen et al. 1990). These results indicate that embryo maturation in vitro is dependent on ABA, which appears to control the biosynthetic capacity of the cotyledons.

Plant regeneration and doubled haploid production

Following full development and maturation of embryos, plants are generally recovered by direct germination of these embryos. In most dicot species such as *Brassica*, fully matured embryos germinate into plantlets at a relatively low frequency (Kott and Beversdorf 1990). In other cases, there may be extensive secondary or tertiary embryo development from the initial embryos (Loh and Ingram 1982). This can give rise to complete plants. The regeneration of plants through secondary shoot formation from the cotyledonary areas has been observed in some species (Siebel and Pauls 1989a; Duijs et al. 1992). The tendency toward this mode of regeneration may be genotype dependent (Duijs et al. 1992).

Conditions for embryo maturation and germination are different, and most protocols require a modification of culture conditions to ensure a high frequency of plant recovery. A period at low temperatures, partial desiccation, and culture agitation improve the quality of *Brassica* microspore-derived embryos and the frequency of plant recovery (Coventry et al. 1988; Gland et al. 1988; Lichter 1989; Kott and Beversdorf 1990; Brown et al. 1993).

Under optimal culture conditions, embryos of *Brassica* species reach the cotyledonary stage by 21 days. However, development is asynchronous and various stages of embryogenesis can be detected at this time. A process of selection is required to remove advanced embryos and allow maturation of less-developed embryos. Initial culture density, competition for nutrients, and production of inhibitory substances by more advanced embryos may contribute to this asynchronous pattern. Established protocols for plant recovery from embryos involve culture on a reduced-sucrose and hormone-free medium in light (Coventry et al. 1988).

The quality of embryos is an important factor in plant recovery. In cereals, albino embryos can occur at a relatively high frequency and may be genotype dependent (Knudsen et al. 1989; Larsen et al. 1991). This phenomenon may be due to cytoplasmic DNA modifications but can be reduced through culture manipulations (Day and Ellis 1985; Knudsen et al. 1989; Kao et al. 1991).

With embryogenesis from gametic cells, a high frequency of haploid chromosome complement is expected and often occurs. However, mixoploids, aneuploids, and polyploids may occur as a result of endoreduplication and chromosomal aberrations during culture. These changes are generally more frequent where there is extensive callusing before embryogenesis (Keller and Armstrong 1981; Canhoto et al. 1990). The use of colchicine is the primary method of doubled haploid production from haploid embryos. Colchicine treatments are usually employed on regenerated plants. In some cases, chromosome doubling is performed either during the early stages of microspore culture to recover diploid embryos (Hannig 1993; Möllers et al. 1994) or before plants are transferred to soil (Mathias and Röbbelen 1991). In such approaches, chromosome doubling is not limited to the use of colchicine, because other antimicrotubule agents, such as oryzalin, trifluralin, and pronamide, are equally effective (Hannig 1993).

In some species, spontaneous diploidization occurs at a high frequency (Charne et al. 1988; Lichter et al. 1988). Cryopreservation of isolated microspores of *Brassica napus* prior to culture also results in a large percentage of diploid embryos (Chen and Beversdorf 1992a). Where such diploids result from chromosome duplication of the haploid complement, homozygosity is assured. However, if diploids originate from nuclear restitution or unreduced gametes, there will be some heterozygosity. In some species, such as *Solanum*, this phenomenon is quite common (Meyer et al. 1993).

Utilization of pollen embryos

Breeding and genetic studies

The most obvious advantage of haploid embryos to crop improvement is the potential to create homozygous plants through chromosome doubling (Keller et al. 1987; Morrison and Evans 1988). This reduces the need to conduct exhaustive and time-consuming inbreeding cycles to achieve homozygosity by conventional means. With doubled haploids, all the loci are homozygous and all functional genes are expressed.

Commercial cultivars have been developed using this technique in wheat and tobacco (Chaplin et al. 1980; de Buyser et al. 1987; Ferrie et al. 1995). The value of doubled haploids is evident in outcrossing populations such as *Brassica oleracea* and *B. rapa*, where it is a reliable method to achieve homozygosity (Duijs et al. 1992). Because of the absence of dominance and recessive interactions, selection is easier because characteristics are more defined (Choo et al. 1985) and population size for selection is smaller compared to conventional F_2 (Siebel and Pauls 1989b, c). In cytoplasmic male sterile lines, microspore development may still proceed to a stage where embryo induction is possible (Gadu et al. 1993) and the production of pollen-derived embryos is a convenient means of obtaining homozygous breeding lines.

A basic assumption relating to the use of microspore-derived embryos in breeding is that they should represent a random array of gametic recombinations and that there is no preferential regeneration from selected microspores. Evidence indicates that this is the case (Chen and Beversdorf 1990; Charne and Beversdorf 1991; Murigneaux et al. 1993). However, there are reports of segregation distortions among doubled haploids, which are indicative of selective regeneration of embryos from microspores (Thompson et al. 1991; Zivy et al. 1992). Segregation distribution patterns may be related to the existence of embryogenesis genes (Foisset et al. 1993).

Mutation and selection

The microspore-derived embryo system is valuable for the isolation of mutants and is a close parallel to microbial systems. Large numbers of embryogenic microspores can be subjected to in vitro selection pressure, or can be mutagenized, in a relatively small space and mutants can be isolated. Mutants are easily selected because all recessives are expressed. The added advantage is the ready production of homozygotes through chromosome doubling. In a diploid system, in contrast, recessives can remain hidden in the heterozygote.

The system has been exploited for the isolation of herbicide tolerant *Nicotiana tabacum* and *Brassica napus* plants by in vitro mutagenesis of isolated microspores or cells derived from haploid embryos (Chaleff and Ray 1984; Swanson et al. 1988, 1989), and, in the latter species, commercial cultivars are under development. The system has been effectively used for the isolation of mutants of *Brassica napus* having altered storage lipid composition (Turner and Facciotti 1990; Huang et al. 1991). In this case, pollen embryos allow screening for fatty acid profile of one cotyledon while the rest of the embryo can be retained for plant regeneration (Huang 1992).

The occurrence of gametoclonal variation among microspore-derived embryos may allow the selection of useful traits without the need for mutagenesis (Morrison and Evans 1988). This has been used for selection of variants for disease resistance, increased alkaloid content, and increased protein levels (Collins et al. 1974; Sacristan 1982; Foroughi-Wehr and Friedt 1984; Schaeffer et al. 1984; Voorrips and Visser 1990; Witherspoon et al. 1991).

Gene transfer

Haploid embryos, or cells derived from such embryos, are potentially useful as recipients for foreign genes, especially where there is a high frequency of embryogenesis (see Chapter 21). A high plant regeneration frequency is required for successful gene transfer and the recovery of transgenics. Microspore-derived embryos generally exhibit a high regeneration potential. Also, chromosome doubling results in duplication of the introduced trait and the homozygote can be evaluated.

A number of gene transfer techniques have been utilized in the transformation of pollen embryos (Huang 1992). *Agrobacterium*-mediated gene transfer is the most frequent technique and fertile homozygous transgenic plants have been recovered (Swanson and Erickson 1989; Oelck et al. 1991; Huang 1992; Sangwan et al. 1993). The use of electroporation and polyethylene glycol (PEG) to facilitate direct DNA uptake by microspores resulted in transient gene expression, but stable transformants were not recovered (Kuhlmann et al. 1991; Fennell and Hauptmann 1992; Jardinaud et al. 1993). Electroporation was very successful in transformation of maize zygotic embryos (D'Halluin et al. 1992) and should be applicable to pollen-derived embryos. The use of the biolistic technique in combination with desiccation and imbibition of DNA resulted in stable transformation of *Brassica napus* microspore-derived embryos with an efficiency of 2% (Chen and Beversdorf 1994). Microinjection of DNA into microspore-derived embryos of *Brassica napus* resulted in the recovery of transgenic plants (Neuhaus et al. 1987). However, this approach has not been widely applicable.

One attractive feature of the microspore embryogenic system is the potential to transform individual cells (Pechan 1989). Because embryos are derived from individual cells, transformants can be recovered without the complication of chimeras.

Protoplasts from embryogenic microspores or cells derived from haploid embryos have been utilized for gene transfer through cell fusion (Chuong et al. 1988a, b). The advantage is the recovery of diploids and completely fertile plants in contrast to diploid fusions where aneuploidy, chromosomal aberrations, and low fertility are common.

Biochemical and physiological studies

Because of their uniformity and developmental similarity to the zygotic embryo, microspore embryos are ideal for studies of the biochemistry and physiology of embryogenesis. In highly responding genotypes, populations of embryos at specific stages of development can be conveniently obtained and the influence of regulatory factors examined. In contrast, the zygotic embryo is relatively inaccessible.

With *Brassica* microspore-derived embryos, storage lipid and protein biosynthesis and accumulation have been extensively studied (Taylor et al. 1990, 1993; Wilen et al. 1990; Pomeroy et al. 1991; Wiberg et al. 1991). These embryos are rich in lipid biosynthetic enzymes and are useful for in vitro screening for oil quality (Taylor et al. 1990; Wiberg et al. 1991). This system has been utilized for studies of chilling tolerance, metabolism of chlorophyll during seed degreening, and glucosinolate metabolism (Cloutier 1990; Johnson-Flanagan et al. 1992; Johnson-Flanagan and Singh 1993; McClellan et al. 1993).

Artificial seeds and germplasm storage technology

The use of somatic embryos as artificial seeds is an emerging technology that has been applied to a few species (Janick et al. 1989; Senaratna et al. 1990; Redenbaugh 1993). The process involves embryo maturation, development of desiccation tolerance, and in some cases encapsulation, to permit convenient handling of seeds. This technology has been extended to pollen-derived embryos (Datta and Potrykus 1989; Senaratna et al. 1991; Brown et al. 1993). Embryos of *Brassica napus* and *B. oleracea* were desiccated and stored at room temperature, and they subsequently germinated at a high frequency (Takahata et al. 1992, 1993; Brown et al. 1993). Important factors in desiccation tolerance were the stage of embryo development and maturation in the presence of exogenous ABA. Once desiccated, these embryos can be handled as seeds. Microspore-derived embryos are potentially more advantageous for artificial seed production because embryogenesis is much more uniform compared to the production of somatic embryos. The other advantage is the potential to diploidize microspores before embryo formation and recover homozygous embryos. In spite of these advantages, the use of microspore-derived embryos as artificial seeds is restricted so far to a few species (Datta and Potrykus 1989; Brown et al. 1993).

With the generation of large numbers of microspore-derived embryos, the embryo population can be viewed as a potential source of germplasm resulting from genetic recombinations and novel variations that may occur as a result of in vitro culture conditions or chromosome doubling techniques. This may be a valuable source of germplasm for crop improvement. Storage of these embryos in a viable state may be necessary to permit convenient screening for novel genotypes. There are two methods currently used for storage: cryopreservation and desiccation storage. Pollen embryos have been cryopreserved and plants successfully recovered (Bajaj 1983). In addition, embryogenic microspores can be stored at ultra low temperature without loss of embryogenic potential (Charne et al. 1988; Chen 1991; Chen and Beversdorf 1992b; Skladal et al. 1992).

Conclusions and future potentials

The most successful protocols for pollen-derived embryos are those developed for members of Solanaceae, Brassicaceae, and Poaceae. A large part of this success is due to the recognition of key factors affecting pollen embryogenesis. These include genotype, stage of microspore development, donor plant physiology, and manipulation of in vitro culture conditions. However, there is a lack of information on the biochemical and physiological characteristics of embryogenic microspores. Since it is now possible to enrich the population of embryogenic microspores through fractionation techniques (Fan et al. 1988; Martensson and Widell 1993), it should be possible to address this deficiency.

Undoubtedly, specific gene expression controls various aspects of microsporogenesis (see Chapter 3). A crucial factor in pollen embryo development may be the continued expression of sporophytic genes in the microspores. The existence of specific embryogenic microspores in association with gametophytic microspores appears to be the case in some species (Heberle-Bors 1985). However, there is a body of evidence to support the view that commitment to the gametophytic pathway is dependent on the stage of microspore development (Sunderland and Huang 1987). The use of in vitro culture manipulations to control gametophytic or sporophytic development from binucleate pollen of *Nicotiana tabacum* suggests a lack of predetermination (Benito-Moreno et al. 1988; Zarsky et al. 1992). So far, success from binucleate pollen has not been reported in other species, although it is possible that they may not respond to the same signals as *Nicotiana tabacum*. For most species, the frequency of pollen embryos is very low and an understanding of the factors governing gametophytic development should allow manipulations to increase this frequency. It appears that pretreatments prior to culture serve to arrest gametophytic development (Foroughi-Wehr and Wenzel 1989).

Attempts are now being made to understand the biochemical and physiological aspects of pollen embryogenesis in vitro with studies of gene expression and biosynthesis of specific proteins (Pechan et al. 1991; Reynolds and Kitto 1992; Binarova et al. 1993; Garrido et al. 1993). In particular, culture conditions that alter the expression of specific genes associated with embryogenesis have been examined.

Even though the developmental stages of microspore-derived embryos are similar to those of the zygote, there are generally more abnormalities and a low frequency of direct embryo-to-plant conversion with microspore-derived embryos. This is indicative of abnormal physiological conditions during in vitro development. Methods are now available to normalize maturation and to improve plant recovery. Also, culture modifications such as microculture techniques should improve embryo quality.

The ability to recover homozygous plants from pollen-derived embryos is extremely useful in plant breeding programs, and the recovery of novel recombinations is recognized. Homozygosity can be readily achieved even in rigidly outcrossing populations. Mutagenesis and mutant selection are important attributes of the haploid embryogenic system. The use of haploid cells and embryos as recipients for foreign genes is gaining prominence because of the high regeneration potential of these embryos and the recovery of the homozygous trait through chromosome doubling. These embryos are excellent model systems for studies of fundamental aspects of embryogenesis and the biochemistry of storage product accumulation. In some species, pollen embryos can be desiccated and handled as artificial seeds. With the development of techniques for chromosome duplication prior to embryogenesis, the use of these embryos as artificial seeds should gain prominence.

The main limitation to the widespread utilization of pollen embryos is the relatively few economically important species in which embryos can be reliably obtained. Because androgenic response can be identified by marker proteins (Vergue et al. 1993) and because very young microspores, or even microsporocytes, can be cultured in vitro to produce viable mature pollen (Heberle-Bors 1989; Mouritzen and Holm 1992; Takacs et al. 1994), it should be possible to analyze the factors regulating the switch from gametophytic to sporophytic development.

References

Afele, J.C., Kannenberg, L.W., Keats, R., Sohota, S. & Swanson, E.B. (1992). Increased induction of microspore embryos following manipulation of donor plant environment and culture temperature in corn (*Zea mays* L.). *Plant Cell Tissue Organ. Cult.* 28:87–90.

Agache, S., de Buyser, J., Henry, Y. & Snape, J.W. (1989). Studies of the genetic relationship between anther culture and somatic tissue culture abilities in wheat. *Plant Breeding* 100:26–33.

Agarwal, P.K. & Bhojwani, S.S. (1993). Enhanced pollen grain embryogenesis and plant regeneration in anther cultures of *Brassica juncea* cv. PR-45. *Euphytica* 70:191–203.

Arnison, P.G. & Keller, W.A. (1990). A survey of the anther culture response of *Brassica oleracea* L. cultivars grown under field conditions. *Plant Breeding* 104:125–133.

Arnison, P.G., Donaldson, P., Ho, L.C.C. & Keller, W.A. (1990a). The influence of various physical parameters on anther culture of Broccoli (*Brassica oleracea* var italica). *Plant Cell Tissue Organ. Cult.* 20:147–155.

Arnison, P.G., Donaldson, P., Jackson, A., Semple, C. & Keller, W. (1990b). Genotype-specific response of cultured broccoli (*Brassica oleracea* var italia) anthers to cytokinins. *Plant Cell Tissue and Organ. Cult.* 20:217–222.

Attree, S.M., Pomeroy, M.K. & Fowke, L.C. (1992). Manipulation of conditions for the culture of somatic embryos of white spruce for improved triacylglycerol biosynthesis and desiccation tolerance. *Planta* 187:395–404.

Baillie, A.M.R., Epp, D.J., Hutcheson, D. & Keller, W.A. (1992). *In vitro* culture of isolated microspores and regeneration of plants in *Brassica campestris*. *Plant Cell Reprod.* 11:234–237.

Bajaj, Y.P.S. (1983). Regeneration of plants from pollen embryos of *Arachis, Brassica* and *Triticum* spp. cryopreserved for one year. *Current Sci.* 52:484–486.

Bajaj, Y.P.S. (1990). *In vitro* production of haploids and their use in cell genetics and plant breeding. In *Haploids in Crop Improvement. I. Biotechnology in Agriculture and Forestry*, ed. Y.P.S. Bajaj, pp. 3–44. Berlin: Springer-Verlag.

Barcelo, P., Cabrera, A., Hagel, C. & Lörz, H. (1994). Production of doubled-haploid plants from *tritordeum* anther culture. *Theor. Appl. Genet.* 87:741–745.

Batty, N. & Dunwell, J.M. (1989). Effect of carbohydrate source on the response of potato anthers in culture. *Plant Cell Tissue Organ. Cult.* 18:221–226.

Benito-Moreno, R.M., Macke, F., Hauser, M.T., Alwen, A. & Heberle-Bors, E. (1988). Sporophytes and male gametophytes from *in vitro* cultured immature tobacco pollen. In *Sexual Reproduction in Higher Plants*, eds. M. Cresti, P. Gori, and E. Pacini, pp. 137–142. Berlin: Springer-Verlag.

Biddington, N.L. & Robinson, H.T. (1991). Ethylene production during anther culture of Brussels sprout (*Brassica oleracea* var. *gemmifera*) and its relationship with factors that affect embryo production. *Plant Cell Tissue Organ. Cult.* 25:169–177.

Binarova, P., Straatman, K., Hause, B., Hause, G. & Van Lammeren, A.A.M. (1993). Nuclear DNA synthesis during the induction of embryogenesis in cultured microspores and pollen of *Brassica napus* L. *Theor. Appl. Genet.* 87:9–16.

Blakeslee, A.F., Belling, J., Farnham, M.E. & Bergner, A.D. (1922). A haploid mutant in the Jimson weed, *Datura stramonium*. *Science* 55:646–647.

Bohanec, B., Neskovic, M. & Vujicic, R. (1993). Anther culture and androgenetic plant regeneration in buckwheat (*Fagopyrum exulentum* Moench.). *Plant Cell Tissue Organ. Cult.* 35:259–266.

Bourgin, J.P. & Nitsch, J.P. (1967). Obtention de *Nicotiana* haploids à partir d'étamines culturées *in vitro*. *Ann. Physiol. Vég Paris* 9:377–382.

Bray, E.A. & Beachy, R.N. (1985). Regulation by ABA of ß-conglycinin expression in cultured developing soybean cotyledons. *Plant Physiol.* 79:746–750.

Brown, D.C.W., Watson, E.M. & Pechan, P.M. (1993). Induction of desiccation tolerance in microspore-derived embryos of *Brassica napus*. *In Vitro Cell. Devel. Biol.* 29:113–118.

Burnett, L., Yarrow, S. & Huang, B. (1992). Embryogenesis and plant regeneration from isolated microspores of *Brassica rapa* L. ssp. Oleifera. *Plant Cell Reprod.* 11:215–218.

Buter, B., Schmid, J.E. & Stamp, P. (1991). Effects of L-proline and post-plating temperature treatment on maize (*Zea mays*) anther culture. *Plant Cell Reprod.* 10:325–328.

Buter, B., Pescitelli, S.M., Berger, K., Schmid, J.E. & Stamp, P. (1993). Autoclaved and filter sterilized liquid media in maize anther culture: significance of activated charcoal. *Plant Cell Reprod.* 13:79–82.

Canhoto, J.M., Ludovina, M., Guimaraes, S., Cruz, G.S. (1990). *In vitro* induction of haploid, diploid, and triploid plantlets by anther culture of *Iochroma warscewiczii* Regel. *Plant Cell Tissue Organ. Cult.* 21:171–177.

Chaleff, R.S. & Ray, T.B. (1984). Herbicide resistant mutants from tobacco cell cultures. *Science* 223:1148–1151.

Chaplin, J.F., Berk, L.G., Gooding, A.V. & Powell, N.T. (1980). Registration of NC744 tobacco germplasm (Reg No GP18). *Crop Sci.* 20:677.

Charne, D.G. & Beversdorf, W.D. (1988). Improving microspore culture as a rapeseed breeding tool: the use of auxins and cytokinins in an induction medium. *Can. J. Bot.* 66:1671–1675.

Charne, D.G. & Beversdorf, W.D. (1991). Comparisons of agronomic and compositional traits in microspore-derived and conventional populations of spring *Brassica napus*. In *Proceedings of the 8th International Rapeseed Congress, Canola Council of Canada*, ed. D.I. McGregor, pp. 64–69.

Charne, D.G., Pukachi, P., Kott, L.S. & Beversdorf, W.D. (1988). Embryogenesis following cryopreservation in isolated microspores of rapeseed *Brassica napus* L. *Plant Cell Reprod.* 7:407–409.

Chen, J.L. (1991). Evaluation of microspore culture in germplasm preservation, lipid biosynthesis and DNA uptake studies in rapeseed *Brassica napus* L. Ph.D. thesis, Dept. of Crop Science, Univ. of Guelph, Guelph, ON.

Chen, J.L. & Beversdorf, W.D. (1990). A comparison of traditional and haploid-derived breeding populations of oilseed rape (*Brassica napus* L.) for fatty acid composition of the seed oil. *Euphytica* 51:59–65.

Chen, J.L. & Beversdorf, W.D. (1992a). Production of spontaneous diploid lines in isolated microspores following cryopreservation of spring rapeseed (*Brassica napus* L.). *Plant Breeding* 108:324–327.

Chen, J.L. & Beversdorf, W.D. (1992b). Cryopreservation of isolated microspores of spring rapeseed (*Brassica napus* L.) for *in vitro* embryo production. *Plant Cell Tissue Organ. Cult.* 31:141–149.

Chen, J.L. & Beversdorf, W.D. (1994). A combined use of microprojectile bombardment and DNA imbibition enhances transformation frequency of canola (*Brassica napus* L.). *Theor. Appl. Genet.* 88:187–192.

Choo, T.M., Reinbergs, F. & Kasha, K.J. (1985). Use of haploids in breeding barley. *Plant Breeding Rev.* 3:219–252.

Chuong, P.V., Beversdorf, W.D., Powell, A.D. & Pauls, K.P. (1988a). The use of haploid protoplast fusion to combine cytoplasmic atrazine resistance and cytoplasmic male sterility in *Brassica napus*. *Plant Cell Tissue Organ. Cult.* 12:181–184.

Chuong, P.V., Beversdorf, W.D., Powell, A.D. & Pauls, K.P. (1988b). Somatic transfer of cytoplasmic traits in *Brassica napus* L. by haploid protoplast fusion. *Mol. Gen. Genet.* 211:197–201.

Cloutier, S. (1990). *In vitro* selection for freezing tolerance using *Brassica napus* microspore culture. M.Sc. thesis, Dept. of Crop Science, Univ. of Guelph, Guelph, ON.

Collins, G.B., Legg, P.B. & Kasperbauer, M.J. (1974). Use of anther-derived haploids in *Nicotiana*. I. Isolation of breeding lines differing in total alkaloid content. *Crop Sci.* 14:77–80.

Coventry, J., Kott, L. & Beversdorf, W.D. (1988). Manual for microspore culture technique for *Brassica napus*. *Dept. of Crop Science Technology Bullettin OAC Publication 0489*. Guelph. ON: Univ. of Guelph.

Custers, J.B.M., Cordewener, J.H.G., Möllen, Y., Dons, H.J.M. & Van Lookeren-Campagne, N.M. (1994). Temperature controls both gametophytic and sporophytic development in microspore culture of *Brassica napus*. *Plant Cell Reprod*. 13:267–271.

D'Halluin, K., Bonne, E., Bossut, M., DeBuckeleer, M. & Leemans, J. (1992). Transgenic maize plants by tissue electroporation. *Plant Cell* 4:1495–1505.

Datta, S.K. & Potrykus, I. (1989). Artificial seeds in barley: encapsulation of microspore derived embryos. *Theor. Appl. Genet*. 77:820–824.

Day, A. & Ellis, T.H.N. (1985). Deleted forms of plastid DNA in albino plants from cereal anther culture. *Current Genet*. 9:671–678.

de Buyser, J., Henry, Y., Lonnet, P., Hertzog, R. & Hespel, A. (1987). Florin: a doubled haploid wheat variety developed by the anther culture method. *Plant Breeding* 98:53–57.

de Jong, A.J., Schmidt, E.D.L. & de Vries, S.C. (1993). Early events in higher-plant embryogenesis. *Plant Mol. Biol*. 22:367–377.

Dore, C. (1989). Obtention de plantes haploides de chou cabus (*Brassica oleracea* L. ssp. capitata) après culture *in vitro* d'ovules pollinisés par du pollen irradiè. *C.R. Acad. Sci. Paris* t. 309 Serie III: 729–734.

Duijs, J.G., Voorrips, R.E., Visser, D.L. & Custers, J.B.M. (1992). Microspore culture is successful in most crop types of *Brassica oleracea* L. *Euphytica* 60:45–55.

Dunwell, J.M. (1986). Pollen, ovule and embryo culture as tools in plant breeding. In *Plant Tissue Culture and its Agricultural Applications,* eds. L.A. Withers and P.G. Alderson, pp. 375–404. London: Butterworth.

Dunwell, J.M. & Cornish, L.M. (1985). Influence of preculture variables on microspore embryo production in *Brassica napus* ssp. *oleifera* cv. Duplo. *Ann. Bot*. 56:281–289.

Dunwell, J.M. & Sunderland, N. (1975). Pollen ultrastructure in anther cultures of *Nicotiana tabacum*. III. The first sporophytic division. *J. Exp. Bot*. 26:240–252.

Dunwell, J.M. & Thurling, N. (1985). Role of sucrose in microspore embryo production in *Brassica napus* ssp. *oleifera*. *J. Exp. Bot*. 36:1478–1491.

Dunwell, J.M., Cornish, L.M. & Decourcel, A.G.L. (1985). Influence of genotype, plant growth temperature and anther incubation temperature on microspore embryo production in *Brassica napus* ssp. *oleifera*. *J. Exp. Bot*. 36:679–698.

Eikenberry, E.J., Chuong, P.V., Esser, J., Romero, J. & Ram, R. (1991). Maturation desiccation, germination and storage lipid accumulation in microspore embryos of *Brassica napus* L. In *Proceedings of the 8th International Rapeseed Congress, Canola Council of Canada*, ed. D.I. McGregor, pp. 1809–1814.

Fabijanski, S., Altossar, I. & Arnison, P.G. (1992). Heat shock response during anther culture of Broccoli (*B. oleracea* var *italica*). *Plant Cell Tissue Organ. Cult*. 26:203–212.

Fan, Z., Armstrong, K.C. & Keller, W.A. (1988). Development of microspores *in vivo* and *in vitro* in *Brassica napus* L. *Protoplasma* 147:191–199.

Fehlinghaus, T., Deimling, S. & Geiger, H.H. (1991). Methodical improvements in rye anther culture. *Plant Cell Reprod*. 10:397–400.

Feng, X.R. & Wolyn, D.J. (1991). High frequency production of haploid embryos in Asparagus anther culture. *Plant Cell Reprod*. 10:574–578.

Feng, X.R. & Wolyn, D.J. (1993). Development of haploid Asparagus embryos from liquid cultures of anther derived calli enhanced by ancymidol. *Plant Cell Reprod*. 12:281–288.

Fennell, A. & Hauptmann, R. (1992). Electroporation and PEG delivery of DNA into maize microspores. *Plant Cell Reprod.* 11:567–570.

Ferrie, A.M.R., Palmer, C.E. & Keller, W.A. (1995). In vitro embryogenesis in plants. In *Plant Embryogenesis,* ed. T.A. Thorpe, pp. 309–344. Dordrecht: Kluwer Acad. Publ.

Finkelstein, R.R. & Crouch, M.L. (1986). Rapeseed embryo development in culture on high osmoticum is similar to that of seeds. *Plant Physiol.* 81:907–912.

Finkelstein, R.R. & Crouch, M.L. (1987). Hormonal and osmotic effects on developmental potential of maturing rapeseed. *Hort. Sci.* 22:797–800.

Finkelstein, R.R. & Somerville, C.H. (1989). Abscisic acid or high osmoticum promotes accumulation of long chain fatty acids in developing embryos of *Brassica napus. Plant Sci.* 61:213–217.

Finnie, S.J., Powell, W. & Dyer, A.F. (1989). The effect of carbohydrate composition and concentration on anther response in barley. *Plant Breeding* 103:110–119.

Foisset, M., Delourme, R., Lucas, M.O. & Renard, M. (1993). Segregation analysis of isozyme markers on isolated microspore-derived embryos in *Brassica napus* L. *Plant Breeding* 110:315–322.

Foroughi-Wehr, B. & Friedt, W. (1984). Rapid production of recombinant barley yellow mosaic virus resistant *Hordeum vulgare* lines by anther culture. *Theor. Appl. Genet.* 67:377–382.

Foroughi-Wehr, B. & Wenzel, G. (1989). Androgenetic haploid production. *IPTAC Newslett.* 58:11–18.

Gadu, S., Procunier, J.D., Ziauddin, A. & Kasha, K.J. (1993). Anther culture–derived homozygous lines in *Hordeum bulbosum. Plant Breeding* 110:109–115.

Garrido, D., Eller, N., Heberle-Bors, E. & Vicente, O. (1993). *De novo* transcription of specific mRNAs during the induction of tobacco pollen embryogenesis. *Sex. Plant Reprod.* 6:40–45.

Ghaemi, M., Sarraji, A. & Alibert, G. (1993). Influence of genotype and culture conditions on the production of embryos from anthers of tetraploid wheat (*Triticum turgidum*). *Euphytica* 65:81–85.

Gland, A., Lichter, R. & Schweiger, H.G. (1988). Genetic and exogenous factors affecting embryogenesis in isolated microspore cultures of *Brassica napus* L. *J. Plant Physiol.* 132:613–617.

Gruber, S. & Röbbelen, G. (1991). Fatty acid synthesis in microspore-derived embryoids of rapeseed (*Brassica napus*). In *Proceedings of the 8th International Rapeseed Congress, Canola Council of Canada,* ed. D.I. McGregor, pp. 1818–1820.

Guha, S. & Maheshwari, S.C. (1964). *In vitro* production of embryos from anthers of *Datura. Nature* 204:497.

Guha, S. & Maheshwari, S.C. (1966). Cell division and differentiation of embryos in the pollen grains of *Datura in vitro. Nature* 212:97–98.

Gürel, A., Kontowski, S., Nichterlein, K. & Friedt, W. (1991). Embryogenesis in microspore cultures of sunflower (*Helianthus annuus* L.). *Helia* 14:123–128.

Hamoaka, Y., Fujita, Y. & Iwai, S. (1991). Effects of temperature on the mode of pollen development in anther culture of *Brassica campestris. Physiol. Plant.* 82:67–72.

Hannig, A. (1993). Antimicrotubule agents for diploidizing haploid tissues in *Brassica napus. 8th Crucifer Genetics Workshop. Abs. #57.* National Research Council of Canada.

Hansen, M. & Svinnset, K. (1993). Microspore culture of Swede (*Brassica napus* ssp *rapifera*) and the effects of fresh and conditioned medium. *Plant Cell Reprod.* 12:496–500.

Hassawi, D.S. & Laing, G.H. (1990). Effect of cultivar, incubation temperature and stage of microspore development on anther culture in wheat and triticale. *Plant Breeding* 105:332–336.

Hause, B., Hause, G., Pechan, P. & VanLammeren, A.A.M. (1993). Cytoskeletal changes in induction of embryogenesis in microspore and pollen cultures of *Brassica napus* L. *Cell Biol. Intl.* 17:153–168.

Heberle-Bors, E. (1985). *In vitro* haploid formation from pollen: a critical review. *Theor. Appl. Genet.* 71:361–374.

Heberle-Bors, E. (1989). Isolated pollen culture in tobacco: plant reproductive development in a nutshell. *Sex. Plant Reprod.* 2:1–10.

Hoekstra, S., van Zijderveld, M.H., Louwerse, J.D., Heidekamp, F. & van der Mark, F. (1992). Anther and microspore culture of *Hordeum vulgare* L. cv Igri. *Plant Sci.* 86:89–96.

Horner, M. & Street, F.E. (1978). Pollen dimorphism-origin and significance in pollen plant formation by anther culture. *Ann. Bot.* 42:763–771.

Hu, H. & Huang, B. (1987) Application of pollen-derived plants to crop improvement. *Intl. Rev. Cytol.* 107:397–420.

Hu, H. & Yang, H. (1986). Haploids of higher plants *in vitro*. Berlin, Heidelberg: Springer-Verlag.

Huang, B. (1986). Ultrastructural aspects of pollen embryogenesis in *Hordeum, Triticum* and *Paeonia*. In *Haploids of Higher Plants*, eds. H. Hu and H. Yang, pp. 91–117. Berlin, Heidelberg, New York, Tokyo: Springer-Verlag.

Huang, B. (1992). Genetic manipulation of microspores and microspore-derived embryos. *In Vitro Cell Devel. Biol.* 281:53–58.

Huang, B. & Keller, W.A. (1989). Microspore culture technology. *J. Tissue Cult. Methods* 12:171–178.

Huang, B., Swanson, E.B., Baszcyznski, C.L., MacRae, W.D., Bardour, E., Armavil, V., Wobe, L., Arnoldo, M., Rozakis, S., Westecott, M., Keats, R.F. & Kemble, R. (1991). Application of microspore culture to canola improvement. In *Proceedings of the 8th International Rapeseed Congress, Canola Council of Canada,* ed. D.I. McGregor, pp. 298–303.

Hunter, C.P. (1988). Plant regeneration from microspores of barley, *Hordeum vulgare* L. Ph.D. thesis, Wye College, University of London.

Janick, J., Kitto, S.L. & Dim, V.A. (1989). Synthetic seed production by desiccation and encapsulation. *In Vitro Cell Devel. Biol.* 25:1167–1172.

Jardinaud, M-F., Souvre, A. & Alibert, G. (1993). Transient GUS gene expression in *Brassica napus* electroporated microspores. *Plant Sci.* 93:177–184.

Jing, J.K., Xi, Z-Y. & Hu, H. (1982). Effects of high temperature and physiological conditions of donor plants on induction of pollen derived plants in wheat. *Ann. Rep. Inst. Genet. Acad. Sinica, Beijing* 1981:67–72.

Johnson-Flanagan, A. & Singh, J. (1993). A method to study seed degreening using haploid embryos of *Brassica napus* cv. Topas. *J. Plant Physiol.* 141: 487-493.

Johnson-Flanagan, A.M., Huiwey, Z., Geng, X-M., Brown, D.C.W., Nykifarutk, C.L. & Singh, J. (1992). Frost, abscisic acid and desiccation hasten embryo development in *Brassica napus* L. *Plant Physiol.* 94:700–706.

Kameya, T. & Hinata, K. (1970). Induction of haploid plants from pollen grains of *Brassica. Japan. J. Breeding* 20:82–87.

Kasha, K.J., Ziauddin, A. & Cho, U.H. (1990). Haploids in cereal improvement: anther and microspore culture. In *Gene Manipulation in Plant Improvement* II, ed. J.P. Gustafson, pp. 213–235. New York: Plenum Press.

Kao, K.M., Saleem, M., Abrams, S., Peeras, M., Horn, D. & Mallard, C. (1991). Culture conditions for induction of green plants from barley microspores by anther culture methods. *Plant Cell Reprod.* 9:595–601.

Keller, W.A. & Armstrong, K.C. (1977). Embryogenesis and plant regeneration in

Brassica anther culture. *Can. J. Bot.* 55:1383–1388.

Keller, W.A. & Armstrong, K.C. (1981). Dihaploid plant production by anther culture in autotetraploid narrow stem kale (*Brassica oleracea* L. var Acephala D.C.). *Can. J. Genet. Cytol.* 23:259–265.

Keller, W.A., Arnison, P.G. & Cardy, B.K. (1987). Haploids from gametophytic cells: recent developments and future prospects. In *Plant Tissue and Cell Culture*, eds. C.E. Green, D.A. Somers, W.P. Hackett, and D.D. Biesboer, pp. 233–241. New York: Allan R. Liss.

Kieffer, M., Fuller, M.P., Chauvin, T.E. & Schlesser, A. (1993). Anther culture of kale (*Brassica oleracea* L. con var. Acephala. D.C. Alef). *Plant Cell Tissue and Organ. Cult.* 33:303–313.

Kimber, G. & Riley, R. (1963). Haploid angiosperms. *Bot. Rev.* 29:480–531.

Knudsen, S., Due, I.K. & Anderson, S.B. (1989). Components of response in barley anther culture. *Plant Breeding* 103:241–246.

Kostoff, D. (1934). A haploid plant of *Nicotiana sylvestris*. *Nature* 133:949.

Kott, L.S., Polsoni, L., Ellis, B. & Beversdorf, W. (1988). Autotoxicity in isolated microspore cultures of *Brassica napus*. *Can. J. Bot.* 66:1665–1670.

Kott, L.S. & Beversdorf, W.D. (1990). Enhanced plant regeneration from microspore-derived embryos of *Brassica napus* by chilling, partial desiccation and age selection. *Plant Cell Tissue Organ. Cult.* 23:187–192.

Kuhlmann, U., Foroughi-Wehr, B., Garner, A. & Wenzel, G. (1991). Improved culture system for microspore of barley to become a target for DNA uptake. *Plant Breeding* 107:165–168.

Kyo, M. & Harada, H. (1985). Studies on conditions for cell division and embryogenesis in isolated pollen culture in *Nicotiana rustica. Plant Physiol.* 79:90–94.

Kyo, M. & Harada, H. (1986). Control of the developmental pathway of tobacco pollen *in vitro*. *Planta* 168:427–432.

Kyo, M. & Harada, H. (1990). Specific phosphoproteins in the initial period of tobacco pollen embryogenesis. *Planta* 182:58–63.

LaCour, L.F. (1949). Nuclear differentiation in the pollen grain. *Heredity* 3:319–337.

Larsen, E.T., Tuvesson, I.K.D. & Andersen, S.B. (1991). Nuclear genes affecting percentage of green plants in barley (*Hordeum vulgare* L.) anther culture. *Theor. Appl. Genet.* 82:417–420.

Lichter, R. (1982). Induction of haploid plants from isolated pollen of *Brassica napus*. *Z. Pflanzenphysiol.* 105:427–434.

Lichter, R. (1989). Efficient yield of embryoids by culture of isolated microspores of different Brassicaceae species. *Plant Breeding* 103:119–123.

Lichter, R., DeGroat, E., Fiebey, D., Schweiger, R. & Gland, A. (1988). Glucosinolates determined by HPLC in the seeds of microspore-derived homozygous lines of rapeseed (*Brassica napus* L.). *Plant Breeding* 100:209–221.

Liu, C-M., Xu, Z-H. & Chua, N-H. (1993). Auxin polar transport is essential for the establishment of bilateral symmetry during early plant embryogenesis. *Plant Cell* 5:621–630.

Lo, K-H. & Pauls, K.P. (1992). Plant growth environment effects on rapeseed microspore development and culture. *Plant Physiol.* 99:468–472.

Loh, C.S. & Ingram, D.S. (1982). Production of haploid plants from anther cultures and secondary embryoids of winter oilseed rape, *Brassica napus* ssp. *oleifera*. *New Phytol.* 91:507–516.

Maheshwari, S.C., Tyagi, A.K., Malhorta, K. & Sopory, S.K. (1980). Induction of haploidy from pollen grains in angiosperms: the current status. *Theor. Appl. Genet.* 58:193–206.

Maheshwari, S.C., Rashid, A. & Tyagi, A.K. (1982). Haploids from pollen grains: retrospect and prospect. *Amer. J. Bot.* 69:865–879.

Mandel, R.M. (1991). The hormone physiology of microspore-derived embryos of *Brassica napus* cv Topaz. M.Sc. thesis, University of Calgary, Calgary, AB, Canada.

Maquoi, E., Hanke, D.E. & Deltour, R. (1993). The effects of abscisic acid on the maturation of *Brassica napus* somatic embryos: an ultrastructural study. *Protoplasma* 174:147–157.

Marsolais, A.A., Wheatley, W.G. & Kasha, K.J. (1986). Progress in wheat haploid induction using anther culture. *Proceedings of the DSIR Plant Breeding Symp. 1986. Agronomy Society of N.Z. Special Publication* 5:340–343.

Martensson, B. & Widell, S. (1993). Pollen from cold-treated *Nicotiana tabacum* buds: embryogenic capacity, peroxidase activity and partitioning in aqueous two phase systems. *Plant Cell Tissue Organ. Cult.* 35:141–149.

Mascarenhas, J.P. (1992). Pollen gene expression: molecular evidence. *Intl. Rev. Cytol.* 140:3–18.

Mathias, R. & Röbbelen, G. (1991). Effective diploidization of microspore-derived haploids of rape (*Brassica napus* L.) by *in vitro* colchicine treatment. *Plant Breeding* 106:82–84.

Mayer, U., Büttner, G. & Jürgens, G. (1993). Apical–basal pattern formation in the *Arabidopsis* embryo: studies on the role of the gnom gene. *Development* 117:149–162.

McClellan, D., Kott, L., Beversdorf, W.D. & Ellis, B.E. (1993). Glucosinolate metabolism in zygotic and microspore-derived embryos of *Brassica napus* L. *J. Plant Physiol.* 141:153–159.

Meyer, R., Salamini, F. & Uhrig, H. (1993). Isolation and characterization of potato diploid clones generating a high frequency of monohaploids or homozygous diploid androgenetic plants. *Theor. Appl. Genet.* 85:905–912.

Möllers, C., Iqbal, M.C.M. & Röbbelen, G. (1994). Efficient production of doubled haploid *Brassica napus* plants by colchicine treatment of microspores. *Euphytica* 75:95–100.

Mordhorst, A.P. & Lörz, H. (1993). Embryogenesis and development of isolated barley (*Hordeum vulgare* L.) microspores are influenced by the amount and composition of nitrogen sources in culture media. *J. Plant Physiol.* 142:485–492.

Morrison, R.A. & Evans, D.A. (1988). Haploid plants from tissue culture: new varieties in a shortened time frame. *Biotechnol.* 6:684–690.

Mouritzen, P. & Holm, P.B. (1992). Microspore embryogenesis and plant regeneration from anthers of barley cultured through meiosis. *Hereditas* 117:179–188.

Murigneaux, A., Baud, S. & Bechert, M. (1993). Molecular and morphological evaluation of doubled-haploid lines in maize. 2. Comparison with single seed–descent lines. *Theor. Appl. Genet.* 87:278–287.

Neuhaus, G., Spangenberg, G., Scheid, O.M. & Schweiger, G. (1987). Transgenic rapeseed plants obtained by the microinjection of DNA into microspore-derived embryoids. *Theor. Appl. Genet.* 75:30–36.

Niell, S.J., Horgan, R. & Rees, A.F. (1987). Seed development and vivipary in *Zea mays* L. *Planta* 171:358–364.

Oelck, M.M., Phan, C.V., Eckes, P., Donn, G., Rakow, G. & Keller, W.A. (1991). Field resistance of canola transformants (*Brassica napus* L.) to ignite (Phosphinothricin). In *Proceedings of the 8th International Rapeseed Congress, Canola Council of Canada*, ed. D.I. McGregor, pp. 293–297.

Olsen, F.L. (1991). Isolation and cultivation of embryogenic microspores from barley (*Hordeum vulgare* L.). *Hereditas* 115:255–266.

Ormerod, A.J. & Caligari, P.D.S. (1994). Anther and microspore culture of *Lupinus albus* in liquid culture medium. *Plant Cell Tissue Organ. Cult.* 36:227–236.

Osolnik, B., Bohanec, B. & Jelaska, S. (1993). Stimulation of androgenesis in white cabbage (*Brassica oleracea* var *capitata*) anthers by low temperature and anther dissection. *Plant Cell Tissue Organ. Cult.* 32:241–246.

Pan, Q.Y., Seguin-Swartz, G., Downey, R.K. & Rakow, G.F.W. (1991). Number of microspores in immature and mature flower buds in *Brassica* species. In *Proceedings of the 8th International Rapeseed Congress, Canola Council of Canada*, ed. D.I. McGregor, pp. 1836–1839.

Pechan, P.M. (1989). Successful cocultivation of *Brassica napus* microspores and proembryos with *Agrobacterium*. *Plant Cell Reprod.* 8:387–390.

Pechan, P.M., Bartels, D., Brown, D.C.W. & Schell, J. (1991). Messenger RNA and protein changes associated with induction of *Brassica* microspore embryogenesis. *Planta* 184:161–165.

Pedroso, M.C. & Pais, S. (1993). Regeneration from anthers of adult *Camellia japonica* L. *In Vitro Cell Devel. Biol.* 29:155–159.

Pescitelli, S.M., Johnson, C.D. & Petolino, J.F. (1990). Isolated microspore culture of maize: effects of isolation technique, reduced temperature, and sucrose level. *Plant Cell Reprod.* 8:628–630.

Phippen, C. & Ockendon, D.J. (1990). Genotype, plant, bud size and media factors affecting anther culture of cauliflowers (*Brassica oleracea* var botrytis). *Theor. Appl. Genet.* 79:33–38.

Piccirilli, M. & Arcioni, S. (1991). Haploid plants regenerated via anther culture in wild barley (*Hordeum spontaneum* C. Koc). *Plant Cell Reprod.* 10:273–276.

Pickering, R.A. & Devaux, P. (1992). Haploid production: approaches and uses in plant breeding. In *Genetics, Molecular Biology and Biotechnology*, ed. P.R. Shewry, pp. 511–539. Wallingford: Barley CAB International Wallingford.

Pomeroy, M.K., Kramer, J.K.G., Hunt, D.J. & Keller, W.A. (1991). Fatty acid changes during development of zygotic and microspore-derived embryos of *Brassica napus*. *Physiol. Plant.* 81:447–454.

Quatrano, R.S. (1987). The role of hormones during seed development. In *Plant Hormones and Their Role in Plant Growth and Development*, ed. P.J. Davies, pp. 494–514. Dordrecht: Kluwer Acad. Publ.

Raghavan, V. (1990). From microspore to embryoid: faces of the angiosperm pollen grain. In *Progress in Plant Cellular and Molecular Biology*, eds. H.J.J. Nijkamp, L.H.W. Van der Plas, and J. Van Aartrijk, pp. 213–221. London: Kluwer Acad. Publ.

Rahman, M.H. (1993). Microspore-derived embryos of *Brassica napus* L.: stress tolerance and embryo development. Ph.D. thesis, University of Calgary.

Rashid, A. & Reinert, J. (1980). Selection of embryogenic pollen from cold-treated buds of *Nicotiana tabacum* L. var Badischer barley and their development into embryos in culture. *Protoplasma* 105:161–165.

Rashid, A., Siddiqui, A.W. & Reinert, J. (1982). Subcellular aspects of origin and structure of pollen embryos of *Nicotiana*. *Protoplasma* 113:202–208.

Redenbaugh, K. (1993). *Syn Seeds: Applications of Synthetic Seeds to Crop Improvement*. Boca Raton, Ann Arbor, London, Tokyo: CRC Press.

Reynolds, T.L. (1990). Ultrastructure of pollen embryogenesis. In *Biotechnology in Agriculture and Forestry, Vol. 12, Haploids in Crop Improvement I*, ed. Y.P.S. Bajaj, pp. 66–82. Berlin, Heidelberg: Springer-Verlag.

Reynolds, T.L. (1993). A cytological analysis of microspores of *Triticum aestivum* (Poaceae) during normal ontogeny and induced embryogenic development. *Amer. Bot.* 80:569–576.

Reynolds, T.L., & Kitto, S.L. (1992). Identification of embryoid-abundant genes that are temporarily expressed during pollen embryogenesis in wheat anther cultures. *Plant Physiol.* 100:1744–1750.

Rivin, C.J. & Grudt, T. (1991). Abscisic acid and the developmental regulation of embryo storage proteins in maize. *Plant Physiol.* 95:358–365.

Roulund, N., Andersen, S.B. & Forestveit, B. (1991). Optimal concentration of sucrose for head cabbage (*Brassica oleracea* L. convar capitata L. Alef.) anther culture. *Euphytica* 52:125–129.

Sacristan, M.D. (1982). Resistance response to *Phoma lingam* of plants regenerated from selected cells and mutagenized embryogenic culture of haploid *Brassica napus*. *Theor. Appl. Genet.* 61:193–200.

Sangwan, R.S. & Sangwan-Norreel, B.S. (1987). Ultrastructural cytology of plastids in pollen grains of certain androgenic and nonandrogenic plants. *Protoplasma* 138:11–22.

Sangwan, R.S., Ducrocq, C. & Sangwan-Norreel, B. (1993). *Agrobacterium*-mediated transformation of pollen embryos in *Datura innoxia* and *Nicotiana tabacum*: production of transgenic haploid and fertile homozygous dihaploid plants. *Plant Sci.* 95:99–115.

Sarvesh, A., Reddy, T.P. & Kavi-Kishor, P.B. (1993). Embryogenesis and organogenesis in cultured anthers of an oil yielding crop niger (*Guizotia absyssinica* Cass.). *Plant Cell Tissue Organ. Cult.* 35:75–80.

Sax, K. (1935). The effect of temperature on nuclear differentiation in microspore development. *J. Arnold Abor.* 16:301–310.

Schaeffer, G.W., Sharpe, F.T. & Cregan, P.B. (1984). Variation for improved protein and yield from rice anther culture. *Theor. Appl. Genet.* 67:383–389.

Scott, P. & Lyne, R.L. (1994). The effect of different carbohydrate sources upon the initiation of embryogenesis from barley microspores. *Plant Cell Tiss. Organ. Cult.* 36:129–133.

Senaratna, T., McKersie, B.D. & Bowley, S.R. (1990). Artificial seeds of alfalfa, *Medicago sativa* L. induction of desiccation tolerance in somatic embryos. *In Vitro Cell. Devel. Biol.* 26:85–90.

Senaratna, T., Kott, L., Beversdorf, W.D. & McKersie, B.C. (1991). Desiccation of microspore-derived embryos of oilseed rape (*Brassica napus* L.). *Plant Cell Rep.* 10:342–344.

Siebel, J. & Pauls, K.P. (1989a). A comparison of anther and microspore culture as a breeding tool in *Brassica napus*. *Theor. Appl. Genet.* 78:473–479.

Siebel, J. & Pauls, K.P. (1989b). Inheritance patterns of erucic acid content in populations of *Brassica napus* microspore-derived spontaneous diploids. *Theor. Appl. Genet.* 77:489–494.

Siebel, J. & Pauls, K.P. (1989c). Alkenyl glucosinolate levels in androgenic populations of *Brassica napus*. *Plant Breeding* 103:124–132.

Simmonds, D.H., Gervais, C. & Keller, W.A. (1991). Embryogenesis from microspores of embryogenic and non embryogenic lines of *Brassica napus*. In *Proceedings of the 8th International Rapeseed Congress, Canola Council of Canada,* ed. D.I. McGregor, pp. 306–311.

Simonsen, Randal, L., & Baenziger, P.S. (1992). The effect of gelling agents on wheat anther culture and immature embryo culture. *Plant Breeding* 109:211–217.

Singsit, C. & Veilleux, R.E. (1989). Intra and interspecific transmission of androgenetic competence in diploid potato species. *Euphytica* 43:105–112.

Skladal, V., Vyvadilova, M. & Lachovo, J. (1992). Cryopreservation of microspore suspension of *Brassica napus* L. and *Brassica oleracea* L. *Biol. Plant.* 34, Suppl. 548.

Stoehr, M.U. & Zsuffa, L. (1990). Induction of haploids in *Populus maximowiczii* via embryogenic callus. *Plant Cell Tissue Org. Cult.* 23:49–58.

Sunderland, N. & Huang, B. (1987). Ultrastructural aspects of pollen dimorphism. *Intl. Rev. Cytol.* 107:175–220.

Swanson, E.B. & Erickson, L.R. (1989). Haploid transformation in *Brassica napus* using an octopine-producing strain of *Agrobacterium tumefaciens*. *Theor. Appl. Genet.* 78:831–835.

Swanson, E.B., Coumans, M.P., Brown, G.L., Patel, J.D. & Beversdorf, W.D. (1988). The characterization of herbicide tolerant plants in *Brassica napus* L. after *in vitro* selection of microspores and protoplasts. *Plant Cell Reprod.* 7:83–87.

Swanson, E.B., Herrgesell, M.J., Arnoldo, M., Sipell, D.W. & Wang, R.S.C. (1989). Microspore mutagenesis and selection: canola plants with field tolerance to the imidazolinones. *Theor. Appl. Genet.* 78:525–531.

Takacs, I., Kovacs, G. & Barnabás, B. (1994). Analysis of the genotypic effect on different developmental pathways in wheat gametophyte cultures. *Plant Cell Reprod.* 13:227–230.

Takahata, Y. & Keller, W.A. (1991). High frequency embryogenesis and plant regeneration in isolated microspore culture of *Brassica oleracea* L. *Plant Sci.* 74:235–242.

Takahata, Y., Brown, D.C.W. & Keller, W.A. (1991). Effect of donor plant age and inflorescence age on microspore culture of *Brassica napus* L. *Euphytica* 58:51–55.

Takahata, Y., Wakui, K. & Kiazuma, N. (1992). A dry artificial seed system for *Brassica* crops. *Acta Hort.* 319:317–322.

Takahata, Y., Brown, D.C.W., Keller, W.A. & Kiazuma, N. (1993). Dry artificial seeds and desiccation tolerance induction in microspore-derived embryos of broccoli. *Plant Cell Tissue Org. Cult.* 35:121–129.

Tanaka, I. (1993). Development of male gametes in flowering plants. *J. Plant Res.* 106:55–63.

Taylor, D.C., Weber, N., Underhill, E.W., Pomeroy, M.K., Keller, W.A., Scowcroft, W.R., Wilen, K.W., Moloney, M.M. & Holbrook, L.A. (1990). Storage protein regulation and lipid accumulation in microspore embryos of *Brassica napus* L. *Planta* 181:18–26.

Taylor, D.C., Ferrie, A.M.R., Keller, W.A., Giblin, E.M., Pass, E.U. & Mackenzie, S.L. (1993). Bioassembly of acyllipids in microspore-derived embryos of *Brassica campestris*. *Plant Cell Reprod.* 12:375–384.

Telmer, C.A. (1992). Embryogenesis from microspores of *Brassica napus* cv Topas. Ph.D. Thesis, Queens University, Kingston, ON.

Telmer, C.A., Simmonds, D.H. & Newcomb, W. (1992a). Determination of developmental stage to obtain high frequencies of embryogenic microspores in *Brassica napus*. *Physiol. Plant.* 84:417–424.

Telmer, C.A., Newcomb, W. & Simmonds, D.H. (1992b). Microspore development in *Brassica napus* and the effect of high temperature on division *in vivo* and *in vitro*. *Protoplasma* 172:154–165.

Telmer, C.A., Newcomb, W. & Simmonds, D.H. (1993). Microspore development in *Brassica napus* and the effect of high temperature on division *in vivo* and *in vitro*. *Protoplasma* 172:154–165.

Thengane, S.R., Joshi, M.S., Khaspe, S.S. & Mascarenhas, A.F. (1994). Anther culture in *Helianthus annuus* L.: influence of genotype and culture conditions on embryo induction and plant regeneration. *Plant Cell Reprod.* 13:222–226.

Thompson, D.M., Chalmers, K., Waugh, R., Forster, B.P., Thomas, W.T.B., Caligari, P.D.S. & Powell, W. (1991). The inheritance of genetic markers in microspore-derived plants of barley (*Hordeum vulgare* L.). *Theor. Appl. Genet.* 81:492–497.

Tiainen, T. (1992). The role of ethylene and reducing agents on anther culture response of tetraploid potato (*Solanum tuberosum* L.). *Plant Cell Reprod.* 10:604–607.

Trottier, M.C., Collin, J. & Comena, A. (1993). Comparison of media for their aptitude in wheat anther culture. *Plant Cell Tissue Org. Cult.* 35:59–67.

Tsay, A.S., Yeh, C.C. & Hsu, J.Y. (1990). Embryogenesis and plant regeneration from anther culture of bamboo (*Sinocalamus latiflora* [Munro] McClare). *Plant Cell Reprod.* 9: 349-351.

Turner, J., Facciotti, D. (1990). High oleic acid *Brassica napus* from mutagenized microspores. In *Proceedings of the 6th Crucifer Genetics Workshop, USDA-ARS. Geneva NY,* eds. J.R. McFerson, S. Kresovich, and S.G. Dwyer, p. 24.

Tuvesson, I.K.D. & Ohuland, R.C.V. (1993). Plant regeneration through culture of isolated microspores of *Triticum aestivum* L. *Plant Cell Tissue Org. Cult.* 34:162–167.

Vergue, P., Riccardi, F., Beckerf, M. & Dermas, C. (1993). Identification of a 32-KDa anther marker protein for androgenic response in maize, *Zea mays* L. *Theor. Appl. Genet.* 86:843–850.

Voorrips, R.E. & Visser, D.L. (1990). Doubled haploid lines with clubroot resistance in *Brassica oleracea*. In *Proceedings of the 6th Crucifer Genetics Workshop, USDA-ARS Geneva, NY,* eds. J.R. McFerson, S. Kresovich, and S.G. Dwyer, p. 40.

Wiberg, E., Rahlen, L., Hellman, M., Tillberg, E., Glimelius, K. & Stymne, S. (1991). The microspore-derived embryo of *Brassica napus* L. as a tool for studying embryo-specific lipid biogenesis and regulation of oil quality. *Theor. Appl. Genet.* 82:515–520.

Wilen, R.W., Mandel, R.M., Pharis, R.P., Holbrook, L.A. & Moloney, M.M. (1990). Effects of abscisic acid and high osmoticum on storage protein gene expression in microspore embryos of *Brassica napus*. *Plant Physiol.* 94:875–881.

Witherspoon, W.D., Jr., Wernsman, E.A., Gooding, G.V., Jr. & Rufty, R.C. (1991). Characterization of a gameto-clonal variant controlling virus resistance in tobacco. *Theor. Appl. Genet.* 81:1–5.

Yang, Q., Chauvin, J.E. & Herve, Y. (1992). A study of factors affecting anther culture of cauliflower (*Brassica oleracea* var Botrytis). *Plant Cell Tissue Organ. Cult.* 28:289–296.

Zaki, M. & Dickinson, H.G. (1990). Structural changes during the first divisions of embryos resulting from anther and microspore culture in *Brassica napus*. *Protoplasma* 156:149–162.

Zaki, M. & Dickinson, H.G. (1991). Microspore-derived embryos in *Brassica:* the significance of division symmetry in pollen mitosis I to embryogenic development. *Sex. Plant Reprod.* 4:48–55.

Zarsky, V., Garrido, D., Rihova, L., Tupy, J., Vicente, O. & Heberle-Bors, E. (1992). Derepression of the cell cycle by starvation is involved in the induction of tobacco pollen embryogenesis. *Sex. Plant Reprod.* 5:189–194.

Zhang, C. & Qifeng, C. (1993). Genetic studies of rice (*Oryza sativa* L.) anther culture response. *Plant Cell Tissue Organ. Cult.* 34:177–182.

Ziauddin, A., Simion, E. & Kasha, K.J. (1990). Improved plant regeneration from shed microspore culture in Barley (*Hordeum vulgare* L.) cv. Igri. *Plant Cell Reprod.* 9:69–71.

Ziauddin, A., Marsolais, P., Simon, D. & Kasha, K.J. (1992). Improved plant regeneration from wheat anther and barley microspore culture using phenylacetic acid (PAA). *Plant Cell Reprod.* 11:489–498.

Zivy, M., Devaux, P., Blaisonneau, J., Jean, R. & Thiellement, H. (1992). Segregation distortion and linkage studies in microspore-derived doubled haploid lines of *Hordeum vulgare* L. *Theor. Appl. Genet.* 83:919–924.

21

Use of pollen in gene transfer

HIROMICHI MORIKAWA and MASAHIRO NISHIHARA

Summary 423
Significance of pollen transformation for crop improvement 423
Successful gene transfer into pollen by particle bombardment 425
Promoters from anther- and pollen-expressed genes 427
Formation of haploid plants from transformed pollen cells 429
Direct pollination with bombarded pollen grains to obtain transgenic seeds 432
References 434

Summary

We review here the results on gene delivery into pollen grains and transformation by particle bombardment. Most of the studies that have been done up to date on bombardment-mediated transformation of pollen are still confined to transient expression of several pollen- or anther-specific promoters. However, these studies do provide evidence that pollen grains can be transformed at least transiently, and suggest the potential use of these "transformed" pollen grains or those bearing foreign DNA for direct pollination to obtain transgenic seeds via natural reproduction system. The alternative way of using pollen transformation for crop improvement is to develop haploid plants via in vitro culture of bombarded immature pollen grains. Included here are the authors' own recent results on successful production of transgenic haploid plants derived from in vitro culture of bombarded pollen. These results provide a basis for a discussion of the usefulness of pollen transformation for crop improvement.

Significance of pollen transformation for crop improvement

Pollen transformation is an attractive approach for plant breeding and crop improvement. There have been reports of various techniques for gene delivery into pollen or microspores, including imbibition of pollen with DNA (Hess 1980), *Agrobacterium*-mediated transformation (Hess 1987; Pechan 1989), electroporation of pollen (Matthews et al. 1990; Fennell and Hauptmann 1992; Jardinaud et al. 1993), and polyethylene glycol–mediated transformation (Fennell and Hauptmann 1992).

However, all these techniques are problematic in their applicability or reproducibility. Electroporation is useful for introducing foreign DNA into germinating pollen (Matthews et al. 1990), but it is not applicable to plant species in which in vitro pollen germination is difficult. Stoger et al. (1992) did not succeed in transformation of tobacco pollen by the imbibition of DNA solutions or cocultivation with *Agrobacterium tumefaciens*. Hydrated pollen has been reported to have high nuclease activity, which is inhibitory to the introduction of exogenous DNA into pollen cells (Matousek and Tupy 1983; Booy et al. 1989). Isolation of pollen protoplasts is difficult, except in some plant species, including members of Liliaceae (Tanaka et al. 1987, 1989; Zhou 1988). *Agrobacterium* is generally thought not to infect pollen cells. Thus, studies on pollen transformation are far behind those on transformation of somatic cells, and hence more reproducible and reliable techniques for gene transfer into pollen are needed.

Since the pioneering study of Klein et al. (1987), particle bombardment has been shown to be a vital method for gene delivery into intact cells of various plant species. Initially, some workers thought that transformation of pollen grains by this method would be difficult because these cells have a thick and rigid exine except for the germpore region. Twell et al. (1989b) were the first to show that pollen grains of tomato and tobacco can be transformed by particle bombardment. Since then, successful transient transformation by this method has been reported for pollen of various plant species, including dicotyledonous species such as tobacco (Twell et al. 1989b, 1991a; McCormick et al. 1991; Stoger et al. 1992; Nishihara et al. 1993), tomato (Twell et al. 1989b), *Nicotiana glutinosa* (Plegt et al. 1992), *Nicotiana rustica*, and peony (Nishihara et al. 1993), and monocotyledonous species such as maize, *Tradescantia* (Hamilton et al. 1992), and lily (Plegt et al. 1992; Nishihara et al. 1993).

There are two strategic approaches to the use of particle bombardment to achieve pollen transformation for crop improvement: (1) formation of transgenic seeds upon pollination with transformed pollen grains, and (2) formation of haploid plants from transformed pollen grains (Figure 21.1).

Theoretically, the first approach, which utilizes the natural reproductive system, is not species specific and does not need in vitro culture of pollen. This method can be a useful tool for those crop plants that are recalcitrant to somatic cell transformation. Transgenic seeds obtained by transformed pollen have, however, not been reported. The key step for this approach is to find highly efficient conditions of gene delivery into, and gene expression in, the pollen grains. This chapter will first focus on the studies on transient transformation of pollen and then discuss the factors that are essential for the successful production of transformed seeds from direct pollination of bombarded pollen.

The alternative approach is to induce in vitro androgenesis of immature pollen grains that have been bombarded with foreign genes (see Figure 21.1). The authors have obtained transgenic haploid plants for the first time from bombarded immature pollen of *N. rustica* (Nishihara et al. 1995). Plant

Direct pollination with bombarded pollen

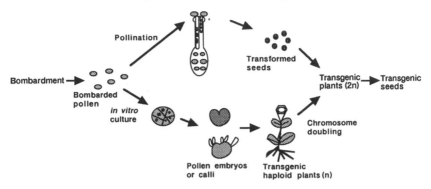

Androgenesis of bombarded pollen

Figure 21.1 Strategies for the use of pollen transformation for crop improvement.

species in which immature pollen can regenerate plants via androgenesis are still limited. But production of haploid plants via androgenesis of bombarded immature pollen is an attractive approach for breeding programs because these plants can easily gain homozygosity by chromosome doubling.

Successful gene transfer into pollen by particle bombardment

Since the report of Twell et al. (1989b), which showed that *LAT52* promoter from tomato (Twell et al. 1989a) can drive expression of the *GUS* gene in tobacco and tomato pollen using particle bombardment, this promoter has been used successfully to drive the expression of foreign genes in the pollen of various species, including tobacco (Twell et al. 1989b, 1991a; McCormick et al. 1991; Nishihara et al. 1993), *Nicotiana glutinosa* (Plegt et al. 1992), *Nicotiana rustica,* and peony (Nishihara et al. 1993). Two other types of anther-specific promoters from tomato (*LAT56* and *LAT59)* and a pollen-specific *PA2* promoter from petunia (Van Tunen et al. 1990) have also been shown to work in the pollen of tobacco (McCormick et al. 1991; Twell et al. 1991a; Stoger et al. 1992). The *Zm13* promoter from maize (Hamilton et al. 1989; Hanson et al. 1989) has also been shown to work in the pollen of maize and *Tradescantia* (Hamilton et al. 1992).

A ß-glucuronidase (*GUS*) gene is a useful reporter gene for studies of gene expression in higher plants because of its low endogenous activity in most types of plant cells (Jefferson et al. 1987). However, pollen cells of various plant species in general have a strong endogenous GUS activity that interferes with detection of the expression of the introduced foreign genes (Plegt and Bino 1989; Alwen et al. 1992).

The *GUS* gene driven by an anther-specific *LAT52* promoter from tomato was reported to be expressed in pollen grains of tobacco and tomato but not of lily (Twell et al. 1989b, 1991b). Plegt et al. (1992) also reported similar results. Nishihara et al. (1993) have presented evidence showing that pollen grains, including those of lily, generally have strong endogenous GUS activity and that this background activity hinders the expression of *LAT52*-driven GUS activity in lily pollen. They have found that the background activity of GUS in these pollen cells can successfully, and almost completely, be suppressed by incubating bombarded pollen samples for less than 12 hours in an enzyme assay solution in the presence of 20% methanol and phosphate buffer of pH 7 (Nishihara et al. 1993). Prior to this study, previous authors had shown that the addition of 20% methanol (Kosugi et al. 1990) and the use of a neutral pH solution (Alwen et al. 1992) are necessary to suppress endogenous GUS activities in various plant cells.

The efficiency of gene delivery and expression in bombardment-mediated transformation in somatic cells is largely influenced by physical and cell physiological factors. Physical parameters, such as type of metal particles (size, shape), accelerating pressure, amount of metal particles per projectile, and the number of shots, affect transient expression frequency of introduced *GUS* gene (Klein et al. 1988). In the case of lily pollen, increasing the accelerating pressure from 84 to 200 kg/cm^2 gave rise to an approximately nine-fold higher GUS expression frequency. Bombardment with gold particles averaging 1.1 mm in diameter (Tokuriki Honten Co., Ltd., Tokyo, Japan) gave approximately a three-fold higher frequency than the particles averaging 2.0 mm in diameter (Alfa Chemical Co., Danvers, Mass.). A change in the amount of gold particles from 0.05 to 0.2 mg/projectile gave approximately a two-fold higher frequency (Nishihara et al., unpublished results). Also, the number of shots greatly influenced the frequency of pollen showing GUS expression.

It has been reported that preculture of sample tissues influences the transient expression of foreign genes in *Arabidopsis* leaf and root tissues (Seki et al. 1991) and in cucumber cotyledons (Kodama et al. 1993). Preculture of pollen also influenced transient GUS activity driven by *LAT52* promoter (Nishihara et al., unpublished results). In *N. rustica*, *LAT52* promoter activity of 3-day-precultured immature pollen in a sucrose-free medium was decreased markedly, that is, 1/150 of that observed in freshly isolated pollen by particle bombardment. The stage of pollen seems to influence the expression efficiency of introduced foreign genes. In the pollen of lily and *N. rustica*, no or very low GUS activities were observed when pollen was bombarded at early binucleate stage, whereas at the late binucleate stage the GUS activity was more than seven-fold that of the early binucleate stage in pollen of both species.

A change in the culture medium from White's modified medium (White 1963) to Murashige and Skoog's medium (Murashige and Skoog 1962) after bombardment did not influence the number of blue spots of GUS-expressing pollen (authors' unpublished results). The effects of other components – such as borate and Ca^{2+}, which are known to stimulate pollen germination and pollen tube growth (Vasil 1987), and osmotic stress, which has recently been shown to influence the efficiency of bombardment-mediated transformation (Armaleo et al. 1990; Vain et al. 1993) – are subjects of future study.

In general, pollen grains have starch granules and a thick exine, which hamper the direct observation of gold particles introduced into the cells. However, as reported by Nishihara et al. (1993), when bombarded pollen grains of lily, after assayed for GUS expression, were fixed in acetic ethanol, hydrolyzed with 1N HCl (to remove the exine and starch granules), and stained with orcein, introduced gold particles became microscopically visible. Gold particles were detected (as black dots) in the vegetative cytoplasm, vegetative nucleus, and generative cytoplasm (Nishihara et al. 1993). Nuclear delivery of DNA-coated gold particles is a prerequisite for successful expression of the introduced foreign gene in cultured tobacco cells (Yamashita et al. 1991). The introduction and expression of the introduced DNA generally seems to be restricted to the vegetative nucleus.

Promoters from anther- and pollen-expressed genes

A number of pollen-expressed genes have been cloned (Mascarenhas 1990, 1992; McCormick 1991; also see Chapter 3), and their 5' upstream regulatory sequences have been isolated and studied using transient expression assay by particle bombardment and transgenic plants obtained by the *Agrobacterium* method. Of the sequences that have been isolated, the *LAT* promoters from tomato anther-expressed genes and the *Zm13* promoter from maize pollen-expressed gene are the most extensively studied.

Interestingly, results obtained by gene expression assay of tobacco pollen that was bombarded with constructs containing deleted upstream sequences of various lengths connected to the reporter (*GUS*) gene are in good agreement with those obtained by the analysis of pollen of transgenic tobacco that contains the same constructs in the genome. Both sets of results indicate that the minimal proximal sequences of the regulatory region of tomato *LAT52* and *LAT59* promoters required for correct temporal and spatial expression during pollen development are approximately 200 bp (–71 to +110 for *LAT52*, and –115 to +91 for *LAT59*) (McCormick et al. 1991; Twell et al. 1991a).

The upstream region of a pollen-expressed gene (*Zm13*) from maize (Hamilton et al. 1989; Hanson et al. 1989) has been characterized by transient expression assay using *Tradescantia paludosa* pollen by particle bombardment (Hamilton et al. 1992). Hamilton and co-workers have shown that sequences necessary for pollen-specific expression are present in a region from –100 to –54 and that those between –260 to –100 are necessary for amplification of the pollen-specific expression of the gene. This result agrees with that obtained from the analysis of pollen of transgenic tobacco; the latter analysis shows that 375-bp sequences between –314 to +61 are necessary for correct temporal and spatial expression of the gene (Guerrero et al. 1990). The results on transient gene expression indicate that particle bombardment is useful for rapid assay for the activity of deletion mutants and to determine essential cis elements in 5' upstream sequences. Tables 21.1 and 21.2 summarize GUS enzyme activities detected by histochemical and fluorometric assay, respectively, in pollen of lily, tulip, tobacco, and N. rustica that were bombarded with three different promoter constructs in which the GUS gene is driven by Zm13 (plasmid –260-GUS), LAT52 (pLAT52-7), and CaMV35S (pBI221) promoters. A single shot was given to each of the target pollen samples. The number of GUS-expressing cells per target sample of pollen highly depends on the type of promoter. In the case of lily pollen, more than 30% of the target pollen sample received gold particle(s) after a single shot; more than 5% of the cells bombarded with –260-GUS expressed the GUS gene, but only 0.6% of the cells bombarded with pLAT52-7 expressed this gene (Table 21.1). Essentially similar results were obtained with tulip pollen. Thus, in lily and tulip pollen, the Zm13 promoter from monocot maize is much more active than the LAT52 one from tomato. Bombardment with pLAT52-7 resulted in an almost two-fold higher number of GUS-expressing cells in mature pollen of tobacco and N. rustica than the number produced with –260-GUS. Thus, this suggests that in tobacco and N. rustica pollen, the LAT52 promoter is more active than the Zm13 one. Accordingly, it is likely that Zm13 and LAT52 promoters have cis-acting elements in their DNA sequences that determine monocot- and dicot-pollen–specific gene expression (McCormick 1991; Mascarenhas 1992). Similar differences in cis-acting elements between monocots and dicots have been reported in rice cab gene promoter with somatic cells (Luan and Bogorad 1992).

It has been reported that the expression of the GUS gene driven by the CaMV35S promoter is histochemically not detectable in pollen grains of lily (Plegt et al. 1992) and N. glutinosa (Plegt et al. 1992) and only rarely detectable in those of tobacco (Twell et al. 1989b; Stoger et al. 1992) and Tradescantia (Hamilton et al. 1992). However, as reported previously (Nishihara et al. 1993), the CaMV35S promoter does induce expression of the GUS gene in pollen and its activity in pollen seems to depend somewhat upon plant species (Table 21.1).

Fluorometric assay showed that –260-GUS gave about 400–500-fold higher GUS activity than did pLAT52-7 in pollen of monocots, whereas pLAT52-7 gave about 15–50-fold higher activity than did –260-GUS in pollen of dicots (Table 21.2). This is in line with the results obtained by histochemical assay of the expression of this gene as shown in Table 21.1, but the difference in the GUS expression of each promoter between monocot and dicot pollen is greater as measured by fluorometric assay compared to histochemical assay. It is conceivable that fluorometric assay reflects the enzyme activity more directly (hence is more "sensitive") than the histochemical assay.

Addition of 20% methanol to the X-gluc solution for GUS enzyme activity assay greatly decreased background endogenous GUS activity (Nishihara et al. 1993). As shown in Table 21.2, addition of the same in the fluorometric enzyme activity assay solution also greatly decreased the background level of GUS activity in the extract from pollen grains that had been bombarded with non-

Table 21.1. *GUS expression frequency in pollen of four plant species bombarded with three different promoter constructs*

Species	Stage	GUS-expressing cells (% of total)		
		-260-GUS	*pLAT52-7*	*pBI221*
Lilium longiflorum	Mature	5.2 ± 1.2a	0.61 ± 0.09a	0.14 ± 0.07a
Tulipa gesneriana	Mature	4.1 ± 1.9	0.10 ± 0.03	0.34 ± 0.08
Nicotiana tabacum	Mature	0.17 ± 0.03	0.34 ± 0.04	0.07 ± 0.02
Nicotiana rustica	Mature	0.33 ± 0.08	0.56 ± 0.11	0.10 ± 0.03
	Immature	0.25 ± 0.09	0.42 ± 0.10	0.13 ± 0.02

a = average of three to five experiments ± standard deviation.
Bombarded pollen was cultured for 24 hours and assayed histochemically (Nishihara et al. 1993). The number of pollen cells in a target sample was 4×10^4 (lily), 10^5 (tulip), and 2×10^5 (tobacco and *N. rustica*).

coated gold particles. The presence or absence of methanol in the enzyme assay solution for fluorometry did not greatly affect the GUS activities detected in the extracts from pollen grains bombarded with DNA-coated gold particles, and the values obtained by assay in the presence or absence of methanol was within experimental error in each of the plasmid/pollen combinations (see Table 21.2).

The *CaMV35S* promoter seemed to have, in general, lower activity than the other promoters tested in pollen grains, but, interestingly, tulip pollen grains are exceptional, and the activity observed in mature pollen grains of tulip was more than 10-fold that of the other plant species. The differences in GUS activities among different constructs (driven by different promoters) were smaller in immature than in mature pollen of *N. rustica* (see Table 21.2), suggesting that these promoters become more specific to pollen during their development.

Formation of haploid plants from transformed pollen cells

There are reports of hundreds of plant species that can produce haploid plants from anther and/or isolated pollen culture (for example, Vasil 1980; Maheshwari et al. 1982; Prakash and Giles 1987; also see Chapter 20). Various means of pollen transformation have been studied. More recently, production of transgenic haploid plants has been reported by electroporation of protoplasts isolated from microspore-derived cultures of maize (Sukhapinda et al. 1993). To our knowledge, however, transgenic haploid plants, in which integration of foreign genes was confirmed by molecular evidence, produced from in vitro androgenesis of transformed pollen cells have not been reported (Heberle-Bors 1991).

Table 21.2. *GUS activity in pollen of four plant species after bombardment with three different promoter constructs*

Species	Stage	Methanol 20%	GUS activity (pmole 4-MUa/h/ug protein)			
			-260-GUS	pLAT52-7	pBI221	Control (Au)
Lilium longiflorum	Mature	+	642.0 ±159.7b	1.31 ± 0.55b	0.36 ± 0.01b	0.01 ± 0.01b
		–	740.1 ± 287.6	1.37 ± 0.55	0.36 ± 0.02	0.10 ± 0.01
Tulipa gesneriana	Mature	+	837.6 ±116.7	2.24 ± 0.87	89.6 ± 21.6	0.24 ± 0.24
		–	868.1 ±136.8	2.88 ± 0.73	82.5 ± 21.6	0.80 ± 0.17
Nicotiana tabacum	Mature	+	61.6 ± 10.4	3222 ± 793	0.99 ± 0.43	~0
		–	60.5 ± 12.7	3164 ± 849	0.99 ± 0.43	0.10 ± 0.09
Nicotiana rustica	Mature	+	120.8 ± 52.7	1810 ± 887	0.77 ± 0.18	0.02 ± 0.01
		–	136.6 ± 67.4	1799 ± 735	0.78 ± 0.34	0.16 ± 0.06
	Immature	+	215.5 ± 19.1	702.4 ± 139.2	12.2 ± 2.2	0.05 ± 0.03
		–	178.7 ± 26.3	640.6 ± 129.6	13.2 ± 4.2	0.09 ± 0.02

a = 4-methyl umbelliferone.
b = average of three experiments ± standard deviation.
Bombarded pollen was cultured for 24 hours and assayed fluorometrically for GUS enzyme activity.
The enzyme assay solution contained (+) or did not contain (-) 20% methanol. Control (Au)
corresponds to the results of pollen cells bombarded with noncoated gold particles.

Figure 21.2 A transgenic haploid plantlet of *Nicotiana rustica* derived from bombarded immature pollen. This plantlet is about 3 months (after bombardment) and it grew well on the medium containing 200 ppm kanamycin. After transfer to a pot, it developed into a mature plant, about 5 months after bombardment.

Figure 21.2 is an example of transgenic haploid plantlets of *Nicotiana rustica* that were obtained from the culture of bombarded immature pollen (Nishihara et al. 1995). After culture in a starvation medium for 3 days (Kyo and Harada 1985), immature pollen of *Nicotiana rustica* was bombarded with gold particles coated with plasmid DNA bearing neomycin phosphotransferase II (*NPTII*) and *GUS* genes. Both these genes are under the control of the cauliflower mosaic virus *CaMV35S* promoter and nopaline synthase *NOS* terminator in the plasmid.

Pollen embryoids derived from bombarded immature pollen cells were selected for kanamycin resistance, and some independent lines of transgenic plantlets were obtained. Transgenic plants had 24 chromosomes, confirming their haploidy (Narajan and Rees 1974). They had aborted pollen grains at the mature stage in their anthers, and the fluorochromatic reaction (FCR) test indicated greatly reduced viability of pollen grains. These results indicate that transgenic haploid plants can be produced from the culture of bombarded immature pollen. Integration of *NPTII* and *GUS* genes and their expression were confirmed respectively by Southern blot analysis and NPTII and GUS enzyme assays.

Transgenic haploid plants will be useful for fundamental genetic studies and for breeding programs of higher plants. They will be used to uncover recessive traits, and also to achieve rapid homozygosity for introduced foreign genes to circumvent segregation problems. They also will be used in gene tagging and mutation studies because of the advantage of having a single genome. Furthermore, in vitro culture techniques for haploid production will not only shorten the breeding cycles but also recover useful types of gene combinations and invaluable materials for plant breeding. Genetic stability of these transgenic haploid plants is a subject of future study.

Direct pollination with bombarded pollen grains to obtain transgenic seeds

Direct pollination with the bombarded pollen to obtain transgenic seeds via the process of sexual reproduction is a potential area for transformation and crop improvement. However, to the authors' knowledge, no reports of this have been published. In plant species that show maternal inheritance of organelle DNA, only genomes in the nucleus of sperm cells formed by division of generative cell are transferred into egg cells upon pollination (Tilney-Bassett 1978; Van Went 1992). Thus, in this case, introduction of foreign genes into the generative or sperm nucleus by particle bombardment is an essential prerequisite for successful use of bombarded pollen for the production of transgenic seeds by direct pollination. At the binucleate stage, the generative cell is surrounded by a cell wall in the pollen cells, and hence the DNA-coated gold particles upon bombardment must penetrate two wall layers (that is, pollen and generative cell) in order to reach the generative nucleus.

We have evidence that gold particle(s) can be successfully introduced into the generative cell cytoplasm (Nishihara et al., unpublished results); some gold particles launched by a particle gun device have enough energy to reach the generative cytoplasm. The frequency of the delivery of gold particles into the generative cell under present conditions is still low (about 10^{-3}–10^{-4} of total) and the improvement of this frequency is a subject of future study. Also, whether DNA introduced into the generative nucleus is functional will need to be studied.

In plant species that show maternal inheritance of organelle DNA, usually only the sperm nuclei carry the paternal DNA to the embryo sac. However, foreign DNA in the pollen tube is also thought to be conveyed into the embryo sac (Tilney-Bassett 1978; Van Went 1992). Thus, foreign genes introduced into the vegetative cytoplasm may also be transferred into egg cells. As shown in Figure 21.3, multiple shots of pollen grains result in the introduction of gold particle(s) into all of the pollen cells. It is possible that all of these cells have DNA in their cellular compartment(s). Pollination using pollen grains after being subjected to multiple shots is an interesting experiment for future study. Note that not all cells expressed the introduced *GUS* gene; at the most, about 30% of the total bombarded cells expressed the gene. The reasons for this are not yet understood.

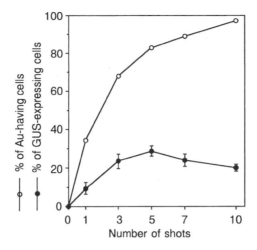

Figure 21.3 Effects of multiple shots on the frequencies of gene delivery and gene expression in mature pollen of lily. About 5×10^4 cells were bombarded with the –260-*GUS* construct (Hamilton et al. 1992) and cultured for 24 hours, after which they were assayed histochemically for GUS expression as described in the text (Nishihara et al. 1993). The number of gold-particle-containing and GUS-expressing cells was determined microscopically as described elsewhere (Nishihara et al. 1993). More than 200 and more than 500 cells, respectively, were counted to determine the frequencies of gold-particle-containing and GUS-expressing cells. The vertical bar indicates standard deviation of three experiments.

As shown previously (Nishihara et al. 1993), lily pollen grains frequently expressed the introduced foreign *GUS* gene even when they had more than one hundred gold particles. This is in contrast to the case of cultured tobacco cells (50–100 mm in diameter), in which introduction of more than three gold particles per cell seemed to be lethal to the cell. Pollen cells in general are rich in cytoplasm and practically lack a vacuole, whereas cultured tobacco cells have a large vacuole. Thus it is likely that the absence of vacuoles may be a cause for the high tolerance of pollen to multiple shots of gold particles.

Because the nucleus of a microspore at the uninucleate stage is surrounded by the nuclear envelope, bombardment at this stage has a much higher probability of delivering foreign genes into the microspore nucleus. In addition, integration of foreign DNA into the microspore genome is also expected, because DNA synthesis occurs prior to microspore mitosis. Successful in vitro culture of microspores at uninucleate stage has been reported in lily (Tanaka and Ito 1980, 1981a), tulip (Tanaka and Ito 1980, 1981b), and tobacco (Benito Moreno et al. 1988). Thus, use of in vitro cultured pollen that has been bombarded at the uninucleate stage for direct pollination is an interesting subject for future study on the production of transgenic seeds.

There have been studies showing that bombarded pollen grains retain their germination ability in tobacco (Twell et al. 1989b; Stoger et al. 1992), *Tradescantia* (Hamilton et al. 1992), and in lily (authors' unpublished results),

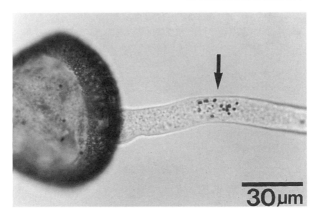

Figure 21.4 Germination of a mature pollen grain that was bombarded with DNA-coated gold particles. Bombarded pollen grains were cultured for 24 hours in White's modified medium at 26°C in the dark. More than 10 gold particles (seen as black dots indicated by an arrow) were detected in the pollen tube.

in which gold particles were observed in the pollen tube (Figure 21.4). The authors' preliminary experiments showed that lily plants that were pollinated with bombarded pollen set seeds normally. In conclusion, these results strongly suggest the possibility of the production of transgenic seeds by direct pollination of bombarded pollen. The authors are currently continuing the study on direct pollination of lily as described.

Acknowledgments

This research was supported by a grant from Iketani Science and Technology Foundation, by CIBA-GEIGY Foundation (Japan) for the Promotion of Science, and by Grants-in-Aid from the Ministry of Education, Science, and Culture of Japan.

References

Alwen, A., Benito Moreno, R.M., Vicente, O. & Heberle-Bors, E. (1992). Plant endogenous ß-glucuronidase activity: how to avoid interference with the use of the *E. coli* b-glucuronidase as a reporter gene in transgenic plants. *Transgenic Res*. 1:63–70.
Armaleo, D., Ye, G.N., Klein, T.M., Shark, K.B., Sanford, J.C. & Johnston, S.A. (1990). Biolistic nuclear transformation of *Saccharomyces cerevisiae* and other fungi. *Curr. Genet*. 17:97–103.
Benito Moreno, R.M., Macke, F., Alwen, A. & Heberle-Bors, E. (1988). *In situ* seed production after pollination with in vitro matured, isolated pollen. *Planta* 176:145–148.
Booy, G., Krens, F.A. & Huizing, H.J. (1989). Attempted pollen-mediated transformation of maize. *J. Plant Physiol*. 135:319–324.

Fennell, A. & Hauptmann, R. (1992). Electroporation and PEG delivery of DNA into maize microspores. *Plant Cell Reprod.* 11:567–570.

Guerrero, F.D., Crossland, L., Smutzer, G.S., Hamilton, D.A. & Mascarenhas, J.P. (1990). Promoter sequences from a maize pollen-specific gene direct tissue-specific transcription in tobacco. *Mol. Gen. Genet.* 224:161–168.

Hamilton, D.A., Bashe, D.M., Stinson, J.R. & Mascarenhas, J.P. (1989). Characterization of a pollen-specific genomic clone from maize. *Sex. Plant Reprod.* 2:208–212.

Hamilton, D.A., Roy, M., Rueda, J., Sindhu, R.K., Sanford, J. & Mascarenhas, J.P. (1992). Dissection of a pollen-specific promoter from maize by transient transformation assays. *Plant Mol. Biol.* 18:211–218.

Hanson, D.D., Hamilton, D.A., Travis, J.L., Bashe, D.M. & Mascarenhas, J.P. (1989). Characterization of a pollen-specific cDNA clone from *Zea mays* and its expression. *Plant Cell* 1:173–179..

Heberle-Bors, E. (1991) Germ line transformation in higher plants. *IAPTC Newslett.* 64:2–10.

Hess, D. (1980). Investigations on the intra- and interspecific transfer of anthocyanin genes using pollen as vectors. *J. Plant Physiol.* 98:321–337.

Hess, D. (1987). Pollen based techniques in genetic manipulation. *Int. Rev. Cytol.* 107:367–395.

Jardinaud, M.F., Souvre, A. & Alibert, G. (1993). Transient GUS gene expression in *Brassica napus* electroporated microspores. *Plant Sci.* 93:177–184.

Jefferson, R.A., Kavanagh, T.A. & Bevan, M.W. (1987). GUS fusions: ß-glucuronidase as a sensitive and versatile gene fusion marker in higher plants. *EMBO J.* 6:3901–3907.

Klein, T.M., Wolf, E.D., Wu, R. & Sanford, J.C. (1987). High velocity microprojectiles for delivering nucleic acids into living cells. *Nature* 327:70–73.

Klein, T.M., Gradziel, T., Fromm, M.E. & Sanford, J.C. (1988). Factors influencing gene delivery into *Zea mays* cells by high-velocity microprojectiles. *Bio/Technology* 6:559–563.

Kodama, H., Irifune, K., Kamada, H. & Morikawa, H. (1993). Transgenic roots produced by introducing Ri-rol genes into cucumber cotyledons by particle bombardment. *Transgenic Res.* 2:147–152.

Kosugi, S., Ohashi, Y., Nakajima, K. & Arai, Y. (1990). An improved assay for b-glucuronidase in transformed cells: methanol almost completely suppresses a putative endogenous b-glucuronidase activity. *Plant Sci.* 70:133–140.

Kyo, M. & Harada, H. (1985). Studies on conditions for cell division and embryogenesis in isolated pollen culture of *Nicotiana rustica*. *Plant Physiol.* 79:90–94.

Luan, S. & Bogorad, L. (1992). A rice cab gene promoter contains separate cis-acting elements that regulate expression in dicot and monocot plants. *Plant Cell* 4:971–981.

Maheshwari, S.C., Rashid, A. & Tyagi, A.K. (1982). Haploids from pollen grains: retrospect and prospect. *Amer. J. Bot.* 69:865–879.

Mascarenhas, J.P. (1990). Gene activity during pollen development. *Ann. Rev. Plant Physiol. Plant Mol. Biol.* 41:317–338.

Mascarenhas, J.P. (1992). Pollen gene expression: molecular evidence. *Int. Rev. Cytol.* 140:3–18.

Matousek, J. & Tupy, J. (1983.) The release of nucleases from tobacco pollen. *Plant Sci. Lett.* 30:83–89.

Matthews, B.F., Abdul-Baki, A.A. & Saunders, J.A. (1990). Expression of a foreign gene in electroporated pollen grains of tobacco. *Sex. Plant Reprod.* 3:147–151.

McCormick, S. (1991). Molecular analysis of male gametogenesis in plants. *Trends Genet.* 7:298–303.

McCormick, S., Yamaguchi, J. & Twell, D. (1991). Deletion analysis of pollen-expressed promoters. *In Vitro Cell. Dev. Biol.* 27:15–20.

Murashige, T. & Skoog, F. (1962). A revised medium for rapid growth and bioassays with tobacco tissue cultures. *Physiol. Plant.* 15:473–497.

Narajan, R.K. & Rees, H. (1974). Nuclear DNA, heterochromatin and phylogeny in *Nicotiana amphiploids. Chromosoma (Berl.)* 47:75–83.

Nishihara, M., Ito, M., Tanaka, I., Kyo, M., Ono, K., Irifune, K. & Morikawa H. (1993). Expression of the b-glucuronidase gene in pollen of lily (*Lilium longiflorum*), tobacco (*Nicotiana tabacum*), *Nicotiana rustica*, and peony (*Paeonia lactiflora*) by particle bombardment. *Plant Physiol.* 102:357–361.

Nishihara, M., Seki, M., Kyo, M., Irifune, K. & Morikawa, H. (1995). Transgenic haploid plants of *Nicotiana rustica* produced by bombardment-mediated transformation of pollen. *Transgenic Res.* 4:341–348.

Pechan, P.M. (1989). Successful cocultivation of *Brassica napus* microspores and proembryos with *Agrobacterium. Plant Cell Reprod.* 8:387–390.

Plegt, L. & Bino, R.J. (1989). b-glucuronidase activity during development of the male gametophyte from transgenic and non-transgenic plants. *Mol. Gen. Genet.* 216:321–327.

Plegt, L.M., Ven, B.C.E., Bino, R.J., Salm, T.P.M. & Tunen, A.J. (1992). Introduction and differential use of various promoters in pollen grains of *Nicotiana glutinosa* and *Lilium longiflorum. Plant Cell Reprod.* 11:20–24.

Prakash, J. & Giles, K.L. (1987). Induction and growth of androgenic haploids. *Int. Rev. Cytol.* 107:273–292.

Seki, M., Komeda, Y., Iida, A., Yamada, Y. & Morikawa, H. (1991). Transient expression of b-glucuronidase in *Arabidopsis thaliana* leaves and roots and *Brassica napus* stems using a pneumatic particle gun. *Plant Mol. Biol.* 17:259–263.

Stoger, E., Benito Moreno, R.M., Ylstra, B., Vicente, O. & Heberle-Bors, E. (1992). Comparison of different techniques for gene transfer into mature and immature tobacco pollen. *Transgenic Res.* 1:71–78.

Sukhapinda, K., Kozuch, M.E., Rubin-Wilson, B., Ainley, W.M. & Merlo, D.J. (1993). Transformation of maize (*Zea mays* L.) protoplasts and regeneration of haploid transgenic plants. *Plant Cell Rep.* 13:63–68.

Tanaka, I. & Ito, M. (1980). Induction of typical cell division in isolated microspores of *Lilium longiflorum* and *Tulipa gesneriana. Plant Sci. Lett.* 17:279–285.

Tanaka, I. & Ito, M. (1981a). Studies on microspore development in liliaceous plants. III. Pollen tube development in lily pollens cultured from the uninucleate microspore stage. *Plant Cell Physiol.* 22:149–153.

Tanaka, I. & Ito, M. (1981b). Control of division patterns in explanted microspores of *Tulipa gesneriana. Protoplasma* 108:329–340.

Tanaka, I., Kitazume, C. & Ito, M. (1987). The isolation and culture of lily pollen protoplasts. *Plant Sci.* 50:205–211.

Tanaka, I., Nakamura, S. & Miki-Hirosige, H. (1989). Structural features of isolated generative cells and their protoplasts from pollen of some liliaceous plants. *Gamete Res.* 24:361–374.

Tilney-Bassett, R.A.E. (1978). The inheritance and genetic behavior of plastids. In *The Plastids*, eds. J.T.O. Kirk and R.A.E. Tilney-Bassett, pp. 251–524. Amsterdam: Elsevier.

Twell, D., Wing, R., Yamaguchi, J. & McCormick, S. (1989a). Isolation and expression of an anther-specific gene from tomato. *Mol. Gen. Genet.* 217:240–245.

Twell, D., Klein, T.M., Fromm, M.E. & McCormick, S. (1989b). Transient expression of chimeric genes delivered into pollen by microprojectile bombardment. *Plant Physiol.* 91:1270–1274.

Twell, D., Yamaguchi, J., Wing, R.A., Ushiba, J. & McCormick, S. (1991a). Promoter analysis of genes that are coordinately expressed during pollen development reveals pollen-specific enhancer sequences and shared regulatory elements. *Genes Dev.* 5:496–507.

Twell, D., Klein, T.M. & McCormick, S. (1991b). Transformation of pollen by particle bombardment. In *Plant Tissue Culture Manual,* ed. K. Lindsey, pp. D2: 1–14. Dordrecht: Kluwer Acad. Publ.

Vain, P., McMullen, M.D. & Finer, J.J. (1993). Osmotic treatment enhances particle bombardment–mediated transient and stable transformation of maize. *Plant Cell Reprod.* 12:84–88.

Van Tunen, A.S., Mur, L.A., Rienstra, J.D., Koes, R.E. & MÒl, J.N.M. (1990). Pollen- and anther-specific chiA and B promoters from *Petunia hybrida*; tandem promoter regulation of chiA gene expression. *Plant Cell* 2:393–401.

Van Went, J.L. (1992). Pollen tube entrance in the embryo sac and fertilization. In *Sexual Plant Reproduction,* eds. M. Cresti and A. Tiezzi, pp. 135–141. Berlin, Heidelberg, New York: Springer-Verlag.

Vasil, I.K. (1980). Androgenetic haploids. *Int. Rev. Cytol. Suppl.* 11A:195–223.

Vasil, I.K. (1987). Physiology and culture of pollen. *Int. Rev. Cytol.* 107:127–174.

White, P.R. (1963). *The Cultivation of Animal and Plant Cells*, 2nd ed. New York: Ronald Press.

Yamashita, T., Iida, A. & Morikawa, H. (1991). Evidence that more than 90% of b-glucuronidase–expressing cells after particle bombardment directly receive the foreign gene in their nucleus. *Plant Physiol.* 97:829–831.

Zhou, C. (1988). A study on isolation and culture of pollen protoplasts. *Plant Sci.* 59:101–109.

Index

abscisic acid, 191, 192, 222, 223, 328, 404–5, 409
abscisic acid mutants, 405
Acacia, 29, 68
Aconitum, 97–8
acriflavin, 278
actin, 45, 47, 366
Actinidia spp., 71, 106, 139
adenine, 190
adenine phosphoribosyltransferase, 190
Adh-1 gene, 43, 52, 53
advertisements, 63, 64
Aesculus, 65
Agalinis, 108
agamospermy, 70, 72
Agrobacterium, 408, 423–4
agrochemicals, 5
agronomic practices, 134
Alachlor, 346
alfalfa, 126
alkali bee, 124
allergens, 44
Allium cepa, 134, 158, 162, 166, 171, 281
    CMS S, 162, 166
allogamy, 60
allophilic, 62, 69
alloplasmics, 7
allopolyploids, 285
allotetraploids, 285
allotropic, 63
Alstroemeria, 279, 280–1, 285–6
Althaea, 94
Amaryllis, 31, 285
Amb a I gene, 48
Amb a II, II-4 genes, 44, 48
Ambrosia artemisifolia, 47
amino-n-caproic acid, 278
aminocyclopropane carboxylate, 225, 228
AMP, 190
amylases, 190, 297
Andrena, 91, 94, 99, 104
androdioecy, 72
androgenesis, 392, 424–5; *see also* pollen
    embryos

andromonoecy, 71
anemophily, 62, 73, 88
aneuploids, 406
Anigozanthos, 285
aniline blue fluorescence, 263
anther, 97, 239
anther culture, 232–3, 393, 395–6
anther dehiscence, 21
anther differentiation, 16
anther walls, 396, 398–9
anther-specific cDNA, 239
anther-specific mRNA, 239
anther-specific promoter, 240, 251, 425, 427
anthophiles, 68
antipodal cells, 26, 365
Antirrhinum, 127, 212, 302
antisense, 243
apetalous male-sterile mutant, 191
Apis mellifera (or *Apis*), 89–90, 93–7, 99, 101–4, 124–5, 140; *see also* honeybee
Apoidea, 61
apple, 319–21, 325, 345
Arabidopsis thaliana (or *Arabidopsis*), 44, 45, 47, 49–50, 53, 190, 208, 249, 337, 340–1, 426
Arachis, 285
Aralia, 98
archesporial cells, 16
Armenica vulgaris, 301
arsenates, 224
artificial seeds, 409–10
Asclepias, 91, 353
ascorbate oxidase, 44, 50
Asimina, 71
Asparagus officinalis, 71
Aspergillus oryzae, 202, 240
atp6 gene, 163
Atta, 102
autogamy, 60
autonomous replicating sequences (ARS), 175
auxins, 222, 277, 399, 404
Averrhoa carambola, 71
azetidine-3-carboxylate, 231, 232

Bacillus amyloliquefaciens, 240, 241, 243
Banksia, 94
barley, 7, 187, 192, 248, 340, 372, 379, 398;
    see also *Hordeum vulgare*
barnase, 240–1, 243–4, 246
barstar, 241, 243, 245
bee colony, 125–6, 131–2, 136
bee color vision, 65
bee emigration, 125
bee pollination, 94–6
bee, polytrophic, 127
bee, population management, 124
bee visitation, 127
beetles, 61–2
benzoic acid analogues, 231
benzotriazole, 225
6-benzylaminopurine, 281
Berberis, 98
Bet v I gene, 48
Bgp1 gene, 52
biotechnological approaches, 7
Bipolaris maydis, 173
Bipolaris maydis Race T, 251
birch pollen, 48
bird pollination, 88
bisexual inflorescences, 70
Bombus, 65, 89–90, 92, 94–101, 104, 110,
    123–5, 136–7; see also bumblebee
Bp4 gene, 45
Bp10/19 genes, 44, 53
Brassica, 4, 7, 8, 102, 108, 209, 268, 279,
    281–2, 285, 398
Brassica campestris, 43, 52, 124, 207–8,
    321, 400
Brassica campestris × *B. oleraceae*, 321
Brassica chinensis, 282
Brassica napus, 44, 45, 47, 65, 67, 71, 124,
    140, 159, 160, 170, 185, 187, 191,
    395, 400–5, 408
    CMS *ogu*, 160, 163, 170, 191
    CMS *pol*, 159–60, 163, 170
    CMS *nap*, 159–60, 163
Brassica napus × *B. campestris*, 282
Brassica napus × *Diplotaxis tenuifolia*, 282
Brassica napus × *Moricandia arvensis*, 282
Brassica oleracea, 186, 206–8, 397, 407
Brassica oleracea × *Diplotaxis tenuifolia*, 282
Brassica pikenensis, 282
Brassica rapa, 407
breeding systems, 59, 70
Brewbaker–Kwack medium, 358
bumblebee, 3, 68, 79, 91–2, 95, 124–5, 137
bumblebee foragers, 136
buzz-pollination, 124

calcium, 381–2
callase, 19, 190

callose, 17, 18, 26, 267, 341
Camellia japonica, 300–1, 398
Campsis, 106, 108
CaMV35S promoter, 428–9, 431
canal cells, 23, 25, 32
cantherophily, 62, 69
Capsella, 280
Capsicum annuum, 223, 354
Cardamine, 94
carpel, closed conduplicate, 69
Cassia, 97–8, 108
cell cycle, 402
central cell, 26–7, 33, 365–7, 385
central cell protoplast, 379–80, 382, 384
Centris, 97–8
Ceratina, 98
chalcone synthase, 247, 337, 340
chalcone synthase mutants, 340
chemical hybridization, 220
chemical hybridizing agents (CHAs), 4,
    219–22, 224, 227, 229, 231–3, 250–1
    Genesis, 229, 231
    MON 21200, 229, 231
    phytotoxicity of, 222
    SC-1058, 226–8
    SC-1271, 226–30
    SC-2053, 226–8
chemical mutagens, 188
chemical thinners, 133
chemotropic substance, 34
chiA gene, 44
chickpea, 28
chimeric genes, 240
chiropterophily, 62, 69
chitinase promoter, 202
Chlorella zofingiensis, 232
2-chloroethyl-trimethyl ammonium chloride,
    223
chloroplast, 250
chlorsulfuron, 347
chromosome doubling, 285
chromosome pairing, 285
chromosome walking, 249
Chrysanthemum, 285
Chrysopsis, 94
Claviceps purpurea, 173
Claytonia, 91
Clintonia, 108, 110
Cnidoscolus, 108
Cochlospermum, 107
Cocos nucifera, 70
coenocytic microspores, 188
Coix lacryma-jobi, 386
colchicine, 285, 401, 406
Coleoptera, 61, 109
Colias, 100, 102
competition, between crops, 126–7, 131

competition, between plants, 125
conifers, 70–1, 73
copper chelators, 224, 225
corbiculae, 95, 96, 103
corn, 168, 173, 184, 188, 193; *see also* maize
    and *Zea mays*
corolla tube, 92
Cosmos, 192, 321
cosuppression, 247
cotton, 28, 224, 365
coumarin, 222
Crinum, 31
Crocus, 25
crop attractiveness, 125, 127
crop maturity, 122
crop quality, 122
crop quantity, 122
crop uniformity, 122
crop yield, 140
cross-incompatibility, 341
crossability barriers, 261–9
crossing barriers, 273–4
cryopreservation, 273, 407, 410; *see also*
    pollen cryopreservation
Cucumis, 96, 321, 328
Cucurbita, 91, 93, 107, 108
Cucurbita pepo, 70
cutinases, 266
cybrids, 174, 175, 184, 386
Cyclamen, 279
cytochrome P450, 250
cytokinins, 191, 277
cytomictic channels, 18
cytoplasmic DNA, 406
cytoplasmic male sterility (CMS), 156–75,
    238, 240, 252
    characterization, 159
    limitations of, 172–4
    maintainers, 172
    mechanisms of action, 159–67
    origins of, 157–9
    restoration, 159
    use in hybrid seed production, 169
cytoskeleton proteins, 47

Datura, 267, 275, 282, 354, 393
Delphinium, 108
dichogamy, 71
dicliny, 77
Dieffenbachia, 106
Diervilla, 108
dihydrozeatin, 191
dioecy, 72, 77
Diplotaxis, 8
Diptera, 61, 108, 109
Dipterocarp, 77
distyly, 71, 72, 210

DNA replication, 402
DNA uptake, 408
DNA-binding protein, 45
dominant male-sterile transgenes, 242
double fertilization, 33, 268, 364, 371
double haploids, 393, 406–7
dry matter yield, 167–8
Duftmale, 67
dystropic, 63
Echium, 96, 98, 103, 107, 108
egg apparatus, 26
egg cell 26–7, 33, 363–73, 379
    isolation, 363, 370–1, 378–9
    manipulation, 371
    protoplast, 378–84, 386
    ultrastructure, 371
Elaeis spp.,70–71
electroporation, 408, 423–4
electrostatic charges, 94
embryo breakdown, 269
embryo culture, 279–82
embryo rescue, 282–5
embryo sac, 25–27, 33, 363–70
    isolation, 359, 363, 368–70, 378–9
    manipulation, 359, 363, 368–70
    protoplast, 381
embryo-specific genes, 402, 407
embryogenesis, 353, 364, 383–5
embryogenic microspores, 397
embryos, 383–5; *see also* pollen embryos
endoplasmic reticulum, 201
endoreduplication, 406
endosperm, 269, 280, 385
endosperm mutants, 337
endothecium, 17, 21
endothelium, 26
Englosini, 67
enpollination, 137
entomophily, 62
ephydrophily, 62
Epibolium, 93
EPSP promoter, 52
Eristalis, 94
Erwinia, 46
Erythronium, 99, 102, 107, 108, 110, 324
esterases, 190
ethrel, 222, 232
ethyl acetate, 278
ethylene, 29, 225, 397
Eulaema, 98
euphilic, 63
eutrophic, 63
exine ornamentation, 19
extracellular matrix, 338–9

F_1 hybrids, 200, 212, 238, 245, 251, 253
F_1 sterility, 285

female fertility, 251
female gametophyte, 364
female germ unit, 27, 365
fenridazon, 225, 226
fenridazon potassium, 232–3
fertility restoration, 243
fertilization, 33–4, 353, 367
filiform apparatus, 27, 33
flavonoids, 184, 247
flavonols, 29, 341
floral colors, 64–5
floral competition, 125, 131
floral constancy, 66, 88, 91, 102, 124
floral density, 88, 105, 125, 134
floral nectars, 69
floral receptivity, 134
floral rewards, 94, 134
floral scents, 66–7
floral size, 66
floral syndromes, 62, 66
floral visitation, 89, 91, 95–7, 105–6,
    108–111, 123, 138, 141
flow cytometry, 355–6, 358, 361
flowering time, 77
fluorescein diacetate test, 299, 305, 356, 358,
    369
fluorochromatic reaction test, 431
fluorochromes, 263, 356, 360
foraging insects, *see* insect foraging
Fourier transform infrared spectroscopy, 303–4
Fragaria, 23
Fragaria × *ananassa*, 132
Freesia, 281, 285
Fritillaria, 276–7
fruit set, 105–6
fruit symmetry, 90
fruit yield, 167
fusiogenic medium, 380–2

G10 gene, 44, 46
β-galactosidase, 43, 327
gametes, 2n, 285
gametic recombinations, 407
gametocides, 4, 247, 250; *see also* chemical
    hybridizing agents
gametoclonal variation, 408
gametophytic competition, *see* pollen compe-
    tition
gametophytic factors, 341
gametostat, 247
gamma radiation, 322
geitonogamy, 60
gene duplication, 209
gene flow, 75
gene tagging, 432
gene transfer, 5, 6
generative cell, 19, 42, 401, 432

genetic engineering, 184, 193, 220, 240–53
genetic transformation, 7, 202, 386, 408
genic male sterility (GMS), 183–94, 241,
    248–9
    biochemical changes, 189–90
    cytological changes, 187–8
    environmental effects, 186–7
    hormonal effects, 191–2
    use in hybrid programs, 192–3
germplasm, 410
gibberellins, 30, 191, 222–3, 277, 327
Gingko biloba, 301
Gladiolus, 25, 285
β-glucuronidase, *see* GUS
glufosinate ammonium, 244
glutamine synthetase gene, 51
Glycine max, 339
glycoproteins, 205, 339
gold particles, 428, 431–3
Gossypium spp., 32, 127, 158; *see also* cotton
growth regulators, 222, 277, 281, 399
    LY195259, 224
    RH-531, 223
    TD1123, 224
growth-inhibitory substances, 219
GUS (β-glucuronidase), 47, 50–2, 425–33
gynodioecy, 72, 184
gynomonoecy, 71

Halictus, 104
halogenated aliphatic acids, 224
haploid chromosome, 406
haploid embryos, 393; *see also* pollen
    embryos
haploid plants, 424, 429–32
heat shock proteins, 399
Helianthus annuus, 71, 91, 95, 160, 163,
    170, 398
    CMS *pet*, 160, 170
Helophilus, 94
hemiphilic, 63
hemitropic, 63
Heracleum, 88
herkogamy, 71
heteromorphy, 71, 210–11
heterosis, 4, 156, 167–9, 200, 238
    high parent, 167–8
    standard, 167–9
heterostyly, 71–2, 211
heterozygosity, 407
Hibiscus moscheutos, 324
histones, 19
homogamy, 70, 71
homozygosity, 407, 410–11
honey, 9
honey production, 139
honeybee, 68, 90, 95, 124–5, 131, 136–8, 140

brood, 136
    queen, 125, 129, 136
Hoplitis, 98
Hordeum vulgare, 32, 399
Howea, 281
hybrid breakdown, 285
hybrid canola, 124, 140
hybrid corn, 4
hybrid cultivars, 156
hybrid maize, 253
hybrid onion, 223
hybrid programs, 192
hybrid promoters, 250
hybrid purity, 184, 227
hybrid seed (hybridity), 3, 4, 156, 169, 184,
    224, 233
hybrid seed production, 137–9, 169, 184,
    192–4, 221, 238, 245
    block method, 169
    mixed method, 169
hybrid vigor, *see* heterosis
hybrid wheat, 225, 227, 229
hybridization barriers, 261–9
hybrids
    intergeneric, 281, 386
    interspecific, 281, 386
hydrophily, 62
Hymenoptera, 61, 93, 108–9

I3 gene, 45
Iberis amara, 206
immunosuppressors, 278–9
    acriflavin, 278
    amino-n-caproic acid, 278
    salicylic acid, 278
Impatiens, 96, 300–1
in situ hybridization, 49, 365
in vitro culture methods, 279, 283, 284
in vitro embryogenesis, 383, 385
in vitro fertilization 8, 335, 366, 371–2,
    377–87,
in vitro pollination, 283, 286
in vitro–produced zygotes, 383, 384
incompatibility, 274
    active versus passive, 264
    barriers, 261, 315
    unilateral, 263–4
incongruity, 35, 265, 267, 274–6
indole acetic acid, 191
indole butyric acid, 222
input costs, 138
insect body hairs, 88, 92–4
insect body size, 89, 107
insect characters in pollination, 66
insect competition 89–90
insect foraging, 75–6, 128
    flight directionality, 88

    flight distance, 105
    insect interactions, 88–90
    nectar vs. pollen collection, 88, 94–6
    scent marking, 90
    selective exclusion, 107
insect pheromones, 67; *see also* pheromones
insect pollen loads, 91, 96
insect pollination, 61, 87–111, 121–2
    sex and castes role of, 88, 90–91
insect population size, 89, 105
insect tongue length, influence on foraging,
    88, 91–3
insect visitation, 122
insects, 61, 62, 68–9
    attractant sprays, 125, 128–9
    isolation distance, 126, 133
    management of, 121
    nest boxes, 123
    nest sites, 131–2
    nesting materials, 123
    nesting tunnels, 136
    repellent compounds, 131
    sex ratio, 136
    spatial and temporal isolation, 126, 133
integumentary tapetum, 26
intercellular matrix, 23, 32
intergeneric crosses, 157, 262–3, 282, 285
interspecific barriers, 261–9
interspecific competition, 66
interspecific crosses, 157–8, 262–3, 265,
    269, 274, 278, 282
interspecific hybridization, 184, 274, 286
intraspecific crosses, 157–8
intraspecific incompatibility, 319
introgression, 285
Ipomoea, 23, 97, 99, 108
irradiation, 275
isolated embryo sacs, 379
isolated gametes, 357–61, 366, 370–1, 377–87
    electrofusion, 359, 370–2, 380–83, 386
    spontaneous fusion, 371–2
isozymes, 347

Japanese cedar pollen, 48
Jepsonia, 105

Kalmia, 95
kanamycin resistance, 431
karyogamy, 384
Kentucky bluegrass, 44
KGB gene family, 44
Kunitz trypsin inhibitor, 44

Lantana camara, 65
Lasthenia, 93
LAT52 gene, 44, 45, 48–9, 51, 425–30
LAT52/GUS construct, 49

LAT56 gene, 46, 51, 339
LAT58 gene, 425
LAT59 gene, 46, 51, 339, 425, 427
Lathyrus, 276
Lavandula, 92–3, 109
leaf-cutter bees, 68, 69, 124, 126, 132, 135, 140
Lepidoptera, 93, 108–9
Leucaena leucocephala, 338
Lilium, 24–5, 32, 267, 274, 276, 279, 281, 283–5, 301, 426, 433
Lilium auratum × *L. henryi*, 286
Lilium henryi, 284
Lilium longiflorum × *L. candidum*, 286
Lilium longiflorum × *L. dauricum*, 284
Lilium longiflorum × *L. henryi*, 284
Lilium longiflorum, 276–7, 282–5, 339
Lilium pumilum, 286
lily, *see Lilium*
LIM motif, 50
Linaria, 109
linkage, 249
Linum, 32, 94, 109, 211, 265
Lol p gene, 44
Lotus corniculatus, 51–2
Lucilia, 94
Lupinus, 97, 99
Lycopersicon esculentum (or *Lycopersicon*) 23, 185, 265, 279, 281, 339; *see also* tomato
Lycopersicum peruvianum, 202
Lythrum salicaria, 71

Macadamia, 90
Magnolia, 71
maintainer, 172–3, 245
maize, 32, 44, 45, 47, 48, 52, 238, 240, 246, 248, 250–1, 253, 268, 269, 299, 304–5, 334–5, 337–8, 340–42, 346–7, 355–60, 364–70, 372–3, 378–9, 381–7, 425, 427; *see also* corn *and Zea mays*
Malaysian dwarf coconut, 70
male gametes, 19
male gametocides, 219
male gametophyte, 183, 219–20, 239, 333, 400; *see also* pollen
male gametophytic fitness, *see* pollen fitness
male gametophytic selection, *see* pollen selection
male germ unit, 20, 360
male sterility, 4, 184, 188, 190–1, 194, 218, 220, 227–8, 231, 233, 237–8, 240, 398
  hormone sensitivity, 191–3
  photoperiod sensitivity, 187, 194
  temperature sensitivity, 186
male-sterility genes, dominant, 240–1
male-sterile flowers, 185–6

male-sterile lines, 137, 140, 156–75, 186–8, 407
male-sterile (*ms*) mutants, 184, 188, 190, 192
  ms1, 187
  ms2, 188
  ms3, 188
  ms4, 188
  ms6, 188
  ms9, 187
  ms10, 188
  ms13, 188
  ms15, 186, 192
  ms22, 188
  ms24, 188
  ms33, 186, 192
  msp, 188
  pop1, 341
  vms, 186
male-sterile plants, 186
maleic hydrazide, 222
Malus domestica, 135
Malus pumila, 301
Malvaviscus, 105
mammal pollination, 88
marker genes, 192–3, 244, 245, 248–9, 252
Mason bees, 132
mate finding, 70
mating systems, 70
Medicago sativa, 67, 123
Megachile spp., 98, 123–4, 140
megaspore, 26
meiosis, 16–18
Melandrium album × *M. rubrum*, 282
Melandrium album × *Silene schafta*, 282
Melissodes, 91
melittophily, 62, 69
mellissopalynology, 9
mentor pollen, 248, 275, 315, 318, 329
  action, 326–8
  effects, 315, 329
  functions, 323
  preparation, 319, 322, 324–5
  promoter of fruit retention, 328
  recognition substances, 327
mentor pollen index, 320
Mercurialis annua, 191
Mertensia, 93, 109
metabolic inhibitors, 222, 224
microinjection, 359, 366, 368–70, 408
micromanipulation, 386
microspore culture, 393, 395–6
microspore development, 18–19
microspore isolation, 17–19
microspore tetrads, 18–19
microspore-derived embryos, *see* pollen embryos
microspore-expressed genes, early, 45

microspore-expressed genes, late, 45
microtubules, 19, 399
mitochondria, 18
mitochondrial genes, 164, 166–7
mitomycin, 159
mixoploids, 406
molecular markers, 347
Monarda fistulosa, 65
monocarboxylates, 231
monoculture, 172
monoecy, 70–1
monolectic, 63
monophilic, 63
Monumetha, 104
mRNA, 339, 402
mutagenesis, 408
mutagens, 158, 184, 188
mutualism, 77–8, 88
    plant/plant, 78
    pollinator/pollinator, 78
myophily, 61, 62, 69
Myosotis, 93
myrmecophily, 62, 69

naphthalene acetic acid, 222
Nasanov gland, 128
nectar, 63, 64, 68
nectar collection, 88, 94–5
nectar foragers, 135
nectar guides, 65, 67
nectar production, 127–8
nectar rewards, 131
nectar robbing, 90–2
Nerine, 279, 285
Nicotiana, 24, 31, 158, 267, 276, 279, 321,
    354, 366, 386
Nicotiana alata, 202, 328
Nicotiana glutinosa, 424–5, 428
Nicotiana rustica, 424–6, 430
Nicotiana tabacum, 53, 398, 402, 408, 410
Nicotiana tabacum × *N. acuminata*, 280
Nicotiana tabacum × *N. oesoebila*, 282
Nomia melanderi, 124
NOS (nopaline synthase), 431
Npg1 gene, 44
NPTII (neomycin phosphotransferase II), 431
NTP303 gene, 44
nuclear magnetic resonance spectrometry,
    299, 355

obturator, 32
Oenethera, 24, 44, 47, 93, 97, 99, 109, 341
oilseed rape, 167, 172, 238, 240, 253; *see
    also Brassica napus* and *B. campestris*
Ole e I gene, 44, 48
oleosin, 45
olfactory cues, 127

oligolectic, 63
oligophilic, 63
olive, 44
onion, 4, 168–9, 174; *see also Allium cepa*
Ophrys, 67
orchard bee, 124
orchids, 30
ornithophily, 62, 69
Oryza sativa, 160, 163, 170, 233, 399
    CMS WA, 160, 163, 170
oryzalin, 285
Osmia spp., 98, 124, 132
outcrossing, 75, 134
ovary, 25, 69
ovary culture, 279, 281
ovary manipulation, 275
overpollination, 133
ovule, 25–7
ovule culture, 279–81, 284

P2 gene family, 44
palynology, 2
Papaver, 103, 212, 377
particle bombardment, 50, 424, 427, 432–3
partitioning, 78
Passiflora, 71, 338
pcf gene, 167, 175
peach, 32
pear, 319–21, 324–5
pearl millet, 4, 168, 173, 251, 340; *see also
    Pennisetum americanum*
pectate lyase, 44
pectin, 205
pectin esterase, 44, 47
Pedicularis, 93, 95, 97
pellicle, 23, 31
Pennisetum americanum, 34, 158, 160, 163,
    170
    CMS A1, 160, 170
Peponapis pruinosa, 70
perianth, color changes, 65
periplasmodium, 20, 21
pest management, 139
Petunia hybrida (or *Petunia*), 24, 44, 31, 161,
    166, 169, 171, 174, 202, 240, 268,
    327, 425
    CMS *axillaris*, 171
Petunia inflata, 202–4
Petunia parodii × *P. inflata*, 282
PG1 gene, 44
phalaenophily, 62, 69
Phaseolus, 279, 285
Phaseolus vulgaris × *P. acutifolius*, 281
phenylcinnoline carboxylates, 226
pheromones, 128–9
    brood-based, 136–7
    mandibular, 129, 136

queen, 136–7
Phleum pratense, 48
phloem mobility, 232
Phlox, 97, 100, 109
phosphatases, 297
Phyllosticta maydis, 174
phytoalaccin, 204
pin flowers, 71, 72
Pinus, 345
Pistacia, 345
pistil, 22, 338–41
plant breeding, 386, 411
plant breeding systems, 59, 70–2
plant density, 134
plant growth substances (plant hormones),
    191–2, 222, 277–8
plant–plant interactions, 77
Plantago lanceolata, 61
plastids during meiosis, 18
plastids in generative cell, 19
pleiotropy, 210
Plumbago, 26, 33, 353, 360, 364, 366
polar nuclei, 365
Polemonium, 67, 100, 109
pollen, 41, 67, 183, 239
    as health food, 31
    assay, 342, 345
    binucleate, 295–7, 307, 334, 401, 410
    2-celled, 19, 353–4
    3-celled, 19, 353–4
    characteristics, 68, 88
    cryopreservation of, 301–3, 305
    cytoskeleton elements in, 300
    echinate, 93, 103
    gene transfer, 423–7
    in animal diet, 9
    in pharmaceutical industry, 9
    irradiated, 321, 322, 324
    methylated, 320
    misplaced, 102
    morphology, 88, 93
    2n, 286
    pioneer, 321, 325
    population effect, 338
    residual, 97, 105
    size, 92–3
    trinucleate, 295–7, 307, 334
    water content, 297–9, 302, 304
    white, 247
pollen adherence, 93
pollen adhesion, 30, 266–7
pollen allelopathy, 77
pollen allergens, 47–8
pollen allergy, 2
pollen banks, 294, 308
pollen baskets, 68
pollen biology, 1

pollen biotechnology, 3, 334, 342
pollen box, 51
pollen capture, 73
pollen carryover, 74–5, 89, 105, 134
pollen coat, 20, 210
pollen collection, 88, 90, 94–7, 102–3, 110
pollen compatibility, 133
pollen competition, 34–5, 77, 323–4, 336–7
pollen consumption, 103
pollen culture, hanging drop technique, 334
pollen dehydration, 304–5
pollen density, 338
pollen deposition, 96, 110–11, 133
pollen development, 16–21, 41, 184, 187,
    221, 225, 228–9
    chemical inhibitors of, 221
pollen dimorphism, 397, 400
pollen discarding, 96, 102
pollen dispenser, 137
pollen dryer, 304–5
pollen embryogenesis, 393
    factors affecting, 396–400
pollen embryos, 2, 392–411, 424
    comparison with zygotic embryos, 403
    development, 400–5
    encapsulation, 409
    germination, 405–6
    growth medium, 398–9
    in mutant isolation, 407
    maturation, 405
    plant regeneration from, 406
    transformation of, 408
    utilization, 407–10
pollen fitness, 336–7
pollen flow, 66, 134
pollen foraging, 136
pollen function, 334, 338
pollen gene expression, timing of, 43–6
pollen genes, numbers of, 42–3
pollen germination, 31, 42, 266–7, 307–8
    in vitro, 307, 334, 335
    in vivo, 336
    semi in vitro, 307
pollen grooming, 90, 95–6, 102–3
pollen growth promoting substances, 327
pollen hydration, 30, 266–7
pollen lethality, 244, 246, 249
pollen loads, 95–6, 103
pollen longevity, 295–6
pollen losses, 88, 96–7, 102–3
pollen membrane, 299
pollen mix, 318
pollen movement, 72–7, 92–4
pollen odor, 130
pollen products, 9
pollen protoplast, 378
pollen purity, 96

pollen quality, 2, 133–4, 355
pollen rewards, 127, 131
pollen screening, 34–5
pollen selection, 2, 335–5, 337, 345–7
pollen storability, 295
pollen storage, 89, 274, 293–308
    freeze or vacuum drying, 301–2
    humidity effects, 300
    in organic solvents, 300–1
    lyophilization of, 301–2
    temperature effects on, 300
pollen suppression, 220
pollen transfer, 122, 132–3, 135, 137, 139
pollen transformation, 423–7, 429
pollen trap, 136
pollen tube, 105–10, 201, 205, 323–4
    entry into ovule, 32–3
    entry into stigma, 31
    growth through style, 32
    guidance, 34
pollen tube inhibition, 265–9
pollen tube growth, 42, 333, 334, 336, 338,
        341
    in vitro, 335
    semi vivo technique, 335
pollen tube competition, 337
pollen tube density, 324
pollen vector, 62, 63, 170–1
pollen viability, 28, 91, 102, 294–7, 307–8
    assays, 305, 307
    biochemical changes, 297
    factors affecting, 295–7
pollen vigor, 308
pollen volatiles, 130
pollen wall, 1, 20, 327
pollen wall proteins, 20, 31
pollen-expressed genes, 42, 43, 45
pollen-limited crops, 122
pollen–ovary dialogue, 323
pollen–pistil interaction, 28–35, 265, 319,
        323, 327, 338–41
pollen–pistil recognition, 327
pollen–pollen interaction, 319, 323, 338
pollen-specific genes, 42, 44, 46
    trans-species expre-sion of, 53
pollen-specific promoters, 50–2, 244, 425, 427
pollenkitt, 20
pollinating agents, 3
pollination, 59, 284
    double, 321, 325
    nocturnal, 67
    placental, 282
    stigmatal, 283
    stump, 276
    value of, 78
pollination biology, 2, 59–60

pollination constraints, 3, 121
pollination control, 238–9
pollination control systems, 4, 156, 238,
        240–51
pollination distance, 89
pollination ecology, dynamics of, 78
pollination efficiency, 87–8
pollination inputs, 130
pollination intensity, 336
pollination management, 138
pollination mechanisms, 64
pollination stimulus, 29
pollination techniques
    cut-style, 275–7, 283–4
    grafted-style, 277–8, 284
pollination thresholds, 140
pollination units, 125, 136
pollinator activity, 140
pollinator breeding, 134
pollinator evaluation, 105–11
pollinator food preference, 135
pollinator guilds, 123
pollinator introduction, timing of, 125–6
pollinator patterns of movements, 76
pollinator visitation, 122, 126, 130–2
pollinator–pollinator interactions, 77
pollinators, 62, 78
    alternative, 79
    behavioral constraints on, 135
    distribution of, 124
    habitat management, 122–3, 139
    native, 122, 139
    social insect, 137
pollinators' fidelity, 64
pollinia, 68
polyads, 68
polyembryony, 385
polyethylene glycol, 334, 408, 423
polygalacturonase, 44, 47
polylectic, 63
polymerase chain reaction, 203
polyphilic, 63
polyploidization, 285–6
polyploidy, 72, 406
Pontederia, 100, 109–10
Populus, 278, 320, 327, 328
postfertilization barriers, 269, 274, 276, 279,
        282
prefertilization barriers, 263–9, 274, 276,
        278, 282
primexine, 18, 19
Primula, 210, 211
proembryos, 282
profilin, 44, 47, 48
promoters, 49–53, 240–1, 244–52, 425–9,
        431

protandry, 28, 71, 76
Protea, 106
protein kinase, 208
proteinase inhibitors, 24
protogyny, 28, 71
Prunus spp., 301, 345
Pseudaugo-chloropsis, 98
pseudocopulation, 67
PSI gene, 44, 48, 51
psychophily, 62, 69
Pyrus ussuriensis, 301

radiations, 184, 188
ragweed, 44
ramet, 60
Raphanobrassica, 7
Raphanus, 101, 109, 206, 208
Raphanus raphanistrum, 65
recombinant DNA technology, 5, 237–53
red clover, 124
repellent compounds, 131
reproductive isolation, 265
resource limitation, 132
restorer genes, 167, 172, 243, 245–6, 252
rewards, 63–4, 67, 75, 131
Rf1 gene, 163, 164
Rf2 gene, 164
Rf3 gene, 166
Rf4 gene, 166
RFLP analysis, 338, 340, 365
Rhyzopus niveus, 202
ribonucleases, 202–3, 240
ribosomes, 18
rice, 44, 168, 172, 187, 191, 252; *see also*
    Oryza sativa
Ricin A, 204
RNases, 204, 286
Rosa, 89, 266
Rubus, 104
rye, 304–5
ryegrass, 44, 231

S-allele, 201, 203, 206–7, 264–5
    high-activity, 208
    low-activity, 208
S-gene, 209–10, 265
S-locus, 200, 206
S-locus glycoprotein (*SLG*) gene, 208–9
S-locus glycoproteins, 207
S-proteins, 201, 202–3
S-related kinase, 209
S-RNase, 204
salicylic acid, 278
Salix babylonica, 301
sapromyophily, 62, 69
Sarcotheca celebica, 71

scopa, 95
seed abortion, 105
seed set, 105–6
seed yield, 167–8, 227
self-incompatibility (SI), 4, 71, 199–212,
    264, 275, 318, 321, 377
    diallelic, 210–11
    gametophytic, 201, 318
    response mechanism, 204
    SI systems, 200–7, 210–11
    sporophytic, 206–7, 318, 321
    two-locus gametophytic system, 204–5
self-pollination, 70, 209
Senecio spp., 75
sexual expression in plants, 70
SF3 gene, 45, 50
shoot meristem, 403
Silene schafta, 282
Simondsia, 71
Solanum, 281, 407
Solanum chacoense, 202
Solanum tuberosum, 202–3
somatic embryos, 405
sorghum, 4, 168, 173, 248, 268, 275; *see also*
    Sorghum bicolor
Sorghum bicolor, 159, 161, 163, 171, 386
    CMS milo, 160, 163, 171
soybean, 185, 187
sperm cell, 19, 33, 385, 387, 400
    dimorphism, 33
    isolation, 352–61, 378
    manipulation, 359–61, 371
    microscopy, 355–8
    protoplasts, 379–84, 386, 371
    viability, 355
sperm nuclei, 366–70
    microinjection of, 359, 366, 368–70
Spinacia oleraceae, 71
Spondias mombin, 71
sporopollenin, 19, 20, 190, 225, 228
*Sta*44-4 gene, 44
stamen, 183
stamenless-1 mutant, 185, 186, 192
stamenless-2 mutant, 185, 186, 188, 190–2
Stapelia, 67
Stellaria, 109
Sternbergia, 23
stigma, 205, 206
    types, 22–3
stigma exertion, 110
stigma exudate, 23–4, 32, 266
stigma papillae, 22
stigma receptivity, 28, 110, 368
stigma surface, 265
stigma touches, 105, 107
stigmatal barriers, 275

stigmatic pollen load, 106–8, 110
Streptomyces griseolus, 250
streptomycin, 159
stylar barriers, 275
stylar canal, 23–4, 32
stylar protein, 339
style manipulations, 275
style, types, 23–5
substrate adhesion molecules, 339
sugarbeet, 252
sulfonylurea compound R7402, 250
sunflower, 7, 45, 167–8, 172, 252; *see also*
       Helianthus annuus
suspensor, 385
Symplocarpus foetidus, 67
syncytium, 17–8
synergids, 26–7, 32, 33, 365–6, 378, 380
Syrphus, 94

T-DNA mutagenesis, 249
TA29 gene, 240–1
TA29 promoter, 250
TA29-barnase, 241, 243–5, 252
TA29-barstar, 241, 243, 252
Tac25 gene, 45
Tagetes, 285
tapetum, 17, 20–1, 41, 156, 159, 163, 164,
       166, 187–8, 222–3, 225–6, 230,
       239–40, 251
Taraxacum officinale, 72
target crop, 125–6
temperature, high, 275
temperature, low, 397
threshold models, 140
thrum flowers, 71, 72
tissue-specific promoters, 286
tobacco, 7, 44, 240, 268, 339, 341, 382, 433
tomato, 4, 44–6, 79, 185–8, 191–4 223, 238,
       268, 339–40, 347, 425; *see also*
       Lycopersicon esculentum
Tpc44/70 genes, 45
Tradescantia, 42, 45, 53, 424–5, 427–8, 434
transactivating protein, 251
transgenic plants, 190, 252, 408–9, 429, 431
transgenic seeds, 424, 432–4
transmitting tissue, 23–4, 32, 34, 327, 339
transposon tagging, 184, 249
Trifolium, 96, 101, 103, 123–4, 281
Trigona, 89, 95
trioecy, 72
triploids, 286
tripping, 28
Tripsacum, 268

tristyly, 72
Triticale, 7
Triticum aestivum (or *Triticum*), 161, 163,
       171, 231, 397, 401
      CMS *timopheevi*, 161, 163, 171
Triticum timopheevi, 229
tryphine, 20, 341
TUA1 gene, 44, 47, 51
α-tubulin, 44, 47
ß-tubulin, 47
Tulipa, 279

Ubisch bodies, 20, 228, 230
unilateral incompatibility, 263–4
unisexual plants, 70
URF13 protein, 165

vegetative cell, 19, 42, 299, 401
Vespa, 94
Vicia, 91, 97, 101, 109–10, 339
viscin, 68, 93
Vitis spp., 31, 71, 216, 279
vitronectin, 339

W2247 gene, 44
watermelon, 29
wheat, 5, 7, 45, 168, 173, 219, 223–4, 226–9,
       231–2, 238, 248, 266, 268, 300, 305,
       372, 379; *see also Triticum aestivum*
wide hybridization, 6–8, 262, 315
wind pollination, 73–4, 88

xenogamy, 60
Xylocopa, 90

Zea m I gene, 44
Zea mays, 41, 158, 159, 161, 164, 171, 208,
       280, 301, 354, 364, 367, 369, 378,
       381, 383, 386
      CMS C, 158, 161, 166, 171, 174
      CMS S, 158, 161, 166, 171, 174–5
      CMS T, 158, 161, 164–5, 173, 175
      CMS *T-urf13*, 164–5, 175
Zea mays x *Z. mexicana*, 282
zeatin nucleotide, 191
Zm13 gene, 44, 48–9, 51, 53, 244, 246, 425,
       427–8
Zm58 gene, 44
ZmPRO 1–3 genes, 44
zoophily, 62, 73
zygote, 372, 384–7, 403
zygote, artificial, 382, 386
zygotic embryogenesis, 384, 385